RE

BASIC ELECT

FOR

ENGINEERS

REED'S
BASIC ELECTROTECHNOLOGY
FOR
ENGINEERS

by

EDMUND G. R. KRAAL
C.Eng., D.F.H. (Hons.), M.I.E.E., M.I. Mar. E.
Formerly Head of Electrical Engineering and Radio Department
South Shields Marine and Technical College

Revised and enlarged by
STANLEY BUYERS
B.Ed.,T.Eng., M.I. Elec.I.E.,
Senior Lecturer, Department of Electrical and Electronic Engineering
South Shields Marine and Technical College

THOMAS REED PUBLICATIONS
A DIVISION OF THE ABR COMPANY LIMITED

First Edition – 1965
Second Edition – 1973 (SI Units)
Reprinted – 1977
Reprinted – 1980
Third Edition – 1985
Reprinted – 1994
Reprinted – 1995
Reprinted – 1996
Reprinted – 2000
Reprinted – 2002

ISBN 0 900335 96 3

THOMAS REED PUBLICATIONS
The Barn
Ford Farm
Bradford Leigh
Bradford-on-Avon
Wiltshire BA15 2RP
United Kingdom

E-mail: sales@abreed.demon.co.uk

Produced by Omega Profiles Ltd, SP11 7RW
Printed and Bound in Great Britain

PREFACE

This book is intended to cover the basic theoretical work in the syllabuses for Electrotechnology in Part B of the Department of Transport Examinations for Class 1 and Class 2 Engineers. It is also suitable for Marine Engineering Cadets studying the Electrical Engineering Principles unit of the Business & Technician Education Council (BTEC) programme.

A similar pattern to other volumes in this series has been adopted, giving emphasis on first principles, referring to numerous illustrations, providing worked examples within the text, and supplying many problems for the student to attempt on his own. Typical examination questions at the end provide the student with the opportunity of finally testing himself thoroughly before attempting the examination. Fully worked out step-by-step solutions are given to every problem thus being particularly useful to the engineer at sea without a college tutor at hand.

In this edition, additional material has been included to cover basic solid state electronics and devices, atomic theory of conduction and the assessment of d.c. machine efficiency.

I wish to acknowledge the help and constructive advice given by colleagues and students of South Tyneside College (formerly South Shields Marine and Technical College). Acknowledgement is also made to the Controller of Her Majesty's Stationery Office for permission to reproduce and use the specimen questions from "Examination of Engineers in the Mercantile Marine" as are made available by the Department of Transport.

S. BUYERS

CONTENTS

THE S.I. SYSTEM

PREFIXES, SYMBOLS, MULTIPLES AND SUB-MULTIPLES

PREFIX	SYMBOL	UNITS MULTIPLYING FACTOR
tetra	T	$\times 10^{12}$
giga	G	$\times 10^{9}$
mega	M	$\times 10^{6}$
kilo	k	$\times 10^{3}$
milli	m	$\times 10^{-3}$
micro	μ	$\times 10^{-6}$
nano	n	$\times 10^{-9}$
pico	p	$\times 10^{-12}$

Examples			
	1 megawatt (MW)	=	1×10^{3} kilowatts (kW)
		=	1×10^{6} watts (W)
	1 kilovolt (kV)	=	1×10^{3} volts (V)
	1 milliampere (mA)	=	1×10^{-3} ampere (A)
	1 microfarad (μF)	=	1×10^{-6} farad (F)

PHYSICAL QUANTITIES (ELECTRICAL), SYMBOLS AND UNITS

The table has been compiled from recommendations in B.S. 1991 and the List of Symbols and Abbreviations issued by the I.E.E.

QUANTITY	SYMBOL	UNIT	ABBREVIATION OF UNIT AFTER NUMERICAL VALUE
Force	F	newton	N
Work or Energy	W	joule or newton metre	J Nm
Torque	T	newton metre	Nm
Power	P	watt	W
Time	t	second	s
Angular Velocity	ω (omega)	radians per second	rad/s
Speed	N	revolutions per minute	rev/min
	n	revolutions per second	rev/s
Electric charge	Q	coulomb	C
Potential difference (p.d.)	V	volt	V

QUANTITY	SYMBOL	UNIT	ABBREVIATION OF UNIT AFTER NUMERICAL VALUE
Electronic force (e.m.f.)	E	volt	V
Current	I	ampere	A
Resistance	R	ohm	Ω (omega)
Resistivity (specific resistance)	ϱ (rho)	ohm metre	Ωm
Conductance	G	siemens	S
Magnetomotive force (m.m.f.)	F	ampere-turn	At
Magnetic field strength	H	ampere-turn per metre or ampere per metre	At/m A/m
Magnetic flux	Φ (phi)	weber	Wb
Magnetic flux density	B	tesla	T
Reluctance	S	ampere-turn or ampere per weber	At/Wb or A/Wb
Absolute permeability of free space	μ_o	henry per metre	H/m
Absolute permeability	μ (mu)	henry per metre	H/m
Relative permeability	μ_r	—	—
Self inductance	L	henry	H
Mutual inductance	M	henry	H
Reactance	X	ohm	Ω
Impedance	Z	ohm	Ω
Frequency	f	hertz	Hz
Capacitance	C	farad	F
Absolute permittivity of free space	ϵ_o (epsilon)	farad per metre	F/m
Absolute permittivity	ϵ	farad per metre	F/m
Relative permittivity (dielectric constant, specific inductive capacity)	ϵ_r	—	—
Electric field strength, electric force	E	volt per metre	V/m

Electric flux	Ψ (psi)	coulomb	C
Electric flux density, electric displacement	D	coulomb per square metre	C/m^2
Active power	P	watt	W
Reactive power	Q	volt ampere reactive	VAr
Apparent power	S	volt ampere	VA
Phase difference	ϕ (phi)	degree	°
Power factor (p.f.)	$\cos \phi$	—	—

CHAPTER 1

ELECTRON THEORY
THE ELECTRIC CIRCUIT
TERMS AND LAWS

THE NATURE OF ELECTRICITY

To enable the student engineer to obtain practice with appropriate problems and to appreciate fundamentals, a start is made in this chapter with basic circuit theory and relevant calculations. However, it is also considered advantageous if an introduction is first made to the nature of electricity through the subject area of atomic physics. A more detailed explanation will be developed as required in later relevant chapters but it is hoped that the student will be dissuaded from the concept that electronics and aspects of electrical engineering are unrelated. The nature of electricity and the many electrical phenomena and effects all have their origin in atomic structure.

THE STRUCTURE OF THE ATOM

It is now universally accepted that the passage of electricity of a current is due to a flow of *electrons* and as there is no observable indication of such a flow in a conductor, we must perforce accept the classical atomic theory on the constitution of matter and the effects of electron movement and rearrangement. Matter may be defined as anything that occupies space. It may be in solid, liquid or gaseous form but basically consists of *molecules* of the substance. A molecule is the smallest particle of a substance that can exist by itself. Thus molecules have the properties of the substance which they form but themselves consist of groups of *atoms*. As an example, a molecule of water, written H_2O, consists of 2 atoms of hydrogen and one of oxygen. The atom is defined as the smallest particle that can enter into chemical action, but is itself a complex structure consisting of sub-atomic particles. A substance that contains only atoms with the same properties is called an *element,* but one containing atoms of different properties is called a *compound*. All atoms of a given element are identical and atoms of different elements differ only in the number and arrangement of the sub-atomic particles contained therein. The sub-atomic particles can be *charged* or *uncharged*. This reference to charge will be made repeatedly, as study progresses, but at this stage, it can be stated that electricity in its minutest form consists of charges and these can be of two kinds, namely, positive (+ve) and negative (−ve). Like charges repel each other and unlike charges attract each other. In general the space in which a physical force exists between charges is referred to as an *Electric Field* and at this stage this definition will suffice but more detailed consideration will be made in Chapter 8 when dealing with Electrostatics.

According to the theory, propounded by eminent scientists like Rutherford and Bohr, each atom has a core or *nucleus* surrounded by a number of *orbital electrons*. The nucleus consists of minute masses of positively charged sub-atomic particles or *protons,* and *neutrons* which have no charge. The sole purpose of the neutrons is to cement the positively charged protons together within the nucleus. The number of protons in an atom determines the *atomic number* (Z) of the element and also the number of negatively charged orbital electrons. In a normal stable atom the number of protons is equal to the number of orbiting electrons. An electron has a mass of 9.04×10^{28} g and possesses a charge of 1.6×10^{-19} Coulomb but the proton has a mass some 1850 times greater than that of an electron whilst the neutron has nearly an equal mass. The conception of the atom is shown by the diagram (Fig 1).

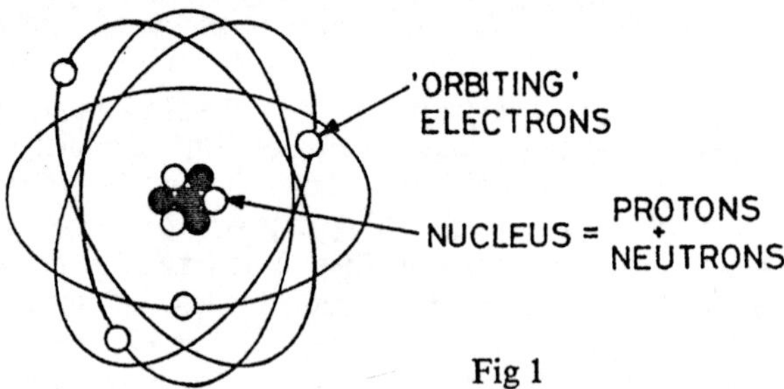

Fig 1

The negatively charged electrons are considered to spin about an axis and also to revolve round the nucleus so constituting a miniature solar system. The nucleus thus represents the sun and the electrons represent the planets. Under normal conditions the atom is said to be stable or unexcited. As stated above the planetary electrons together neutralise the positively charged protons of the nucleus, so the complete atom itself has no electrical charge. The diagram (Fig 2) shows examples of the atomic structure of different elements but the illustrations are schematic being drawn for one plane only. The simplest atom is that of the element hydrogen, consisting of a nucleus with one proton (having a +ve charge) around which travels one electron in an orbit. The electron with its −ve charge neutralises that of the proton. In the diagrams, the electrons are denoted by circles, with their charges shown, and are considered to be moving on dotted orbits. The nucleus is shown with a full circle, has a net positive charge attributed to the protons contained therein and these are shown by + marked circles. The neutrons are shown by small circles with no charge sign.

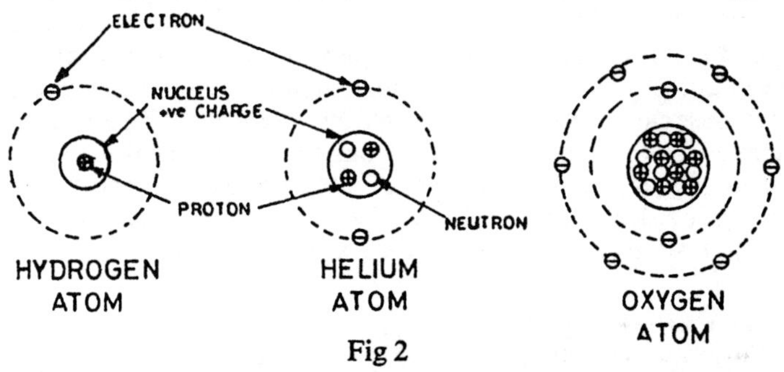

Fig 2

The next element considered is helium. This has 2 planetary electrons and the nucleus consists of 2 protons and 2 neutrons. The planetary electrons of most atoms are associated with the nucleus in a definite manner *ie* the electrons are in groups termed *'shells'*, such that the planetary path of each shell is different. This is shown if an oxygen atom is considered. This has a nucleus of eight protons and eight neutrons. The planetary electrons are eight in two orbits or shells — six in the outer shell and two in the inner shell. For any one atom, the electrons in the first shell can be less than, but never more than two electrons and in the second shell more than eight.

The diagram (Fig 3) represents the atomic structure of two metals: lithium and sodium. In each case, and if other metals are considered, it will be seen that all have one or two electrons in the outermost shell. This feature is thought to be the reason for metals having good electrical conducting properties. It is suggested that for metals in their normal crystalline state, the

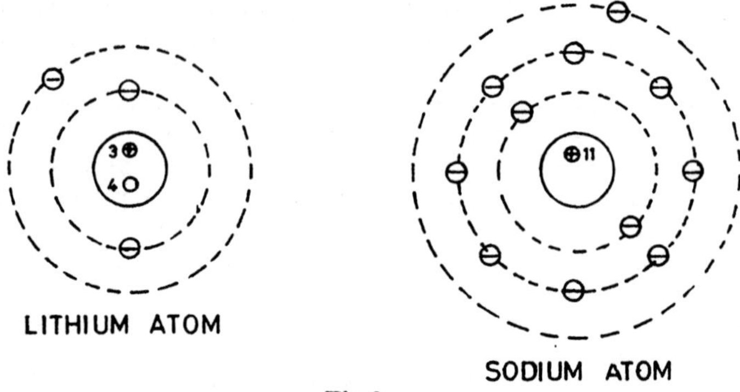

Fig 3

atoms are so lined up that their outermost electrons are partially screened from the +ve attractive effect of the nucleus and are thus not so closely bound. Thus they can move comparatively freely between one atom and its neighbours. Such outer orbital electrons, called *'mobile or valence electrons'* move in a random manner from atom to atom and constitute a 'pool' of moveable negative charges, the existence of which is used to explain the passage of electricity or current in a circuit. Note that valency is a chemical term to which mention will be made later.

CURRENT AS ELECTRON MOVEMENT

Current, according to the electron theory, is due to the movement of electrons from one atom to the next, each electron carrying with it a −ve charge. As explained above, since the mobile electrons move in a random manner between atoms, the transference of charge and therefore passage of electricity, in any one particular direction, does not occur and no current is considered to flow. If an electrical force, in the form of an electromotive force or potential difference is applied across a good conductor then the mobile electrons will move under the influence of this force towards the higher potential or +ve terminal. The required electrical force can be produced by a battery or generator which can be regarded as a pump moving the electrons round the circuit. A stream or movement of electrons is said to constitute an electric current but, it is stressed here that, attention must be paid to the difference between the direction of conventional current flow and electron flow. Thus if a length of wire is connected to two terminals, between which an electromotive force or potential difference exists, then a current will flow from the +ve terminal through the wire to the −ve terminal. Electron flow will be, however, from the −ve terminal to the +ve terminal. This fundamental difference between conventional current and electron flow must always be remembered and is illustrated by the diagram (Fig 4a and 4b). It is also stressed here that the electrical generator or battery, which maintains an e.m.f. or potential difference (p.d.) between the ends of a conductor, does not itself make electricity but merely causes a movement of the charges or electrons which are already present in the circuit.

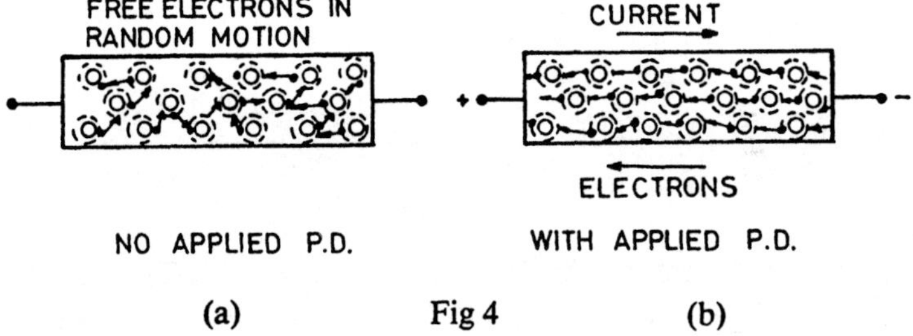

(a) Fig 4 (b)

IONISATION

An atom may lose or gain an electron as the result of a disturbing action. It then becomes electrically unbalanced having acquired a charge and is called an *ion*. Thus an atom minus an electron, exhibits a +ve charge and is a +ve ion. Similarly an atom which gains an electron, exhibits a −ve charge and is a −ve ion. When an electron is made to leave a parent atom by the application of some effect, such as the force due to an electric field, or by the application of heat or light, it may acquire sufficient energy to detach further electrons from any other atoms with which it may collide. Such action causes the struck atoms to become +ve ions and, if electrons leave such atoms faster than they can be regained, the state of ionisation continues. Electronic apparatus such as the fluorescent lamp and cathode ray tube depend on ionisation for satisfactory operation. Such action will be described in detail when such devices are considered in later chapters.

THE ELECTRIC CIRCUIT

A circuit can be defined as the path taken by an electric *current*. Such a current will flow through a circuit if (i) a source of electrical energy such as a battery or generator is connected and (ii) the circuit is continuous or conducting throughout its whole length. The diagram (Fig 5) represents a simple circuit in which current is flowing. It shows the source, from which energy is transmitted through the medium of the current, the conducting path or cables along which the current flows and the 'load'. The load is the point at which the energy is required to be released or work is to be done through the agency of the current flowing.

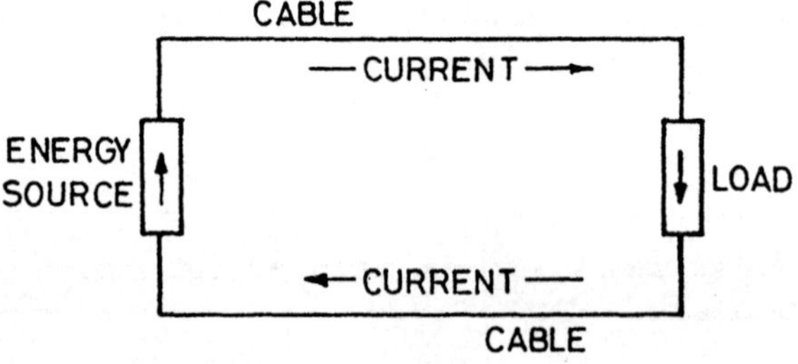

Fig 5

The conditions of Fig 5 are better represented by a circuit diagram as in Fig 6, which illustrates the energy source as a chemical cell, the conducting path as the leads or wires and the load. A switch is shown as a pivoted link which, when opened, interrupts the continuity of the circuit and thus stops current flowing.

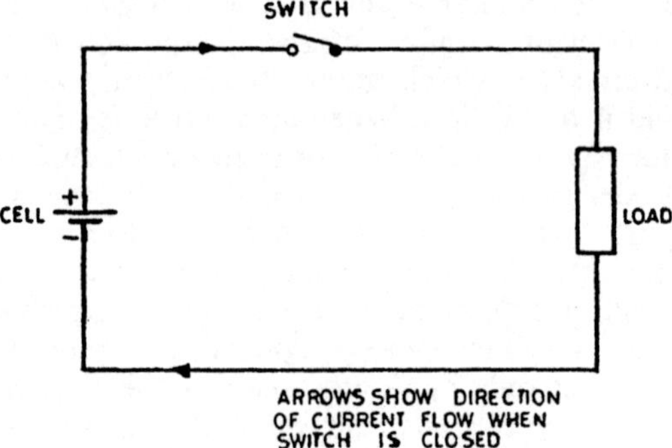

Fig 6

Consideration of the simple circuit introduces more fundamental terms and the practical units as used in electrical engineering. Flow of electricity or current is the result of a pressure built up within the energy source which manifests itself, at the circuit connecting-points or terminals, as a pressure difference. One terminal, called the positive, is considered to be at a higher pressure or potential than the other terminal, called the negative. A *potential difference* (p.d.) is said to exist between these terminals. The direction of current is from the positive (+ve) terminal through the circuit external to the energy source, back to the negative (−ve) terminal and thence through the source to the +ve terminal. Thus for the load, current is from +ve to −ve terminal, but for the energy source in the form of a cell, battery or generator, current is from the −ve to the +ve terminal.

The electrical pressure generated by the energy source is termed its *electromotive force* (e.m.f.). The symbol used is *E*, and e.m.f. is measured as a *voltage*. The unit is the *Volt*, which will be defined later, but any voltage value can be represented by the letter V attached to the numerical value. Thus a voltage of two hundred and twenty volts would be written as 220V. For reasons which will be explained when the mathematics of the circuit is considered, the whole generated e.m.f. of a cell, battery or generator does not appear at the terminals, when current is flowing. The p.d. across the terminals is also measured in terms of the potential or voltage dropped round the external circuit. The symbol used for the terminal p.d. is *V* and it is measured as a voltage, *ie*, in volts.

CIRCUIT LAWS

1. For any circuit, current strength is found to be proportional to the voltage applied across its ends. Current strength is denoted by the symbol *I* and is measured in *Amperes*. The ampere will be defined later by consideration of the electromagnetic effect of a current flow, but any current value can be represented by the letter A appended to the numerical value. Thus 200A means two hundred amperes.

Any electrical circuit is found to offer opposition to the flow of current. This opposition is termed the *resistance* of the circuit and is denoted by the symbol *R*. The practical unit of resistance is the *Ohm*, but any value is represented by the Greek letter capital Ω (omega) appended to the numerical value. Thus 100Ω means one hundred ohms. The ohm can be defined in terms of the volt and ampere thus: a resistor has a value of one ohm resistance, if a current of one ampere passes through it when a potential difference of one volt is applied across its ends. An alternative definition will be given in Chapter 2.

2. The current in any circuit, for a constant voltage, is found to vary inversely with the resistance; for instance, the greater the resistance, the smaller the current and vice versa.

OHM'S LAW

The relationships set out above, are summarised by the first law of the electrical circuit, which is called Ohm's law and can be expressed thus: the current in a circuit is directly proportional to the voltage and inversely proportional to the resistance. This can be written

$$\text{Current} = \frac{\text{Voltage}}{\text{Resistance}}$$

$$\text{or } I \text{ (amperes)} = \frac{V \text{ (volts)}}{R \text{ (ohms)}} \text{ or } \frac{\text{E}}{\text{R}}$$

$$\text{Other forms are } V = IR \text{ or } R = \frac{V}{I}$$

When using the Ohm's law formula it is essential to pay due regard to the correct magnitudes of the units used. Reference should be made to the appropriate table of conversions.

Example 1. An e.m.f. of 6V is applied across a 300Ω resistor. Find the current which will flow.

$$I = \frac{E}{R} = \frac{6}{300} = 0.02\text{A} = 20\text{mA}.$$

Example 2. A current of 10mA passes through a 60kΩ resistor. Find the voltage drop across the ends of the resistor.

$$V = IR = (10 \times 10^{-3}) \times (60 \times 10^{3}) = 600\text{V}.$$

SERIES AND PARALLEL CIRCUITS

Study of the electrical circuit shows that in its simplest form it may be built up as (i) a series circuit, (ii) a parallel circuit. For convenience, resistance is considered to be concentrated in a resistor, as for Fig 6, or in more than one resistor; while the connecting leads are assumed to have negligible resistance, unless a definite resistance value for these is stated. Similarly the cell, battery or generator can be assumed to have no resistance unless otherwise stated.

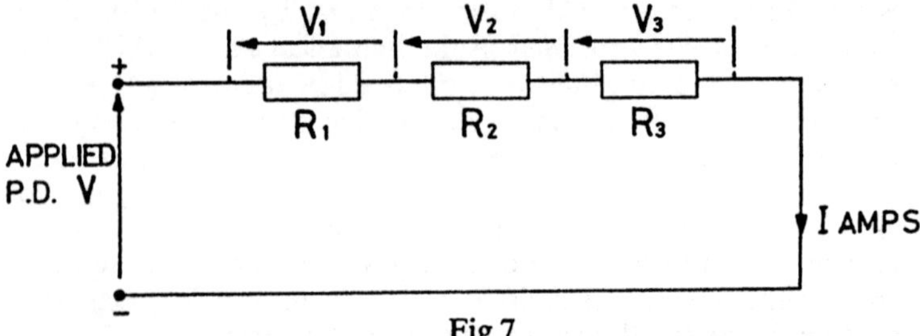

Fig 7

The diagram (Fig 7) shows a Series Circuit. It will be seen that only one current path is provided and that the same current passes through all the resistors. The current is thus common for such a circuit but the applied potential is dropped progressively as the current flows along the circuit.

The diagram (Fig 8) shows a Parallel Circuit. Here the main current is made up of a number of branch currents, but the applied p.d. is the same or common for all branches. It will be noted that at any junction point there is no accumulation of current. *ie* the total current entering that point is the same as the total current leaving the point. Simple laws based on the voltage conditions for the series circuit and the current conditions for the parallel circuit have been evolved which allow the solution of associated problems for these simple sircuits, and also for those of the more complicated series-parallel arrangements and electrical networks.

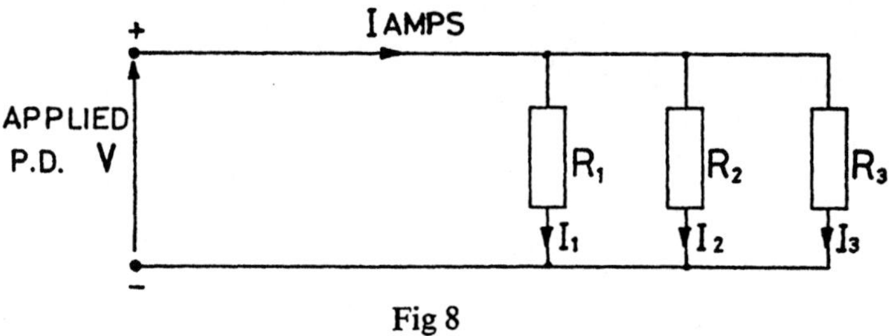

Fig 8

KIRCHHOFF'S LAWS

1. VOLTAGE LAW. The sum of the potential or voltage drops taken round a circuit must be equal to the applied potential difference. Thus for Fig 7:

$$V_1 + V_2 + V_3 = V$$

2. CURRENT LAW. The current flowing away from a junction point in a circuit must equal the current flowing into that point. Thus for Fig 8:

$$I_1 + I_2 + I_3 = I$$

The above laws are used to deduce simple formulae for the series and parallel circuits in terms of the equivalent resistances of the circuits.

THE SERIES CIRCUIT. For Fig 7, let I amperes be the common current flowing round the circuit. Then from Ohm's law, the voltage dropped across resistor R_1 is V_1 volts $= IR_1$. Similarly the voltage dropped across R_2 is $V_2 = IR_2$ and so on. If R is taken as the equivalent resistance of the whole circuit then as V is the applied voltage and it will be dropped over this equivalent resistance, we can write $V = IR$.

Using Kirchhoff's voltage law then

$$V = V_1 + V_2 + V_3$$

$$\text{or } IR = IR_1 + IR_2 + IR_3 = I(R_1 + R_2 + R_3)$$

$$\therefore R = R_1 + R_2 + R_3$$

THE PARALLEL CIRCUIT. For Fig 8, let V volts be the common voltage applied to all the parallel branches and let it cause a total main current of I amperes. Voltage V would also cause a current of I amperes through an equivalent circuit of resistance R ohms.

Thus $I = \dfrac{V}{R}$ and using Kirchhoff's current law then

$$I = I_1 + I_2 + I_3$$

But for branch 1 $V = I_1R_1$ or $I_1 = \dfrac{V}{R_1}$

Similarly for branch 2 $I_2 = \dfrac{V}{R_2}$ and so on

Thus $I = I_1 + I_2 + I_3$ can be written:

$$\frac{V}{R} = \frac{V}{R_1} + \frac{V}{R_2} + \frac{V}{R_3} = V\left(\frac{1}{R_1} + \frac{1}{R_2} + \frac{1}{R_3}\right)$$

$$\therefore \frac{1}{R} = \frac{1}{R_1} + \frac{1}{R_2} + \frac{1}{R_3}$$

Note. The reciprocal of resistance is frequently referred to as *Conductance*. Symbol $G = \dfrac{1}{R}$

The unit is the *Siemens;* the symbol S being appended to the numerical value.

So for a parallel circuit $G = G_1 + G_2 + G_3$ *etc.*

Example 3. Three resistors of values 2, 4 and 8 ohms are connected in series across a supply of 42 volts. Find the current taken from the supply and the voltage dropped across each resistor.

Here $R = R_1 + R_2 + R_3 = 2 + 4 + 8 = 14\Omega$

So supply current $= \dfrac{42}{14} = 3\text{A}$

Voltage dropped across 2Ω resistor $= 3 \times 2 = 6\text{V}$
Voltage ,, ,, 4Ω ,, $= 3 \times 4 = 12\text{V}$
Voltage ,, ,, 8Ω ,, $= 3 \times 8 = 24\text{V}$

Check. 6V + 12V + 24V = 42V (the applied voltage).

Example 4. The above resistors are connected in parallel across the same supply voltage, find the total current and the current in each branch.

Here $\frac{1}{R} = \frac{1}{R_1} + \frac{1}{R_2} + \frac{1}{R_3} = \frac{1}{2} + \frac{1}{4} + \frac{1}{8} = \frac{7}{8} = 0.875$ S

or $R = \frac{8}{7} = 1.14\Omega$ and $I = \frac{42}{1.14} = 36.75$A

The current in Branch 1 $= \frac{42}{2} = 21$A

The ,, ,, 2 $= \frac{42}{4} = 10.5$A

The ,, ,, 3 $= \frac{42}{8} = 5.25$A

Check. 21A + 10.5A + 5.25A = 36.75A (the total supply current).

INTERNAL RESISTANCE OF SUPPLY SOURCE

Up to now the energy supply source has been considered as having negligible resistance. In practice a cell, battery or generator, no matter how well designed, has an internal resistance which results in an internal voltage drop when current is being supplied. Thus the e.m.f. generated appears at the supply terminals, shown as AB in the diagram (Fig 9), only when the circuit switch is open, *ie* the cell is on 'open circuit' (O.C.). When current I is supplied, an internal voltage drop given by IR_i occurs, R_i being the internal resistance of the cell. The potential difference V appearing at the energy source terminals is $E - IR_i$. Thus V is less than the generated e.m.f. E by the p.d. required to drive the current through the resistance of the cell itself.

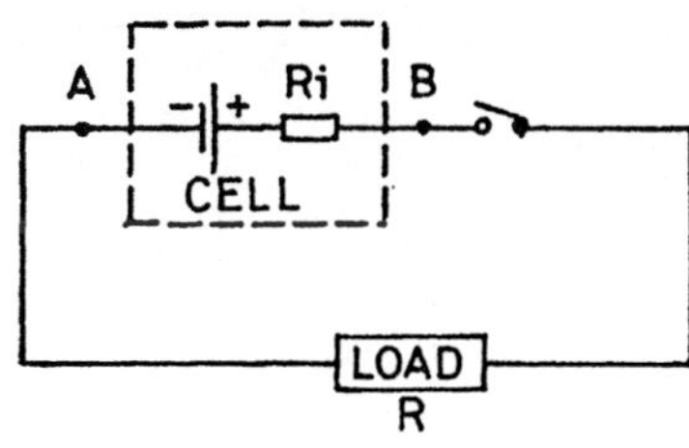

Fig 9

ELECTROMOTIVE FORCE AND TERMINAL P.D. OR VOLTAGE

In Fig 9, R_i represents the internal resistance of the cell as shown external to the cell itself. This is diagrammatic only;

sometimes this resistance is not shown, being written as a figure only beside the cell e.m.f. If given, however, internal resistance must be taken into account. The foregoing paragraph describes the fundamental difference between Electromotive Force E, and Potential Difference V. On open-circuit (O.C.) the terminal p.d. of an energy source equals the e.m.f. generated; but 'on load', *ie* when current is supplied, the terminal p.d. equals the e.m.f. minus the internal voltage drop. This can be summarised mathematically thus:

On Open Circuit (O.C.) $V = E$

On Load $V = E - IR_i$

We can also deduce that since $V = IR$ where R is the load resistance, then

Then $IR = E - IR_i$ and $E = IR + IR_i$ or $E = I(R + R_i)$

Expressed another way:

On O.C. Cell terminal voltage $V = E$

On Load Cell terminal voltage $V = E - IR_i$

also Cell terminal voltage $V = IR$

Problems can be treated as a simple series circuit, if E is used as the circuit voltage and R_i is included in the series resistance.

Example 5. A battery of e.m.f. 42V and internal resistance 7Ω is used to supply the series circuit of Example 3, *ie* three resistors of 2, 4 and 8Ω in series. Find the current and the terminal voltage and by how much the cell voltage 'sits down' when supplying the load.

Note. It is appropriate at this stage to explain that a battery is an arrangement of more than one cell. The methods of connecting the cells will be discussed in Chapter 2, but for the present a battery can be considered to be an arrangement of cells in series. Thus the battery e.m.f. is the sum of the cell e.m.f.s and the battery internal resistance is the sum of the cell internal resistances.

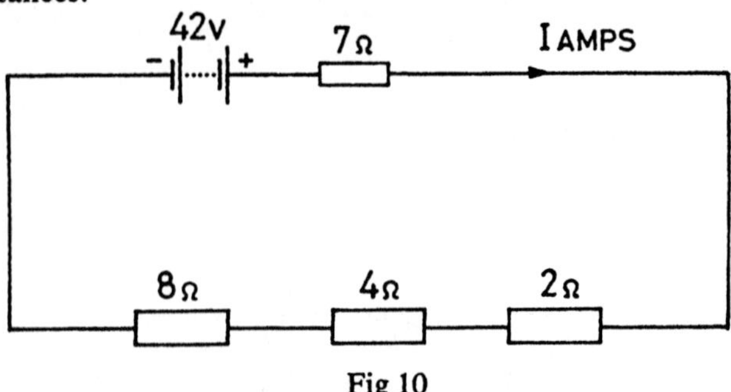

Fig 10

External resistance of the circuit $R = 2 + 4 + 8 = 14\Omega$
Battery resistance $= 7\Omega$
Total resistance of circuit $= 14 + 7 = 21\Omega$
Circuit current $I = \frac{E}{R} = \frac{42}{21} = 2A$
Terminal voltage $V = IR = 2 \times 14 = 28V$
Voltage drop in cell $IR_i = 2 \times 7 = 14V$
Check Terminal voltage $V = E - IR_i = 42 - 14 = 28V$
Cell voltage 'sits down' by 14 volts.

THE SERIES-PARALLEL CIRCUIT. As the title implies, circuits may be built up from series and parallel grouping of resistors. The solution of associated problems, though not straightforward, follows a logical sequence of operation based on the methods used for the simple series and parallel arrangement of resistors. It cannot be too strongly urged that here is a case of 'practice making perfect' and the reader should work a variety of appropriate problems. A method of solution for any particular problem will thus readily become apparent once its form is recognised and both time and labour will ultimately be saved. The more difficult network problems are solved by reverting to fundamental principles such as the direct application of Kirchhoff's laws or using a Theorem or Method of Solution based on these laws. Such problems will be introduced at a very much later stage in the book, when the reader should feel sufficiently confident to appreciate the finer points of theory and may even enjoy the methods and techniques necessary to allow solution.

The diagram (Fig 11a) shows a straightforward series-parallel circuit. It consists of a series circuit made up of two sections,

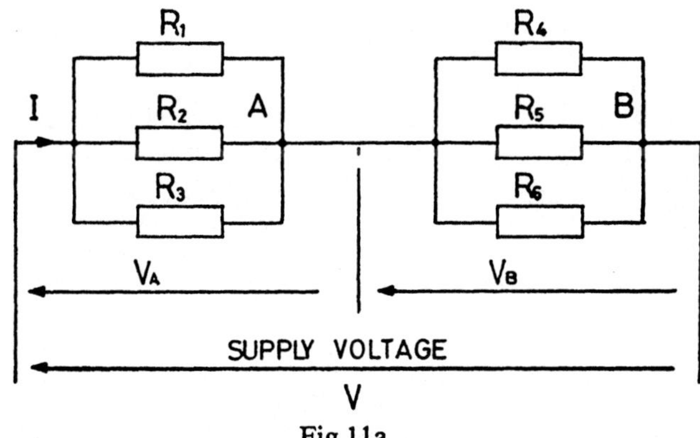

Fig 11a

each comprising of a group of resistors in parallel. Since the main circuit or supply current may require to be found together with the current in each resistor, solution can only be obtained by a simplification of the problem. It will be noted that the parallel groups or banks of resistors have been called sections A and B respectively. The voltage dropped across these sections is not known and since such knowledge is essential, procedure would be as set out below.

The circuit is simplified by finding the equivalent resistance values R_A and R_B of the parallel banks from

$$\frac{1}{R_A} = \frac{1}{R_1} + \frac{1}{R_2} + \frac{1}{R_3} \text{ and } \frac{1}{R_B} = \frac{1}{R_4} + \frac{1}{R_5} + \frac{1}{R_6}$$

Once R_A and R_B have been found the total supply current can be obtained since, as is shown in the diagram (Fig 11b), the equivalent circuit is now of the simple series type.

The supply current is

$$I = \frac{V}{R_A + R_B}$$ and the voltage drops across groups A and B are

$$V_A = IR_A \text{ and } V_B = IR_B$$

$$\text{or } V_A = \left(\frac{V}{R_A + R_B}\right) R_A \text{ and } V_B = \left(\frac{V}{R_A + R_B}\right) R_B$$

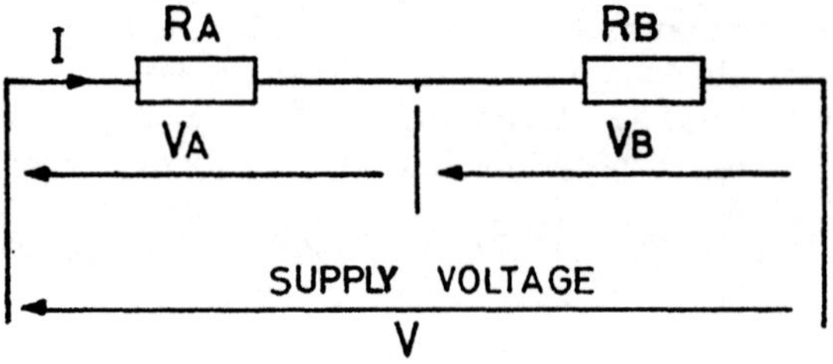

Fig 11b

Once V_A and V_B are known, the individual currents in each resistor can be found by reverting to the original circuit.

$$\text{Thus } I_1 = \frac{V_A}{R_1} \text{ and } I_2 = \frac{V_A}{R_2} \quad \text{Also } I_4 = \frac{V_B}{R_4} \text{ etc}$$

It is not intended that the above method of solution should be memorised. It is given to explain the solution of Example 6 and to illustrate the step-by-step procedure. There is no short-cut for problems of the series-parallel type. The reader should work only with the data given and should not make any assumptions. Methods of solution using proportions for currents of voltages across parallel or series sections of the circuit are discouraged, since seldom in practice are the resistance ratios simple, and only good can come from adhering to and following the straightforward, though sometimes more tedious methods.

Example 6. A circuit is built up from five resistors. Resistors of values 4Ω, 6Ω and 8Ω are connected in parallel to form a group, whilst resistors of 3Ω and 6Ω are connected in parallel to form another group. The two parallel groups of resistors are connected in series across a 10V supply. Find the voltage dropped across each parallel group, the main supply current and the current in each resistor.

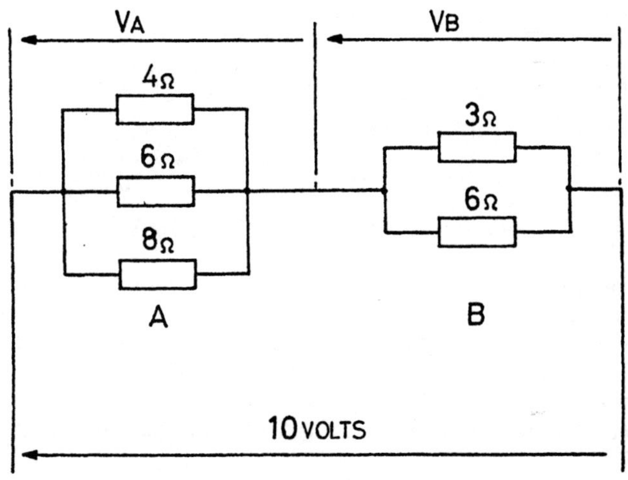

Fig 12

Let R_A be the equivalent of the first group.

$$\text{Then } \frac{1}{R_A} = \frac{1}{4} + \frac{1}{6} + \frac{1}{8} = \frac{13}{24} = 0.54\text{S}$$

$$\text{or } R_A = \frac{24}{13} = 1.85\Omega$$

Similarly let R_B be the equivalent resistance of the second group.

$$\text{Then } \frac{1}{R_B} = \frac{1}{3} + \frac{1}{6} = \frac{3}{6} = 0.5\text{S}$$

$$\text{or } R_B = \frac{6}{3} = 2\Omega.$$

For the equivalent series circuit, total resistance is R

$$\text{or } R = R_A + R_B = 1.85 + 2 = 3.85\Omega$$

Main supply current $I = \frac{V}{R} = \frac{10}{3.85} = 2.6\text{A}$

Voltage drop across R_A or the first parallel group

$= 1.85 \times 2.6$
$= 4.8\text{V}$

,, ,, ,, R_B or the second ,, ,, $= 2 \times 2.6$
$= 5.2\text{V}$

Check. Total supply voltage is (4.8 + 5.2) = 10V.

Current in 4Ω resistor $= \frac{4.8}{4} = 1.2\text{A}$

,, 6Ω ,, $= \frac{4.8}{6} = 0.8\text{A}$

,, 8Ω ,, $= \frac{4.8}{8} = 0.6\text{A}$

Check. Total current is (1.2 + 0.8 + 0.6) = 2.6A

Similarly: Current in 3Ω resistor $= \frac{5.2}{3} = 1.73\text{A}$

,, ,, 6Ω ,, $= \frac{5.2}{6} = 0.87\text{A}$

Check. Total current is (1.73 + 0.87) = 2.6A.

Example 7. A battery of e.m.f. 42V and internal resistance 7Ω feeds a circuit consisting of three resistors connected in parallel. The resistors have values of 2Ω, 4Ω and 8Ω. Find the battery current, the battery terminal voltage and the current in each resistor.

Let R be the equivalent resistance of the parallel-connected load.

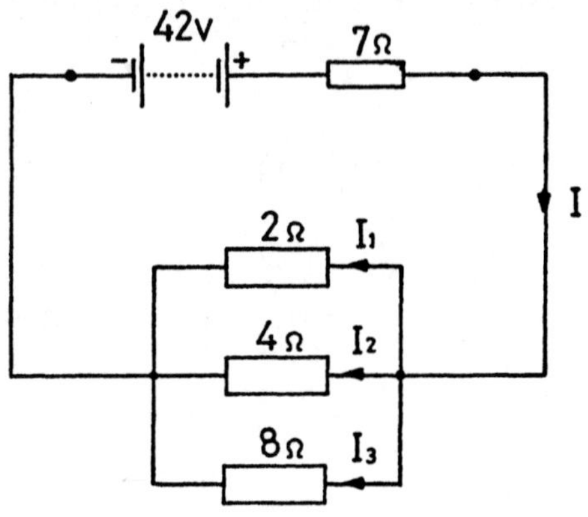

Fig 13

Then $\frac{1}{R} = \frac{1}{2} + \frac{1}{4} + \frac{1}{8} = \frac{7}{8} = 0.875$ S and

$$R = \frac{8}{7} = 1.14\Omega$$

The circuit can now be considered to have a total resistance of 8.14Ω made up from 1.14Ω and 7Ω in series.

The battery current I is given by $\frac{42}{8.14} = 5.16$A

The terminal voltage will be $5.16 \times 1.14 = 5.88$V
or ,, ,, ,, ,, $42 - (7 \times 5.16) = 42 - 36.12$
$= 5.88$V

Current I_1 in 2Ω resistor is $\frac{5.88}{2} = 2.94$A

Current I_2 in 4Ω ,, ,, $\frac{5.88}{4} = 1.47$A

Current I_3 in 8Ω ,, ,, $\frac{5.88}{8} = 0.74$A

Check. Total current $I = 5.16$A.

AMMETERS AND VOLTMETERS

These are the primary instruments used for electrical work and the diagram (Fig 14), shows how they are connected into the circuit. Ammeters are used for measuring current and voltmeters for measuring potential difference or voltage. Both instruments operate on the same principle, but ammeters must be of very low resistance since they are in series with the load and must not result in any appreciable voltage drop. Voltmeters on the other hand must be of high resistance, since they may be connected across points which could be at a high potential difference. For most circuit purposes, the ammeter is considered to have

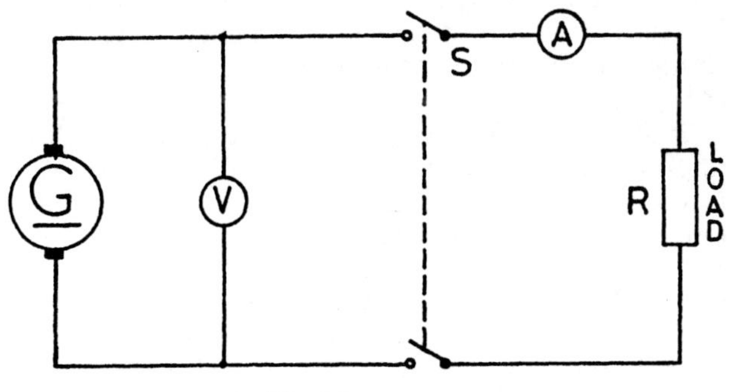

Fig 14

negligible resistance and the voltmeter to have infinite resistance, *ie* to take no current.

In Fig 14 a generator is shown as the energy source, S may be a single-pole or double-pole switch, as is shown here, and R is the load resistance. As a practical example, the generator may have an internal resistance of 0.02Ω, the cable leads may have a total resistance of 0.03Ω and R may have a value of 5Ω. If the generator is set to 220V on open-circuit, *ie* with the switch open, then when the switch is closed a current of $\frac{220}{5 + 0.02 + 0.03} = \frac{220}{5.05} = 43.56$A would flow round the circuit.

The terminal voltage of the generator would 'sit down' to 220 − (43.56 × 0.02) volts = 220 − 0.87 = 219.13V. This would be shown by the voltmeter, while the ammeter would show 43.56A. If the voltmeter was disconnected and then connected directly across R it would indicate 219.13 − (43.56 × 0.03) volts = 219.13 − 1.3 = 217.83V or voltage across R = IR = 43.56 × 5 = 217.83V. The voltage drop in the cables would be 1.3V. It will be seen that the example of a simple distribution system has been worked as a simple series circuit and that the instruments perform their required functions. The ammeter shows the series circuit current, whilst the voltmeter indicates the potential drop across any particular portion of the circuit. It also can record the e.m.f. built up by the generator when the switch is open, since this is the only condition when the e.m.f. appears at the terminals of this energy source.

RANGE OF EXTENSION OF AMMETERS AND VOLTMETERS

For practical work it may not be possible to pass all the circuit current through the ammeter. It may be difficult to construct a suitable instrument because of size or other limitations, and in order to introduce a certain amount of standardisation, it may be easier to use the ammeter with a *shunt* in order to measure the circuit current. Before considering the applications of a shunt it is appropriate here to point out that there are various types of electrical measuring instruments which are described by their 'movements'. Such 'movements' utilise different operating forces and a shunt is normally only used with the 'moving-coil' type since this can be constructed to the highest degrees of accuracy and sensitivity and is ideal for working with various forms of transducer. Transducers are devices which can be made to register both mechanical and electrical quantities. It can be assumed that for subsequent work in this chapter, a moving-coil ammeter or voltmeter is being considered.

A shunt is a specially constructed resistor of low ohmic value and, in order to make an ammeter capable of measuring a current greater than that which can be passed through it, a parallel arrangement of the ammeter and the shunt is used. The ammeter is designed to carry a definite but small fraction of the main current and the rest of the current is made to by-pass the ammeter through the shunt, which is accurately made and set to a definite resistance value. It is calibrated with the ammeter instrument and must always be used with it. The calibrated leads between instrument and shunt form part of the arrangement and must not be cut or substituted for by pieces of ordinary copper wire. The diagram (Fig 15) shows the normal arrangement of instrument and shunt and the example shows the form of calculation necessary. It will be seen that the calculation follows the pattern set for parallel resistance circuits.

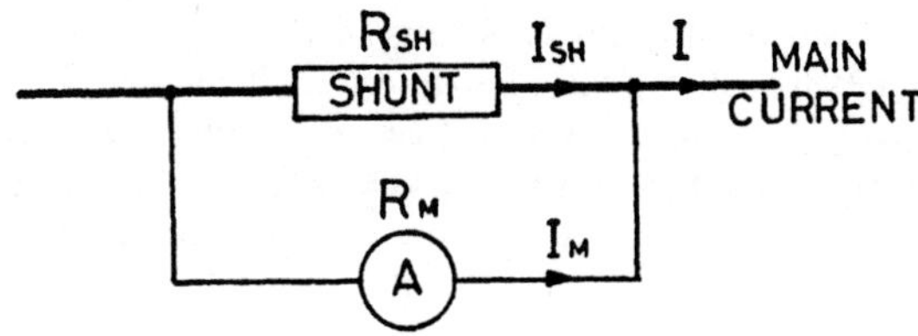

RM INCLUDES RESISTANCE OF INSTRUMENT AND LEADS

Fig 15

Example 8. Calculate the resistance of a shunt required to operate with a moving-coil milliammeter, which gives full-scale deflection for a current of 15mA and which has a resistance of 5Ω. (*Note.* 5Ω can be taken to include the resistance of the connecting leads, since no specific mention of lead resistance has been made.) The combination of meter and shunt is required to read currents up to 100A.

Voltage drop across instrument when giving full-scale deflection = current causing full-scale deflection × resistance of instrument circuit

$= I_M \times R_M = (15 \times 10^{-3}) \times 5 = 75 \times 10^{-3}$ volts

$= 0.075$V or 75mV

Now the voltage drop across the instrument is the same as the voltage drop across the shunt.

Thus $I_{SH} \times R_{SH} = 0.075$ volts.

But the shunt current I_{SH} would be 100 − meter current

$$= 100 - 0.015 = 99.985\text{A}$$

$$\text{So } R_{SH} = \frac{75 \times 10^{-3}}{99.985}\ \Omega$$

$$= 0.000751\Omega$$

It is important to note the low resistance value of the shunt which is designed to carry the current without 'heating up'. The shunt is usually mounted on the switchboard, behind the ammeter and in the main current circuit. The 'light' calibrated leads are coiled to take up any 'slack' and then brought out to the instrument. Thus the ammeter may be marked 0-100 amperes, but in actual fact only a minute current, some 15mA, passes through the instrument itself. The remainder and by far the largest proportion of the current, passes through the shunt. The reason for always using the instrument with its own calibrated shunt and leads is thus obvious.

To measure voltages higher than that for which the instrument movement is designed a *series* or *range* resistor must be used. This resistor is designed to drop the excess voltage and dissipates a certain amount of heat. It consists of special fine-gauge wire wound on a porcelain spool or on a mica card, the whole being mounted inside a ventilated case. Here again the arrangement may be mounted behind the switchboard, if it cannot be contained in the case of the instrument. Thin leads for carrying the small instrument current connect the range resistor unit and the instrument to the main supply terminals, usually through fuses. Thus the voltmeter may be scaled 0–250 volts, but in fact only 0.075V may be dropped across it, when full-scale deflection occurs. By far the major voltage drop occurs across the range resistor, which is always high in ohmic value: thousands of ohms. This fact should be noted. The diagram (Fig 16) shows the

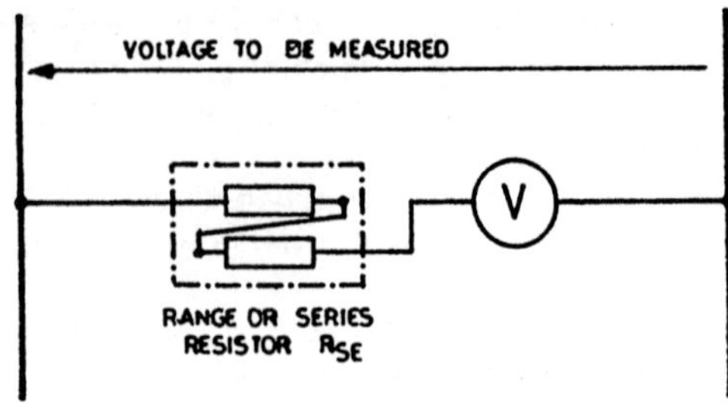

Fig 16

arrangement and Example 9 shows how the value of an appropriate range resistor is calculated.

Example 9. Calculate the resistance of the range resistor required to be placed in series with the instrument of Example 8 to make it into a voltmeter reading 0–250V. (The instrument has a resistance of 5Ω and gives full-scale deflection with a current of 15mA.)

The current through the complete voltmeter circuit must be limited to 15mA, otherwise the instrument would be 'burnt out'.

Thus resistance of the voltmeter circuit must be

$$\frac{250}{15 \times 10^{-3}} = 16.667 \times 10^3\Omega = 16.667\ \text{k}\Omega$$

But the instrument already has a resistance of 5Ω, therefore the series or range resistor R_{SE} must have a value of (16 667 − 5) ohms = 16 662Ω.

The actual 'movement' or working unit of an ammeter or voltmeter is much the same and it is the use of a shunt or range resistor which decides whether current or voltage can be measured. Multi-purpose portable test instruments are available which can make a range of measurements. A range switch, termed a *'range multiplier'* is provided which makes the appropriate connection of shunt or range resistors.

Consider an instrument movement in which 15mA at a p.d. of 75mV gives full-scale deflection. Its resistance $= \dfrac{75 \times 10^{-3}}{15 \times 10^{-3}}$

$= 5\Omega$.

If a voltage range 0–15V is required, the instrument circuit resistance must be $\dfrac{15}{15 \times 10^{-3}} = 1000\Omega$ and a range resistor of 1000 − 5 = 995Ω must be switched in.

Similarly if a voltage range of 0–150V is required the range resistor must be $\dfrac{150}{15 \times 10^{-3}} - 5 = 10\,000 - 5 = 9995\Omega$.

If a current range 0–5A is required, a shunt is used whose value can be obtained thus:

P.D. across shunt = p.d. across instrument movement for full-scale deflection

or ,, ,, = 75mV = 75 × 10^{-3} volts

The current through the shunt = 5 − 0.015 = 4.985A

and the resistance of the shunt would be

$$\frac{75 \times 10^{-3}}{4.985} = 0.015\ 05\Omega$$

SENSITIVITY

The term is used to denote the suitability of a measuring instrument for a particular purpose. If, for example, a voltmeter is so poorly constructed that it requires a comparatively large current for full-scale deflection (f.s.d.), it will be apparent that the overall circuit current would be adversely affected when such an instrument is connected across any particular part of the circuit. This is of the greatest importance for electronic circuitry. Consider a component of resistance value 1kΩ forming part of a series circuit drawing 1mA. A voltmeter of resistance 5kΩ connected across such a component would lower the resistance of the parallel arrangement to 0.803kΩ.

Note $\frac{1}{R} = \frac{1}{5} + \frac{1}{1} = \frac{6}{5} = 1.25$ or $R = \frac{5}{6} = 0.803\text{k}\Omega$

Accordingly the circuit current would also rise appreciably and the overall circuit conditions would be altered — an undesirable effect. The higher the resistance value of the voltmeter, the less the effect and voltmeters are therefore often given a 'sensitivity' figure of ohms per volt. Thus a meter rated at 20kΩ/v would require a current of $\frac{1}{20 \times 10^3}$ amperes or $\frac{1}{20}$ milliamperes or $\frac{1}{20} \times 10^{-3} = 50\mu\text{A}$ for full-scale deflection and the range resistors required would be calculated on this basis. Such a voltmeter connected across the component of the example would have little effect on the circuit current and should be the instrument used.

CHAPTER 1

PRACTICE EXAMPLES

1. A circuit is made up from four resistors of value 2Ω, 4Ω, 5Ω and 10Ω connected in parallel. If the current is 8.6A, find the voltage drop across the arrangement and the current in each resistor.

2. One resistor group consists of 4Ω, 6Ω and 8Ω connected in parallel and a second group consists of 3Ω and 6Ω in parallel. The two groups are connected in series across a 24V supply. Calculate (a) the circuit current, (b) the p.d. across each group, (c) the current in each resistor.

3. If the resistor arrangement of Q1 is connected to a 12V battery of internal resistance 0.65Ω, find the circuit current and the battery terminal voltage. Find also, the current in the 5Ω resistor.

4. A moving-coil instrument has a resistance of 10Ω and requires a current of 15mA to give a full-scale deflection. Calculate the resistance value of the resistor necessary to enable it to be used to measure (a) currents up to 25A, (b) voltages up to 500V.

5. Two resistors of 60kΩ and 40kΩ value are connected in series across a 240V supply and a voltmeter having a resistance value of 40kΩ is connected across the 40kΩ resistor. What is the reading on the voltmeter?

6. When a 10Ω resistor is connected across a battery, the current is measured to be 0.18A. If similarly tested with a 25Ω resistor, the current is measured to be 0.08A. Find the e.m.f. of the battery and its internal resistance. Neglect the resistance of the ammeter used to measure the current.

7. Two groups of resistors A and B are connected in series. Group A consists of four resistors of values 2Ω, 4Ω, 6Ω and 8Ω connected in parallel and group B consists of two resistors of values 10Ω and 15Ω in parallel. If the current in the 4Ω resistor is 1.5A, calculate, (a) the current in each of the remaining resistors, (b) the supply voltage, (c) the voltage drop across the groups A and B.

8. The voltage of a d.c. generator, when supplying a current of 75A to a load, is measured to be 108.8V at the switchboard. At the load, the voltage recorded is 105V and when the load is switched off the voltage rises to 110V. Find the internal resistance of the generator, the resistance of the supply cables and estimate the fault current if a 'short-circuit' of negligible resistance occurred at the load terminals.

9. The ammeter on a switchboard, scaled 0–300A is accidentally damaged. The associated shunt is marked 300A, 150mV. A small ammeter, scaled 0–1A with a resistance of 0.12Ω, is available, and the possibility of using this is considered. Find if such an arrangement is possible, and if so, how it could be achieved using surplus resistors which are also available.

10. Five resistors AB, BC, CD, DE and EA are connected to form a closed ring ABCDEA. A supply of 90V is connected across AD, A being positive. The following is known about the resistors: AB is 10Ω, BC is of unknown value R_1 ohms, CD is of unknown value R_2 ohms. DE is 6Ω and EA is 9Ω. A high resistance voltmeter (taking negligible current) when connected across BE reads 34V with B positive and when connected across CE reads 6V with E positive. Find the values of R_1 and R_2, the current in branch ABCD and the main supply current.

CHAPTER 2

THE ELECTRIC CIRCUIT (CONTINUED): UNITS

All engineering studies stress the need for units and an introduction to some of these will have been made when the subjects of mechanics and heat were being covered. Units allow measurements to be taken and calculations to be made. They are essential to the derivation of formulae from the basics of theory and enable presentations of related principles to be evolved. Thus for electrical engineering, even at the Chapter 1 stage, the ampere, volt and ohm were considered, and although these units have yet to be defined, their importance in relation to the basic electric circuit will have been appreciated. The student will also recognise these units as being amongst those in common everyday usage. If, however, the impression has been given that the study of electrotechnology will involve the knowledge of an entire range of new units, then it is stressed that this is not the case. The whole modern concept of engineering technology is based on the universal adoption of SI units and, since some of these have been encountered in earlier work, it will not be long before, in this study of electrical engineering, common ground is being covered and the relevant relationships with associated units, already treated from the mechanical engineering aspect, are being stressed.

Before proceeding with any further study of units of the SI system, it would be useful to introduce an historical note and consider the situation in engineering as it has developed. Towards the end of the last century two systems of units began to be employed in engineering; the British or foot-pound-second (fps) system and a metric or centimetre-gramme-second (cgs) system. The British or Imperial system had no merits since all units of the same kind, such as those of length, area, volume etc, bore no relation to each other, indeed there were also additional units such as the calorie and horsepower which were arbitrarily and sometimes differently defined. The metric system was primarily devised as a benefit to industry and commerce but physicists soon realised its advantages and it was adopted prior to 1870 in scientific and technical circles. In 1873 the British Association for the Advancement of Science selected the centimetre and the gramme as basic units of length and mass for physical purposes. Measurement of other quantities called for a base-unit of time and the adoption of the second, for this purpose, gave the centimetre-gramme-second (cgs) system.

The metric system, in the cgs form, was adopted to a large extent for electrical engineering in the early days of development. The system had the advantage that all the same kind of quantities are multiples of ten and it was also international. The sizes of the absolute unit of the centimetre and the gramme were found to give rise to difficulties for the desired electrical units which became either too large or too small for practical working. The use of these absolute units for essential formulae in engineering work also proved difficult and thus more workable or practical units had to be devised. Such practical units were to include the volt, ampere and ohm. In about 1900 practical measurement in metric units began to be based on the metre, kilogramme and the second and the aforementioned electrical practical units. These constituted the unrationalised MKS system.

The next development came from a fact, which was repeatedly pointed out over the first half of the present century, that a system of units could be devised to make the practical units of volt, ampere and ohm the absolute units of such a system. If, in addition, suitable adjustments of certain constants encountered in electromagnetism and electrostatics are accepted, then a more workable system of units would result. This system was known as the rationalised MKS system and its adoption was recommended by the International Electrotechnical Commission of 1950. The change to the MKS caused some little inconvenience to the older electrical engineer and necessitated the revision of many of the better known works of reference and text-books. The student was required to appreciate however, that the new system did not upset the course of learning in any way and that, if anything, the units introduced made matters easier and formulae more manageable.

Prior to 1970, conditions did exist when both the older Imperial and the newer rationalised MKS systems of units were in use. The latest extension of metric units into all branches of commerce and industry has enabled engineering to evolve the SI system, the units of which are used throughout this book. Thus from the electrical viewpoint, it can be said that the SI system is the rationalised MKS system with units in all the other fields of measurement being fully metricated.

THE SI SYSTEM

All measurement consists in comparison with some standard or unit. The three fundamental units are those of length, of mass and of time. In the SI system the metre is taken as the fundamental unit of length (s), the kilogramme as the unit of mass (m) and the second as the unit of time (t). From the fundamental units, can be built up the derived units, which can be further classified as mechanical or electrical units. Thus Force is a derived mechanical unit involving a fundamental unit and a derived unit, *ie* mass and acceleration. For the SI system, a unit of force, called the *Newton*, has been introduced. Velocity is similarly a derived unit involving distance and time. So also is acceleration a derived unit, involving velocity and time. Both velocity and acceleration are mechanical units. The ampere is really a derived unit involving force and length but as stated previously it is used as a fundamental electrical unit. Other electrical units are the volt and the ohm which are also derived units, but the *Joule* and *Watt*, although used principally in the past in connection with electrical engineering, can be derived from mechanical relationships and will be defined later under this heading.

Once the units are recognised and understood, the reader is advised to discontinue their classification as mechanical or electrical units and to accept them as general engineering units. This applies particularly to the units of work and power. Both the mechanical and electrical engineering fields are concerned with common appliances or associated problems and a ready use of the appropriate units, with a correct appraisal of the magnitudes of the quantities involved, is essential to the practising engineer.

MECHANICAL UNITS

The fundamental units require little definition since they are accepted standards. Thus the metre is the absolute standard, taken as the distance between two marks on a certain metal bar. Similarly the kilogramme is the mass of an accepted standard lump of metal. The time unit is the second, which is defined as $\frac{1}{86\,400}$ of a mean solar day.

Most of the principal SI derived units will already have been introduced to the student but a revision is made here to allow an extension into the field of electrical engineering.

UNIT OF FORCE

Force can be defined as that which tends to cause an object to move, to change that motion or to keep that object at rest.

THE NEWTON. This is the force required to accelerate a mass of one kilogramme at a rate of one metre/second^2 (*ie* F = ma NEWTON). However, if the acceleration of the mass is due to gravity alone, the rate of acceleration is a constant 9.81 m/s^2. This acceleration due to gravity is given the symbol 'g',

$$\therefore \mathrm{F} = \mathrm{mg\ NEWTON.}$$

This force due to gravitation is referred to as *weight.* Thus a mass of 1kg has a weight of 1kg × 9.81m/s^2 = 9.81N

ie 1kg = 9.81N

The symbol for force is F but any value of newtons can be represented by the letter N after the numerical value.

UNIT OF WORK AND ENERGY

THE JOULE. This can be defined as the *work* done or *energy* stored when a force of one newton acts through a distance of one metre in the direction of the force.

The symbol for work or energy is W but any value in joules can be represented by the letter J after the numerical value.

From the definition, it follows that a force of F newtons, acting through a distance of s metres, does $F \times s$ newton metres of work or $F \times s$ joules.

Thus: W (joules) = F (newtons) × s (metres)

UNIT OF POWER

THE WATT. *Power* is the rate at which work is done or energy is converted and the unit is the watt. A watt is the power resulting, when a joule of energy is expended in a second.

The symbol for power or rate of doing work is P but any value in watts can be represented by the letter W after the numerical value.

The definition can be more generally written as

$$P\,(\text{watts}) = \frac{W\,(\text{joules})}{t\,(\text{seconds})}$$

The joule and the watt are the units originally used in electrical engineering and they will be encountered constantly in electrical problems. Example 10 is set out here to serve as an introduction to electro/mechanical relationships.

Example 10. A pump is required to lift 1200 litres of water through 10 metres in 6 minutes. Calculate the work done in joules and the power rating of the pump. Assume 1 litre of water to have a mass of 1 kilogramme.

Work done = force of gravity × distance lifted

$$\text{Thus } W = Fs = mgs$$
$$= (1200 \times 9.81) \times 10$$
$$= 12 \times 9.81 \times 10^3 \text{ newton metres}$$
$$= 117.72 \times 10^3 = 117\,720\text{Nm or } 117\,720\text{J}$$

$$\text{Power} = \frac{\text{work done}}{\text{time}} = \frac{117\,720}{6 \times 60} = 327\text{W}$$

Power rating of pump would be 327 watts.

Note. In the above problem no account has been taken of machine efficiency. This will be introduced in due course, but for the example, the practical rating figure of the electric motor, assuming this to be the means of driving the pump, would be larger.

ELECTRICAL UNITS

The same fundamental units are used as for the mechanical units namely: the metre, kilogramme and second. The primary derived unit is the ampere, which has been adopted as the basic electrical unit of current and as a fourth fundamental unit. Before considering the definition for the ampere, it is necessary at this stage, to describe two associated effects, which would be observed when a current flowed in a circuit.

(1) If the resistance of the circuit was concentrated in a short length of conductor, then a temperature rise of the wire in this region would be noted, showing a conversion of electrical energy into heat energy.

(2) If the circuit was supplied through two wires laid together, then especially if the current is large and the wires flexible, a mechanical effect would be noted. When the current is switched on, the wires would be observed to move and this electromagnetic effect, as it is called, has been used to define the ampere for the SI system. The factors governing the magnitude and direction of the force on the wires will be described in the chapter on Electromagnetism.

UNIT OF CURRENT

THE AMPERE. This is that current which, when maintained in each of two infinitely long, straight, parallel conductors situated in a vacuum and separated by a distance of one metre between centres, produces on each conductor a force of 2×10^{-7} newtons per metre length of conductor.

As stated in Chapter 1, the symbol for current is I and any value in amperes is represented by the letter A after the numerical value. The reader is reminded that practical circuit currents may range from thousands of amperes to minute values of micro-amperes and attention is drawn to the Table of Prefixes of Magnitudes as given at the front of this book. Full consideration *must be* given to the correct use of the abbreviation which follows the numerical value.

When a current flows for a given time, a quantity of electricity is said to be conveyed round the circuit. The quantity which passes can be shown to be related to the work done in the circuit, but before this relationship is considered further, it is necessary to define quantity of electricity in terms of current and time.

UNIT OF QUANTITY

THE COULOMB. The usual unit—sometimes called the ampere second. For practical purposes a larger unit, for everyday electrical engineering is used. This is the *Ampere hour* as used in connection with the capacity of batteries and for accumulator charging.

The symbol for quantity of electricity is Q and any value in coulombs can be represented by the letter C after the numerical value. Any value in ampere hours is represented by the letters Ah after the numerical value. Since the quantity of electricity which is conveyed round a circuit would vary with the strength of the flow of electricity and with time, a simple definition for the coulomb can be deduced thus:

A coulomb is the quantity of electricity conveyed by a steady current of one ampere flowing for a time of one second.

Thus Q (coulombs) $= I$ (amperes) $\times\ t$ (seconds)
or Q (ampere hours) $= I$ (amperes) $\times\ t$ (hours)

From the above, the following can be deduced:

$$\begin{aligned} 1 \text{ ampere hour} &= 1 \text{ ampere} \times 1 \text{ hour} \\ &= 1 \text{ ampere} \times 3600 \text{ seconds} \\ &= 3600 \text{ ampere-seconds} \\ &= 3600 \text{ coulombs} \\ \text{Thus } 1\text{A h} &= 3600 \text{ C} \end{aligned}$$

Example 11. Consider Example 5, where a battery of e.m.f. 42V and internal resistance 7Ω is used to supply a circuit of three resistors 2, 4 and 8Ω in series. If the current is switched on for 30 minutes, find the quantity of electricity which would have been conveyed.

Total resistance of circuit $= 7 + 2 + 4 + 8 = 21\Omega$

Circuit Current $= \dfrac{42}{21} = 2\text{A}$

Quantity of Electricity = current × time in seconds

$= 2 \times 30 \times 60 = 3600\text{C}$

or Quantity of Electricity = current × time in hours

$= 2 \times \dfrac{30}{60} = 1\text{A h}.$

The passage of an electric current results in energy being expended. This energy may appear as the work done by the rotation of an electric motor, as the action of heating up a furnace element or as the agency responsible for the electrolytic dissociation of a salt solution. The relation between conveying a quantity of electricity round a circuit by an applied voltage and the resulting work done can be used to derive the units of voltage and resistance in terms of the coulomb and the joule which have already been defined.

UNIT OF VOLTAGE

THE VOLT. This is the unit of electromotive force and potential difference and can be defined as the e.m.f. to be applied, or the p.d. available between two points in a circuit, if one joule of work is to be done when passing one coulomb of electricity between the points.

As stated in Chapter 1, the symbol for voltage or e.m.f. is V and any value in volts is represented by the letter V after the numerical value. In accord with the remarks made by concerning the representation of current, the reader's attention is drawn to the Table of Prefixes of Magnitudes, and to the correct use of the Abbreviations.

From the definition set out above it is stated that the work done by part of an electric circuit equals the voltage applied across that part of the circuit times the quantity of electricity conveyed.

Thus: W(joules) = V(volts) × Q(coulombs)

or W(joules) = V(volts) × I(amperes) × t(seconds)

$$W = VIt = I^2Rt$$

Example 12. Consider Example 11. A battery of e.m.f. 42V and internal resistance 7Ω is used to supply a circuit of three resistors, 2, 4 and 8Ω in series. If the current is switched on for 30 minutes, find the energy converted (as heat) by each resistor and inside the battery itself.

Circuit current was found to be 2 amperes.
Using form $W = I^2Rt$ then energy converted in
2 ohm resistor $= 2^2 \times 2 \times 30 \times 60 = 14\,400$ joules
4 ohm resistor $= 2^2 \times 4 \times 30 \times 60 = 28\,800$ joules
8 ohm resistor $= 2^2 \times 8 \times 30 \times 60 = 57\,600$ joules
7 ohm battery $= 2^2 \times 7 \times 30 \times 60 = 50\,400$ joules
Total energy converted by the circuit =
$14\,400 + 28\,800 + 57\,600 + 50\,400 = 151\,200$ joules

Check. The total energy converted by the entire circuit may be found from $W = VIt$ joules
$= 42 \times 2 \times 30 \times 60 = 151\,200$ joules.

The definitions of Power and Energy have already been considered, but it would be as well to summarise the points of importance, namely that power is the rate at which work is done and the unit is the Watt. Thus: $P = \frac{W}{t}$ or $W = Pt$

From deductions set out above $W = VIt$ and it follows that $P = VI$ or P (watts) $= V$ (volts) $\times$ I (amperes).

The above is a most important relationship, It can also be expressed in the following forms:

$$P = I^2 R \text{ or } P = \frac{V^2}{R}$$

The attention of the reader is drawn to the following which must also be known.

Since Energy = Power × time
1 Joule = 1 Watt-second

Now a joule is a small unit of energy and for practical purposes a much larger electrical unit of energy is used. This is the *kilowatt hour*, abbreviated to ***kW h*** and is also known as the commercial unit of electricity or more commonly as 'a unit'.

Since one kilowatt hour = one kilowatt × one hour
= 1000 watts × 3600 seconds
So one kilowatt hour = 3 600 000 joules.

Example 13. A 220V electric fire is rated at 2kW. Find the current taken when the fire is switched on and also how much it would cost to use the fire for 5 hours with electricity being charged at 6p per unit.

Current taken $= \frac{2 \times 1000}{220} = 9.09$A

Electricity used $= 2 \times 5 = 10$ kW h $= 10$ units
Cost $= 10 \times 6 = 60$p.

UNIT OF RESISTANCE

THE OHM. This was defined in Chapter 1 as the unit of resistance and in terms of the volt and ampere thus: a resistor has a value of one ohm resistance, if one ampere passes through it when a potential difference of one volt is applied across its ends. Now that the relations between the ampere, volt, joule and watt have been defined, it is possible to give a further definition for the ohm which is associated with power or energy dissipation. Thus the ohm can be defined as: that resistance which when one ampere passes through it dissipates energy at the rate of one joule per second (*ie* one watt).

Alternatively, the ohm is that resistance in which a current of one ampere flowing for one second generates a joule of energy.

For a resistor the energy produced by current flow appears as heat, and the following is of importance.

Since $P = VI$ and $V = IR$ then $P = (IR)I$

or $P = I^2R$ as developed earlier.

Power dissipated in a resistor is thus proportional to the current squared. If the current was doubled by raising the voltage, the power dissipation would be four times as large. The temperature would rise in proportion and assuming the resistor was capable of carrying its normal current only and had very little capacity for working at a higher temperature, then a 'burnout' would occur. The same limitations apply to cables, electrical machines and switchgear. Electrical equipment is assigned a rating which, on full load, enables it to operate with a safe temperature rise. An increase of the normal rated current, brought about by overloading or by an overvoltage, results in a temperature rise proportional to the new current squared. The total temperature will rise very rapidly as the overcurrent occurs and if this is maintained then damage will result. Damage to electrical insulating materials can occur because of sustained overloads and overheating should be regarded as the main cause of failures of electrical machines.

Example 14. A hot-plate of a ship's electric galley is fitted with a control marked High, Medium and Low. The heating element consists of two equal sections, which are connected in parallel for High and in series for Low. Only one section is used for Medium. If the plate when set at high is rated at 2kW on 220V, find the wattage rating when the control is set at Low and at Medium.

Current taken at High. Two sections in parallel

$$= \frac{2000}{220} = 9.09\text{A}$$

$$\text{Current taken by 1 section} = \frac{1000}{220} = 4.545\text{A}$$

$$\text{Resistance of 1 section} = \frac{220}{4.545} = 48.41\Omega$$

Resistance of 2 sections in series = 96.82Ω
Current taken at Low. Two sections in series

$$= \frac{220}{96.82} = 2.27\text{A}$$

Power dissipated = 220 × 2.27 = 499.4W
or wattage rating = 500W approximately.
Current taken at Medium. One section only across 220V
or I = 4.545A
Wattage rating is one half that of the High setting = 1kW or wattage rating = 220 × 4.545 = 1000 Watts = 1kW.

EXAMPLES RELATING MECHANICAL AND ELECTRICAL ENERGY

The best understanding of the various units as discussed earlier, is achieved, by considering examples where mechanical work is performed by electrical means or vice versa. The relations between the units will need to be introduced at all times, but it is necessary before proceeding to such examples, to stress that no machine is perfect and that its overall performance is measured by its efficiency.

EFFICIENCY

The symbol usually used is η—the Greek letter eta. In all apparatus and machines, losses of energy occur due to bearing and brush friction, air turbulence, unwanted electrical currents, escape of heat, etc. These losses result in the output of such apparatus or machines, when measured as work, being always less than the input when this is measured in the same work units. The ratio of the output to the input is termed the efficiency. Thus:

$$\text{Efficiency} = \frac{\text{output}}{\text{input}} \text{ or } = \frac{\text{input} - \text{losses}}{\text{input}}$$

$$\text{or} \quad \text{,,} \quad = \frac{\text{output}}{\text{output} + \text{losses}}$$

Efficiency is usually expressed as a percentage.

Example 15. A diesel engine has a measured indicated power of 7.5kW and a mechanical efficiency of 85 per cent. It drives a generator which supplies a lamp load at 110V. How many 60W lamps can be supplied, if the efficiency of the generator is measured to be 88 per cent? Find the total load current.

The output of the engine = input × efficiency

$$= 7.5 \times \frac{85}{100} = 6.375\text{kW}$$

At the coupling between engine and generator, it can be assumed that there is no loss of energy, so the power input to the generator must be the power output of the engine and it follows that:

Output of engine = input to generator = 6.375kW

Thus generator output = input × efficiency

$$= 6.375 \times \frac{88}{100} = 5.61\text{kW}$$

$$\text{Number of lamps} = \frac{5610}{60} = 93.5 \text{ say } 93$$

$$\text{Load current} = \frac{93 \times 60}{110}$$

$$= 50.73\text{A say } 51\text{A}$$

or alternatively;

$$\text{Load current} = \frac{5610}{110} = 51\text{A.}$$

Example 16. A pump is required to lift 12 tonnes of water through 10m in 2 minutes. Calculate the power required to drive the pump, the current taken if driven by a 220V motor and the cost of pumping at 5p per unit. Assume the efficiency of the pump to be 60 per cent and the efficiency of the driving motor is 85 per cent.

Work to be done = Force opposing gravity × distance lifted

Note 1 tonne = 10^3kg

Thus work to be done = $(12 \times 10^3 \times 9.81) \times 10$ newton metres

$$= 117.72 \times 10^4\text{Nm}$$

Also 1 177 200Nm = 1177.2kJ

This is the output of the pump. The input would be greater, *ie*

$$1177.2 \times \frac{100}{60} = 1962\text{kJ}$$

Since the pumping is to be accomplished in 2 minutes or 120 seconds, the power input during this time $= \dfrac{1\,962\,000}{120}$

$$= 16\,350\text{W}$$

Thus power required to drive the pump is 16.35kW.

The output power rating of the motor must be 16.35kW and the input power would be $16.35 \times \dfrac{100}{85} = 19.24\text{kW}$

$$\text{Current taken by motor} = \frac{19\,240}{220} = \frac{962}{11} = 87.45\text{A}$$

$$\text{Energy used} = 19.24 \times \frac{2}{60} = \frac{19.24}{30}$$

$$= 0.641\text{kW h}$$

$$\text{Cost} = 0.641 \times 5 = 3.2\text{p.}$$

Example 17. The electric motor used to drive a ship's winch has an efficiency of 86 per cent. The winch can lift a mass of 0.5 tonnes through a distance of 22m in 22 seconds. The winding gear of the winch has an efficiency of 60 per cent. Calculate the power rating of the motor and also the current taken from the 220V ship's mains.

Work done by the winch = $500 \times 9.81 \times 22$ newton metres
= 107 910Nm or 107 910J

This is the output of the winch or the output of the winding gear. The input to the winding gear would be

$$107\,910 \times \frac{100}{60} = 179\,850\text{J}$$

The input to the winding gear would also be the motor output = 179 850J.

Since the lifting is done in 22 seconds, the motor would give out power during this time.

$$= \frac{179\,850}{22} = 8175\text{W}$$

For a motor output of 8175W, the input power would be

$$8175 \times \frac{100}{86} = 9505.8\text{W}$$

Thus power rating of motor is 9.51kW

$$\text{Input or motor current} = \frac{9506}{220} = 43.21\text{A.}$$

Example 18. A storage battery is provided for emergency use aboard a ship. The battery is arranged to supply certain essential services during the period of time taken to start-up the 'stand-by' generator. The principal load to be supplied by the battery is the 'emergency' motor for an electric-hydraulic steering gear. This motor is rated at 220V, 15kW, and has an efficiency of 88 per cent. The battery is to be of a capacity sufficient to operate this motor and an additional lighting load of twenty 60W lamps for a period of 30 minutes. Estimate the size of the battery and also its discharge current.

Output of motor = 15kW

Input to motor $= 15 \times \frac{100}{88} = \frac{187.5}{11} = 17.045\text{kW}$

Input current to motor $= \frac{17\,045}{220} = \frac{852.3}{11} = 77.5\text{A}$

Lighting load = 60 × 20 = 1200W

Lighting current $= \frac{1200}{220} = \frac{60}{11} = 5.45\text{A}$

Total current = 77.5 + 5.45 = 82.95A
or Discharge current = 82.95A

Size of battery $= 82.95 \times \frac{30}{60} = 41.475\text{A h.}$

GROUPING OF CELLS

Ohm's law states that the current in a circuit can be increased by raising the potential difference applied across the circuit or by decreasing the circuit resistance. If the supply source is a generator, the applied p.d. can be varied by controlling the e.m.f. being generated in the machine, but if a battery is the source of energy then the applied voltage cannot be varied easily. Since a battery consists of a group of cells and since the e.m.f. of any cell is fixed, being decided by its chemical composition, then a larger e.m.f. or a greater current can only be obtained by appropriate arrangement of the cells. The cells can be connected in series, parallel or series-parallel arrangements.

SERIES CONNECTION. For this arrangement the −ve terminal of a cell is connected to the +ve terminal of the adjacent cell as shown in the diagram (Fig 17a). The arrangement is more simply depicted by Fig 17b. A battery of 3 cells in series is shown.

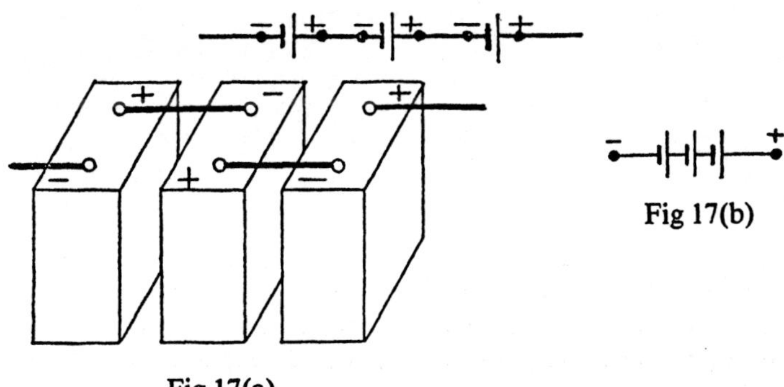

Fig 17(b)

Fig 17(a)

From Kirchhoff's voltage law, the e.m.f. of the source is equal to the sum of the e.m.f.s taken round the circuit and thus for a battery of n cells in series, the e.m.f. = e.m.f. of 1 cell × n. Also since this is a series circuit, then the current in any 1 cell is the circuit current. The internal resistances of the cells are also in series and should be treated in accordance with the deductions already made for the resistance of a series circuit. These points are illustrated by the example.

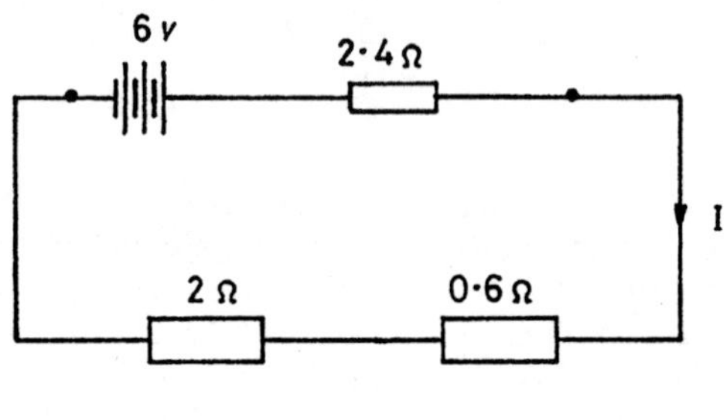

Fig 18

Example 19. A battery consists of 4 cells in series, each of e.m.f. 1.5V and internal resistance 0.6Ω. Find the current flowing, if the battery is connected to two resistors of 2Ω and 0.6Ω which are connected in series. The arrangement is shown by the diagram (Fig 18).

Total e.m.f. = 4 × 1.5 = 6V

Total battery internal resistance = 4 × 0.6 = 2.4Ω

Total circuit resistance = 2.4 + 2 + 0.6 = 5Ω

So circuit current $= \frac{6}{5} = 1.2$A

Other values of interest would be

Battery terminal voltage = 1.2 × 2.6 = 3.12V

or Battery ,, ,, = 6 − (1.2 × 2.4)
= 6 − 2.88
= 3.12V

Voltage drop across each resistance = 1.2 × 2 = 2.4V
and 1.2 × 0.6 = 0.72V

Current in 1 cell = circuit current = 1.2A.

PARALLEL CONNECTION. For this arrangement, the +ve terminals of all the cells are connected together as are all the −ve terminals. The arrangement is as shown in the diagram (Fig 19).

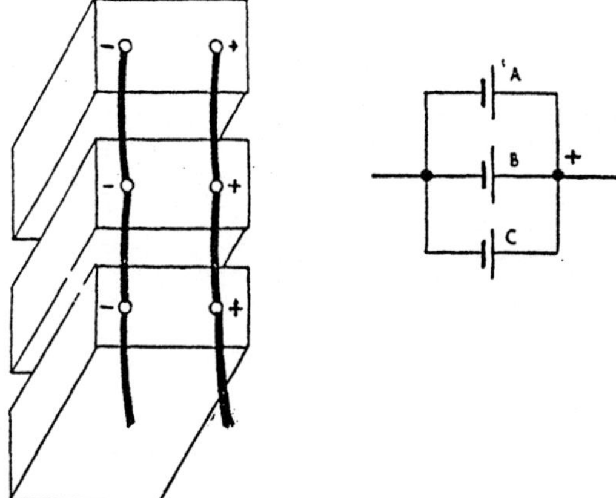

Fig 19

From Kirchhoff's current law, the total current is the sum of the currents in each branch. Thus the total current from the battery is equal to the sum of the currents available from each cell. For correct working, the e.m.f. of each cell should be the same. So also should the internal resistance although this is not essential. If n cells are in parallel, the total current is n times that given by one cell, but the battery e.m.f. is that of any one cell. This latter point can be reasoned from the fact that if the +ve terminal of A is 2V above its −ve terminal and the +ve terminal of B is 2V above its −ve terminal, then the +ve connection between A and B is 2V above the −ve connection. If this is carried on for cell C and any further number of cells then it is seen that the whole +ve connection is 2V above the −ve connection, *ie* the battery voltage is 2V.

The battery internal resistance is obtained from the parallel-resistance formula, *ie* it is $\frac{1}{n}$ th of a cell resistance. The battery resistance once determined, is added to the external resistance to give the total circuit resistance as in the following example.

Example 20. A battery consists of 4 cells in parallel, each of e.m.f. 1.5V and internal resistance 0.6Ω. Find the current flowing if connected to a resistance of 2.6Ω. The arrangement is shown in the diagram (Fig 20).

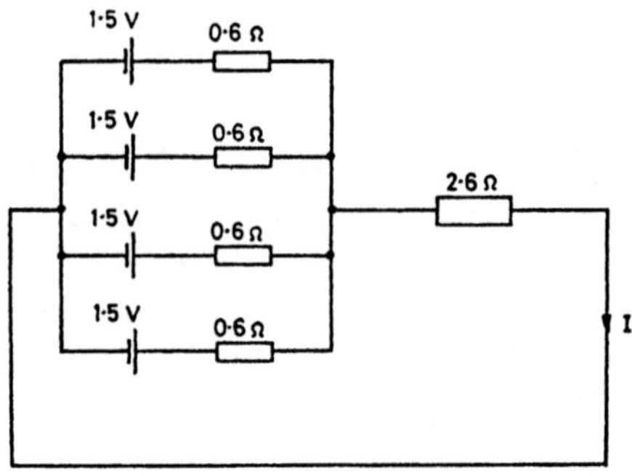

Fig 20

E.m.f. of battery = e.m.f. of 1 cell = 1.5V

Internal resistance of battery = $\frac{0.6}{4}$ = 0.15Ω

Total resistance of circuit = 2.6 + 0.15 = 2.75Ω

Current = $\frac{1.5}{2.75}$ = 0.545A. Other information would be

Terminal voltage = 0.545 × 2.6 = 1.418V

or ,, ,, = 1.5 − (0.545 × 0.15) = 1.5 − 0.082 = 1.418V

Current of 1 cell = $\frac{0.545}{4}$ = 0.136A.

SERIES-PARALLEL CONNECTION. To build up this arrangement a number of cells which are connected in series are then connected in parallel with a similar number of cells in series. The arrangement is as shown in the diagram (Fig 21) and is used to provide both increased voltage and current for a circuit. The cells in series provide the increased e.m.f. and the parallel banks of cells supply the extra current.

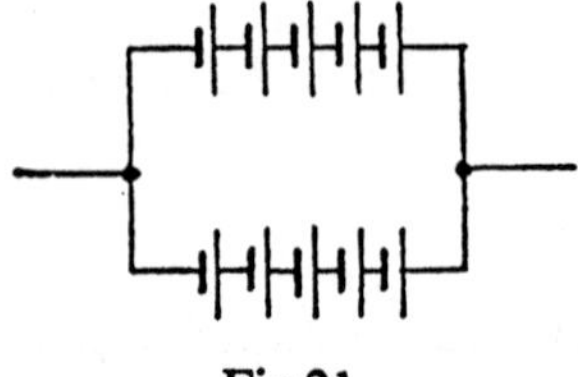

Fig 21

The procedure for solving problems follows the reasoning already covered for the series and parallel arrangements, but should be taken step by step.

Example 21. Ten cells each of internal resistance 3Ω and e.m.f. 2V are connected in two banks of 5 cells per bank. They are then connected to an external load resistance of 20Ω. Find the load current and the p.d. across the battery terminals. The arrangement is shown in the diagram (Fig 22a).

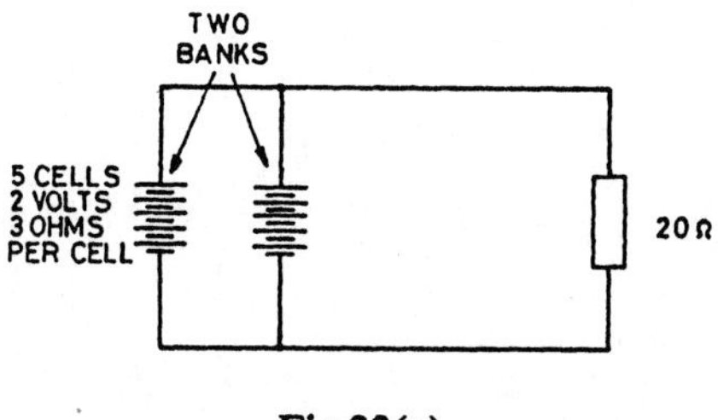

Fig 22(a)

E.m.f. of a bank = 5 × 2 = 10V = battery e.m.f.

Resistance of 1 bank = 5 × 3 = 15Ω

$$\text{Resistance of battery} = \frac{15}{2} = 7.5\Omega$$

Total circuit resistance = 7.5 + 20 = 27.5Ω

$$\text{Circuit or load current} = \frac{10}{27.5} = 0.364\text{A}$$

P.d. or terminal voltage = 0.364 × 20 = 7.28V

$$\text{Current per cell} = \text{current of 1 bank} = \frac{0.364}{2} = 0.182\text{A}$$

Example 21 (continued). If the battery is rearranged with 5 banks of 2 cells in each, find the new current and voltage. The arrangement is shown in the diagram (Fig 22b).

E.m.f. of a bank = 2 × 2 = 4V = battery e.m.f.

Internal resistance of a 1 bank = 2 × 3 = 6Ω

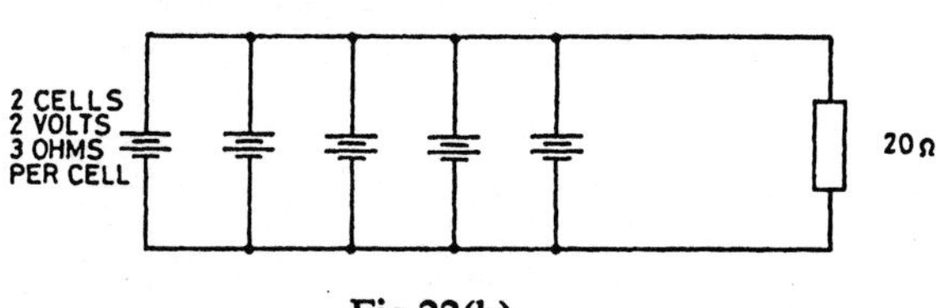

Fig 22(b)

Internal resistance of battery $= \frac{6}{5} = 1.2\Omega$

Total circuit resistance $= 20 + 1.2 = 21.2\Omega$

Circuit or load current $= \frac{4}{21.2} = 0.188\text{A}$

Terminal voltage $= 20 \times 0.188 = 3.77\text{V}$

or $4 - 1.2 \times 0.188 = 4 - 0.23 = 3.77\text{V}$

Current/cell = current of 1 bank $= \frac{0.188}{5} = 0.0376\text{A}$.

MAXIMUM POWER CONDITION FOR THE LOAD.

Consider the simple circuit of Fig 5. If I was the circuit current then the power P supplied to the external load R would be given by I^2R, where $I = \frac{E}{R + R_i}$. Note here R_i is the internal resistance of the supply source, thus:

$$P = \left(\frac{E}{R + R_i}\right)^2 R = \left[\frac{E^2R}{(R + R_i)^2}\right]$$

By differentiating with respect to R, the maximum power condition can be obtained when $\frac{d}{dR}\left[\frac{E^2R}{(R + Ri)^2}\right] = 0$

Such differentiation results in a maximum for P when $R = R_i$, that is, when the resistance of the load is equal to the internal resistance of the supply source.

The efficiency (as a percentage) of the supply for the condition $R = R_i$ is

$$\frac{\text{Output Power}}{\text{Total Power}} \times 100 \text{ or } \eta = \frac{I^2R}{I^2(R + R_i)} \times 100$$

But $R = R_i$

$$\therefore \eta = \frac{I^2\text{R}}{I^2 2R} \times 100 = \frac{1}{2} \times 100 = 50 \text{ per cent}$$

CHAPTER 2

PRACTICE EXAMPLES

1. An electric hoist is required to lift a load of 2 tonnes to a height of 30m. The cage has a mass of 0.25 tonnes and the lifting operation is timed to be completed in 1.5 minutes. If the 220V motor is metered to take a current of 50A, find the efficiency of the installation.

2. Thirty cells each having an e.m.f. of 2.2V and an internal resistance of 0.3Ω are so connected to give a supply e.m.f. of 22V. If the arrangement is then connected to three 20V, 10W lamps in parallel, calculate (a) the terminal voltage of the battery, (b) the current taken by each lamp, (c) the power wasted in each cell.

3. A pump delivers 12 700 litres of water per hour into a boiler working at 15 bars. The pump which is 82 per cent efficient is driven by a 220V motor, having an efficiency of 89 per cent. Calculate the current taken by the motor. Assume 1 litre of water to have a mass of 1kg and 1 bar = $10^5 N/m^2$.

4. A resistor of 5Ω is connected to a battery made up of four similar cells in series. Each cell has an e.m.f. of 2.2V and the current which flows is 1.4A. If the cells were connected in parallel, find the current which would flow through the 5Ω resistor.

5. A five-tonne cargo winch is required to lift a load of 5 tonnes at 36.5m/min. Calculate the power rating of the 220V driving motor if the efficiency of the winch gearing is 75 per cent and that of the motor can be taken as 85 per cent. Calculate also the current taken from the ship's 220V mains.

6. A 220V diesel-driven generator is required to supply the following on full load. (a) Lighting load comprising one hundred 100W and two hundred 60W lamps. (b) A heating load of 25kW. (c) Miscellaneous small loads taking a current of 30A. Calculate the required power output of the diesel engine when the generator is supplying all the loads at the same time. Assume a generator efficiency of 85 per cent.

7. A battery is made up from three similar correctly connected dry cells in series. The open-circuit e.m.f. is measured to be 4.3V. When the battery is connected to an unknown resistor the current is metered to be 0.4A and the battery terminal voltage as 4.23V. If one of the cells of the battery is reversed and the circuit made up as before, estimate the new current value.

8. A 150W, 100V lamp is to be connected in series with a 40W, 110V lamp across a 230V supply. The lamps are required to operate at their rated power values. Determine the values of suitable resistors to be used with the lamps and make a sketch showing how they would be connected.

9. A resistor of 0.525Ω is connected to the terminals of a battery consisting of 4 cells, each of e.m.f. 1.46V joined in parallel. The circuit current is found to be 0.8A. Find the internal resistance of each cell.

10. Twelve cells, each of e.m.f. 1.5V and internal resistance 0.225Ω, are arranged four in series per row or bank, with three banks in parallel. The battery so formed is connected to a load consisting of a series-parallel resistor arrangement, made up of a 2Ω resistor connected in parallel with a 3Ω resistor, these in turn being connected in series with a 2.5Ω resistor. Find the battery terminal voltage, the power ratings of the resistors and the energy converted into heat in the complete circuit if the arrangement is switched on for 1 hour.

CHAPTER 3

CONDUCTORS AND INSULATORS

The reasons as to why certain materials are good conductors of electricity while others are not, will be considered in detail later when the electron theory is studied; here it can be stated that a substance which freely allows the passage of electricity is classed as a conductor. Examples are metals, certain grades of carbon and certain liquids — chiefly solution of salts, acids or alkalis. An insulator can be defined here as a substance which will not allow the free passage of electricity. Examples are rubber, porcelain, slate, mica, some organic materials and certain liquids — notably oils.

RESISTANCE OF A CONDUCTOR

VARIATION OF CONDUCTOR RESISTANCE WITH DIMENSIONS AND MATERIAL

The resistance of 'ohmic' value of a conductor, such as a coil of wire, can be altered in different ways. Thus if coils of different lengths of the same wire, *ie* same material and same cross-sectional area, are measured for resistance, their ohmic values would be found to vary in direct proportion to their lengths. Again if coils of wire of the same material and length, but of different cross-section are measured, their resistance values would be found to vary in inverse proportion to the areas of the wires with which they are wound.

A similar series of comparative measurements with coils of wire of the same length and cross-sectional area but different material, would show that the resistance value varied with the conductor material.

The elementary tests described above, indicate that the resistance of a conductor or resistor can be altered by varying its dimensions or the nature of material used, and the relation of these factors to the actual conductor resistance will now be examined in detail.

1(a) DIMENSIONS. Resistance of a conductor is proportional to its length or, more simply and using an example, it can be said that the conductor resistance of a 100m length of cable will be double that of a 50m length of the same cable. This can be readily shown as follows:

Let R_A ohms = the resistance of a 50m length. Then two 50m sections in series would have a resistance of R ohms

Hence $R = R_A + R_A = 2R_A$

But the length has been doubled

So 2 × Length = 2 × Resistance of original length.

Summarising:

Resistance if proportional to Length or $R \propto l$.

1(b) DIMENSIONS. Resistance of a conductor is inversely proportional to its area or, more simply and using an example, it can be said that the conductor resistance of a 1mm² cross-sectional area cable will be twice that of the same length of cable, of the same conductor material but of 2mm² or twice the cross-sectional area. This can also be shown thus:
Let the resistance of 1mm² cable be R_A and suppose an identical cable to be connected in parallel with it. The resistance of the combination would be R ohms.

$$\text{Hence } \frac{1}{R} = \frac{1}{R_A} + \frac{1}{R_A} = \frac{2}{R_A} \text{ or } R = \frac{R_A}{2}$$

Thus the resistance of the combination is half the original cable resistance, but the area of the combination is twice that of the original cable

or 2 × Area = ½ × Resistance of original length.

If the reasoning was repeated for four 1mm² cables in parallel, the area would be four times the original and the new resistance would be $R = \frac{R_A}{4}$. The resistance of a 4 × 1.0 or 4mm² cable would be quarter of the original cable resistance and the area would be 4 times that of the original cable.

From the above, it follows that doubling the area halves the resistance of a conductor of the same length and material and quadrupling the area reduces the resistance to a quarter of the original value. Thus:

Resistance is inversely proportional to Area or $R \propto \frac{1}{A}$.

2. MATERIAL. The resistance of a conductor depends upon the material from which it is made. Thus the resistance of a length of iron wire is approximately 7 times greater than the resistance of a piece of copper wire of exactly similar dimensions, *ie* same length and cross-sectional area. If therefore resistance varies

with the nature of the material, it is necessary to define this property in such a way as to allow a comparison of resistance value for accepted standard dimensions of the conductor material. The term *resistivity* or *specific resistance* (symbol ϱ — the Greek letter rho), is now introduced. This was originally expressed in ohms or microhms per cm³. The modern tendency is to specify it as — ohm-metre or — microhm-millimetre.

The Resistivity or Specific Resistance of a material is the resistance as measured between the opposite faces of a cube of unit dimensions.

Thus for the diagram (Fig 23) a cube of copper of sides 1 metre is taken and the resistance is measured between the faces as shown by the arrows. The resistivity would thus be determined and would be expressed as $1.725 \times 10^{-8}\Omega$m or $17.25\mu\Omega$mm.

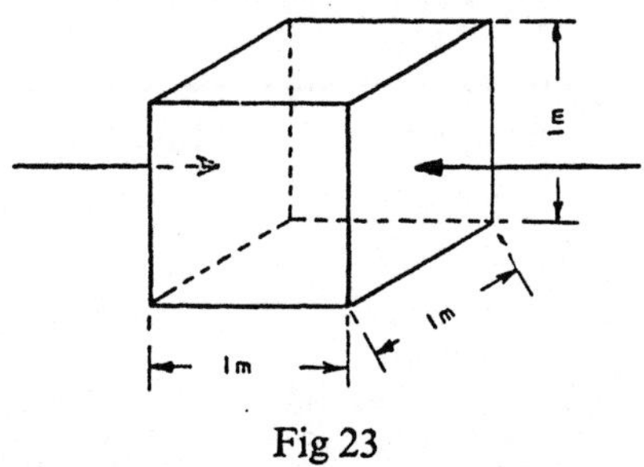

Fig 23

It should be noted that the temperature of the material is measured at the time the test is made and is frequently specified with the resistivity figure. Thus ϱ for copper is given as 1.725×10^{-8} ohm-metre at 20°C. The reasons for specifying the temperature in connection with the resistivity figure will be explained in due course.

It has been seen that $R \propto l$ and $R \propto \frac{1}{A}$. Therefore $R \propto \frac{l}{A}$ or $R = k\frac{l}{A}$ where k is constant. If ϱ is taken as this constant k, then the foregoing can be written as:

$$R = \frac{\varrho l}{A}$$

The use of ϱ as the constant k is acceptable since resistance depends upon resistivity and if in the formula, unit length of 1m and unit area of 1m² is taken, then $R = \varrho$, *ie* the definition of resistivity is satisfied. It is the resistance of a unit cube measured across opposite faces. Problems involving resistivity are solved by use of the above expression $R = \frac{\varrho l}{A}$, but it is essential to note the units used for the resistivity figure. If l and A are not in these units, they should be converted first as is shown by Example 23.

Example 22. If 5m of manganin resistance wire, 0.1mm diameter, have a resistance of 267.5 ohms, find the resistivity of the material.

$$A = \frac{\pi \times d^2}{4} = \frac{3.14 \times (1 \times 10^{-4})^2}{4}$$

$$= 0.785 \times 1 \times 10^{-8} \text{ square metre}$$

$$l = 5\text{m}$$

$$\text{Since } R = \frac{\varrho l}{A} \text{ then } \varrho = \frac{RA}{l} = \frac{267.5 \times 0.785 \times 1 \times 10^{-8}}{5}$$

$$\text{or } \varrho = \frac{535 \times 0.785 \times 10^{-8}}{10}$$

$$= 41.998 \times 10^{-8} \text{ ohm-metre or } 42 \times 10^{-8}\Omega\text{m}$$

$$= 420 \times 10^{-9} = 420 \times 10^{-6} \times 10^{-3}$$

$$= 420\mu\Omega\text{mm}.$$

Example 23. Find the length of wire required to make a 10Ω resistor, if the diameter is 1mm and the resistivity is 450μΩmm.

Here $R = 10\Omega$. $\qquad \varrho = 450 \times 10^{-6} \times 10^{-3}$

$$= 45 \times 10^{-8} \text{ ohm-metre}$$

$$d = 1\text{mm} = 1 \times 10^{-3} \text{ metre}$$

$$\textit{Since } R = \frac{\varrho l}{A} \text{ hence } l = \frac{RA}{\varrho}$$

$$\text{or } l = \frac{10 \times \pi \times (1.0 \times 10^{-3})^2}{45 \times 10^{-8} \times 4} \text{ metres}$$

$$\therefore l = 17.44\text{m}.$$

Occasionally a problem involving the formula $R = \frac{\varrho l}{A}$ can be worked by a method of proportion.

This makes for easier working than finding the resistivity values and resubstituting in the formula to obtain the answer. This is illustrated by the example.

Example 24. If the resistance of 1.6km of copper wire of 0.5mm diameter is 170Ω, calculate the resistance of 1km of iron wire of 1.0mm diameter, assuming that the resistivity of iron is 7 times that of copper.

Resistance of 1600m of copper wire 0.5mm diameter is 170

ohms, then resistance of 1000m of copper wire 0.5mm diameter is

$$\frac{170 \times 1000}{1600} \text{ ohms}$$

and resistance of 1000m of copper wire 1.0mm diameter is

$$\frac{170 \times 1000}{1600} \times \frac{1}{4} \text{ ohms}.$$

Note. The iron wire is of double the diameter, so copper wire of double the diameter has been considered.

Since Area $= \frac{\pi d^2}{4}$, it follows that wire of twice the diameter will have an area four times as great and the resistance therefore will be reduced by 4.

Proceeding with the solution.

Resistance of 1000m of copper wire 1.0mm in diameter is

$$\frac{170 \times 1000}{1600 \times 4} \text{ ohms}$$

So resistance of 1000m of iron wire 1.0mm diameter

$$= \frac{170 \times 1000 \times 7}{1600 \times 4} = 186\Omega.$$

VARIATION OF CONDUCTOR RESISTANCE WITH TEMPERATURE

Most conductors are found to show a variation of resistance when their temperature is changed. Usually this variation follows a straight line relation as is shown by the diagram (Fig 24). If the resistance of a resistor is measured each time its temperature is altered and the results are plotted, a graph such as (1) or (2) or (3) would be obtained. These graphs would cover the main types of conductor and would show that

(1) For pure metals, resistance increases regularly with temperature.

(2) For certain metal alloys used for making resistors, such as Manganin or Constantan, the graph is horizontal. The resistance is substantially unaffected by temperature.

(3) For certain partial conductors, such as carbon, the resistance is found to decrease with temperature.

TEMPERATURE COEFFICIENT OF RESISTANCE. Because of the straight-line relationship between resistance and temperature, illustrated in Fig 24, a simple law is evident and an equation can be deduced, which within normal ranges of temperature, allows the resistance R of a conductor at any temperature T to be obtained in terms of the resistance R_0 and at 0°C and a

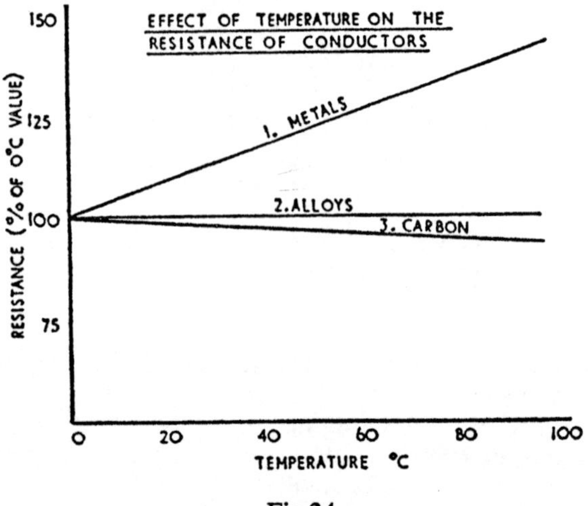

Fig 24

coefficient (symbol α — the Greek letter alpha), known as the *temperature coefficient* of the conductor material. The appropriate equation is:

$$R = R_0 (1 + \alpha T)$$

For the diagram (Fig 25) the graph for a pure metal (copper) is illustrated and enlarged to show the above formula has been deduced. It cuts the R axis to give a value of R_0, the resistance at 0°C. If now, the resistance has a value R_0 at 0°C then at 1°C it will be increased by a small amount x. The fraction $\frac{x}{R_0}$ is taken as the temperature coefficient α of the metal or $\frac{x}{R_0} = \alpha$ and $x = \alpha R_0$.

α is thus the fraction of the resistance at 0°C by which the resistance increases for 1°C rise in temperature.

If x = increase in resistance for 1° rise in temperature, then
xT = ,, ,, $T°$,, ,,

Thus $R = R_0 + xT = R_0 + R_0\alpha T$

or $R = R_0 (1 + \alpha T)$

For copper, α has a value $\frac{1}{234.5} = 0.004\,265$.

The temperature coefficient of resistance α, is usually based on average conditions obtaining from 0°C to 100°C and is thus the ratio of the increase in resistance per °C rise in temperature to the resistance at 0°C. Alternatively, the temperature

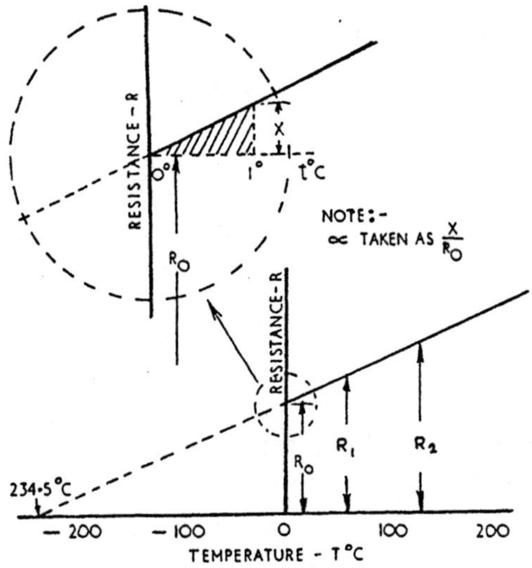

Fig 25

coefficient of resistance can be defined as the increase of resistance of 1 ohm at 0°C for 1°C rise of temperature.

If the graph, as shown in Fig 25, is rising as temperature increases, then the material is said to have a 'positive' temperature coefficient of resistance; whereas if the graph falls, the material has a 'negative' temperature coefficient. Usually for practical work and problems, the ohmic value as given for a resistor will be at a temperature other than 0°C and the following expression has been deduced to facilitate the obtaining of the resistance value at any particular temperature, if its resistance value is given for any other temperature condition.

Let R_1 be the resistance at a temperature T_1°C and R_2 be the resistance at a higher temperature T_2°C.

Then $R_2 = R_0 (1 + \alpha T_2)$ and $R_1 = R_0 (1 + \alpha T_1)$

Dividing, we have

$$\frac{R_2}{R_1} = \frac{R_0 (1 + \alpha T_2)}{R_0 (1 + \alpha T_1)} \text{ or } R_2 = R_1 \frac{(1 + \alpha T_2)}{(1 + \alpha T_1)}$$

Example 25. The cold resistance of a coil of wire is 20Ω at 15°C. It is heated to give a hot resistance of 23Ω. Find its temperature rise, if the temperature coefficient of the resistance material is 0.42 per cent per °C.

Note. Temperature coefficient can be expressed as given above for the example, but is more usually written as 0.0042 from 0°C or more fully 0.0042 ohm per ohm per 0°C at °C *ie* 0.0042Ω/Ω/°C at 0°C. — Sometimes written as 0.0042 Ω/°C at 0°C.

The problem can be solved indirectly as

$$20 = R_0[1 + (0.0042 \times 15)] \text{ or } 20 = R_0\,(1.063)$$

$$\text{whence } R_0 = \frac{20}{1.063} = 18.8\Omega$$

Resubstituting—

$$23 = 18.8\ [1 + (0.0042 \times T)]$$

$$\text{or } 1.223 = 1 + (0.0042 \times T)$$

$$\text{and } T = \frac{2230}{42} = 53.1°\text{C}$$

The temp. rise = 53.1 − 15 = 38.1°C

The following is a more direct solution.

$$\frac{23}{20} = \frac{R_0[1 + (0.0042 \times T_2)]}{R_0[1 + (0.0042 \times 15)]}$$

$$\text{Thus } 1.15 \times 1.063 = 1 + 0.0042 \times T_2$$

$$\text{or } 1.222\,45 - 1 = 0.0042 \times T_2$$

$$\text{and } 0.223 = 0.0042 \times T_2 \text{ or } T_2 = 53.1°\text{C}$$

As before, the temperature rise is 38.1°C.

Example 26. The filament of a 230V lamp takes a current of 0.261A when working at its normal temperature of 2000°C. The temperature coefficient of the tungsten filament material can be taken as 0.005Ω/Ω/°C at °C. Find the approximate current which flows at the instant of switching on the supply to the cold lamp, which can be considered to be at a room temperature of 20°C.

$$\text{Resistance of lamp (hot)} = \frac{230}{0.261} = 882\ \Omega \text{ at } 2000°\text{C}$$

$$\text{Here } R_2 = R_0\,(1 + \alpha T_2)$$

$$\text{So } R_0 = \frac{882}{[1 + (0.005 \times 2000)]} = \frac{882}{11} = 80.2\Omega$$

Again

$$R_{20} = 80.2\ [1 + (0.005 \times 20)]$$

$$\text{or } R_{20} = 80.2\ (1 + 0.1) = 80.2 \times 1.1 = 88.22\Omega$$

$$\text{and current taken when cold} = \frac{230}{88.22} = 2.6\text{A}.$$

The example shows how change of resistance with temperature can affect practical working conditions and allowances to this end should be made. The 'tripping' of a circuit-breaker or 'blowing' of a main-fuse, could mean the loss of supply to a large lighting or resistance load. Before any attempt is made to restore the supply, sections of the load should be isolated, so that when the main switching-on takes place, only part load is applied to the supply and this load is then gradually built up to its full value, by closing the individual circuit switches. The reason for 'shedding load' in this manner is related to the fact that the lamps will have cooled when the supply was off and when supply is restored, a current of seven to ten times the full load value will be taken as a surge. This current will fall as the lamps heat up and may only last for a few milli-seconds, but it could be sufficient to re-trip the circuit-breaker or blow the main-fuse. Thus a fault condition would be suspected, but in fact the cause of the current surge can be explained and appropriate action taken.

RESISTANCE OF AN INSULATOR

VARIATION OF INSULATION RESISTANCE WITH DIMENSIONS AND MATERIAL

An insulator has been defined as a substance which will not allow the free flow of electricity. In electrical apparatus, machines and cables, insulation is used to confine the flow of electricity to the required circuit and to prevent the current from taking 'leakage paths'. Leakage currents are minimised by making the resistance of their paths as large as possible. Thus material with a high resistivity (ϱ) is used for the insulation, the length of the leakage path is kept as large as possible and the area as small as possible. It should be realised at this stage that insulation does allow current to pass and that by measuring this current, the soundness of the insulation can be judged. Instruments such as high-resistance ohmmeters or insulation testers — one trade name being 'The Megger', operate on this principle.

The simple diagram of Fig 26 shows the path of the leakage currents in the cables forming part of the circuit feeding a load and it will be noted that the leakage currents flow radially from or towards the conductors of the cable. The cables can be looked upon as steam pipes, radiating heat. The longer the pipe the more the heat radiated, *ie* the greater its radiating area. Similarly for the cables, the thickness of the insulation, shown as t is actually the length l in the formula $R = \frac{\varrho l}{A}$ while the area A is given by the curved surface of the insulation. The larger this surface the easier it is for leakage current to flow.

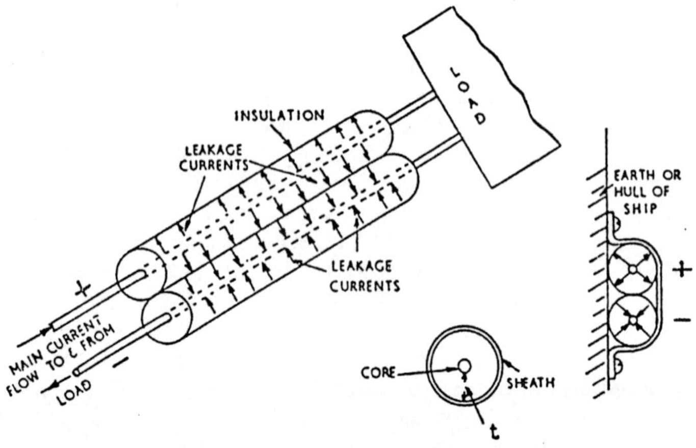

Fig 26

To summarise the foregoing, it is pointed out that insulation resistance of cable would be measured between core and sheath, or 'earth' and would be given by an approximation of the formula $R = \frac{\varrho l}{A}$. Here ϱ would have an extremely high value; for vulcanised rubber it is $10^{15}\Omega$m. or 10^{9}MΩm. l would be the insulation thickness t and surface area A would be proportional to the length of the cable. Thus if the insulation resistance of 100m of cable was measured as being 180 MΩ, then 200m of the same cable would have a resistance value of 90 MΩ. The basic point is that cable-conductor resistance is doubled for double the length, but the insulation resistance is halved. Doubling the length has doubled the area of the leakage paths and since $R \propto \frac{1}{A}$, A if A is doubled R is halved.

It should now be understood why a large electrical cable installation or machine when tested for insulation resistance may give a low figure, whereas the value obtained for a small installation or machine may be considerably larger. Insulation resistance is also affected by other factors, besides the size of the installation or machine. Site conditions such as temperature, humidity, cleanliness together with age must be taken into account and the resistance value means little unless compared with that obtained for a comparable new installation or machine. Acceptable insulation-resistance values for installations and machines are set out in the appropriate

Regulations and the points made above have been stressed to show that test results should be treated with due consideration. Conductor-resistance measurements are more straightforward, although here again, special testing techniques should be employed depending on the type of resistor or apparatus being measured.

VARIATION OF INSULATION RESISTANCE WITH TEMPERATURE

For electrical apparatus, machines and cables, the allowable working temperature and hence the current-carrying capacity of the equipment is limited almost wholly by the restrictions imposed by the insulation. The insulation is usually made up from cotton, silk, rubber, or plastics, and as a general rule, if they are subjected to excessive temperatures their electrical and mechanical properties are impaired. Even if insulation such as mica or porcelain is not damaged by excessive temperatures, it is seen from the attached graph (Fig 27) that, like the partial conductor carbon, the insulation resistance falls with temperature rise, but here the relationship is not straight line. The graph can be shown to follow a logarithmic law and thus insulation resistance falls rapidly as temperature rises An increasing leakage current flows through the insulation as its temperature rises and such a current generates more internal heat which may eventually cause 'breakdown' of the insulation. The allowable

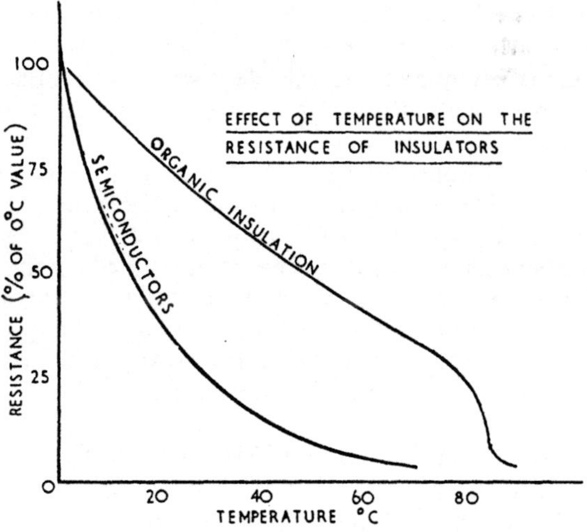

Fig 27

temperature rise for any electrical equipment which gives a safe insulation-resistance value, has been determined by experience and the power rating of an appliance is in accordance with accepted specification. For example, BS Specifications or Lloyd's Regulations may specify a working temperature rise of 50°C for a particular motor when performing a certain duty. This would be when it was developing its rated output in an ambient or room temperature of 30°C. Thus a total temperature of 80°C would be allowed. This figure varies for the type of insulation with which the machine is constructed, but for the example, if the same motor is to work in an ambient of 50°C, then the allowable temperature rise will be reduced to 30°C. The motor would now only be capable of giving a reduced output and would have to be derated. Alternatively a larger machine must be used, if the full original power output was still required. Derating of equipment is necessary to ensure a maximum *safe* working temperature for the insulation and for this condition, the insulation resistance will reach an acceptable minimum value.

Since the insulation-resistance value alters as the temperature of the equipment alters, and it is also affected by other load factors already considered earlier such as, size of installation, humidity, cleanliness, age and site conditions, then a true indication as to the state of the installation or machine can only be gained by reference to a record or log of readings, built up over a period of time. Reference to such practice has already been made earlier and it should be accepted that the keeping of such a log is essential for large electrical installations. Many ships are now fitted with insulation-resistance indicators which record leakage current and thus the state of the insulation resistance. Such indicators assist the keeping of a log which will show comparative readings for the same temperature rise, taken when the installation or machine was new, dry and clean. The difference between the readings can be used to assess the state of the equipment at the time of checking, and if an improvement in readings is deemed essential for safe working, then appropriate arrangements can be made for cleaning down, drying out or for a more thorough inspection and overhaul.

RESISTANCE OF A SEMICONDUCTOR

Electronic devices utilising semiconductor materials will be considered in more detail in Chapter 14 and studied in depth in Volume 7. However, it is necessary, at this stage, to make a reference to the important relationship between the resistance of a semiconductor and its temperature. A semiconductor can be described as a material which, for given dimensions, has a resistance value midway between that of a conductor and an insulator of the same dimensions. The main usage of semiconductor materials is in connection with solid-state devices such as rectifier diodes and transistors, but here we consider the resistance/temperature property in relation to *thermistors.*

VARIATION OF SEMICONDUCTOR RESISTANCE WITH TEMPERATURE

Semiconductor materials have resistance values which alter appreciably when heated. Germanium and silicon, as typical examples, have a negative temperature coefficient which is not constant, but increases as the material is heated. The relationship of resistance with temperature is in accordance with an inverse variation and gives a graph similar to that shown for Fig 27. It will be seen from the graph that as a semiconductor material is heated, its resistance falls and if a piece of this material is used as a resistor then the current passed will increase as the piece heats up. The semiconductor, when used as described here, is known as a thermistor. It can be adapted for use as a measuring or regulating device. As an example of the former, it has been developed for marine work as the detecting element of an electrical temperature-indicating instrument. The original thermometer head consisted of a coil of platinum wire which, when heated, altered the resistance of an indicator circuit so that the latter could be calibrated to indicate temperature. A thermistor element is now being used instead of the platinum wire, being more robust, of smaller dimensions — can be located nearer the 'hot spot', and gives a greater resistance change for a given temperature change. The instrument is thus more sensitive and accurate.

The device can be used as a regulator since it can alter the operating current to a controlling circuit when its temperature is varied. Thus if a thermistor is buried in the windings of an electric motor, any overheating adjacent to its situation will result in the thermistor-circuit current increasing until the connected motor protective device is operated. Such thermistor operated units are now being offered for marine usage in conjunction with motor starters but care must be taken to ensure that the thermistors are correctly located and connected.

The use of thermistors is now becoming so common for electronic circuitry that mention must be made of the fact that research and development have resulted in units being produced which have a positive temperature coefficient in contrast to the more usual negative coefficient characteristic. Such posistors are obviously special and suited to particular applications.

HEAT AND ELECTRICAL ENERGY

Energy can exist in several forms. The mechanical, electrical, thermal and chemical forms are those most used for modern industry and the work done when energy is expended can be put to use in various ways. Although the term 'energy being expended' is commonly used, it should be remembered that energy canot be destroyed or lost. It can only be changed from one form to another and the obvious convertibility between mechanical and electrical energy is seen in a machine like the electrical generator or electric motor. For the former, mechanical energy is passed in at the shaft and electrical energy is obtained and utilised in a circuit connected across the machine terminals. The utilisation may be effected by converting the electrical energy into heat, light or mechanical energy. For the electric motor electrical energy is passed in at the terminals and mechanical energy is passed out at the shaft. The relations between the mechanical and the electrical units of energy and power have already been deduced and our studies now continue with the consideration of the relation between mechanical, electrical and heat energy.

RELATION BETWEEN MECHANICAL AND HEAT ENERGY

The fact that heat is a form of energy is probably the most obvious to the practical engineer, who is only too well aware of the dangers associated with a 'hot bearing', 'slipping belt' or 'clutch'. In these instances mechanical energy is made available by the prime mover and is being converted into unwanted heat through the medium of friction. If this conversion into heat is allowed to continue, the temperature of the associated machine parts may rise to a dangerous level, when a 'seize-up', 'burn-out' or fire may result. The examples have been quoted to show that an elementary deduction can be made showing that the heat energy produced is proportional to the mechanical energy being expended.

SPECIFIC HEAT CAPACITY. This can be found by a comparatively simple mechanical test. The laboratory apparatus would consist of a hollow brass cylinder, which can be rotated by a belt drive. The cylinder can be filled with a known quantity of water and made to rotate against a friction surface applied with a known tension. By simple calculation, the work put in at the driving pulley can be related to the heat produced at the cylinder. James Joule, an English scientist, by careful experimental work showed that 4.187 joules of work are required to produce sufficient heat to raise the temperature of 1 gramme of water by 1 degree Celsius or 1 Kelvin. Since we are now concerned with SI units, then if the mass of water is taken as 1 kilogramme it follows that 4187 joules (4200J approximately) would be required. The joule is now also an SI unit of heat and thus we have the condition where this constant of 4200 must be taken into account by introducing the term *'specific heat capacity'*. This is defined as the quantity of heat required to raise unit mass of a material through a temperature interval of 1 degree Celsius or 1 Kelvin. Different materials would require differing amounts of heat to produce the same temperature rise on the same mass. The units of specific heat capacity (symbol *c*), are heat units per unit mass per unit temperature. Since, for SI units, the most convenient unit of mass is the kilogramme then the kilojoule would be the appropriate size of heat unit to give specific heat capacity in kilojoules per kilogramme per Kelvin or kJ/kgK. In terms of the Celsius temperature scale, this would be kJ/kg°C. Because the relation between the energy and heat is most readily determined for water and has been taken as 4200 joules, it follows that the specific heat capacity value for water would be 4.2kJ/kg°C. The values for other materials are also determined by experiment and can be compiled into the usual tables of physical constants. The following examples illustrates a conversion from mechanical to heat units which involves the use of differing specific heat capacity values.

Example 27. A motor brake-testing rig consists of a water-cooled, cast-iron pulley and a fixed frame which is made to carry two spring balances to which are fastened the ends of a rope which passes round the pulley. Both spring balances hang from screwed rods which are arranged to be adjustable to alter the tension on the rope. Tests made on a small motor running at a full-load speed of 750 rev/min gave the following readings. Spring balances 16.89kg and 0.55kg. The pulley is hollow 102mm long, 380mm in diameter (these are outside dimensions). It has an average wall thickness of 6.4mm. It has a mass of 2.72kg and is designed to be half-filled with water. Estimate the output power of the motor being tested and the time for which the motor can be tested before the water commences to boil. The temperature of the pulley and water is 15°C at the start of the test and the rope diameter is 25mm. Take the specific heat capacities of water and cast iron as 4.2 and 0.42kJ/kg°C respectively.

Power output of the motor is given by the expression $\frac{2\pi NT}{60}$

Effective load on brake = (16.89 − 0.55) = 16.34kg

Hence restraining force F = 16.34 × 9.81 newtons

$$\text{Effective radius} = \frac{380 + 25}{2} = 202.5\text{mm}$$

Restraining torque = 16.34 × 9.81 × 202.5 × 10^{-3} Nm

$$\text{Output power} = 2 \times \pi \times \frac{750}{60} \times (16.34 \times 9.81 \times 202.5 \times 10^{-3})\ \text{W}$$
$$= 2560\text{W} = 2.56\text{kW}$$

Heat energy available per minute = 2.56 × 60 = 153.6kJ

$$\text{Volume of water} = \frac{\pi d^2}{4} \times \text{width} \times \frac{1}{2}$$
$$= \frac{\pi}{4}(380 - 12.8)^2 \times (102 - 12.8) \times \frac{1}{2}\text{mm}^3$$
$$= 4.72 \times 10^{-3}\ \text{m}^3$$

Since 10^3kg is the mass of 1 cubic metre of water

Then mass of water = 4.72 × 10^{-3} × 10^3 = 4.72kg

Mass of pulley = 2.72kg

$$\text{Temp rise/min} = \frac{153.6}{(4.72 \times 4.2) + (2.72 \times 0.42)}$$
$$= \frac{153.6}{20.97} = 7.3°\text{C}$$

Allowable temperature rise = 85°C

$$\text{Heating time} = \frac{85}{7.3} = 11.7 = 12 \text{ min (approximately).}$$

RELATION BETWEEN ELECTRICAL AND HEAT ENERGY

In Chapter 2 mention was made of an associated effect which would be noted when current flowed in a circuit. This effect would be a temperature rise in any part of the circuit, where resistance was concentrated and one definition of the ohm relates the unit of resistance to a joule of energy being generated, when a current of 1 ampere is flowing. Since this energy cannot be destroyed, this is obviously another instance of energy conversion from one form to another, and a simple test can be made to deduce the relation between heat and electrical energy. Such a test would determine the specific heat capacity of a material by an electrical method and, since water is the most convenient substance, an appropriate rig-up is described.

The experimental apparatus consists of a glass flow-tube which is surrounded by a glass water-jacket — spaced some little distance from it. The whole is sealed so that the space between flow-tube and jacket can consist of a vacuum. A heating wire runs centrally along the tube and thermometers are placed at either end of the tube. The water, the specific heat capacity of which is to be found, is arranged to flow steadily through the tube and is heated electrically by a known current passing through the wire. After a period the inlet and outlet temperatures of the water become constant and this difference in temperature is noted. The constant rate of flow of the water is metered and thus the mass of liquid being heated in a given time can be measured. The voltage drop across the heater is also measured and the quantity of heat absorbed by the water equals the electrical energy expended by the heating element:

$$VIt = mcT$$

Where m is the mass of liquid, c the specific heat capacity and T the temperature rise.

The value of c would be found to be 4.2kJ/kg°C. It will be noted that the specific heat capacity of water, when determined either by mechanical or electrical means, is found to be the same. Different forms of apparatus have been developed to find the c values of various materials, the electrical method being usually favoured because of the accuracy with which control of the test can be effected and measurements taken.

The following examples show how the specific heat capacity value is used in electrical problems.

Example 28. A brass calorimeter was found to have a mass of 67 grammes. It was filled with water when the new mass was 131.7 grammes. The temperature of water and container was 18°C. A heater coil was next immersed in the calorimeter, which was suitably lagged to minimise heat loss. Find the time taken to heat the water and calorimeter to a temperature of 33°C, if the heating was done by passing a current of 2A through the coil, the voltage drop across which was 7.5V. Take the specific heat capacity of brass as 0.39kJ/kg°C and that of water as 4.2kJ/kg°C.

Mass of water = 131.7 − 67 = 64.7g or 0.0647kg

Mass of calorimeter = 67g = 0.067kg

Temperature rise of water and calorimeter = 33 − 18 = 15°C

Heat required by water (mcT) = 0.0647 × 4.2 × 15 kJ

" " " calorimeter = 0.067 × 0.39 × 15 kJ

Total amount of heat required

$= 15\,[(0.0647 \times 4.2) + (0.067 \times 0.39)]$ kJ

$= 15 \times 0.2978 \times 10^3$ J

Input power $= 2 \times 7.5 = 15$W

$\therefore$ Time taken to produce the temperature rise

$$\frac{15 \times 297.8}{15} \text{ seconds} = 4.96 \text{ minutes.}$$

$= 5$ min (approximately).

Example 29. A 220V electric kettle has an efficiency of 90 per cent. Calculate the resistance of the heater-coil and the current necessary to raise the temperature of 1 litre of water from 15°C to boiling point in 9 minutes. Take a litre of water to have a mass of 1 kilogramme and the specific heat capacity as 4.2kJ/kg°C.

Since no information is given about the kettle, the effect of heating it must be neglected.

Heat received by water $= 1 \times (100 - 15) \times 4.2$

$= 357$kJ

Heat energy put out by the heater (kettle is only 90 per cent efficient)

$$= \frac{357}{0.9} \text{ kJ}$$

Time of heating $= 9 \times 60$ seconds

Power rating of heater coil $= \dfrac{357 \times 10^3}{0.9 \times 9 \times 60}$

$= 735$W

Heater current $= \dfrac{735}{220} = 3.34$A

Heater resistance $= \dfrac{220}{3.34} = 65.9\Omega$.

Example 30. A 120W electric soldering-iron is plugged into the 120V ship's mains for five minutes, the ambient temperature being 15°C. The mass of the copper bit is 133g and 50 per cent of the heat generated can be assumed to be lost in radiation and heating the other parts of the iron. Find whether the iron has reached working temperature in the time specified. Take the specific heat capacity of copper as 0.39kJ/kg°C and the temperature of melting solder as 310°C.

Heat required by the iron $= 0.133 \times 0.39 \times (310 - 15)$

$= 15.34$kJ

Heat produced by the element $= \dfrac{15.34}{0.5} = 30.68\text{kJ}$

$= 30\,680$ J

Power rating of the element $= 120$W

Time for 30 680 joules to be expended

$$= \frac{30\,680}{120 \times 60} \text{ minutes}$$

$= 4.27$ min.

Since only 4.27 min are required to achieve working temperature, then the time of five minutes as specified would be sufficient.

Example 31. A resistance unit consists of 500 turns of nickel-chrome wire, 0.5mm diameter. It is wound on a former 30 × 100mm and its resistivity can be taken as 1060$\mu\Omega$ mm at 15°C. At 100°C, its resistance is 2 per cent greater than at 0°C. Determine the current taken at a temperature of 300°C, when the resistance is connected across a 250V supply.

Length of a turn $= (2 \times 30) + (2 \times 100) = 260$mm

No of turns $= 500$

Total length of wire $= 260 \times 500 = 130\,000$mm $= 130$m

Resistance at 15°C $= \varrho \dfrac{l}{A}$

$$= \frac{1060 \times 10^{-9} \times 130}{\dfrac{\pi(0.5 \times 10^{-3})^2}{4}}$$

$$\text{or } R = \frac{106 \times 13 \times 10^{-7} \times 4}{\pi \times 0.25 \times 10^{-6}}$$

$= 702\Omega$

Also since

$R = R_0(1 + \alpha\text{T})$ or $1.02 = (1 + \alpha 100)$

then $\alpha = \dfrac{1.02 - 1}{100} = 0.0002\Omega/°\text{C}$ at 0°C

Similarly $R_1 = R_0(1 + \alpha\text{T}_1)$ and $R_2 = R_0(1 + \alpha\text{T}_2)$

$$\text{or } R_2 = R_1 \frac{(1 + \alpha\text{T}_2)}{(1 + \alpha T_1)}$$

where $R_1 = 702\Omega$ at 15°C.

$$\text{Giving } R_2 = 702\,\frac{1 + (300 \times 0.0002)}{1 + (15 \times 0.0002)} = \frac{702(1 + 0.06)}{(1 + 0.003)}$$

$$= \frac{702 \times 1.06}{1.003} = 742\Omega$$

Current taken $= \dfrac{250}{742} = 0.337$A.

ATOMIC THEORY OF CONDUCTION

In Chapter 1 reference was made to the fact that atoms can be assembled to result in the matter being in either a solid, liquid or gaseous form. Returning to this subject area of atomic physics, we now consider in more detail the electron shells — called *quantum shells*. The shells are considered to be concentric and to be seven in number. They are identified by the letters K, L, M, N, O, P and Q, the K shell being closest to the nucleus. A natural law states that the appropriate numbers of electrons in each shell are 2, 8, 18, 32, 18, 13 and 2. Figure 28 illustrates the arrangement of shells in a copper atom (Z = 29) which is the most commonly used metal in electrical work.

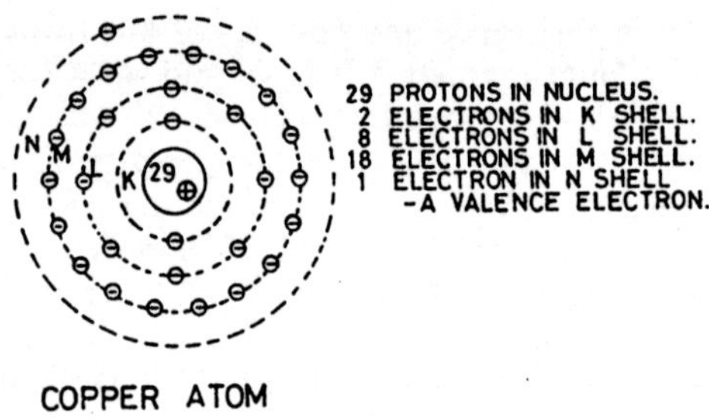

Fig 28

The quantum theory postulates that the shells are identified by specific energy potential levels relating to the orbital distance from the nucleus. These energy levels are important factors in the conduction process and an understanding of them, in relation to the atom is necessary to distinguish between conductors and insulators.

ENERGY LEVELS

When an atom is excited by some extraneous means such as heat, light or some other form of agitation, energy is acquired which has the effect of moving an electron from an inner to an outer shell. Electrons can only exist in the definite energy levels of a shell and cannot exist with energies between these levels. A sudden change in energy level can however result in a move from one shell to another. Electrons with low energy lie in the inner shells *i.e.* nearer the nucleus. Thus if the move or jump is made to a lower or inner shell, the energy given out is emitted as light radiations. The effect of an electron gaining energy has already been mentioned earlier. When the inner shells are filled, additional electrons can only exist in the outer shells and increase until the total −ve charge of the electrons equals the +ve charge of the nucleus, when the atom becomes stable. The unit for measuring electron energy level is the electron-volt (eV).

ENERGY BANDS

A number of fundamental laws have been determined by research which can be seen to control atomic behaviour. One such law states that not more than two electrons can be in the identical energy state and these electrons should have opposite directions of spin. Thus other electrons, whilst within the energy range for the shell, will occupy orbits at progressively higher energy levels. Consider a single isolated atom of copper (Z = 29) Fig 28. There are 2, 8 and 18 electrons respectively in the K, L and M shells. These are filled with electrons which are very nearly, for each shell, at the same energy level. There is also one outer or *Valence* electron in the N shell with its higher energy value. The energy levels are sharp and can be depicted by the diagram (Fig 29a). Consider next a number of adjacent copper atoms whose electrons, if excited, can be given slightly higher energy levels. The original atom is now influenced by the electric fields of the nuclei of adjacent atoms as well as of its own. The shell electrons are in close proximity and would be at slightly different levels. The definite energy lines of Figure 29a are now broadened to a band, shown by two lines spaced from each other, as shown by Figure 29b.

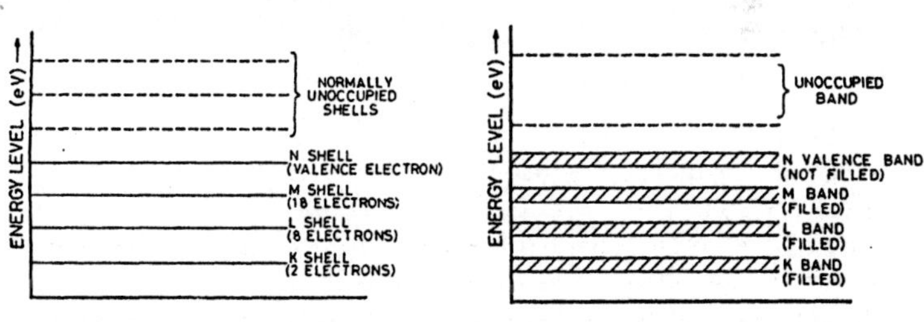

Fig 29

(a) (b)

It will be seen that the K, L and M bands though broadened represent filled shells with no free electrons and thus play no part in electrical conduction. The partially filled valence band indicates the presence of such free electrons with their ability to migrate into the unfilled bands. The unfilled bands constitute a

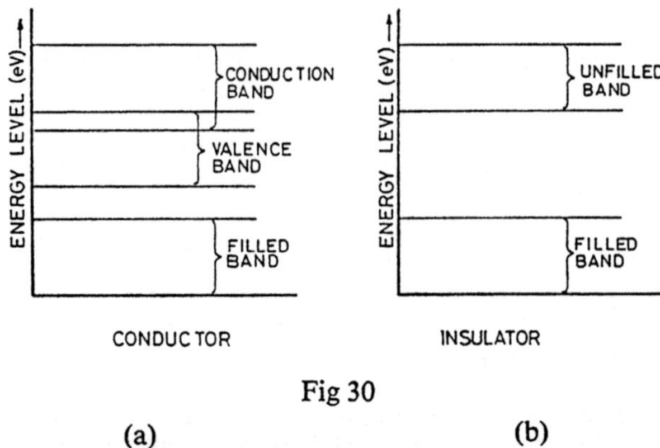

Fig 30

(a) (b)

conduction band since, as described earlier, the electrons are remote from the nucleus and can move readily. Thus for Figure 30a the conduction condition is shown by the upper bands where the valency and conduction bands overlap. This is for a conductor.

For an insulator (Fig 30b) the lower band is filled and remote from the upper band — thus conductivity is very poor. The possibility of an electron gaining enough energy to cross the gap is small and could only be achieved by the application of so high a potential difference that the material would be destroyed by such an electron movement. Such movement would manifest itself in a spark or puncture of the insulating material which could then result in "tracking" — a burnt conducting path or area.

CRYSTAL LATTICE

Some substances, including metals, in their smallest elemental form are found to consist of a large number of *crystals* joined together. The crystals are built up from a regular structure of atoms which repeats itself continuously to form a *lattice.* The simplest crystal to examine is that of carbon but the term *co-valent bonding* must be introduced first to allow a full understanding of the lattice concept. The idea of a covalent bond is simply shown if a molecule of hydrogen is considered, which though not a metal, exhibits many similar characteristics in chemical reactions.

One way in which an atom may combine with other atoms and bring about a change in the number of atoms in its valence shell is by co-valent bonding. This means the sharing of a pair of electrons by two atoms, each atom contributing one electron to form a shared pair. A powerful bond between atoms is the result. Consider a hydrogen atom which has one valence electron. The K shell electrons should be 2 thus a hydrogen molecule contains 2 atoms with the nuclei linked by the valence electrons to form a pair. The arrangement is illustrated by the diagram (Fig 31).

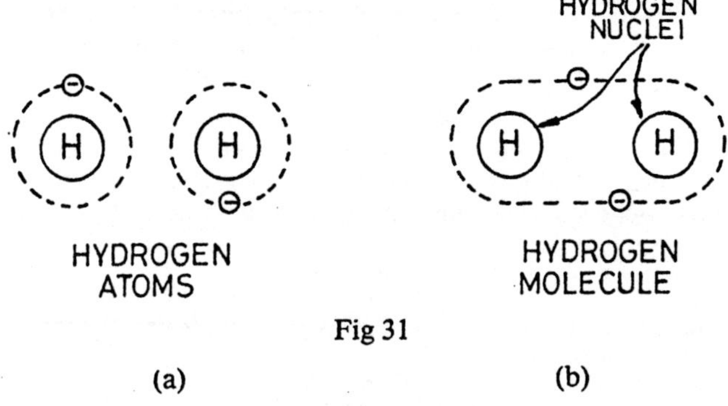

Fig 31

(a) (b)

Now consider the carbon atom, shown by the first diagram of Figure 32a. The K shell is full but the L shell is incomplete, containing only 4 electrons. For a carbon crystal, we see that coupling occurs between the outershells of neighbouring atoms (Figure 32b). An atom is thus produced which has 8 electrons in the L shell. This means that the shell is full having its maximum number of electrons. There are therefore no free electrons and thus pure carbon is an indifferent conductor of electricity. The

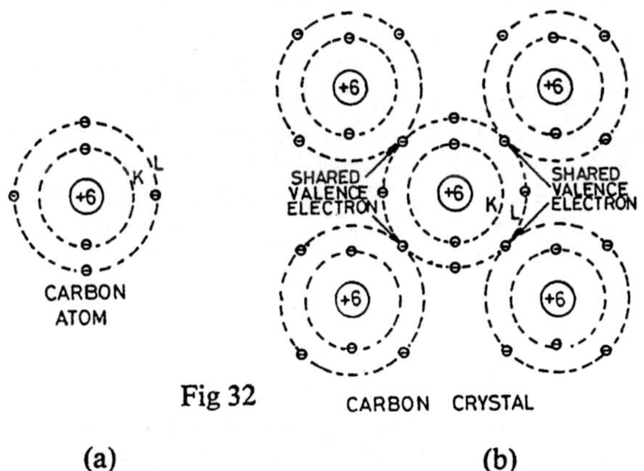

Fig 32

(a) (b)

carbon brushes, as used for electrical machines, are the result of processing to produce certain operational characteristics. A controlled heat treatment results in grades of either natural graphite or amorphous carbon which improves the conductivity.

In crystals, the atoms are arranged in an orderly geometrical pattern and all pure crystals of the same element have the same lattice structure. For the carbon or diamond crystal being considered here, each atom is equidistant from four neighbours with each valence electron being shared between the parent atom and one of the neighbouring atoms. Two electrons, one from each neighbouring atom, thus form a covalent bond and the arrangement can be represented by the diagram of Figure 33a or by the diagram of Figure 33b which depicts the crystal structure.

There are several basic patterns of crystals and for metals it is generally believed that these consist of a lattice of +ve ions

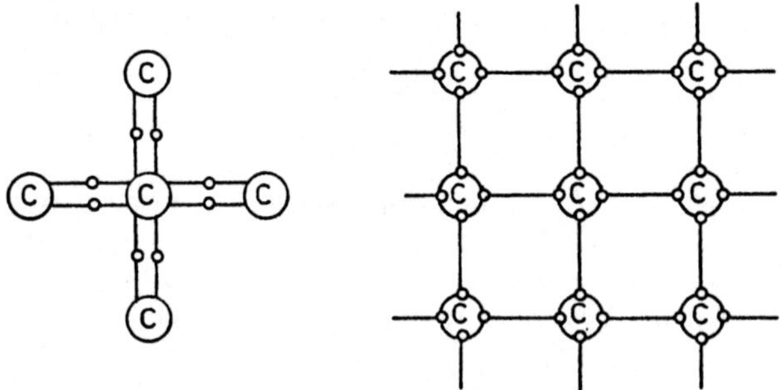

Fig 33

through which a cloud of electrons move. The electrons are the valency electrons of the metal. The crystal lattice will be given further consideration when semiconductor theory is considered in Chapter 15, but it is appropriate here to point out that the crystalline form of metals is responsible for their strength, elasticity and relative resistance to chemical reactions. Strong electrical bonds are provided by the mutual attractive forces between atoms and considerable work must be done to distort the equilibrium arrangement, irrespective of whether the metal is in tension or compression.

CONDUCTIVITY

As described previously, atoms make up a crystal lattice and in such atoms the outermost electrons are influenced by the adjacent atoms of the lattice, the result being that such valence electrons may move continuously from atom to atom. When a potential difference is applied to the conductor, these electrons pass through the crystal towards the positive potential in a somewhat random and erratic manner. This action, representing an opposition to the passage of current, is termed conductor resistance. It is also known that all atoms vibrate about a mean position, the amplitude of this vibration being dependent on the temperature. The magnitude of the vibrations are also known to increase with temperature until the melting point of the material is reached when the crystal lattice breaks up into either individual or groups of atoms. Electronic theory also considers that since the vibrations increase with temperature, extra force is necesary to move the free electrons through the lattice and this explains the fact that the resistance of a pure metal rises with temperature rise. It should also be noted that since extra force is needed to move the electrons then extra energy is absorbed into the conductor to appear as heat. By similar explanation resistance is found to decrease as temperature falls, the relationship being linear down to very low temperature values. Extrapolation of this linear law shows that as the temperature approaches absolute zero, conductor resistance reduces to a negligible value and hence the conductivity rises to an immense value. This condition brings about "super-conductivity" which is at present, being actively investigated by practical research in the electrical power industry.

METALLIC CONDUCTION

Consider a short length of copper wire connected to two terminals or electrodes which are at different potentials. One electrode has thus been given a +ve charge or is deficient in electrons and the other a −ve charge which has a surplus of electrons. The +ve electrode attracts electrons from the nearest copper atoms of the wire and these atoms will be left with a +ve charge *i.e.* will become positive ions. These ions attract the mobile electrons from the next atoms down the wire which in turn are ionised to attract electrons from the adjacent atoms down the line and this process continues until the atoms at the other end of the wire become +ve ions. These ions now accept electrons from the other electrode which, being negatively charged, carries an excess of electrons — these having been positioned there by the energy source. It can be assumed that this 'jumping' action of the electrons is so fast as to be almost instantaneous. It should also be noted that as an electron leaves an atom it occasions the formation of a *hole*. This term is frequently used in explaining semiconductor operation and the opportunity is taken here to introduce it. When an electron with its −ve charge leaves an atom, a gap or space is left in the atom's structure and the atom will now acquire a +ve charge. This gap or hole can be credited with this +ve charge and we can now visualise a 'hole' as an entity in itself having a +ve charge. Any other mobile electron can be captured by the hole to neutralise its charge and make the original atom stable since its structure is now complete. As a result of this action, an electron will have left a neighbouring atom and a hole is created there. This action continues between adjacent atoms and, as we are considering metallic conduction, we see that with the application of a potential difference, electrons move towards the +ve terminal and holes move towards the −ve terminal *i.e.* in the same direction as the current. The similarity with a row of occupied seats is often used. If a person (electron) is induced to leave and everyone is asked to 'move up one seat', the unoccupied space or hole can be reasoned to move down in the direction opposite to the movement of the people. The concept of a hole is useful, it is often referred to as a 'current carrier' and will be considered again later with electronic theory.

The conduction action described above has been treated in terms of a few atoms. It is estimated that in a cubic millimeter of copper there are as many as 10^{20} atoms so, even if we consider a wire one tenth of a millimeter in diameter, it is obvious that the number of mobile electrons and holes available for conduction is considerable.

LIQUID CONDUCTION

Unlike a metal the molecules in a liquid are less closely packed and have little cohesion between each other. There is no crystal lattice and the electrostatic bonding between atoms is also very loose. Conduction of electricity is dependent on the nature of the liquid and also whether it is capable of electrolytic dissociation. For the passage of current ionic action is involved and some basic explanation of this will assist in the appreciation of the conduction process. The theory involved is given detailed attention in Chapter 4.

GASEOUS CONDUCTION

Consider a container evacuated completely and then filled with a gas at normal atmospheric pressure. The gas consists of molecules of a simple structure, which are mainly stable, except for the occasional one which is ionised. Such ionisation means the existence of some free electrons and molecules which have lost an electron and have thus acquired a +ve charge. The ionisation is the result of some action by light, cosmic rays or radio-activity. Gas molecules are known to oscillate at high speed in random directions resulting in frequent collisions. The distance between the gas molecules is small here. The spacings between collisions are small but if the gas pressure is reduced the distance between the collisions becomes larger. When the free electrons move through the gas and meet an ionised molecule, an immediate recombination results and stability is restored. With the reduction of gas pressure, the +ve and −ve charged molecules and electrons travel a relatively longer distance before being normalised. This distance is known as the 'mean free path' and as it increases, the period during which ionisation continues becomes longer. Continued reduction of the gas pressure increases the free path until this latter approximates to the dimensions of the enclosing container. The pressure is now so low that the enclosed space can be considered to be a vacuum and collisions with gas molecules are not now possible. Since no molecules are present in a vacuum, no obstacle to the movement of electrons exists if these are injected into the space. Alternatively, no ionisation is possible and a vacuum, without injected electrons, can be considered to be the perfect insulator.

Reverting to the gas-filled condition, we see that conduction is affected by the pressure. If two electrodes, between which there is a potential difference, are introduced into the gas-filled space, a directional motion of the charges can be expected. Because of the low pressure, the electrons have a long free path and can attain high velocities and gain motional energy on collision. The effect of the collisions will depend upon the velocity on impact with a gas molecule. The high velocity electrons could cause the molecules to merely bounce away from each other. On the other hand the electron velocity on impact could be large enough to cause another electron to be jolted from one shell orbit of the neutral atom to another, thus exciting it and giving rise to the emission of light due to some of the collision energy having been absorbed. If the velocity on impact is very high the collision energy could cause an electron to be knocked out of a neutral gas atom thus ionising the molecule. The process results in the production of an additional free electron and an ionised molecule. The new atom is now available to join the collision action and could shock-ionise a further gas molecule. The effect is a cumulative one and, since there is a potential difference between the electrodes, a current would result, electrons making their way to the +ve plate or *anode* and ionised molecules to the −ve plate or *cathode*. The current would increase rapidly leading to a condition which could be disastrous unless stabilised by the inclusion of a resistor in the circuit. The above action forms the basis of discharge lamp operation and will be discussed further in the appropriate sections.

When considering the conditions for ionisation it was pointed out that if electrons could be made available, then ionisation would continue, especially if electrodes, at a potential, were inserted into the evacuated or gas-filled container. Such electrons could be obtained by either *cold-emission* — tearing them away from the electrodes or by *thermionic emission*. Thermionic emission will be dealt with in detail in Chapter 14.

COLD ELECTRON EMISSION

A vacuum, although a perfect insulator, will nevertheless allow current flow if a potential high enough is applied across the electrodes. Electrons contained in the atoms of the metallic electrodes can be extracted by the electric forces exerted on them by the applied field or potential, provided this is strong enough to overcome the internal electrical bonds of the metal atoms. The movement of the electrons constitutes a current, made evident in the form of a spark. Such a spark may develop into an arc if a supply source is connected to the electrodes. If the discharge is maintained, sufficient to heat the electrodes, thermionic emission is encouraged and a copious supply of electrons will be made available and the discharge would continue. If the current resulting from the initial discharge is limited by the resistance of the circuit then a working arrangement can be contrived to result in a source of illumination. If gas at a low pressure is introduced into the container then, once the initial electron emission has been achieved by the high applied potential, ionisation of the gas would occur and the electrode potential can be reduced to maintain the discharge. This arrangement gives rise to the cold-cathode discharge lamp.

THE COLD-CATHODE DISCHARGE LAMP

The basic principle concerning the operation of this lamp is the use of a high voltage gradient across the gas to produce electron flow and the necessary collisions to cause ionisation. The electrons are initially extracted from the cathodes by the large applied voltage between anode and cathode and, to produce the initial ionisation, a 50 per cent higher voltage than the running voltage is required. When operating, a potential gradient of 1 to 1.3kV per metre is used, and since the total supply voltage for some lamps and signs can be in the region of 10kV, a mains transformer is used. This transformer is designed to have a high internal reactance so that the voltage 'sits down' to the required running value, once the lamp has struck.

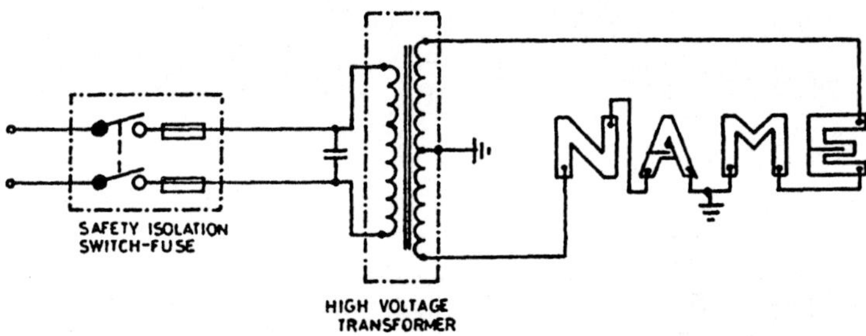

Fig 34

The lamp, in its simple form, is used for advertising purposes and the colour of the discharge is governed by the type of gas used. For illumination purposes, mercury vapour is used together with a fluorescent coating on the inside of the tube. The long tubes (2.4m) with three or four in series are difficult to accommodate and the lamp has only limited applications for marine work. It is however, used for the illumination and decoration of saloons and public-rooms on some of the larger passenger ships.

A more familiar type of cold-cathode lamp is the small pigmy or sign lamp used for indicator purposes. This is arranged to work at mains voltage with a resistance in series, the value of which is some 10 000Ω. Neon gas at high pressure is used and the discharge is usually in the form of a glow which surrounds the electrodes.

CHAPTER 3

PRACTICE EXAMPLES

1. Ten thousand cubic millimetres of copper are (a) drawn into a wire 100 metres long, (b) rolled into a square sheet of 100mm side. Find the resistance of the wire and the resistance between opposite faces of the plate, if the resistance of the copper is 17μΩmm or 1.7×10^{-8} ohm-metres.

2. A coil of copper wire has a resistance of 90Ω at 20°C and is connected to a 230V supply. By how much must the voltage be increased to keep the current constant, if the temperature of the coil rises to 60°C? Take the temperature coefficient of resistance of copper is 0.004 28Ω/Ω/°C at 0°C.

3. An electric kettle is fitted with a heater unit of 120Ω resistance. The efficiency is 84 per cent and the voltage is 220V. How long will it take to heat 0.75 litre of water from 6°C to 100°C? Take the specific heat capacity of water as 4.2kJ/kg°C.

4. Find the length of manganin wire required to make a 15.7Ω resistor, if the diameter is 0.315mm and the resistivity is 407μΩmm.

5. The cold resistance at 15°C of the field coil of a motor is 200Ω and the hot resistance is 240Ω. Determine the temperature rise assuming the temperature coefficient of resistance to be 0.0042Ω/Ω/°C at 0°C.

6. A 2-core cable, each core of which is 300m long and of uniform cross-sectional area of 150mm² is fed from one end at 240V. A load of 200A is taken off from the centre of the cable and a load of 100A from the far end. Calculate the voltage at each load. A single-core cable of similar material 880m in length and of uniform cross-sectional area of 50mm² has a resistance of 0.219Ω.

7. The resistance of a 100m length of copper conductor 1mm diameter, is 2.47Ω. Calculate the resistance of a cable 800m in length composed of 19 similar strands of

conductor, but each 1.5mm diameter. Allow 5 per cent increase in length for the 'lay' (twist) of each strand of the completed cable.

8. A wire has a resistance of 10Ω at 0°C and 15Ω at 100°C. What is the temperature coefficient of the resistance of the material? At what temperature will its resistance be 30Ω?

9. A 200V, 200kW electric furnace is required to raise 500kg of brass from an initial temperature of 15°C to the melting point of 910°C. If the overall efficiency is 0.8, calculate the time of the operation. Specific heat capacity of brass is 0.39kJ/kg°C.

10. A 230V electric water heater takes water at a mean temperature of 16°C and the mean temperature of the outlet water is 82°C. The cost at 2p per unit of the energy it consumes in a given period is £7.44. Determine the quantity of water used if the efficiency of the heater is 80 per cent. Take the specific heat capacity of water as 4.2kJ/kg°C.

CHAPTER 4

ELECTROCHEMISTRY

Mention has already been made of the three main effects of current flow, namely those producing heat, magnetism, and chemical action. In Chapter 3 the heating effect was discussed in some length, whilst the magnetic effects will be covered in Chapter 5. This chapter will be devoted to electrochemistry, which was the subject of many basic investigations concerned with the science of electricity. Such studies soon revealed the chemical action associated with current flow and the reversibility of this action and the electric cell was the principal source of electrical energy before the principles of electrodynamic induction were discovered and the electric generator was developed.

The existence of static electricity, namely electricity at a high potential and exhibiting itself as a stationary charge, had been known for centuries, but this was associated with the lightning flash and friction effects such as the attraction of paper by a piece of amber when the latter was rubbed. The early and if somewhat accidental experiments of men like Galvani and Volta showed that electricity could be produced and controlled by chemical means and led to the first 'voltaic piles' or batteries being constructed. These consisted of a number of plates of zinc and copper alternated by absorbent material such as felt, soaked with an acid or salt solution.

At the commencement of the nineteenth century the only practical means of producing electricity was by chemical means and, it is therefore not surprising that, this somewhat separate, self-contained and special branch of electrical engineering science was the first to be thoroughly investigated and developed. The *laws of electrolysis*, first propounded by Faraday in 1834, summarise the basics of theory sufficiently well, to allow modern ideas on the relation between electrical and chemical action to be accepted without upsetting earlier, well known and hard won knowledge.

The conversion of electrical energy into chemical energy and the reverse action is an important example of the principle of the conservation of energy. The theory involved is best studied from the first action, namely, the conversion of electrical to chemical energy by the conduction of a current through a liquid.

Conduction in liquids is dependent on the presence of ions as current carriers. Pure distilled water is un-ionised and is a poor conductor but a few drops of sulphuric acid added to the water produces a conducting solution when subjected to a potential difference. This conduction is brought about by *electrolytic dissociation* — a process which will now be considered.

ELECTROLYTIC DISSOCIATION

Consider common table salt (NaCl), when in the dry crystalline form. A molecule of such a crystal consists of one atom of sodium (Na) and one atom of chlorine (Cl). In its normal state the sodium atom has only one electron in its valence shell instead of the possible eight. The chlorine atom has seven electrons instead of the possible eight in its outer shell and the combination (NaCl) is achieved by the one electron leaving the sodium atom and entering the chlorine atom. The chlorine atom thus absorbs the valence sodium electron so filling its outer shell. In this condition the chlorine atom becomes a −ve ion and the sodium atom a +ve ion. The salt molecule can be described as an ionic compound and is stable because of the attraction between the reformed atoms. The crystal is composed of the +ve and −ve ions arranged in regular patterns — the crystal lattice.

When the salt is dissolved in water, electrolytic dissociation occurs and some molecules break up allowing the chlorine and sodium atoms to separate. The chlorine atom however still keeps the extra electron taken from the sodium. The solution has now become an electrolyte, both the chlorine ions and the sodium ions being able to wander. The ions in the electrolyte are continuously recombining with other oppositely-charged free ions whilst different molecules are breaking down elsewhere. The dissociation and recombination can be shown by

$$NaCl \rightleftarrows Na^+ + Cl^-$$

Note: the representation of the electron distribution *ie* chlorine has gained one electron hence the Cl^-. Sodium on the other hand has become a +ve ion, hence Na^+. It should also be noted that the electron interchange number is one *ie* the chemical valency number.

ELECTROLYSIS

The passage of current through a solution of an acid, alkali or salt produces a chemical change, explained by the theory of *dissociation*. The solution is called the *electrolyte* and the process is termed electrolysis. The general theory involved is as follows. When an electrolyte is first made up, as would result from dissolving copper sulphate crystals in water, some molecules split into two, independent of any external assistance, to form *ions*, carrying +ve and −ve *charges*. Such ions are extremely mobile. If two plates, termed the *electrodes*, are immersed in the electrolyte and a potential difference applied across them, a current will pass through the solution. The +ve ions migrate, under the influence of the electric field due to the potential difference, to the *cathode*, namely the electrode by which the current leaves the electrolyte. Such ions are called *cations*. The −ve ions, called *anions*, migrate to the *anode*, namely the electrode by means of which the current enters the solutions. Metal or hydrogen ions always carry +ve charges and travel with the current to appear at the cathode, whereas non-metallic ions travel in the opposite direction to the current and may appear at the anode or may engage in secondary reactions, some of which will be described shortly. If the electrodes are made from platinum, which is chemically inert, gas bubbles are produced on the electrodes. Consideration is now given to electrolysis with various liquids.

For the salt solution, ionic action is explained by the fact that Cl^- ions move to the anode and the Na^+ ions move to the cathode and current flows. The Cl^- anions revert to their normal atomic structure by giving up their surplus electrons which are transferred round the external circuit. They are now neutral but in a very active (nascent) state. Two such atoms combine to form a chlorine molecule and some of these rise to the surface as gas bubbles whilst others dissolve in the water. The Na^+ cations meanwhile drift to the cathode, are neutralised on reaching it, but being also in a nascent state combine with the water to form sodium hydroxide and the hydrogen so liberated will appear as bubbles at the cathode. The latter chemical action is shown by the equation; $2Na + 2H_2O = 2Na(OH) + H_2$.

Consider next the case of dilute sulphuric acid (H_2SO_4). The dissociation and recombination can be shown by.

$$H_2SO_4 \rightleftarrows H^+ + H^+ + SO_4^{--}$$

The hydrogen cations $2H^+$ move to the cathode and after giving up their charge are neutralised and rise to the surface as gas. The SO_4^{--} radicals are anions which move to the anode to give up their charge, but cannot exist in this form. They combine with hydrogen atoms of the water to reform acid molecules and liberate the oxygen as gas. Thus $SO_4 + H_2O = H_2SO_4 + O$. Note that the acid is not consumed but allows easy conduction by providing a good supply of ions.

If a copper sulphate ($CuSO_4$) solution is subjected to electrolysis with copper plates being used as the electrodes, action occurs as shown below. When the electrolyte is made up, dissociation and recombination can be shown by

$$CuSO_4 \rightleftarrows Cu^{++} + SO_4^{--}$$

With the application of a potential across the electrodes, the cations (Cu^{++}) reach the cathode, give up their charges, are neutralised, and the copper atoms are deposited on this copper electrode. The anions (SO_4^{--}) being radicals, cannot exist in this form and move to the anode and combine with the metal of this electrode to form a copper sulphate which goes into solution. Thus no gas is given off for this cell and copper is seen to be transferred from the anode to the cathode.

For the salt and acid cells, it would be observed that once electrolysis proceeds, gas bubbles are released slowly and tend to blanket the electrodes, said to be causing *polarisation*. Under this condition, if the applied potential is removed and a sensitive voltmeter is connected across the electrodes, a potential of some 1.5V would be indicated, caused, it is thought, by the slow progress of ions through the electrolyte. The voltmeter would also show a potential difference value between each platinum electrode and the electrolyte.

This polarising e.m.f. would rapidly become ineffective if used to supply current to an external circuit. The cell appears to have an internal back e.m.f. which is confirmed by (1) Ohm's Law not being followed, (2) electrolysis only starts satisfactorily if the applied voltage is of the order of 2V. More is said about polarisation in the larger general context of this chapter but it is appropriate here to point out that if tests are made on a number of different electrolytic cells, graphs can be drawn from the results, the ordinates being voltage applied across the electrodes (V) and current passing (I). The characteristics of a composite graph are seen to follow distinct types (Fig 35). Characteristic 1 shows that Ohm's Law is followed closely and is applicable to cells where no gas is given off. Characteristic 2 is followed for cells where gas is given off. It will be noted that little current flows until the applied voltage exceeds a critical value E_b — usually 1 to 2 volts and is linear after this. The explanation is also associated with polarisation.

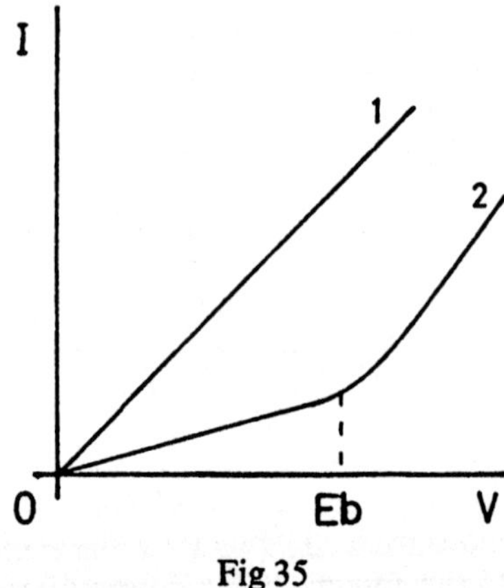

Fig 35

To finalise this introduction to electrolysis, a test condition can be investigated for the acid cell with different metals being used for the electrodes. Consider zinc and copper with no external applications of a potential difference. Dissociation of the acid occurs as before, giving H^+ and SO_4^{--} ions. Chemical action is now involved in that sulphate ions readily combine with zinc atoms but not with copper atoms. Thus the Zinc dissolves to zinc sulphate and leaves behind two electrons *ie* $Zn + SO_4^{--} \rightarrow Zn^{++}SO_4^{--} + 2$ electrons. The zinc electrode then acquires a −ve potential or becomes the cathode for an external circuit. The H^+ ions on reaching the copper electrode acquire two electrons to leave this electrode with a +ve charge and thus is able to act as the anode for an external circuit.

Here then we have the fundamental difference between an electrolytic and voltaic cell. For the latter, as a result of electrolytic dissociation and because electrodes of different metals are used, chemical action is possible if the circuit is completed. The chemical energy made available by the reaction is converted into electrical energy. For the former, with electrodes of similar material and no applied e.m.f., chemical action is not possible since there is no apparent preference for a combination between the electrolyte and one electrode or the other. As a consequence, chemical action is only possible if an external voltage is applied.

ELECTROLYTIC CELLS

The whole arrangement consisting of electrodes and electrolyte as described above, is frequently called an *electrolytic cell* to distinguish it from a *voltaic cell* which will be described later in this chapter. Electrolysis does not occur with solids or gases and is only possible for certain liquids. Some, like oils, are non-conductors, whereas others, like mercury, conduct without decomposition. The remaining liquids are electrolytes, which can therefore be defined as liquids which decompose when current is passed through them. The electrolytic cell can be constructed to enable experiments and measurements to be made with great accuracy. In this form it is frequently referred to as a *Voltameter.*

THE WATER VOLTAMETER (Sulphuric acid solution)

The diagram (Fig 36) shows the construction of the apparatus which is made of glass, with platinum electrode plates placed at A and C. The lead-in wires, passed through rubber corks, are not exposed to the solution to prevent corrosion. The voltameter is filled with acidulated water and the platinum electrodes are connected to a battery of 2 volts or more. Current passes from the anode to the cathode and bubbles of gas are given off which rise into the graduated tubes.

If care had been taken before passing current, to fill both tubes with the acidulated water by opening the taps and then closing them after all air had been expelled, then certain deductions can be made from the experiment. After a period of time, the gas collected would be found to be Hydrogen at the

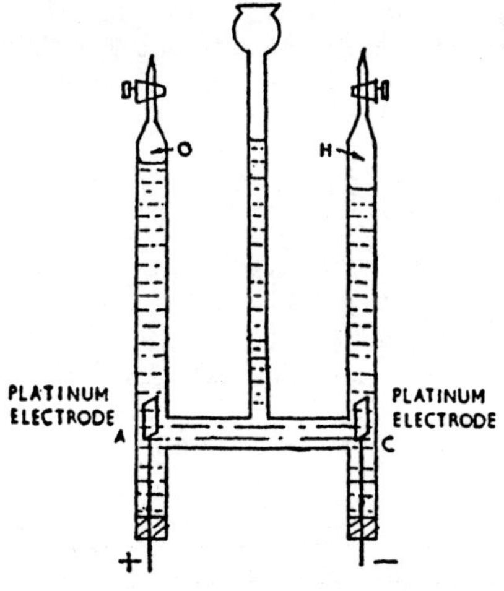

Fig 36

cathode and Oxygen at the anode. The ratio of the volumes of H to O would be 2:1 and the amount of gas collected would be proportional to the strength of the current and the time for which it flowed or more generally to the quantity of electricity passed.

In the acidulated water there are considered to be sulphuric acid molecules which divide into three ions, two of hydrogen carrying +ve charges H^+, H^+ and one with −ve charge represented by SO_4^{--}. Note the total +ve hydrogen charge equals the −ve sulphate or sulphion charge, but the ions migrate under the influence of the electric field. Thus the H^+ ions give up their charges at the cathode and are liberated as hydrogen gas. The sulphions proceed to the anode, but as they cannot exist in a free state they combine with two hydrogen H^+ H^+ ions or atoms from the water thus liberating oxygen as in the equation. Thus: $2SO_4 + 2H_2O = 2H_2SO_4 + O_2$. The oxygen rises from the anode and collects in the tube above it. The H_2SO_4 goes into solution and thus the electrolyte is decomposed in that the water appears to be used up, but the acid content remains the same and the solution gets stronger; that is, its specific gravity rises.

THE COPPER VOLTAMETER (Copper sulphate solution)

The diagram (Fig 37) shows the usual arrangement. A copper sulphate solution, made from crystals dissolved in pure water, is contained in a glass or glazed earthenware tank. The electrodes are made from pure copper sheet. The $CuSO_4$ molecule is considered to split into two ions, Cu^{++} and SO_4^{--}. When a p.d. is applied to the electrodes and current is passed, the copper ions migrate to the copper cathode to combine with it and give up their charges. The sulphions give up their charges at the anode

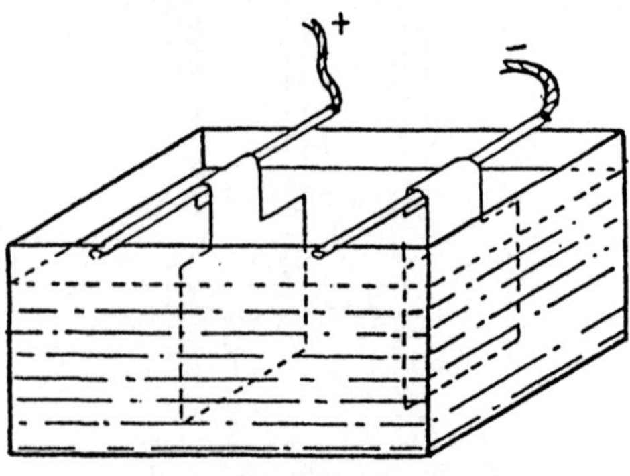

Fig 37

and combine with the copper from this electrode to reform copper sulphate. Thus copper appears to be taken from one electrode and deposited on the other.

The chemical equations for the electrodes are:

(1) Cathode. $CuSO_4 = Cu + SO_4$.

(2) Anode. $Cu + SO_4 = CuSO_4$.

During electroysis a small amount of gassing may be noted at the plates — this would be due to decomposition of water in the solution as described for the water voltameter. Furthermore some complex action may occur in the electrolyte due to sulphions combining with hydrogen in the water to form H_2SO_4. Oxygen from the water is then released to combine with anode copper to give copper oxide. This oxide will then dissolve in the H_2SO_4 to give $CuSO_4$. Irrespective of the action the final result is a simple one, in that the loss in mass of the anode equals the gain in mass of the cathode.

Various forms of voltameter can be constructed to allow research into electrolysis. Thus a silver voltameter may be used consisting of silver (Ag) plates and a silver nitrate ($AgNO_3$) solution. The examples described could be connected in series and the same quantity of electricity passed through all voltameters. If the electrodes were washed and carefully weighed before electrolysis and then washed and weighed again after electrolysis, certain conclusions would be reached which were first enunciated by Faraday in 1834 by his laws of electrolysis.

QUANTITATIVE LAWS OF ELECTROLYSIS (Faraday's Laws)

1. The mass of an element liberated from or deposited on an electrode is proportional to the quantity of electricity which has passed.

2. The masses of elements liberated from or deposited on electrodes by a given quantity of electricity are proportional to their Chemical Equivalent Weight.

Consider the first law. It is found by experiment that the mass of any material deposited or liberated always depends on the quantity of electricity which has passed. Thus $m \propto Q$ coulombs or $m \propto It$. This proportion can be modified to:

$$m = zIt$$

where z is a constant depending on the substance deposited. z is termed the *electrochemical equivalent* of the element.

ELECTROCHEMICAL EQUIVALENT (E.C.E.)

The mass in grammes or kilogrammes liberated by one coulomb of electricity is called the E.C.E. of a substance. Thus,

10 amperes flowing through a copper voltameter for 1000 seconds would result in 10 000 coulombs having passed and 3.3g of copper would be deposited. Thus the E.C.E. of copper $= \frac{3.3}{10\,000} = 0.000\,33$g/C. Similarly that for hydrogen would be 0.000 010 4, for oxygen 0.000 082 9 and for silver 0.001 118g/C.

In line with metrication and the use of SI units, it is more appropriate to think in terms of the kilogramme and the E.C.E. can be defined as the mass (in kilogrammes) of a substance liberated by the passage of one coulomb. Thus the E.C.E. of copper would be 330×10^{-9}kg/C. Since the milligramme is also an accepted SI unit, the E.C.E. can be given as mg/C. Thus for copper it would be 0.33mg/C.

The first law of electrolysis leads to a method of stating the unit of current, which was considered accurate enough to allow an original definition for the International Ampere. This then, was defined as, that unvarying current which deposits silver at the rate of 1118×10^{-9}kg per coulomb when passed through a solution of silver nitrate in water.

The formula already deduced above allows the solution of problems associated with electrolysis and practical electroplating. The unit in which the E.C.E. is given should be noted.

Example 32. Find the time taken to deposit 11.4g of copper when a current of 12A is passed through the copper sulphate solution contained in a copper voltameter.

The E.C.E. of copper can be taken as 330×10^{-9} kilogrammes/coulomb.

$$\text{Since } m = zIt \text{ then } t = \frac{m}{Iz}$$

$$\text{so } t = \frac{11.4 \times 10^{-3}}{12 \times 330 \times 10^{-9}}$$

$$t = 2880 \text{ seconds} = 48 \text{ minutes.}$$

ATOMIC WEIGHT, VALENCY, CHEMICAL EQUIVALENT WEIGHT

The second law of electrolysis can also be deduced by examining the results of tests made with a number of different voltameters in series, having been subjected to the passage of the same quantity of electricity. The results of the experiment would show that the mass of the substances deposited or liberated at the electrodes would be proportional to the chemical equivalent of the substances. It would also be concluded that if the atomic weight of any substance is known, its E.C.E. can be found provided the valency is known and the E.C.E. of hydrogen is assumed. Thus if the chemical equivalents of hydrogen, oxygen, copper and silver were 1, 8, 31.8 and 107 respectively, the masses of H, O, Cu and Ag liberated by the same quantity of electricity would be in the same proportion and therefore the E.C.E. of a substance is the E.C.E. of hydrogen multiplied by the chemical equivalent of the substance. Thus taking the E.C.E. of hydrogen as 0.0104 mg/C, that of silver would be $0.0104 \times 107 = 1.118$ mg/C.

To conclude our deductions from the second law, it would be well to define the following terms.

ATOMIC WEIGHT. Atoms are extremely small and determination of their absolute masses present considerable difficulties. The mass of an atom of hydrogen is believed to be 1.67×10^{-24}g and it is still therefore customary, even in SI units, to refer to the relative weights of the atoms of various substances in terms of the atom of hydrogen. Thus the term atomic weight is still used and is the weight of an atom of the substance in relation to the mass of an atom of hydrogen. Thus the value for oxygen is 16, *ie* it has 16 times the weight of an atom of hydrogen.

VALENCY. This can be described as the combining ratio of a substance. As an example that of oxygen is 2, whereas that of hydrogen is 1, so water is represented by the symbol H_2O. The valency of an elemental substance can also be defined as the number of atoms of hydrogen with which one atom of the element can combine. Thus the valency of a sulphate is 2 since, for example, in sulphuric acid H_2SO_4, 2 atoms of hydrogen are required to combine with the sulphate.

CHEMICAL EQUIVALENT WEIGHT. This is the ratio of atomic weight to valency or Chemical Equivalent Weight $= \dfrac{\text{Atomic Weight}}{\text{Valency}}$.
It is thus the weight of a substance which would combine with one part by weight of hydrogen or eight parts by weight of oxygen. Thus the C.E.Wt. or *combining weight* of hydrogen is 1, that of oxygen is 8, silver 31.8 and so on, as would be found in chemical tables.

From the second law we have:

$$\frac{\text{Mass of Material X liberated}}{\text{Mass of Hydrogen liberated}} = \frac{\text{C.E.Wt. of Material X}}{\text{C.E.Wt. of Hydrogen}}$$

or $\frac{m_X}{m_H} = \frac{\text{C.E.Wt. of X}}{\text{C.E.Wt of H}}$ also from the expression $m = zIt$

$$\frac{z_X It}{z_H It} = \frac{\text{C.E.Wt of X}}{\text{C.E.Wt. of H}} \text{ or } Z_X = z_H \times \frac{\text{C.E.Wt. of X}}{\text{C.E.Wt. of H}}$$

But the chemical equivalent of hydrogen = 1

so $z_X = z_H$ (Chemical equivalent Wt. of substance X)

$$\text{or } z_X = z_H \times \frac{\text{Atomic weight of substance X}}{\text{Valency of substance X}}$$

Example 33. How many amperes would deposit 2g of copper in 15 minutes, if the current is kept constant. Given the E.C.E. of hydrogen as 0.0104mg/C, the Atomic Weight of copper as 63.56 and the Valency of copper as 2.

$$\text{Then } z_{Cu} = z_H \times \frac{63.56}{2} = 0.0104 \times 31.8$$

$$= 0.33\text{mg/C}$$

$$\text{whence } I = \frac{m}{zt} = \frac{2}{0.33 \times 10^{-3} \times 15 \times 60}$$

$$\text{or } I = 6.72\text{A}.$$

Example 34. A voltameter consists of a solution of zinc sulphate and electrodes of zinc and carbon. If current is passed in at the carbon electrode, zinc is found to be deposited on the zinc electrode and oxygen is given off at the carbon plate. If a current 3.5 amperes is passed for 1 hour, find the mass of zinc deposited and oxygen liberated from the solution. The E.C.E. of zinc can be taken as 338×10^{-9}kg/C, the Atomic Weight as 65.38 and the Valency as 2. Take the Atomic Weight of oxygen as 16 and the Valency as 2.

From relationships already deduced:

$$m_{Zn} = z_{Zn}It \text{ or } m_{Zn} = 338 \times 10^{-9} \times 3.5 \times 3600$$

$$\text{or Zinc deposited} = 425.88 \times 10^{-5}$$

$$= 4.26 \times 10^{-3}\text{kg or } 4.26\text{g}$$

$$\text{Also } z_{Zn} = z_H\left(\frac{\text{At wt of Zn}}{\text{Valency of Zn}}\right)$$

$$\text{and } z_O = z_H\left(\frac{\text{At wt of O}}{\text{Valency of O}}\right)$$

$$\text{so } \frac{z_O}{z_{Zn}} = \frac{\text{At wt of O}}{\text{Valency of O}} \Big/ \frac{\text{At wt of Zn}}{\text{Valency of Zn}}$$

$$= \frac{\text{At wt of O}}{\text{Valency of O}} \times \frac{\text{Valency of Zn}}{\text{At wt of Zn}}$$

$$= \frac{\text{At wt of O}}{\text{At wt of Zn}} \times \frac{2}{2}$$

$$\text{Thus } z_0 = z_{Zn} \frac{\text{At wt of 0}}{\text{At wt of Zn}} = 338 \times 10^{-9} \times \frac{16}{65.38}$$

$$= \frac{338 \times 10^{-9}}{4.086}$$

$$\text{or } z_0 = 82.5 \times 10^{-9}\text{kg/C or } 82.5 \times 10^{-6}\text{g/C}$$
$$\text{and } m_0 = 82.5 \times 10^{-6} \times 3.5 \times 3600$$
$$= 1.05 \text{ grammes.}$$

BACK E.M.F. OF ELECTROLYSIS

The circuit laws enunciated in Chapter 1 govern the conditions for the majority of practical circuits. The loads of such circuits are mainly resistive such as, the coils of resistance wire in appliances like electric heaters and filament lamps. Such loads, termed *passive* loads, are recognised by the fact that they conform to Ohm's Law. For other types of loads such as the electric motor and accumulator or storage battery when being charged, Ohm's law is not directly applicable and they represent *active* loads, of which some electrolytic cells are also an example. The difference between pure ohmic resistance and that offered to the passage of current by such electrolytic cells will now be considered.

The diagram (Fig 38) shows a simple circuit for which the source of supply is a battery made up of three similar voltaic cells in series. Assume that the current through an electrolytic cell made up as shown, is adjusted and maintained at 3 amperes

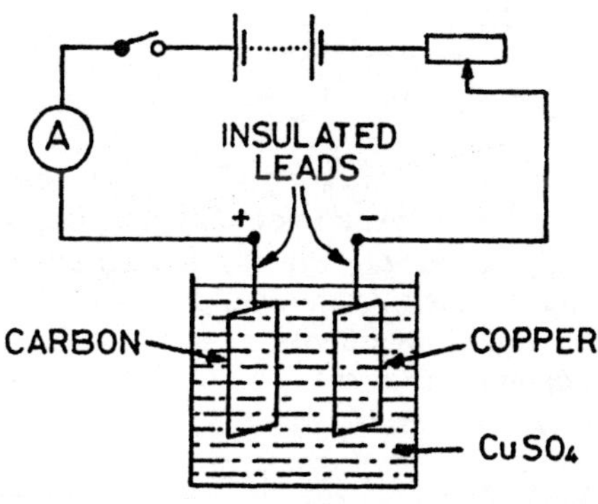

Fig 38

by the variable resistor provided for this purpose. If the circuit is set and the supply potential is then reduced by removing one of the voltaic cells, it may well be assumed that, as the e.m.f. has been reduced to $\frac{2}{3}$ of the original value, the current will fall to $\frac{2}{3} \times 3 = 2$ amperes. In actual fact the new current strength will be well below this value, and if the experiment is repeated by removing another voltaic cell so as to make the supply e.m.f. $\frac{1}{3}$ of the original, then the final current value will be well below that of 1 ampere, the value expected by an application of Ohm's Law.

The experiment shows that an extra current controlling factor is present in a circuit involving an electrolytic cell and the results can be explained by considering that a *back e.m.f.* is produced by the cell, so that the following equation represents the conditions:

$$V = E_b + IR_i$$

Here V represents the voltage applied to the cell, E_b is the back e.m.f. of the cell, I the current causing electrolysis and R_i the internal resistance of the cell.

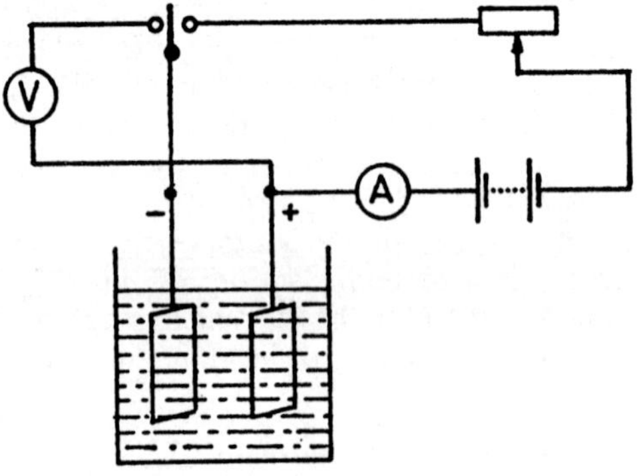

Fig 39

The magnitude of the back e.m.f., for any electrolytic cell, can be found by further experiment and is important, in that, it is involved with *polarisation*, which will be discussed later in this chapter in connection with primary cells. Here the basic action is described, as it would be noted for the simple water voltameter shown in the diagram (Fig 39).

The electrodes are immersed in an electrolyte of sulphuric acid. When the switch is closed current passes from anode to cathode and both electrodes become coated with bubbles of gas, oxygen and hydrogen respectively. Once this occurs the switch is moved to a second position so as to connect a sensitive voltmeter across the cell only. The main circuit current will have stopped, but the voltmeter will register a voltage across the plates, which gradually falls as the bubbles disperse, due to the flow of the small current through the voltmeter. The voltage or e.m.f. is due to the presence of the gas bubbles, and it would be noted that the cell polarity remains the same, namely that the current flows through the meter from anode to cathode in the external circuit so as to be in the opposite direction in the cell to the current flow which caused electrolysis. The value of the back e.m.f. is obviously important in that, if the applied voltage is less than this e.m.f., electrolysis cannot take place. At start a small current would flow but once polarisation begins, the back e.m.f. would rise to equal the applied voltage and the current would cease.

In the case of water the value of the back e.m.f. E_b can also be calculated thus. It is known that 1 gramme of water when formed by the combustion of hydrogen in oxygen produces approximately 15.96kJ of heat. If now we assume that the energy required to separate H and O in 1 gramme of water is the same, then the electrical energy required would also be 15.96kJ.

Since both H and O are released by electrolysis it follows that the total mass of gas or water released by 1 coulomb would be equal to the E.C.E. of hydrogen + E.C.E. of oxygen = 0.000 010 4 + 0.000 082 9 or z_{HO} = 0.000 093 3 gramme ie 93.3 $\times$ 10^{-9}kg are released.

Thus 1 coulomb releases 0.000 093 3g of gas or liberates this mass of water and the electrical energy required to decompose this mass of water = 15 960 $\times$ 0.000 093 3 = 1.49 joules.

If this electrical energy is produced by the work done against the back e.m.f. E_b, then the applied voltage would have a value V where $V = E_b$ and the energy produced by the passage of 1 coulomb = $V \times 1$ joules. Thus $V \times 1 = 1.49$ or $V = 1.49$ and $E_b = 1.49$ volts.

During electrolysis, the decomposition of the liquid produces ions, which having been dissociated, try to recombine and due to their slow progress through the electrolyte a back e.m.f. is produced. For water the back e.m.f. is 1.5 volts approximately and would be the value which would be substituted for E_b in the formula: $V = E_b + IR_i$. Thus for current flow the applied voltage must be greater than the back e.m.f. by the voltage drop due to resistance of the electrolyte.

A back e.m.f. of appreciable value exists for electrolytic cells made up with electrodes of dissimilar materials, but if both electrodes are of the same material, as for the copper voltameter, then the back e.m.f. value is so small that it can be neglected. This is explained by the fact that no difficulty is experienced in enabling the dissociated ions to recombine. They can readily combine with the electrolyte as was stated in the description of the copper voltameter, and the all important result is that the passage of current does not produce an overall chemical change. All that happens is that copper is transferred bodily from anode to cathode. It should be noted that for the water voltameter, although the electrodes are of the same material, namely platinum, yet a back e.m.f. appears when these are coated with H and O gas bubbles. The bubbles have the effect of insulating the electrodes and retarding the passage of ions. This condition results in an appreciable back e.m.f. of 1.5V approximately.

General observations show that when the products of electrolysis possess chemical energy, then the equivalent electrical energy must have been supplied through electrolysis and a back e.m.f. of appreciable value must exist. An example of this reasoning would be the electrolysis of water. The hydrogen and oxygen formed would recombine in an explosive manner to form water with evidence of heat and light. The energy latent for this recombination was derived from the electrical energy put in during electrolysis and a back e.m.f. must therefore have been present. For an electrolytic cell, such as the copper voltameter, since the product of the process possesses no chemical energy it can be assumed that a cell using electrodes of the same material has negligible e.m.f.

RESISTANCE OF ELECTROLYTES. The resistance R_l of a liquid conductor is proportional to the length and inversely proportional to the cross-sectional area. It also varies with the nature of the electrolyte and the concentration but it should be noted that the temperature coefficient is a negative one. Because of the back e.m.f. effect already discussed earlier, the resistance is difficult to measure. The value obtained by dividing the voltage drop across a cell, by the current flowing will give a resistance value for the electrolyte much greater than the true figure and would be erroneous because of neglecting the existence of E_b.

POWER EXPENDED DURING ELECTROLYSIS. If the voltage equation for an electrolytic cell is $V = E_b + IR_l$ then for a current flow of I amperes the power equation would become $VI = E_bI + I^2R_l$.

Here the VI represents the power applied to the cell, E_bI represents the power required to produce chemical dissociation and I^2R_i represents the heat energy produced in the cell which results in a temperature rise.

Example 35. Find the voltage required to pass a current of 4 amperes through a copper voltameter which has an internal resistance of 0.014 ohms and a back e.m.f. of 0.25 volts. Find the power utilised to produce the electrolysis and that wasted in heating the electrolyte. Find also the overall efficiency of the voltameter as a plating vat.

Since $V = E_b + IR_i = 0.25 + (4 \times 0.014) = 0.306\text{V}$

Power utilised for electrolysis $= 0.25 \times 4 = 1\text{W}$

Power wasted $= I^2R_i = 16 \times 0.014 = 0.224\text{W}$

$$\text{Efficiency of the cell for electrolysis} = \frac{1}{1.224}$$

$$= 0.82 \text{ or } 82 \text{ per cent}$$

PRIMARY AND SECONDARY CELLS

Many of the fundamentals of voltaic or galvanic action, as it used to be called originally, have already been mentioned in the earlier pages of this chapter, but we now consider in detail, the theory concerned with the conversion of chemical energy into electrical energy or, in other words, the production of an electromotive force by chemical action. The generation of such an e.m.f. is best studied by describing the action of a simple cell.

THE SIMPLE VOLTAIC CELL

If a piece of commercial zinc is dropped into a glass jar containing dilute sulphuric acid, the zinc is seen to be corroded away and hydrogen gas bubbles are given off. The jar is also found to get warm and it can be deduced that heat is given off by the chemical action. If next, a piece of pure zinc is similarly experimented with, none of the effects mentioned earlier are seen to occur nor do they happen if a dissimilar piece of pure metal such as copper is substituted for the pure zinc.

The chemical action noticed for the commercial zinc is considered to be caused by 'local action' and is explained by the presence of impurities in the zinc, the chief of which are iron and lead. A local closed circuit is made, say between a particle of iron and the zinc, since both are in contact and the acid is common to both dissimilar metals. A small cell is considered to be formed and since current flow is possible, then the generation of an e.m.f. is thought to accompany the chemical action. The truth of the supposition is borne out by the following experiments on a simple cell.

Consider two electrodes to be immersed in a solution of dilute sulphuric acid as shown by the diagram (Fig 40). The electrodes should be plates of pure dissimilar metals, such as zinc and copper and should be placed so as not to touch each other. No action will be seen to take place for the arrangement described but if an ammeter and a resistance load are connected as illustrated, then a current will be seen to flow when the switch is closed. The current through the external circuit will be from the copper +ve pole or anode to the zinc −ve pole or cathode, and it will be seen that hydrogen bubbles are given off from the copper plate while the zinc plate is slowly eaten away.

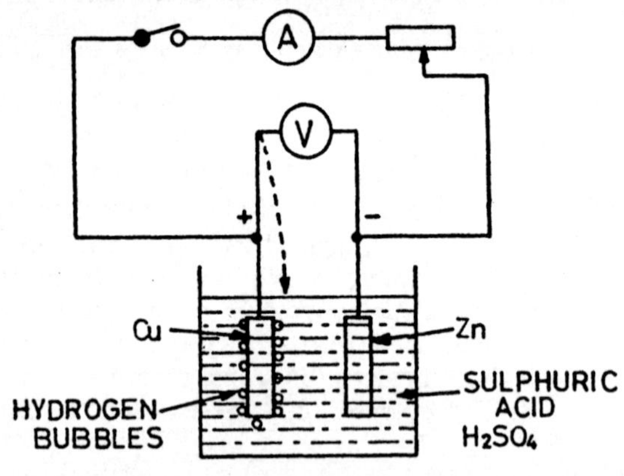

Fig 40

A further point of the experiment is to note that the ammeter shows the current to fall slowly and that it finally ceases after a while. The copper plate will be seen at this time to be covered completely with hydrogen bubbles and if these are wiped off with a glass rod, the current would be found to restart and the cycle of action will repeat itself. The hydrogen bubbles appear to control the chemical action of the cell which is said to have 'polarised' when the current ceases.

Further simple experiments with different combinations of electrodes and electrolytes will be seen to produce different e.m.f.s, as recorded by the voltmeter, which should be a sensitive instrument. When supplying current, the negative pole of each arrangement will corrode away and for every simple cell polarisation will be found to occur. A more complete explanation of the observation made can now be developed.

THE CELL E.M.F. The voltmeter used in Fig 40 will show that an e.m.f. or voltage exists between the electrodes of a cell even when this is an open circuit. When the circuit is completed, current will flow because of this e.m.f. and chemical action will be observed which continues to generate an e.m.f. and maintain the current. The chemical action of a voltaic cell results in a definite e.m.f. which is the result of the action of both electrodes or plates with the electrolyte. If the voltmeter was connected on open circuit between the zinc and the solution, the former would be found to be negative to the latter by some 0.63 volts. When connected between the electrolyte and the copper plate, the voltmeter would record 0.47 volts, with copper positive to the sulphuric acid. With the instrument connected across both plates a reading of 1.1V would result as could be expected. The production of a cell e.m.f. is explained by the electrolytic theory already discussed under electrolysis. It has been said earlier that when an electrolyte is made up, the molecules split into ions which are electrically charged and very mobile. An electrolyte is thus an ionised solution and when a metal is immersed into it, some of the metal appears to enter into solution, in that there is an immediate merging of surface ions of the metal with the ions of the electrolyte. Thus the barrier between the metal and electrolyte is not the surface of the former but along some layer of electrical potential equilibrium, which has caused ion interchange to cease. The action on immersing various metals into an electrolyte differs for the different metals. Thus for zinc in dilute sulphuric acid there is a greater tendency for +ve ions to pass to the solution than for +ve ions of the electrolyte to pass to the zinc. The zinc plate thus becomes deficient in +ve ions and becomes negative to the solution by a voltage of 0.63V. For copper in sulphuric acid, a different action takes place. This metal becomes positive to the solution because there is a greater tendency for the +ve charged hydrogen ions of the electrolyte to move to the copper than for the +ve ions of the metal plate to pass to the solution and the copper rises to a potential of 0.46 volts with respect to the electrolyte.

The interchange of ions as described, results in a potential being set up between the electrodes and solution which gradually opposes the interchange until this finally ceases and equilibrium is established. Thus the arrangement of zinc-copper electrodes in sulphuric acid as described, results in a potential difference of 1.1V between the electrodes. Continuing our investigation of this type of cell on open circuit, we see that the initial +ve Zn ions migrating to the solution combine with the sulphions to form zinc-sulphate, liberating +ve hydrogen ions which move to and accumulate on the copper plate. Thus the chemical action is explained by the formula $Zn + H_2SO_4 = ZnSO_4 + 2H$. The potentials build up within the cell, quickly bring the ion migrations to an end and thus chemical action ceases. If now the open-circuit condition is changed to that of a closed-circuit, by joining the copper electrode to the zinc through an external circuit, the chemical action is immediately noted to recommence. Current is seen to flow and the formation of zinc-sulphate continues with the liberation of hydrogen at the copper plate.

The action of the cell on closed circuit conforms to the following reasoning. The open-circuit e.m.f. of 1.1V, which is the result of the initial ion migration, can now cause a current, the magnitude of which is determined by the circuit resistance. A flow of current means a movement of electrons has resulted, passing from the zinc cathode to the copper anode. The electrons move round the external circuit from the zinc cathode to neutralise the +ve anode charges, thus making way for further migrations of +ve hydrogen ion charges in the cell. The initial cell action, as described for the open-circuit condition, can now continue and the cell functions by maintained chemical action; provided polarisation is avoided. We can now conclude our study of cell action by saying that all metal electrodes produce an e.m.f. as a result of the ion interchange action with the electrolyte, and that they can be arranged in a table, in order of the value of their e.m.f.s.

THE ELECTROCHEMICAL SERIES. If any two elements shown in the table are used for a cell, the element lowest in the series is the +ve terminal, when considered with respect to the external circuit. The list comprises the more usual elements which are mainly metals, but hydrogen and carbon are found to behave like metals and are included. To illustrate the use of the table, the ordinary dry cell as used for a hand-torch can be considered. This cell uses carbon for the +ve electrode and zinc for the −ve electrode.

Materials well spaced apart in the table are usually used for practical cells. For the torch battery the e.m.f. is about 1.5 volts per cell.

Potassium	Lead
Sodium	Tin
Magnesium	Hydrogen
Aluminium	Copper

Manganese	Mercury
Zinc	Silver
Cadmium	Platinum
Iron	Gold
Nickel	Carbon

POLARISATION. It will be remembered that when electrolysis was described, polarisation was introduced but an explanation was postponed until voltaic action had been covered. For the electrolysis of water, it was noted that gas was liberated at the electrodes and this has with its ions resulted in a decrease of the current and ultimately in a change in cell action. Thus once polarisation (collection of gas on the electrodes) occurs, the electrodes have, in effect, changed — these now being hydrogen and oxygen. Voltaic action can now result to set up an e.m.f. operating in the reverse direction to the voltage applied to cause electrolysis. This has already been introduced as the cell back e.m.f.

When the simple voltaic cell supplies current polarisation can be expected to occur. The circuit current gradually falls, even though the chemical action of the cell appears to proceed. Close examination will reveal that as the hydrogen bubbles make their way to the copper plate, not all are liberated here and rise to the surface. Some bubbles stick to the plate and this tendency increases until the whole plate is covered with bubbles to result in the cell becoming increasingly ineffective as a source of e.m.f. The layer of gas surrounding the +ve plate causes a polarising effect because (1) gas has a high resistance, so that any area of the plate covered with bubbles is almost insulated and cannot allow the passage of current. Thus the internal resistance of the cell rises as the gas layer increases and the circuit current falls as a direct result. (2) As hydrogen covers the copper plate, it begins to make its presence felt in that it effectively replaces the +ve copper electrode by a hydrogen electrode and thus reduces the e.m.f. of the cell. It will be seen from the table of the electromotive series that the spacing between zinc and hydrogen is smaller if compared with that for zinc and copper. The cell e.m.f. is thus much reduced giving the final result as described.

Once the cause of polarisation became known it was apparent that, in order to make the simple cell an effective source of electrical energy, a method of preventing the collection of the hydrogen bubbles was necessary.

The simplest forms of *depolarisers* which have been developed, operate chemically, in that they combine with the liberated hydrogen to convert it into water, thus preventing the gas from reaching the +ve electrode and blanketing it. The methods by which this is accomplished will be seen when the examples of *primary* cells are studied.

THE PRIMARY CELL

Under this heading are considered practical cells, which are suitable for providing a constant e.m.f. when operating under everyday conditions. They are however, cells which obtain their electrical energy from chemical energy, the active material being used up in the process. They differ from *secondary* cells in that the latter utilise materials which are not consumed when the cells provide electrical energy. The secondary cell can be electrically 'charged' so that its electrodes are chemically converted into materials which enable the cell to provide an active e.m.f. for supplying electrical energy. In this condition the cell discharges and the electrode materials again change chemically, reverting back to those of the uncharged state. The whole cycle of charge and discharge can then be repeated.

Primary cells suffer from the two main disadvantages of the simple cell, (1) polarisation and (2) local action.

Polarisation is overcome by the use of a suitable chemical depolariser which is therefore an essential component of cell construction. Local action is minimised by using pure metal, such as zinc free from impurities like iron and lead. In its basic form the primary cell is a *wet* cell, which is not used to any extent nowadays. In the *dry* form the Leclanche cell is the most common and particular attention should be paid to its construction and action.

THE LECLANCHE CELL (Wet type). The diagram (Fig 41) shows a cell used for supplying small amounts of electricity in remote locations where cells of the dry type have no particular advantage. Up to quite recently such cells were used for railway signalling in places where no electricity mains were available.

The +ve electrode consists of a carbon rod which is placed in a porous pot and surrounded by small pieces of carbon mixed with powdered manganese dioxide (MnO_2) which serves as the depolariser. The pot is sealed with a layer of bitumen compound or pitch. The negative electrode is a zinc rod and the electrolyte is a solution of ammonium chloride (NH_4Cl).

The action of the cell may be summarised as follows. Zinc ions migrate into the electrolyte leaving the metal electrode with a negative charge. When the external circuit is made, the further

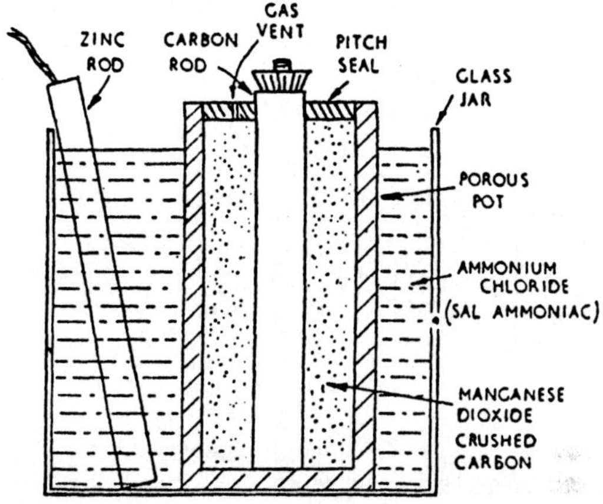

Fig 41

chemical action results in zinc being dissolved into the electrolyte and zinc-chloride ($ZnCl_2$), ammonia (NH_3) and hydrogen are produced. it is the conversion of zinc into zinc chloride which provides the energy of the cell. The chloride and ammonia dissolve in the water of the electrolyte and the hydrogen, as +ve ions, migrate through the porous pot and reaches the carbon. The hydrogen ions, while passing their charges to the carbon electrode, combine with the manganese dioxide, taking from it some of its oxygen to form water. The manganese dioxide is thus reduced to a simpler oxide (Mn_2O_3), frequently known as sesquioxide of manganese.

The solid depolariser is comparatively slow and polarisation takes place if the cell is used continuously. When the circuit current is switched off, the depolarising action continues and the cell can be used again after a little while. It is best suited for intermittent duty such as bell ringing. The e.m.f. is 1.55V, and the following chemical formula defines the chemical action.

Action at negative electrode:

$$Zn + 2NH_4Cl = ZnCl_2 + 2NH_3 + H_2$$

Action at positive electrode:

$$H_2 + 2MnO_2 = Mn_2O_3 + H_2O.$$

THE LECLANCHE CELL (Dry type). One form of construction is illustrated by the diagram (Fig 42) which is a cross-sectional view of a typical practical cell.

The depolariser of manganese dioxide is mixed with powdered carbon and packed round a central carbon rod. This whole

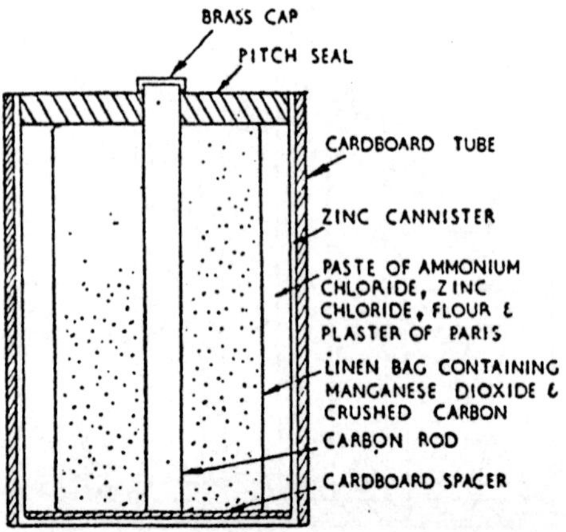

Fig 42

assembly is then placed in a linen bag which serves as the porous pot of the wet cell. The negative electrode is a pressed zinc cannister which contains the linen bag assembly and the electrolyte, which is made up as a paste of ammonium chloride (sal-ammoniac), zinc chloride, flour and plaster of paris. One method of closing the cannister is by sealing it with pitch. Because the cell uses the same materials as the wet type its action is identical and the chemical formulae as already given also apply.

The form of cell as described, is in most general service, but other forms have been developed for incorporating into the layer type of battery, as is used for portable radio sets, calculators etc. The reader should complement the information given here by referring to books specialising in the practical treatment of electrical equipment.

THE SECONDARY CELL (or Accumulator)

Because of the importance of this cell as a means of storing electricity (it is sometimes called a storage cell), the reader is again advised to consult a book giving more details of modern constructional methods, applications and maintenance requirements. The diagram (Fig 43) shows only the basic construction and the description sets out only the elementary principles. The modern accumulator uses 'pasted' plates to allow the maximum use of the available material and the process used in 'forming' the cell is somewhat involved for setting out in a book of basic theory. It is hoped however, that the information given below will provide sufficient knowledge to enable the action of the lead-acid accumulator to be understood. The nickel-iron, or nickel-cadmium alkaline battery also functions on similar principles, although the plate materials and electrolyte differ. This type of cell is also important and should be thoroughly investigated.

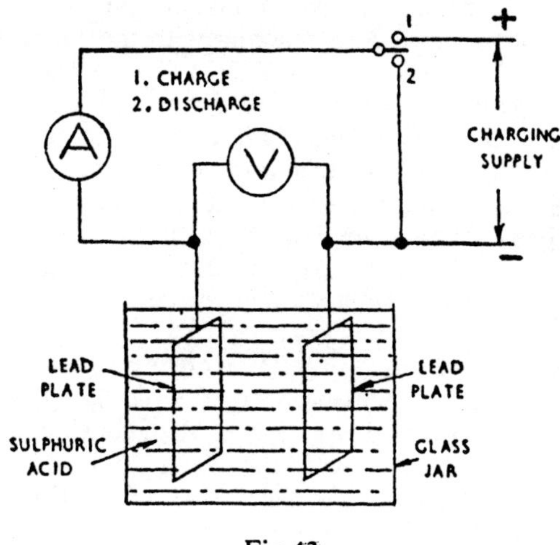

Fig 43

THE LEAD ACID CELL

The simple accumulator consists of two lead (Pb) plates immersed in dilute sulphuric acid, the whole assembly being contained in a glass or moulded ebonite container. The cell has to be worked into a suitable condition before it can be used for storing electricity and the process is carried out by alternatively 'charging' and then 'discharging' the cell. If a d.c. supply is connected to the plates as in the diagram (Fig 43, switch position 1) and the cell is subjected to electrolysis by passing current through it, oxygen and hydrogen gas is given off at the electrodes. As for the water voltameter, the first stage of the reaction would be decomposition of the acid (H_2SO_4). A molecule of acid dissociates to produce hydrogen ions and sulphions (SO_4). The sulphions move to the +ve plate, reacting with the water to form sulphuric acid and oxygen. The latter attacks the +ve plate only to form lead-dioxide (PbO_2), which causes the original lead electrode to assume a dark brown colour. The hydrogen as ions is discharged at the cathode and liberated in the gaseous state. The first chemical action is thus at the +ve plate only but if the supply is switched off (switch—intermediate position), the cell is now found to have the properties of a voltaic cell and will provide an electromotive force.

If the cell is next short-circuited (switch—position 2), it will behave like a primary cell, passing a current for a short time during which period it discharges. The solution is electrolysed in the reversed direction and the original negative plate now acts as the anode with its lead ions reacting with the sulphions of the electrolyte to form lead sulphate. Thus at −ve plate $Pb + SO_4 = PbSO_4$ (lead sulphate). The hydrogen ions from the electrolysis during discharge, move to the original +ve plate, now the cathode. The hydrogen now reduces the lead dioxide to lead oxide which in turn reacts with the acid to form lead sulphate. Thus at the +ve plate the chemical action is $PbO_2 + H_2 + H_2SO_4 = PbSO_4 + 2H_2O$. Both plates are converted into lead sulphate and assume a whitish colour.

If the charging cycle is repeated (switch—position 1), the direction of current flow in the electrolyte is again reversed and the $PbSO_4$ on the +ve plate becomes lead dioxide (PbO_2). This is a complex result of the electrolysis of the acid. Sulphions move to the +ve plate, react with the water to form H_2SO_4 and oxygen. The latter attacks the +ve plate to form PbO_2 and more sulphuric acid. The chemical action at the +ve plate is $SO_4 + H_2O = H_2SO_4 + O$ and $O + H_2O + PbSO_4 = PbO_2 + H_2SO_4$. At the negative plate, lead is produced by the hydrogen ions liberated by the acid decomposition, moving to this electrode and reducing the lead sulphate to spongy lead. The chemical action at the −ve plate is:

$$PbSO_4 + H_2 = H_2SO_4 + Pb$$

After a number of cycles of charging and discharging the plates become porous and the capacity of the cell is increased. When a cell is fully charged, chemical conversions are completed and hydrogen is freely given off, resulting in 'gassing'—the accepted term indicating a full charge.

The lead-acid accumulator in its practical form is provided with 'pasted' plates. Here the active material is applied to plates in the form of a paste, the back-bone of the plate being a lead-antimony grid. One 'forming' charge converts the paste into lead dioxide on the +ve plate and spongy lead on the −ve plate. Irrespective of the method of production the charge and discharge action can be summarised by the following chemical equation.

Charged			Discharged		
+ve Pole		*−ve Pole*	*+ve Pole*		*−ve Pole*
Lead Dioxide	Sulphuric Acid	Lead	Lead Sulphate	Water	Lead Sulphate
PbO_2 +	$2H_2SO_4$ +	Pb =	$PbSO_4$ +	$2H_2O$	+ $PbSO_4$

It will be seen that during discharge water is formed, thus diluting and reducing the specific gravity of the electrolyte. During charge, acid is formed and the tests to check a fully charged cell include:

(1) S.G. of cell charged (1.20 to 1.27); discharged (1.17 to 1.18).
(2) Voltage on open-circuit, charged 2.2V per cell or higher.
(3) 'Gassing' on charge.
(4) Positive Plate—rich dark brown colour. Negative Plate—slate grey.

THE ALKALINE CELL (NICKEL IRON OR NICKEL CADMIUM)

Two variations of this cell are in common usage, being the results of some ninety years developments and patents were taken out simultaneously in America by Edison and in Sweden by Jungner. Both men devised a cell with reversible action which used iron or cadmium for one plate and nickel-hydrate for the other. The construction of both types is basically the same. The active materials are enclosed in steel tubes or flat interlocking pockets which are perforated over the whole surface area with a large number of minute holes. The tubes (Edison form) or pockets (Jungner form) are assembled into steel retaining frames to form the +ve plate. The −ve plate, for both forms of cell, consists of a steel frame into which are assembled the flat pockets. Groups of plates of the same polarity are bolted together to steel terminal pillars. Separators of sheet ebonite are used and the plate groups are assembled in a steel container and, when the plates expand after the initial 'forming' process, no internal movement is possible even under the most severe vibration. The terminals of a cell, for each plate group, are brought through the lid in suitable insulated glands. Batteries of cells are built up in hardwood crates, the cell containers being at a potential. Other forms of enclosure in plastic cases have been developed in recent years leading to a lighter battery and dispensing with the need for insulated wooden crates.

Both forms of cell use the same electrolyte of dilute (21 per cent) potassium hydroxide and have the same e.m.f. of 1.25V. The electrolyte takes no active part but functions merely as a conductor, transferring the hydroxyl (OH^-) ions from one plate to the other when charging or discharging occurs. The cell is inert on open circuit and the electrolyte S.G. (about 1.18) does not alter. The battery is freely described in Britain by the tradename Nife (nickel-iron) even though the nickel-cadmium version is favoured and has been developed in this form.

The chemical reactions in the alkaline cell are complex and involved but as a guide the following equation shows the operation. This equation has been given for the nickel-cadmium version. For the nickel-iron version it would be similar except that Fe would replace Cd and $Fe(OH)_2$ the $Cd(OH)_2$. For a fully charged cell the nickel hydrate is at a high degree of oxidation and the −ve material is reduced to pure cadmium. On discharge the nickel triple-hydrate is reduced to a lower degree of oxidation (double hydrate) and the cadmium of the −ve plate is converted to a hydroxide.

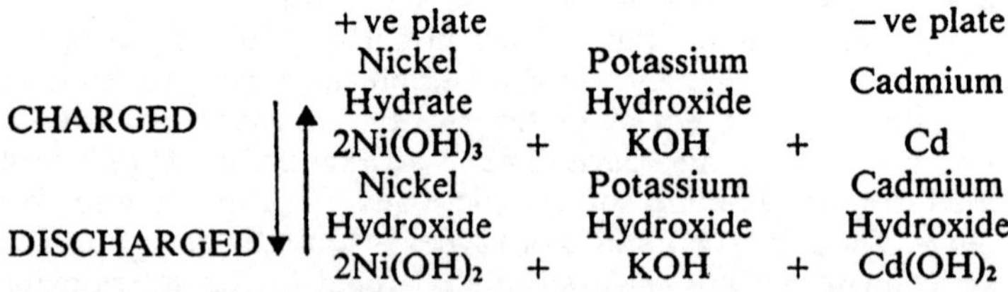

When the alkaline cell is charged the e.m.f. is about 1.5V, decreasing rapidly to 1.3V then more slowly to 1.0V on discharge. The average is 1.25V so for a given voltage the number of cells is 1.7 times the number of lead-acid cells required. The cell is extremely robust and can be 'short-circuited' without damage. The chemical action is completely reversible and no fumes are given off. Provided the plates are covered with electrolyte, the cells will stand almost indefinitely in any state of charge without significant deterioration. The internal resistance of an alkaline cell is 10-50 per cent higher than that of the lead-acid type but this depends upon design factors. The conductivity is adjusted by mixing finely powdered carbon with the nickel hydrate in the pockets of the +ve plate whilst iron powder can be mixed with the cadmium powder in the −ve plate pockets. These additions also prevent the active materials solidifying and the porous requirement of the overall plate construction being lost.

CAPACITY OF A CELL. This is the *ampere hour* figure it can yield on a single discharge, until the e.m.f. falls to about 1.8V per cell (for the lead-acid cell). Generally the capacity is based on a 10 hour rate of discharge, since it decreases as the rate of discharge increases. Research has shown that the performance of a cell can also be improved by working the +ve plate at a higher current density than the negative. This is achieved by keeping the plates of equal area for convenience, and providing an extra −ve plate, *ie* always making the outside plates negative.

EFFICIENCY OF A CELL. This may be expressed in terms of (1) the Ampere hour input and output, (2) the Watt hour input and output. Thus:

$$\text{Ampere hour Efficiency} = \frac{\text{Ampere hours of discharge}}{\text{Ampere hours of charge}}$$

The ampere hour efficiency neglects the varying voltages during charge and discharge. Since this is important, we thus can have an energy efficiency compared to a quantity efficiency, and

$$\text{Watt hour Efficiency} = \frac{\text{Watt hours of discharge}}{\text{Watt hours of charge}}$$

$$= \frac{\text{Average Discharge Volts} \times \text{Amperes} \times \text{Hours}}{\text{Average Charge Volts} \times \text{Amperes} \times \text{Hours}}$$

Example 36. A battery is charged with a constant current of 16 amperes for 11 hours after which time it is considered to be fully charged, its voltage per cell being recorded as 2.2V. Find its ampere hour efficiency if it is (1) discharged at a rate of 16 amperes for 10 hours, and (2) 28 amperes for 4 hours. In either case discharge was discontinued when the voltage per cell fell to 1.8V.

(1) Ampere hour input $= 16 \times 11 = 176$
Ampere hour output $= 160$

$$\therefore \text{Efficiency} = \frac{160}{176} = 0.91 \text{ or } 91 \text{ per cent.}$$

(2)Ampere hour input $= 16 \times 11 = 176$
Ampere hour output $= 28 \times 4 = 112$

$$\therefore \text{Efficiency} = \frac{112}{176} = 0.63 \text{ or } 63 \text{ per cent.}$$

Example 37. A 12V accumulator is charged by means of a constant current of 16A passed for 11 hours. The p.d. during charging varies as shown. The battery is then discharged at a constant current of 16A for 10 hours, the p.d. again varying as shown. Calculate the Watt hour Efficiency of the battery.

	Reading No	Time	Charge	Discharge
Start	1	—	10.8V	12.6V
	2	1 hr	11.0V	12.4V
	3	2 hr	11.5V	12.2V
	4	3 hr	11.8V	12.0V
	5	4 hr	12.0V	11.8V
	6	5 hr	12.2V	11.6V
	7	6 hr	12.4V	11.4V
	8	7 hr	12.6V	11.2V
	9	8 hr	12.8V	11.0V
	10	9 hr	13.0V	10.9V
	11	10 hr	13.1V	10.8V
	12	11 hr	13.2V	127.9
			146.4	

$$\text{Average Charging Voltage} = \frac{146.4}{12} = 12.2 \text{ volts}$$

$$\text{Average Discharge Voltage} = \frac{127.9}{11} = 11.62 \text{ volts}$$

$$\text{Efficiency} = \frac{11.62 \times 16 \times 10}{12.20 \times 16 \times 11} = 0.867 \text{ or } 86.7 \text{ per cent}$$

CHARGING PROCEDURE. British practice uses the 'constant current' method. This is also American practice but on the Continent the 'constant voltage' method is favoured. For this latter method the charging supply voltage is kept constant and is substantially higher than the battery e.m.f. for the discharged condition. The charging current is high initially but falls as the back e.m.f. of the battery rises. This method gives a lower charging time than the 'constant current' method, but due to the violent chemical action and heat generated in the battery there is danger of 'buckling' the plates, unless the battery is specially constructed.

For the 'constant current' method arrangements must be provided for increasing the voltage applied to the battery as charging proceeds and the back e.m.f. rises. If a generator is used and I is the charging current, R_i the internal resistance of the battery and E_b the battery e.m.f. at start of charge, then the applied voltage must be $V = E_b + IR_i$ . . . start of charge (1).

If E_{b1} is the battery e.m.f. at the end of charge, the applied voltage would have to be $V_1 = E_{b1} + IR_i$ . . . end of charge (2).

Thus subtracting (1) from (2) variation of voltage would be $V_1 - V = E_{b1} - E_b$ or the applied voltage must be increased by an amount equal to the rise of the battery e.m.f.

If a constant supply voltage is used for charging, then a variable resistor is required to obtain the necessary current control, and its value will be reduced as charging proceeds.

Let V be the supply voltage, I the charging current, R_i the internal resistance of the battery, E_b the battery e.m.f. at start of charge and R the control resistor. Then:

$$V = E_b + IR_i + IR \ldots \text{start of charge (1)}$$

If E_{b1} is the battery e.m.f. at end of charge and R_1 the new value of the control resistor. Then:

$$V = E_{b1} + IR_i + IR_1 \ldots \text{end of charge (2)}$$

Subtracting (1) from (2) $0 = E_{b1} - E_b + IR_1 - IR$

or $(E_{b1} - E_b) = I(R - R_1)$. Thus the control resistance must be reduced from R to R_1 as the battery voltage rises from E_b to E_{b1}.

Example 38. A 24V emergency battery is to be charged from the 110V ship's mains when the e.m.f. per cell has fallen to a minimum value of 1.8V. The battery consists of 12 cells in series, has a capacity of 100A h at a 10h rate and the internal resistance is 0.03Ω/cell. If charging continues until the voltage/cell rises to 2.2V, find the value of the variable resistor needed to control the charging. The charging current can be assumed to be equal to the maximum allowable discharge current.

$$\text{Discharge current} = \frac{100}{10} = 10 \text{ amperes} = \text{charging current}$$

At start of charge, battery voltage $= 12 \times 1.8 = 21.6\text{V}$

Battery internal resistance $= 12 \times 0.03 = 0.36\Omega$

Then $110 = 21.6 + (10 \times 0.36) + (10 \times R)$

$$\text{or } R = \frac{110 - 25.2}{10} = \frac{8.48}{10} = 8.48 \text{ ohms}$$

At end of charge, battery voltage $= 12 \times 2.2 = 26.4\text{V}$

Then $110 = 26.4 + (10 \times 0.36) + (10 \times R_1)$

$$\text{or } R_1 = \frac{110 - 30}{10} = \frac{80}{10} = 8 \text{ ohms.}$$

Thus the variable resistor should have a value of 8.48 ohms and be capable of being reduced to 8 ohms. In practice a unit of 9 ohms would be used which would be reduced by adjusting the sliding contact until the charging ammeter recorded the correct current. Further adjustments would be made periodically as charging proceeds. It is important to note that besides the ohmic value of the resistor, the wattage rating must be specified. For the unit in the example, a rating of $I^2R = 10^2 \times 9 = 900$ watts is required. The control resistor must be capable of dissipating up to this power as heat during the charging, although this waste of power will decrease slightly as charging proceeds. For example at the end of the charge the power wasted in the resistor would be $10^2 \times 8 = 800$ watts.

The most important point to stress is the correct connecting up of the battery for charging, *ie* +ve terminal of battery to +ve of mains; −ve terminal of battery to −ve of mains. It is surprising how many times this elementary requirement is overlooked through carelessness. For incorrect connection, no control of the current would be possible with the equipment provided and damage of the ammeter, control resistor, or battery could result.

THE MEANING OF pH

Introductory ionic theory dealt with electrolytic dissociation which, for a solution, results in the formation of two separately charged ions (anions and cations). Such ionisation is assisted by a liquid which has a high 'dielectric constant' and thus can separate and support unlike charges. Pure water, being a poor conductor, serves as such a dielectric. The cations of an electrolyte are derived from the metallic part of the molecule, having a +ve charge. The anions are from a non-metallic element or radical and have a −ve charge. (A radical is a fundamental group of atoms, such as a sulphate (SO_4), a nitrate (NO_3) or a carbonate (CO_3) etc. that behave as individual entities and remain unchanged by most chemical reactions.) All acids produce hydrogen ions (H^+) and all alkalis produce hydroxil ions (OH^-). Sulphuric acid (H_2SO_4) ionises to $H_2SO_4 \rightleftarrows 2H^+ + SO_4^{--}$. Since each hydrogen ion (H^+) can carry only one +ve charge and as each H_2SO_4 molecule is electrically neutral, the sulphate ion must have two negative charges (SO_4^{--}). Similarly sodium chloride or common salt (NaCl) splits into $NaCl \rightleftarrows Na^+ + Cl^-$. The double arrows indicate the splitting up is not complete and only a percentage of the solution is ionised; the amount depending on the physical conditions—strength of solution, temperature, type of salt etc. The unionised portion of the solution is assumed to consist of neutral molecules. Caustic soda (NaOH) is an alkali and ionises thus: $NaOH \rightleftarrows Na^+ + OH^-$. A hydroxyl ion ($OH^-$) is produced whereas the sulphuric acid molecule produces two hydrogen ions $2H^+$. Note that hydroxyl ions (OH^-) are in fact radicals.

For an electrolyte, the percentage ionisation is extremely high. It is known that the concentration of hydrogen ions present in the solution determines its properties. The greater this concentration the more acid the solution. Conversely the smaller the concentration the more alkaline the solution. Water is a special case since it ionises only slightly to give both hydrogen and hydroxyl ions. For many modern industrial processes, a knowledge of the acidity of the materials being used is most important. The pH value of a substance or solution is a measure of its acidity or alkalinity. A Swedish scientist Sorensen devised a scale to indicate the hydrogen ion content. It uses the pH symbol which stands for minus the logarithm to the base 10, for the ion concentration. This latter is expressed by $[H^+]$.

Thus if $[H^+] = 10^{-3}$. pH $= 3$ (*ie* $-\log_{10}10^{-3} = 3$)

Note that the letter 'p' comes from the German potenz, meaning 'a power', as the word is used in mathematics. Example $100 = 10^2$ — ten to the power two. The hydrogen ion concentration of a solution is the number of gram-equivalents of hydrogen contained in a litre of the solution. The actual extent of the ionisation of water has been determined and for pure water the ion concentration (H^+) is 0.000 000 1 gram-equivalents per litre or 10^{-7} normal. This power of 10 with the sign changed was taken by Sorensen to provide a scale up to 14 — neutrality being given as 7 (pure water). Thus by measuring its hydrogen ion concentration, it is possible to show where a solution lies on the acid/alkali scale. A strong acid will register near 1 whilst a strong alkali will register towards 14.

An electrical method for determining the pH value of a solution involves measuring the d.c. voltage between two special electrode assemblies. Such electrode assemblies, when immersed in the solution, result in e.m.f.s being produced by voltaic action. The electrode assemblies are known as the *Reference Electrode* and the *Measuring Electrode*. The former is so constructed that a constant potential is produced, irrespective of the pH of the solution under test. The Measuring Electrode assembly is constructed to allow its potential to vary with the solution under test and since the Reference Electrode potential is constant, then the resulting potential variation between the two electrode assemblies can be measured by a sensitive millivoltmeter. Such an instrument can be suitably calibrated to provide a direct indication of the pH value of the sample or solution being tested.

ELECTROCHEMICAL CORROSION

The two main causes of electrochemical corrosion are due to (a) *galvanic* action and (b) *electrolytic* action. In each case the electrolyte may be water with impurities or moist earth. These corrosion causes are now considered separately.

(a) *Galvanic Action.* This results in currents through the

electrolyte from an anode, such as a metal structure, to some adjacent cathode. Metal is lost from the structure which can be both costly and/or dangerous. In the case of a ship such results in rusting by oxidation of the hull when immersed in the conductive salt water. Galvanic action is caused by 'local cells' set up by slight differences in the surface composition of the hull metal, pitting of the plating, welds, rivets and millscale (Fig 44). The corrosion occurs at local anodic areas from which currents flow through the sea to the local cathodic areas. The rate of corrosion is in proportion to the currents which in turn are affected by the composition of the metal, by the temperature of the electrolyte and even by the ship's speed.

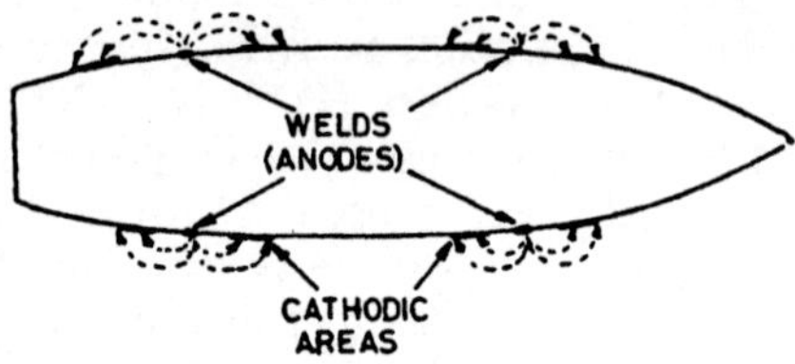

Fig 44

(b) *Electrolytic Action.* This is caused by stray currents due to leakage from some electricity system, such as a d.c. distribution network — examples being an electric dockside crane rail or tramway track. Such corrosion was a severe hazard in the early days of electric traction when the 'live rail' or cable followed a curve of the track and some conductor such as a metal pipe or different cable run was situated within the bend. This condition could provide a leakage current path. The current could leave the live conductor, enter the pipe or adjacent cable sheath and then leave the latter at an appropriate point, to rejoin the live conductor. Where the current leaves the pipe or cable sheath, an anode is formed and as a result metal is dissolved at this point with possible serious results.

CATHODIC PROTECTION METHODS. In both the above cases, corrosion can be prevented by introducing an anode adjacent to the structure, pipe or cable sheath. In this way current is forced to enter the original anode point so that this now becomes a cathode. This current being opposed to the original local currents neutralises or reverses them. For marine work two systems are in general use (a) the 'sacrificial' anode method, (b) the cathodic protection or 'impressed current' method. The latter is most favoured but both systems should be complementary to a good paint system.

(a) *The Sacrificial Anode Method.* Counter currents are encouraged by galvanic action and the method is most useful for protecting smaller structures or ships because the current adjustment range is restricted by constant potential effect. Also, as the name implies, the anodes waste away and require periodic replacement. Examples of materials which readily corrode away are magnesium or zinc. Magnesium gives a potential of some 1.8V positive with respect to iron and thus, provided the resistance of the electrolyte path is sufficiently small, current will enter the cathode (structure or hull). This is shown by Fig 45.

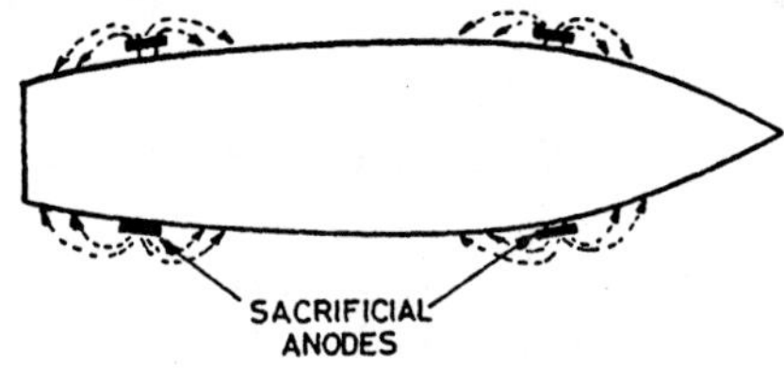

Fig 45

(b) *The Impressed current method.* For a ship, this is achieved by forcing a current through the hull, of a magnitude sufficient to nullify the effect of the local cells already existing. Such a current passes from the external anodes to the hull below the water-line so as to make the latter wholly cathodic. The anodes are supplied from a direct current source (generator or rectifier) and conduct the protective current into the water. The anodes are mounted on insulating plates thus isolating them from the hull (Fig 46). They are connected to the positive terminal of the supply through insulating feeder cables, whilst the hull is connected to the negative terminal. For shore work, the structure to be protected is connected to the negative terminal. The anodes are buried in the ground and suitably sited to give the desired current distribution. For a ship, the current required to give protection may vary from 20 to 300 amperes, the variation being dependent on the paint condition, the ship's speed, the water temperature etc. For shore work, the variation depends mainly on the soil moisture content. The anode material for ship work may be a metal such as raw lead or platinum-plated titanium, which is affected only a little or not at all by the discharge current. It is pointed out that systems do operate where some anode dissolving is inherent such as the aluminium wire system. Such a wire is trailed from the ship's stern. It is suitably insulated from the hull and the wastage is compensated for at regular intervals by paying out a suitable amount of wire. For shore work, the anodes can be made of steel scrap, graphite or ferro-silicon.

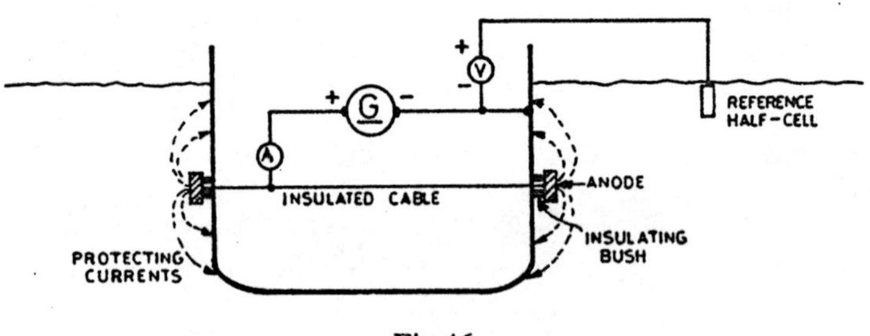

Fig 46

If the protecting current is too high, electrolysis will cause excessive development of alkali and hydrogen resulting in blistering of paint. If the current is too low, corrosion continues and in order to monitor and control protection at the correct value the potential, between the electrolyte (the sea) and the hull, is measured with a sensitive voltmeter and a reference electrode. In actual practice, a measuring half-cell is used for checking purposes — *silver-silver chloride* or *copper-copper sulphate* cells are favoured. Unprotected iron has a potential of some −0.55V against the copper-copper sulphate measuring cell but, if reduced by the impressed current to some −0.85V full protection is achieved. Control can be effected manually or automatically by altering the potential of the motor-generator or transformer-rectifier unit. Modern automatic systems employ a hull-mounted reference electrode and an electronic amplifier to adjust the cathodic potential of the impressed protection current.

CHAPTER 4

PRACTICE EXAMPLES

1. An accumulator is charged at the rate of 6 amperes for 18 hours and then discharged at the rate of 3.5 amperes for 28 hours. Find the ampere hour efficiency.

2. The mass of the cathode of a copper voltameter before deposit was 14.52g, and after a steady current was passed through the circuit for 50 min, its mass was 19.34g. The reading of the ammeter was 5.1A. Find the error of the ammeter, taking the E.C.E. of copper as 330×10^{-9}kg/C.

3. A 90V d.c. generator is used to charge a battery of 40 cells in series, each cell having an average e.m.f. of 1.9V and an internal resistance of 0.0025Ω. If the total resistance of the connecting leads is 1Ω, calculate the value of the charging current.

4. Nickel is to be deposited on the curved surface of a shaft 100mm in diameter and of length 150mm. The thickness of deposit is to be 0.5mm. If the process takes 8h, calculate the current that must flow. The E.C.E. of nickel is 302×10^{-9}kg/C. The density of nickel is 8600kg/m^3.

5. A nickel-alkaline battery is discharged at a constant current of 6A for 12h at an average terminal voltage of 1.2V. A charging current of 4A for 22h, at an average terminal voltage of 1.5V is required to re-charge the battery completely. Calculate the ampere hour and watt hour efficiences.

6. A battery of 80 lead-acid cells in series is to be charged at a constant rate of 5A from a 230V, d.c. supply. If the voltage per cell varies from 1.8 to 2.4V during the charge, calculate the maximum and minimum values of the required control resistor. If the ampere hour capacity of the cells is 60, state the probable charging time required, assuming that the cells were in a completely discharged condition at the commencement of the charge.

7. A metal plate measuring 50mm by 150mm is to be copper-plated in 30min. Calculate the current required to deposit a

thickness of 0.05mm on each side (ignore the edges). The E.C.E. of copper is 330×10^{-9}kg/C and its density is 8800 kg/m^3.

8. A battery of 40 cells in series delivers a constant discharging current of 4A for 40h, the average p.d. per cell being 1.93V during the process. The battery is then completely recharged by a current of 8A flowing for 24h, the average p.d. per cell being 2.2V. Calculate the ampere hour and the watt hour efficiencies for the battery.

9. Thirty lead-acid accumulators are to be charged at a constant current of 10A, from a 200V d.c. supply, the e.m.f. per cell at the beginning and end of charge being 1.85V and 2.2V respectively. Calculate the values of the necessary external resistor required at the beginning and end of charge, assuming the resistance of the leads, connections, etc to be 1Ω and that the internal resistance is 0.01Ω per cell.

10. When a current of 3.5A was passed through a solution of copper sulphate, 4.2g of copper were deposited. If the E.C.E. of copper is 330×10^{-9}kg/C and the chemical equivalent of copper is 31.8, find the time for which the current was passed through the solution and also the mass of hydrogen liberated.

CHAPTER 5

MAGNETISM – ELECTROMAGNETISM

NATURAL MAGNETS

From very early times it was known to ancient civilisation, such as those of the Greeks and Chinese, that pieces of certain types of iron ore have magnetic properties. Pieces of the ore were known, not only to be capable of attracting and repelling other such pieces but could also pass on this property of magnetism. One further known fundamental property of a piece of the ore, called Magnetite or Lodestone, was that if it is freely suspended, as shown in the diagram (Fig 47), then it would come to rest in an approximate geographic North-South direction. The end pointing north is called a north-seeking or simply a North Pole, whilst the other end is a South Pole. The piece of ore constitutes a natural magnet and if brought into contact with a quantity of iron filings, these would be found to adhere mainly to its ends or poles.

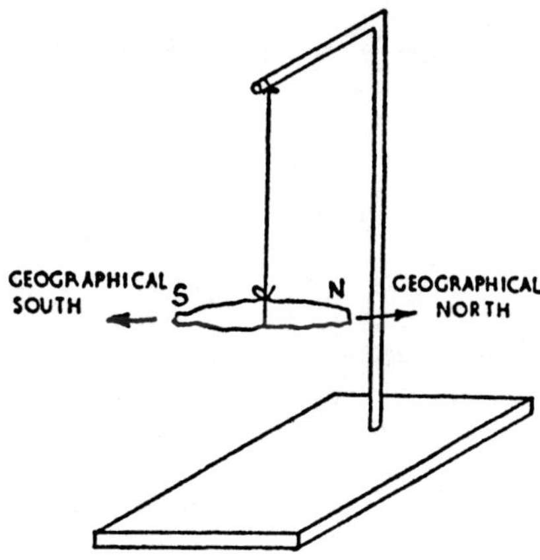

Further simple investigations made with pieces of the magnetic ore would show that, if two such magnets are each suspended as described above and their polarities are determined and marked, then when the *N* pole of one suspended magnet is brought near the *N* pole of the other suspended magnet, repulsion of the poles will result. Two *S* poles brought near each other would behave in a similar manner whereas, a *S* pole brought near the *N* pole of the other magnet will produce an attractive effect. Thus every magnet is seen to have two poles of unlike polarity and that like poles repel whereas unlike poles attract.

ARTIFICIAL MAGNETS

A piece of iron can be converted into a magnet and made to exhibit properties similar to that of the iron ore described above. Such a piece of iron is an artificial magnet and is said to be magnetised. A simple method of magnetising a specimen is by stroking it in one direction from end to end with one pole of an existing magnet, but the most effective method is by electromagnetism, which will be considered later in the chapter.

Certain materials such as copper, aluminium, lead, brass, wood, glass, rubber, etc cannot be magnetised. Thus all known materials can be classified under the heading of magnetic or non-magnetic substances. Some metals such as nickel, cobalt and magnesium exhibit very slight magnetic properties, but it is of interest to record that, when alloyed with iron very strong magnetic properties result.

An artificial magnet is usually made in bar or horse-shoe form. When tested, the tips are found to constitute poles of opposite polarity and, if suspended, a bar magnet will lie on an approximate *N-S* line. The magnetic compass makes use of this principle and consists of a short highly magnetised bar magnet which is pivoted at its centre. A card, calibrated in degrees and/or geographic points, is mounted below and is used with the magnet to obtain a 'bearing'. It is necessary to mention here that the *N—S* direction as indicated by such a compass is not exactly geographic *N* and *S*. The angle between the lines of magnetic and geographic *N—S*, is called the 'variation' and varies for different parts of the world. If the magnetic compass is being used, due allowance must be made for the variation, before a map can be truly orientated and used correctly.

Before proceeding with further study of magnetism, it would be as well to explain why a compass needle lies in the *N—S* direction. The earth itself behaves as though it contains a magnet having its *S* pole in the region of the geographic north and its *N* pole near the geographic south. A compass needle placed on the earth's surface will lie so that its *N* pole will be attracted to the magnetic south (geographic north) pole of the earth and its *S* pole will be attracted to the magnetic north (geographic south). Further mention of the earth's magnetism will be made later on in this chapter.

Summarising the facts deduced so far about natural or artificial magnets, we know that every magnet has two poles of unlike polarity and that *like* poles repel whereas *unlike* poles attract.

THE MAGNETIC FIELD

This is the space around a magnet where its magnetic effects can be felt. If a bar magnet is covered by a sheet of paper and iron filings are sprinkled on the paper, then on tapping the latter, the filings would be seen to align themselves as shown in the diagram (Fig 48). The filings would form a pattern which, if examined closely, would show that lines could be traced from the *N* pole of the magnet to the *S* pole through the space outside and from the *S* to *N* poles inside the magnet.

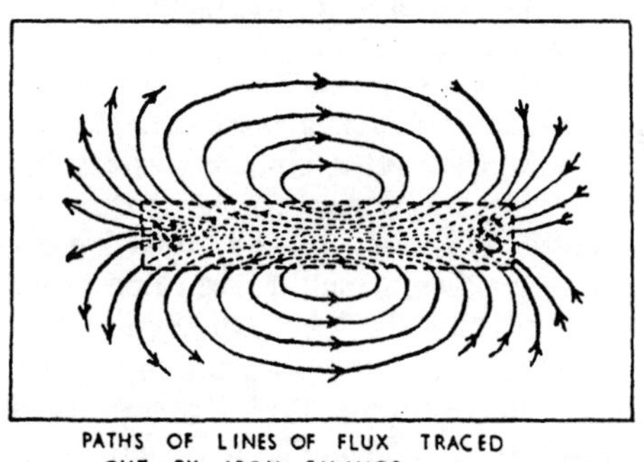

PATHS OF LINES OF FLUX TRACED OUT BY IRON FILINGS

Fig 48

The field can also be plotted by using a small compass needle as shown in the diagram (Fig 49).

Field plotting with the aid of a compass needle is undertaken as follows. Place the magnet on a sheet of paper and draw its outline. Set the compass needle against the *N* pole of the magnet and, with a pencil, mark a dot at the point in line with and adjacent to the *N* pole of the compass needle. Move the compass until the *S* pole of the needle is coincident with the original dot.

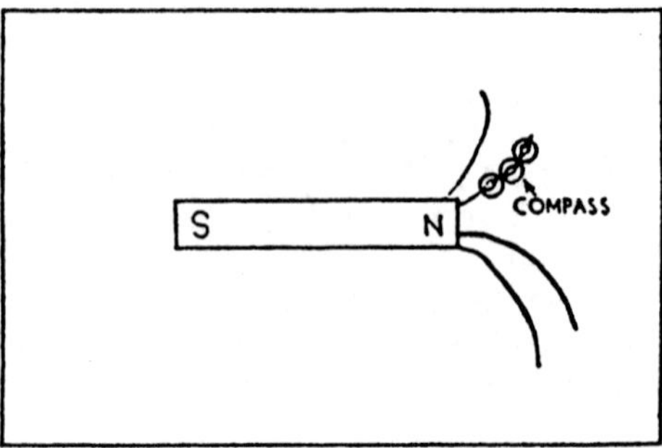

Fig 49

Mark the new point in line with and adjacent to the *N* pole of the needle. Repeat this procedure until the *S* pole of the magnet is reached. Join the dots together to give a *'line of force'* or a *'line of flux'* which can be described as the line which, when drawn through any point in a magnetic field, shows the direction of the magnetic force at that point. Using the compass needle the field can be mapped out for a considerable distance around a magnet and the following deductions can be made.

1. Lines of flux never cross.
2. The lines are always continuous.

If various magnetic field arrangements are plotted as shown in the diagram (Fig 50) then other conclusions can be deduced.

3. Lines of flux are like stretched elastic threads and tend to shorten themselves. This explains the attractive effect between two unlike magnetic poles, which if free to do so will move into contact, thereby reducing the length of the lines of flux.

4. Lines of flux which are parallel and in the same directions repel each other. This deduction is clearly seen for the condition where two magnets are brought together, with like poles adjacent to each other. There would be a force of repulsion between the magnets and if the field is plotted between two like poles a neutral point would be found where the effects of the two repulsion forces balance each other and the total effect is as shown by the absence of control on a compass needle placed at this neutral point.

The strength of the magnetic field around a magnet will vary from point to point, but before this can be measured and methods devised for making such measurements, a system of magnetic units and terms must be introduced. Faraday conceived the idea of the line of flux, as already introduced, and further suggested the use of these lines to depict the strength of the magnetic field.

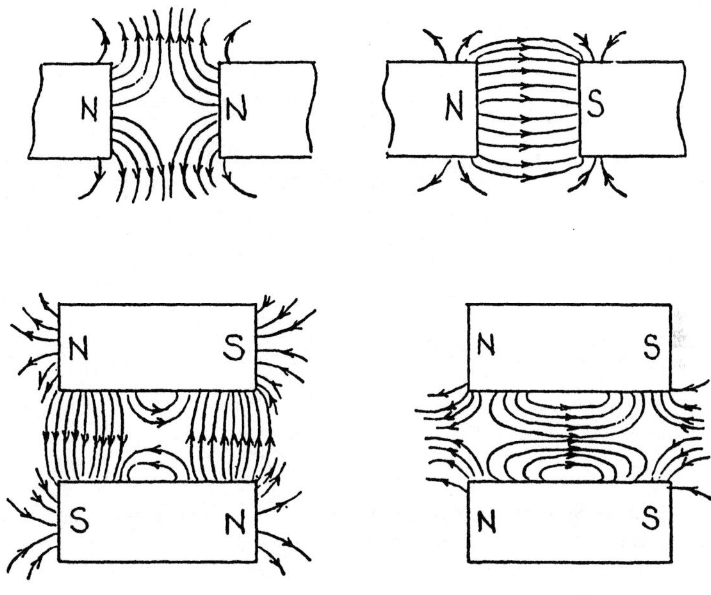

Fig 50

If a unit area at right angles to the lines of flux is considered then further definitions and terms can be introduced.

A number of lines of flux collectively are said to constitute the magnetic *Flux* (symbol Φ—Greek letter phi) which is passing through the area being considered.

Another unit of importance is *Flux Density*—and the value, at any point, is obtained from the expression:

$$\text{Flux Density} = \frac{\text{Flux}}{\text{Area}}$$

The diagram (Fig 51) illustrates the SI unit of flux or the *Weber*. Thus if 50 lines of flux are shown passing through the area of 1 square metre, then for the plane being considered, the magnetic flux is 50 Webers. The symbol for flux density is B and the unit is the *Tesla*. Thus for any point P in the plane being considered, the flux density is 50 teslas.

Note. The tesla is a name introduced for the SI system. The original unit was the weber per square metre *ie* Wb/m².

We now have Flux = Flux Density × Area
or Φ (Webers) = B (teslas) × A (square metres).

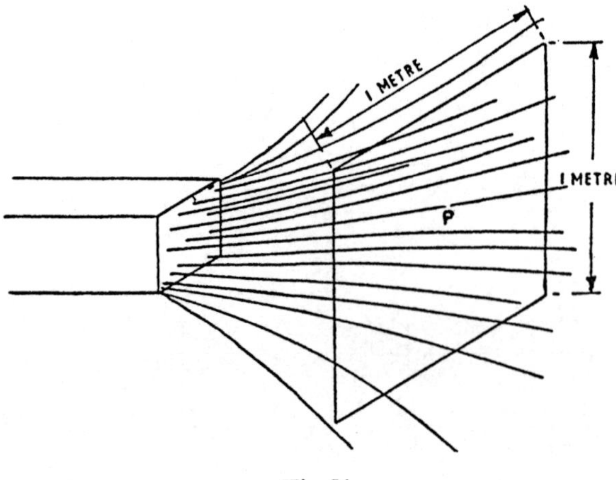

Fig 51

The above relationship will be used continually when the study of electromagnetism and magnetic circuits is made and should be considered a basic and important formula. It is well to stress here the obvious, namely that lines of flux do not exist but the properties of magnets and magnetic fields can best be assessed by assuming their existence and their having definite physical properties. It should also be remembered that the field of a magnet exists in all direections and is not confined to one plane.

MOLECULAR THEORY OF MAGNETISM

A molecule is defined as the smallest particle of a substance that can exist separately and in any magnetic material every molecule is thought to be a complete magnet. In a piece of unmagnetised magnetic material the molecules are considered to

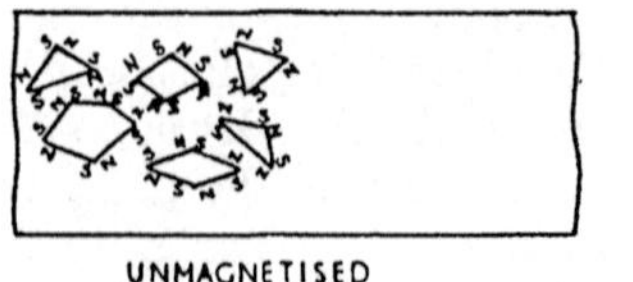

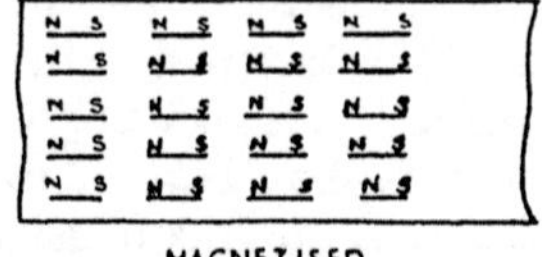

Fig 52

arrange themselves in closed magnetic chains or circuits as shown in the diagram (Fig 52). Under this assumption it is considered that each molecular magnet is neutralised by adjacent molecular magnets so that no magnetism is apparent in the material. The process of magnetising the material is considered to be achieved by simply arranging the molecular magnets so that their axes point in the direction of the magnetising force. The proof of this hypothesis is supported by the following known observations. (1) There is a limit to the amount of magnetism that can be imparted to any one sample of material. This is explained by the supposition that, once all the molecules had been 'lined up', no amount of extra magnetising force can increase the strength of the magnet. (2) When a magnet is broken, the ends of the molecular magnets are exposed and the broken pieces are found to be magnets themselves. (3) If heated to about 100°C and allowed to cool a magnet is weakened. If the magnet is heated to red heat, the magnetic properties are lost altogether. Similarly if a magnetic material like hard steel, is cooled in a strong magnetic field then it will set as a permanent magnet. It is considered that during heating, energy is transferred to the magnet which causes oscillations of the molecular magnets which tend to break the 'lining up' and results in these magnets taking up random directions. Similarly for the cooling process, as energy is passed from the hot material, the oscillations decrease in magnitude and violence and the molecular magnets are allowed to settle in the direction of the magnetising field.

A more modern theory of magnetism is based on the electron theory and the conception of the atom. A chapter considering the electron theory in detail is introduced at a later stage but here it can be stated that an electron, the smallest known -ve charge, when rotating in an elliptical path, constitutes a circular current which sets up a magnetic force along the axis of gyration. In a molecule the magnetic effects of the electrons of the atoms may neutralise each other giving little resultant effect. Again a spinning electron also sets up a magnetic field along its axis of spin. If the fields due to the effects of spin balance out, due to electrons spinning in opposite directions, then the material is non-magnetic. A magnetic material is the result of the fields not balancing out, but to explain the overall apparent effect, it is thought that rather than single atoms or molecules being concerned, it is a group of molecules which act together. Such a group is called a 'domain' and is considered to function like the more elementary molecular magnet already described.

ELECTROMAGNETISM

Earlier theory has referred to an association between magnetism and electricity and this was more specifically mentioned in Chapter 2 when the electrical units were defined. The discovery of a relation between an electric current and magnetism was made in 1820 by the scientist Oersted, when he accidentally noted that a wire arranged above and parallel to a compass needle, caused deflection of the latter when a current was passed through the wire. Reversal of the current caused a reversal of the deflection. Further experiments on the shape, direction and strength of the magnetic fields associated with current-carrying conductors arranged in the form of loops and solenoids were the subject of much work by famous scientists such as Faraday, Maxwell and Gilbert. The result of the discoveries made then led to the deduction of certain fundamental relationships which are now part of accepted basic theory. The shape of the magnetic fields due to simple arrangements of current-carrying conductors will now be considered.

(1) FIELD DUE TO LONG STRAIGHT CURRENT-CARRYING CONDUCTOR

The field associated with such a conductor may be determined by the use of iron filings or a compass needle as was described earlier in the section on magnetism—Figs 48 and 49. Assuming that the current is kept constant during such a test, a field consisting of concentric lines of flux would be confirmed. The diagram (Fig 53) shows a vertical wire passing through a sheet of cardboard. The directions of the current and lines should be particularly noted since this is fundamental knowledge.

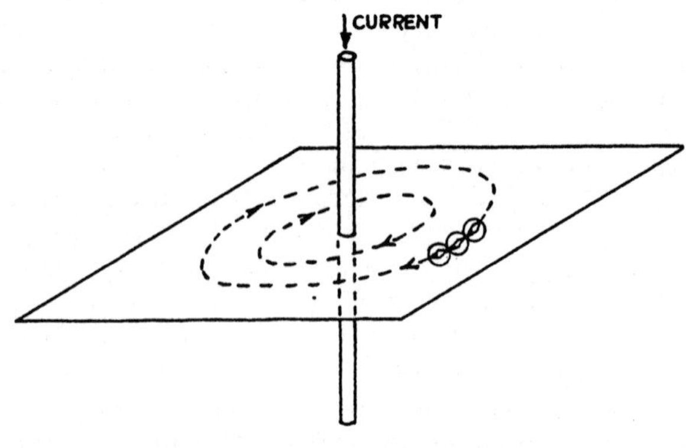

Fig 53

Further tests would show that if the current is reversed, the field would reverse and if the strength of the field was measured by an appropriate sensitive instrument, then consideration of the results would give a graph as illustrated by the diagram (Fig 54a), which shows Flux Density *(B)* plotted to a base of distance (*s*) from the centre of the conductor.

It will be seen that inside the circular conductor, the strength of field or flux density varies from zero at the centre to a maximum on the circumference. Outside the wire the flux density varies inversely as the distance from it.

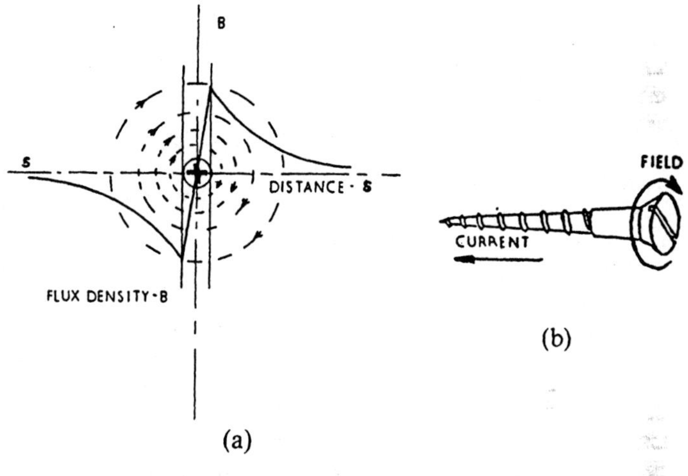

Fig 54

The diagrams (Figs 54a and 54b), make use of the conventional method of indicating current direction. Consider an arrow *ie* current entering the surface of the paper and receding from the viewer, then the feathered end would be seen. This would be shown with a cross. Similarly current flow towards the viewer would be shown with the tip of the arrow *ie* a point or dot. The relation between the direction of the lines of flux and the current is best summarised by Maxwell's 'Right-Hand Screw' Rule. This depicts the association that, if current flows in the direction in which a right-handed screw moves forward when turned clockwise, then the resulting field will be in the direction of turning the screw. If the current is reversed, the screw should be unscrewed and the field would be reversed, or would be in the direction of turning the screw *ie* anti-clockwise.

(2) FIELD DUE TO A CURRENT-CARRYING CONDUCTOR BENT TO FORM A SINGLE LOOP

The diagrams (Figs 55a and 55b) show the loop, the current and the lines of flux which tend to encircle the conductor as deduced from condition (1) above. The resulting field can be plotted by locating the loop in a sheet of cardboard as shown. The result can be considered as the field taken through the section XY of the loop and the similarity with the field of a short bar magnet will be recognised. Thus the loop can be considered to set up a magnetic polarity which can be determined from first principles.

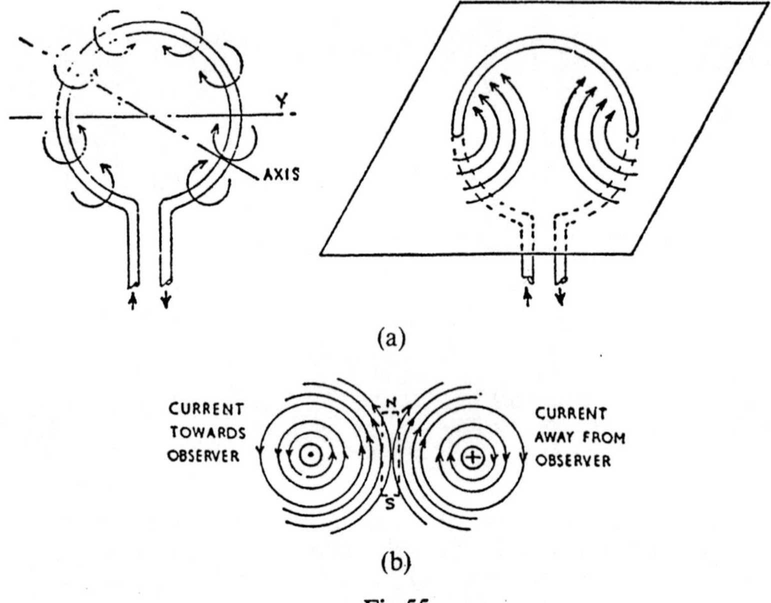

Fig 55

(3) FIELD DUE TO A CURRENT-CARRYING CONDUCTOR WOUND AS A SOLENOID

The next logical step in electromagnetic field investigations is for a coil of wire, which is basically a collection of several loops. A solenoid is a form of a multi-turn coil where the axial length is much greater than its diameter. The turns of wire can be wound in an open spiral or placed close together so that they touch, provided insulated wire is used. The insulation most commonly used is either one of the modern synthetic enamels or a fibrous material such as cotton or silk in the form of thread, tape or braid. The turns of a solenoid may be arranged in several layers provided the current travels through the turns in the same direction. When the field is investigated by plotting with a compass, it is found to be as shown in the diagram (Fig 56). It will be seen that all the turns tend to produce a magnetic field in the same direction, so that this can be deduced by considering the field of a single turn or loop. The turns unite to send a straight field up the centre which comes out at the ends, opens and spreads out to return to the other end, giving the same distribution of lines of flux as would be obtained from a bar magnet.

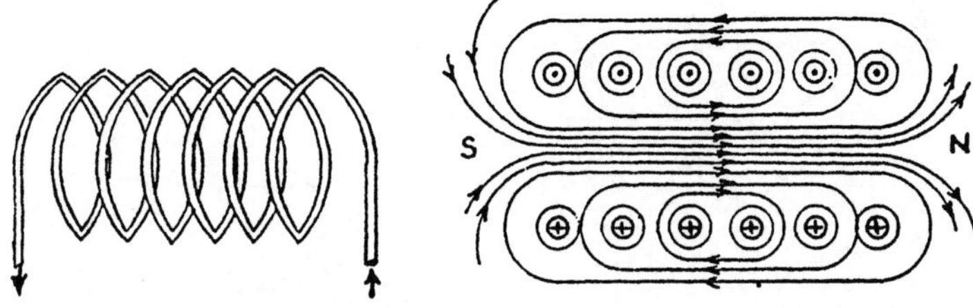

Fig 56

Here again a definite polarity can be attributed to the solenoid when carrying current. The polarity can be determined by finding the direction of the lines for any one turn by applying the Right-Hand Screw Rule but additional aids are useful, the easiest of which being the Right-Hand Rule. This is explained as follows, and is shown in the diagram (Fig 57). Place the right hand on the coil with the fingers pointing in the direction in which current flows. Then the thumb will point in the direction of the *N* pole.

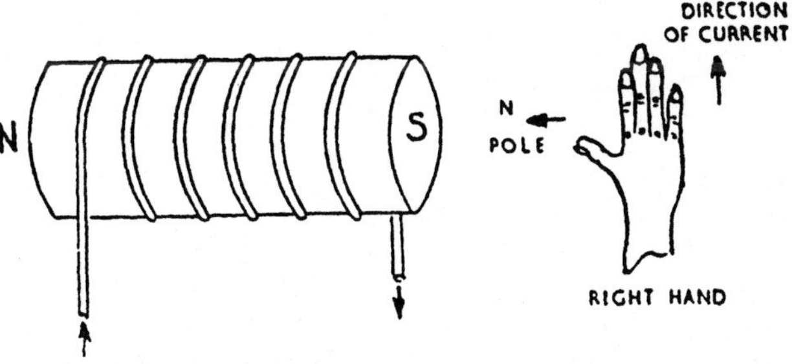

Fig 57

INTRODUCTION OF AN IRON CORE

By iron is meant, at this stage, a magnetic substance. The iron, if made the core of a solenoid, strengthens the field by concentrating the flux and more clearly defining the poles. A magnetic core appears to allow the passage of flux more readily than does air. The reason for this will be introduced in later studies, but experiment shows that the most perfect type of flux path is where the whole of the magnetic circuit is formed from magnetic material. Where this is not practicable the air gaps or air paths are kept as short as possible and good examples are found in the electromagnetic paths for the flux in the electric bell and the electric motor or generator, as illustrated by the diagram (Figs 58a and 58b).

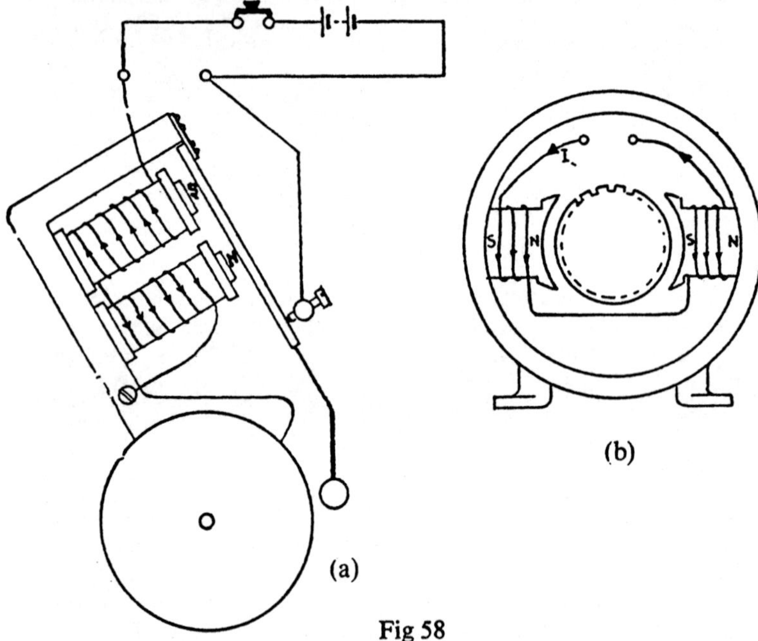

Fig 58

Electromagnets are preferred to permanent magnets in industry for two main reasons. (1) They can be made more powerful than permanent magnets by providing the desired magnetising force, *ie* solenoid coils with sufficient turns and energising current. (2) The magnetism can be controlled, *ie* it can be switched on and off or varied gradually by controlling the current. Although much more will be said about the material used for the core of electromagnets, general practice can be summarised thus:

Permanent magnets are made of *hard steel* because this material retains its magnetism. The material is said to have a high 'retentivity'.

Electromagnets have a core of *soft iron* as it is more readily magnetised but loses its magnetic properties more quickly. The material is said to have a high 'susceptibility'. Thus soft iron is more susceptible than steel.

FORCE ON A CURRENT-CARRYING CONDUCTOR IN A MAGNETIC FIELD

Oersted's experiment with the compass needle and current-carrying wire showed that, a force must be produced when the current is switched on, to bring about deflection of the needle. Converse action can be assumed, *ie* if the needle was fixed and the wire was sufficiently flexible, then movement of the latter would be noted when the current was switched on. Further investigations, to discover the nature, action and magnitude of the effect detected, lead to an accepted fundametal—that a force acts upon a conductor when it is carrying current and situated in a magnetic field provided it is lying at right angles to the lines of flux. The importance of this fundamental truth requires fuller consideration and it would be useful to reconsider the electromagnetic effects which allow the ampere to be defined as a fundamental unit of the SI system.

In Chapter 2 the phenomena leading to the definition of the ampere were mentioned and the points made previously are revised here in the light of electromagnetic theory. If a circuit is supplied through two wires laid together, then especially if the current is large and the wire flexible, a mechanical effect would be noted. This would be particularly noticeable when the current is switched on and off, since the wires would be seen to move. The action noted can be explained with our knowledge of the field associated with a long straight conductor.

Consider the diagram (Fig 59), which shows two conductors carrying current as shown. When the current in both conductors is in the same direction then the resultant magnetic field is such as to embrace both conductors. If the current in each conductor is of the same magnitude then, by Maxwell's right-hand screw rule, the fields between the wires will cancel and the outside lines of flux will unite to make a field which encircles both conductors. If the lines of flux are likened to elastic threads, tending to contract, then the lines of flux are stretched and forces act tending to move the conductors together.

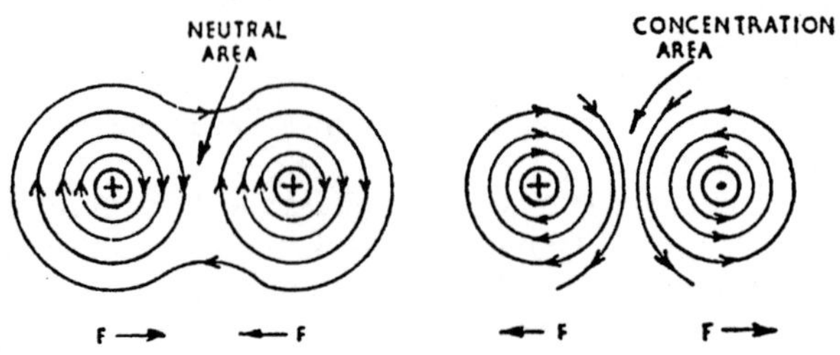

Fig 59

The likening of lines of flux to elastic threads is an acceptable analogy, since lines of force through air will keep to the shortest paths possible and will not elongate readily.

Fig 59 also shows currents in opposite directions in the two parallel conductors. Here the resultant flux is seen to be a concentration between the conductors tending to force them apart.

THE AMPERE

This is defined in accordance with the electromagnetic effects described above and is now the accepted definition for the practical unit of current. Thus the ampere is that value of current which, when flowing in each of two infinitely long parallel conductors, situated in a vacuum and spaced one metre between centres, causes each conductor to have acting on it a force of 2×10^{-7} newton per metre length of conductor.

MAGNITUDE OF FORCE (on a current-carrying conductor in a magnetic field)

The diagram (Fig 60) shows a conductor situated in and at right angles to a magnetic field. Assume that the conductor carries current in the direction shown and that the arrangement can be further illustrated by diagrams (a) and (b). It will be seen that there is a magnetic field due to the current which interacts with the main field tending to distort it so that a strong field exists on one side of the conductor and a weak field on the other side. The lines of flux appear to be stretched and in tending to return to their shortest length, a force is exerted on the conductor pushing it out of the way as shown. This action forms the basis of operation of the electric motor and the reader should pay careful attention to the points being made here.

It can be shown that the force acting on the conductor varies directly with (1) the strength of the magnetic field, (2) the

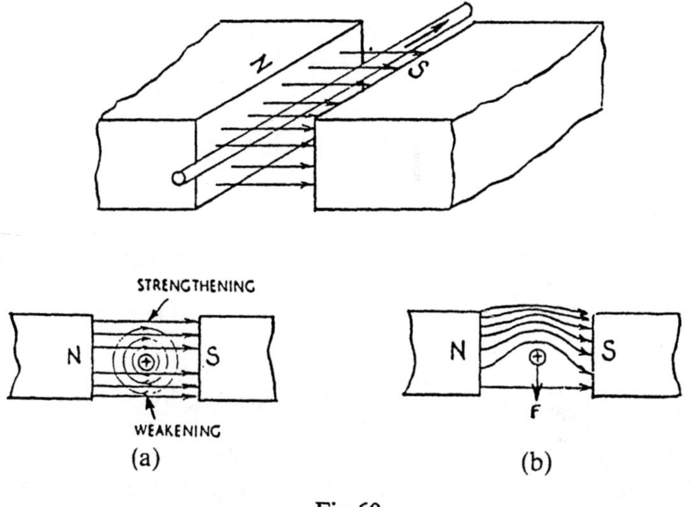

Fig 60

strength of the current in the conductor, (3) the length of the conductor in the magnetic field.

Summarising: Force $\propto$ Strength of field $\times$ Current in conductor $\times$ length of conductor in the magnetic field

or $F \propto BIl$ where F is the force on the conductor in newtons, B is the flux density in teslas, I is the current in amperes, l is the length of the conductor in the field in metres.

The above relationship can be converted to the expression $F = BIl$ if the correct unit of flux density is chosen to allow this equality. This unit is the tesla and is defined below. Thus we have the important formula:

$$F\,(\text{newtons}) = B\,(\text{teslas}) \times I\,(\text{amperes}) \times l\,(\text{metres}).$$

It will be noted from now on that the expression flux density will be used to an increasing extent in preference to strength of magnetic field. This is because, if the idea of using lines of flux to depict a magnetic field is accepted then, the strength of the magnetic field can be represented by the density of the lines. Lines well spaced apart would suggest a weak field, whereas a strong field could be represented by lines closely packed together. Thus field strength can be measured by the density of these lines or by the flux density which in turn can be defined in terms of other accepted SI units.

UNIT OF FLUX DENSITY

This unit is readily defined in accordance with the relationship $F = BIl$ because the units for F, I and l are already known. Thus B is defined in terms of the other three factors and the unit of flux density or the tesla is the density of magnetic field such that a conductor carrying 1 ampere at right angles to the field experiences a force of 1 newton/metre length acting on it.

UNIT OF FLUX

The terms flux and flux density were introduced earlier in this chapter when it was mentioned that the expression flux density was determined by dividing the total flux by the area through which it passed. Thus:

$$\text{Flux Density} = \frac{\text{Flux}}{\text{Area}} \text{ or Flux} = \text{Flux Density} \times \text{Area.}$$

Using our definition for flux density, it follows that the weber is the unit of flux and it is the flux contained in an area of 1 square metre where the flux density is of value 1 tesla. A further and more complex definition of the weber will follow in Chapter 6, which will emphasise the importance of this unit in the study of electromagnetism. Meanwhile it is well to revise the already deduced relationship:

$$\Phi \text{ (webers)} = B \text{ (teslas)} \times A \text{ (square metres)}$$

Example 39. If the flux density inside a solenoid coil is measured to be 140mT and the inside diameter of the solenoid is 40mm, find the value of the total flux produced.

Note. The main purpose of this example is to stress the importance of correct substitution in the formula, with attention being given to the correct numerical magnitude. Thus 140mT is 140 milliteslas. This equals 140×10^{-3}T. Similarly 40mm must be converted to metres before substitution.

$$\text{Thus } A = \frac{\pi d^2}{4} = \frac{3.14 \times 40^2 \times 10^{-6}}{4}$$
$$= 12.56 \times 10^{-4} \text{ m}^2$$
$$\Phi = 140 \times 10^{-3} \times 12.56 \times 10^{-4}$$
$$= 175.8 \times 10^{-6} \text{ Wb}$$
$$\text{or } \Phi = 175.8\mu\text{Wb.}$$

Example 40. Find the force exerted on a conductor 160mm long when carrying 125A and placed at right angles to the lines of flux of a magnetic field of flux density 4×10^{-3} teslas.

Substituting in $F = BIl$ we have

$$F\text{(newtons)} = 4 \times 10^{-3} \times 125 \times 160 \times 10^{-3}$$
$$= 0.08N$$

The force would be one of 0.08 newtons or 0.08N.

THE MAGNETIC CIRCUIT

MAGNETISING FORCE, MAGNETIC FIELD STRENGTH OR MAGNETIC FIELD INTENSITY

Since a magnetic field is produced by a coil of wire carrying a current, it now becomes necessary to deduce a relationship which correlates the flux density produced at any point with the electromagnetic effort required to produce it at that point. To allow this derivation, the electromagnetic effort is defined as the *magnetising force, magnetic field strength* or *magnetic field intensity* (symbol H) and is measured in terms of the factors producing it namely, the current and number of turns of the coil. The magnetising force H at any point in the field is taken as the result of the *Ampere-turns* (symbol IN) of the coil acting over one metre length of the flux path and is considered to cause a flux density of B teslas. IN/l ampere-turns/metre length of the magnetic circuit is therefore a measure of H but, in line with SI unit requirements, it is considered that the same value of H would be caused by a current of IN amperes passing through 1 turn and another method of showing a numerical value of ampere-turns could be with an A as an alternative to At. Thus H can be measured in At/m or in A/m. In this book the original and more common method will be used *ie* 200 ampere-turns, for example, will appear as 200At rather than 200A. In line with this acceptance of duality, a magnetising force value should be read as ampere-turns per metre although it may, in a problem, be given as amperes per metre. Summarising we have:

$$H\text{(magnetising force)} = \frac{IN}{l}\text{(ampere-turns/metre or amperes/metre).}$$

A magnetic circuit is considered to be the complete length of the path through which the flux produced by the coil traverses. Since we are concerned, in practical engineering, with the flux path of machines and electromagnetic devices, it is as well to consider a simple path and the diagrams (Fig 61) show the magnetic system for a simple two-pole generator or motor.

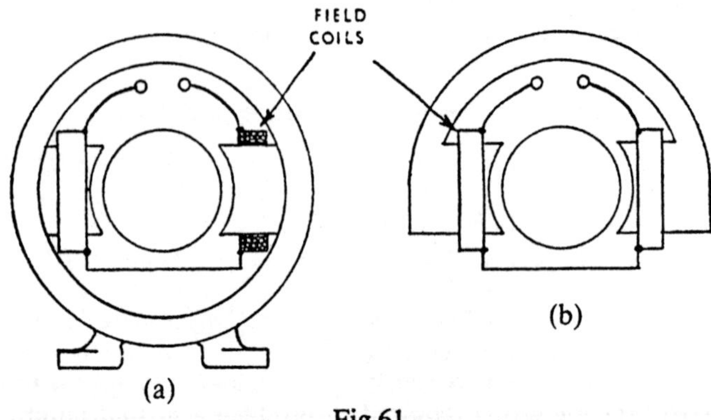

Fig 61

In the diagrams it is seen that the field coils are the energising ampere-turns, spread over the poles of the machine and wound so as to produce a continuous solenoid effect. From the theory point of view the magnetic circuit would be unaffected if both coils were placed on one pole and connected in series, but the practical arrangement gives the more symmetrical layout. Consider each coil to have 2000 turns of thin wire and the coil current to be 1.5 amperes. Then the *total* magnetising force producing the flux for this machine would be (2000 × 2) turns × 1.5 amperes or 6000 ampere turns. By symmetry the flux through the poles and armature splits (Fig 61a) and returns through both halves of the yoke of the machine. This distribution of flux can be seen to be identical with the arrangement in the diagram (Fig 61b), where all the flux returns through one yoke—a lay-out much used in the earlier types of electrical machines employing a horse-shoe shaped electromagnet system.

From the work done so far it will be noted that we consider a certain flux density to exist at a point by virtue of a magnetising force producing it. An analogy can now be made with the electrical circuit and such reasoning will allow a clearer understanding of the magnetic circuit and associated problems.

MAGNETOMOTIVE FORCE OR M.M.F.

A complete path is followed by a group of lines of magnetic flux and this path is called the magnetic circuit. In an electric circuit current is due to an e.m.f. and in a magnetic circuit, flux is considered to be due to a *magnetomotive force* (Symbol *F*) caused by current flowing through a coil of wire. Thus the *m.m.f.* is the total magnetising force produced by a solenoid coil and is measured in ampere-turns (IN). It is stressed at this stage that, from now on, the terms magnetising force, magnetic field strength or magnetic field intensity are used for the force or m.m.f. acting over one metre length of the circuit and that the total force for the circuit is called the magnetomotive force. Thus magnetising force H is the m.m.f./metre length.

$$\text{or } H = \frac{F}{l} = \frac{IN}{l}$$

The passage of the flux through the magnetic circuit is considered to be restriced by the *reluctance* of the circuit. Reluctance (symbol S) can be likened to the resistance in an electrical circuit and is found to be proportional to the length of the magnetic circuit and inversely proportional to the area and the absolute *permeability* (symbol μ).

Electric Circuit		Magnetic Circuit	
Quantity	Unit	Quantity	Unit
e.m.f. (E) Current (I) Resistance (R)	Volt Ampere Ohm	m.m.f. (F) Flux (Φ) Reluctance (S)	Ampere-turn Weber Amp-turn/ Weber or A/Wb
Also $I = \frac{E}{R}$		Also $\Phi = \frac{F}{S}$	
Other Comparables are			
$R = \frac{\varrho l}{A}$		$S = \frac{l}{\mu A}$	
Electric Force (E) $= \frac{V}{d}$	Volts/metre	Magnetising Force (H) $= \frac{F}{l}$	Amp-turns/ metre
Current Density (J) $= \frac{I}{A}$	Ampere/ metre^2	Flux Density (B) $= \frac{\Phi}{A}$	Tesla

The above conception of the magnetic circuit also allows formulae to be deduced, for the magnetising force in the fields of various current-carrying conductor arrangements, such as the long straight conductor, single loop and multi-turn coils like solenoids and toroids.

PERMEABILITY

We now can say that a magnetising force (H) will produce a flux density (B), but the magnitude of this flux density will depend upon the type of material in the magnetic circuit (e.g. air, steel, soft iron, etc). For any material the ratio of flux density to magnetising force is referred to as the absolute permeability (μ) and is measured in Henry per metre (H/m).

Thus: $\mu = \frac{B}{H}$

PERMEABILITY OF FREE SPACE (μ_o).

For a vacuum and most non-magnetic materials the ratio between B and H is a constant value. This can be shown by considering a long straight current carrying conductor situated in a vacuum. Consider the diagram (Fig 62) illustrating a conductor of infinite length carrying a current of 1 ampere.

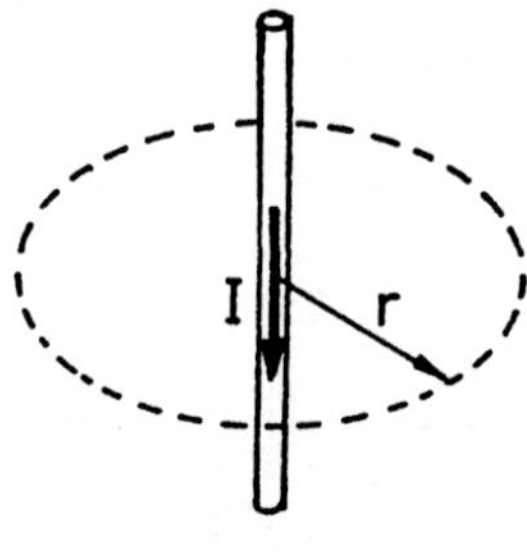

Fig 62

The conductor forming the return path to the supply source, is considered to be at such an infinite distance away that its current will not affect the magnetic field near the conductor being considered.

The conductor arrangement constitutes a single turn and the m.m.f. F is then 1 turn × I amperes or F = I ampere-turns. Consider any point on a line of flux distant r metres from the centre of the conductor. Then the magnetising force H at this point will be the m.m.f./metre length of flux or $H = \frac{F}{l} = \frac{F}{2\pi r}$ whence $H = \frac{I}{2\pi r}$ ampere-turns/metre or amperes/metre.

The above result can now be used to find the flux density resulting from a certain magnetising force and the permeability (μ) of the medium in which the field is established can be deduced.

Consider now Figure 63 which is a plan view of our previous diagram.

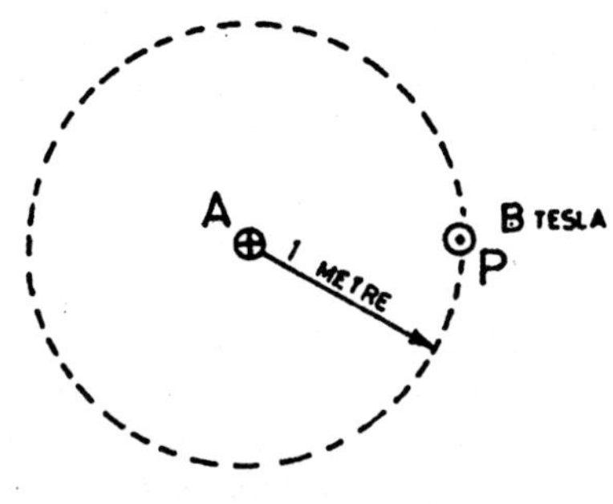

Fig 63

The conductor, in a vacuum, is represented by A and now carries a current of 1 ampere which flows away from the observer, The magnetising force, at any point P distant 1 metre from A is given by H = $\frac{1}{2\pi}$ ampere-turns/metre since both I and r are unity in the formula derived earlier.

Next assume the flux density at point P to be B tesla. Then (a) the force on a metre length of conductor placed at P, parallel to A and carrying a current of 1 ampere would be 2×10^{-7} newtons. This is known from the definition of the ampere. Also (b) the force on a metre length of this conductor is given by BIl newtons, or is

$$B\,(\text{teslas}) \times 1\,(\text{ampere}) \times 1\,(\text{metre}) = B\,(\text{newtons}).$$

Thus equating expressions (a) and (b) for the force on the conductor we see that the value of B for the condition considered would be 2×10^{-7} newtons

Hence:

$$\mu = \frac{\text{Flux density at point P}}{\text{Magnetising force at point P}} = \frac{B}{H} = \frac{2 \times 10^{-7}}{1/2\pi}$$

$$\therefore \mu = 4\pi \times 10^{-7}$$

In this case we have considered a vacuum as the medium in which the field is established therefore:

Permeability of free space = $4\pi \times 10^{-7}$ Henry/metre

or $\mu_0 = 4\pi \times 10^{-7}$ H/m

Example 41. It is required to produce a flux of 0.018Wb across an airgap 2.54mm long and having an effective area of 24 $\times$ 10^{-3} square metres. Find the ampere-turns required.

Area of Gap = 24×10^{-3} square metres

Required flux density $B = \frac{0.018}{24 \times 10^{-3}} = 0.75$T

Also $H = \frac{B}{\mu_o} = \frac{0.75}{4 \times \pi \times 10^{-7}}$

$= 59.7 \times 10^4$At/m

The length of the air gap = 2.54mm = 0.254×10^{-2}m

So total ampere-turns needed = $59.7 \times 10^4 \times 0.254 \times 10^{-2}$

= 1515At.

Example 42. A wooden ring having a mean diameter of 200mm and a cross-sectional area of 400mm² is wound uniformly with a coil of 300 turns. If the current passed through the coil is 5A calculate the value of flux produced in the coil.

The m.m.f. of the coil = 5×300 = 1500At

The mean circumference = $\pi \times 200$ = 628mm = 0.628m

The magnetising force H = At/m = $\frac{1500}{0.628}$ = 2380 At/m

The flux density

$B = \mu_o H$

$= 4 \times \pi \times 10^{-7} \times 2380$

= 0.003T

Total flux $\Phi = BA$

$= 0.003 \times 400 \times 10^{-6}$ Wb

= 1.2μWb.

Example 43. The magnet system of a moving-coil instrument provides a flux density in the air gap of 0.25T. The moving coil, of 120 turns, is carried on a former of (active side) length 25mm and width 18mm (between air-gap centres). If the coil carries a current of 2mA, calculate the turning moment on it.

$F = BIl$ newtons

$= 0.25 \times 2 \times 10^{-3} \times 120 \times 2 \times 25 \times 10^{-3}$

$= 3 \times 10^{-3}$ N

Torque = $F \times$ radius of coil

$= 3 \times 10^{-3} \times 9 \times 10^{-3}$

= 27μN m.

CHAPTER 5

PRACTICE EXAMPLES

1. A conductor carrying a current of 100A is situated in and lying at right angles to a magnetic field having a flux density of 0.25T. Calculate the force in newtons/metre length exerted on the conductor.

2. A coil of 250 turns is wound uniformly over a wooden ring of mean circumference 500mm and uniform cross-sectional area of 400mm^2. If the current passed through the coil is 4A find (a) the magnetising force (b) the total flux.

3. A current of 1A is passed through a solenoid coil, wound with 3200 turns of wire. If the dimensions of the air core are length 800mm, diameter 20mm, find the value of the flux produced inside the coil.

4. Two long parallel busbars, each carry 2000A and are spaced 0.8m apart between centres. Calculate the force per metre acting on the conductors.

5. A moving-coil permanent magnet instrument has a resistance of 10Ω and the flux density in the gap is 0.1T. The coil has 100 turns of wire, is of mean width 30mm and the axial length of the magnetic field is 25mm. If a p.d. of 50mV is required for full-scale deflection, calculate the controlling torque exerted by the spring.

6. An air gap of length 3mm is cut in the iron magnetic circuit of a measuring device. If a flux of 0.05Wb is required in the air gap, which has an area of 650mm^2, find the ampere-turns required for the air gap to produce the necessary flux.

7. A straight horizontal wire carries a steady current of 150A and is situated in a uniform magnetic field of 0.6T acting vertically downwards. Determine the magnitude of the force per metre length and the direction in which it acts.

8. An armature conductor has an effective length of 400mm and carries a current of 25A. Assuming that the average flux density in the air-gap under the poles is 0.5T, calculate the force in newtons exerted on the conductor.

9. In an electric motor the armature has 800 conductors each carrying a current of 8A. The average flux density of the magnetic field is 0.6T. The armature core has an effective length of 250mm and all conductors may be taken as lying on an effective diameter of 200mm. Determine the torque and mechanical power developed when the armature is revolving at 1000 rev/min.

10. Two long straight parallel busbars have their centres 25mm apart. If each carries current of 250A, calculate the mutual force/metre run.

CHAPTER 6

ELECTROMAGNETIC CIRCUITS

In Chapter 5 the fundamental concepts, terminology and relationships of the electromagnetic circuit were introduced and developed. Before proceeding to consider further the effects of ferromagnetic material it will be useful to revise some of the basic relationships.

The magnetomotive force (m.m.f.) F is the force which causes magnetic flux Φ to be established in a magnetic circuit having a reluctance S.

$$i.e.\ \Phi = \frac{F}{S}\ \text{Wb}$$

The m.m.f. is usually established by passing a current through a number of coil turns

$$i.e.\ F = IN\ \text{Ampere-turns}$$

Reluctance S depends upon the dimensions of the magnetic circuit and its permeability

$$i.e.\ S = \frac{l}{\mu A}\ At/Wb$$

Flux Density B is a measure of the magnetic flux Φ in a given area A.

$$B = \frac{\Phi}{A}\ \text{Tesla}$$

The Magnetising Force (Magnetic Field Strength) is a measure of the m.m.f. per metre length of magnetic circuit required to maintain flux in that circuit.

$$H = \frac{IN}{l}\ \text{At/m}$$

Permeability is the ratio of flux density to the magnetising force producing it.

$$\mu = \frac{B}{H} \text{ H/m}$$

For air, vacuum and most magnetic materials we use the permeability of free space μ_o which is a constant value

$$\mu_o = 4\pi \times 10^{-7} \text{ H/m}$$

$$\text{Thus for air } \mu_o = \frac{B}{H} \text{ and } B \propto H$$

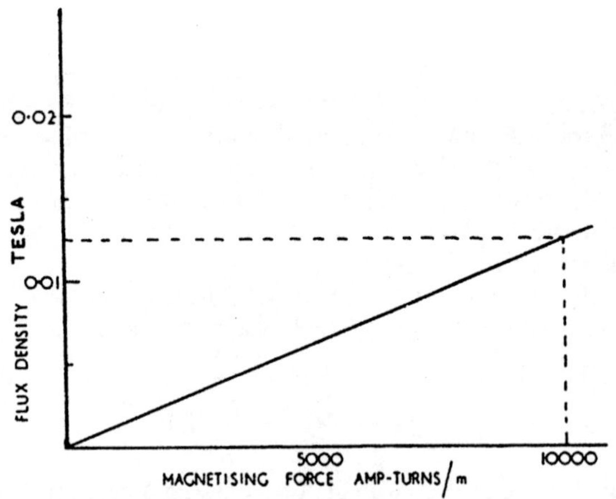

Fig 64

If therefore for air, values of B are plotted to a base of H, a graph, as shown in the diagram (Fig 64), would be obtained. This would obviously be a straight line and if measurements of flux density B were made, at a point outside, but near to, a long straight current-carrying conductor, for various values of magnetising force H, by varying the current; it being remembered that $H = \dfrac{I}{2\pi r}$ where r is the radius from the point to the centre of the conductor, then the straight-line B/H relationship would be confirmed.

1. MAGNETISING FORCE DUE TO A LONG, STRAIGHT CURRENT-CARRYING CONDUCTOR

As mentioned above, the magnetising force outside, but adjacent to, a current-carrying conductor, is given by $H = \dfrac{I}{2\pi r}$.

This expression was deduced earlier (in Chapter 5) and should be revised. It should also be remembered that, H is the m.m.f./metre length. Magnetomotive force—F, is measured in ampere-turns and thus the total m.m.f. for any magnetic circuit outside the conductor can be found from $F = Hl$.

2. MAGNETISING FORCE INSIDE A SOLENOID

If a parallel field of flux lines is assumed inside a solenoid as illustrated by the diagram (Fig 65), then its length can be taken as l metres, the number of turns on the coil as N and the current passed as I amperes. The lines of flux are known to fan out at the ends and for their return path they spread out into space. This external return path has negligibly small magnetic reluctance and the whole m.m.f. of the coil can be regarded as being utilised in setting up the field inside the solenoid. Thus the m.m.f. per unit length is, by definition, H—the magnetising force

$$\text{Thus } H = \frac{F}{l} = \frac{IN}{l} \text{ ampere-turns per metre.}$$

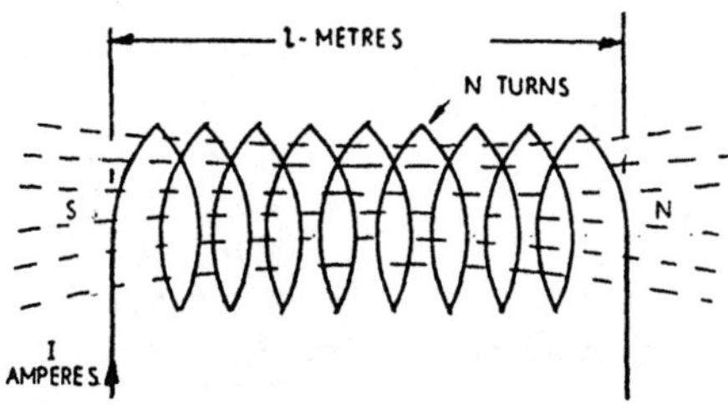

Fig 65

3. MAGNETISING FORCE INSIDE A TOROID

The diagram (Fig 66) shows this simple electromagnetic arrangement. It consists of a solenoid bent back upon itself so that the lines of flux are confined to the inside of the coil. We can now consider a non-magnetic ring to be wound uniformly with a coil of N turns, carrying a current of I amperes. The mean circumference is l metres and since the flux is confined to the inside and the path is uniform, the magnetising force or m.m.f. per unit length is given by:

$$H = \frac{IN}{l} \text{ ampere-turns per metre.}$$

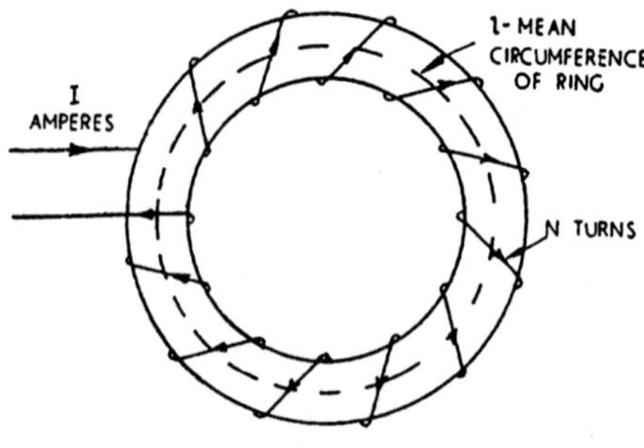

Fig 66

Example 44. A wooden ring having a mean circumference of 300mm and a uniform cross-sectional area of $400mm^2$, is wound uniformly overall with 300 turns of insulated wire. If the current is 3A, calculate (a) the magnetising force (b) the flux density inside the toroid and (c) the total flux produced.

(a) The total m.m.f. produced $= 3 \times 300 = 900$At

The mean circumference is 300mm $= 0.3$ metres

$$\therefore \text{The magnetising force } H = \frac{F}{l} = \frac{900}{0.3} = 3000\text{At/m}$$

(b) The flux density is given

$$\text{by } B = \mu_0 H = 4\pi \times 10^{-7} \times 3000 = 3.768 \times 10^{-3}\text{T} = 3.768 \text{ mT}$$

(c) The total flux produced

$$= B \times A = 3.768 \times 10^{-3} \times 400 \times 10^{-6} \text{ Wb} = 1.5 \times 10^{-6} \text{ Wb}$$

$$\text{or } \Phi = 1.5\mu\text{Wb}$$

FERROMAGNETISM

It is known that, when iron is used as the core of an electromagnet, the field is intensified or that a very much greater flux results from the magnetising ampere-turns of the energising coil. Since the only change in the relation $\Phi = \dfrac{F}{S}$, must be due to the reluctance S and if the dimensions of the core l and A are kept the same as for the air path, it follows that the permeability of the iron must be very much greater than that of air. Thus we can now make reference to the permeability of a magnetic material which is termed *relative permeability*.

RELATIVE PERMEABILITY (μ_r). This is the ratio of the flux density which is produced in a magnetic material to the flux density which would be produced in air by the same magnetomotive force.

$$\therefore \quad \text{Relative Permeability} = \frac{\text{Absolute Permeability}}{\text{Permeability of Free Space}}$$

$$\mu_r = \frac{\mu}{\mu_o}$$

$$\mu = \mu_o\mu_r = \frac{B}{H}$$

$$\therefore B = \mu_o\mu_r H.$$

For materials such as iron, nickel, cobalt, etc this value of μ_r can be very large, ranging from 1000 to 2000 or even more for some special electrical steels. It can be quoted as the permeability figure for the material but is not constant and varies with the flux-density value at which the material is being worked. It is of interest to mention that materials such as bismuth have a relative permeability value of less than 1 and that for air and other non-magnetic materials $\mu_r = 1$.

THE B-H OR MAGNETISATION CURVE

If a specimen of magnetic material is made up in the form of a ring and is wound with an energising coil, measurements of flux density for various values of magnetising force can be made by winding on a secondary coil and using the principle of transformer action. It is not proposed at this stage to discuss the manner in which the tests are made but it can be stated that this is an accepted industrial method for determining the magnetic properties of various materials.

It has already been seen that; if the flux density *B* is plotted against the magnetising force *H* for air, a straight line is obtained, but for magnetic materials, curves as shown in the diagram (Fig 67) will result.

It will be noticed that, at first, the graphs are approximately straight lines, showing *B* to be proportional to *H*. Then the curves begin to turn over forming a 'knee' and finally they become horizontal and exhibit little increase in *B* for a large increase in *H*, In this state, the material is said to 'saturate' — an explanation for this has already been given when dealing with the molecular theory of magnetism.

If the permeability (μ_r) is plotted to a base of *B*, curves as shown in the diagram (Fig 68) would result. The permeability curve has a peak corresponding to the point on the *B-H* curve

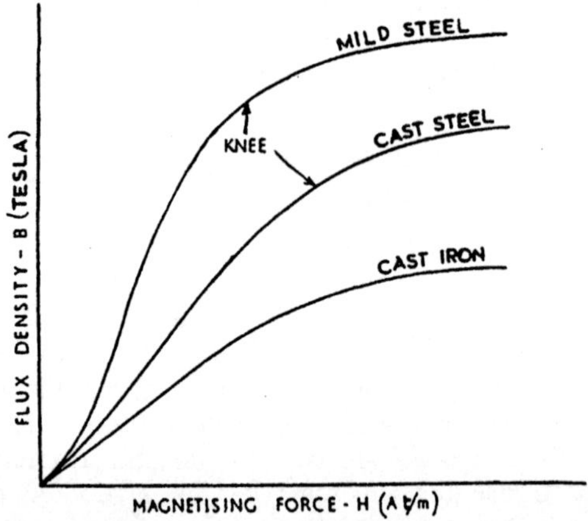

Fig 67

where the tangent goes through the origin. Beyond this peak the permeability value drops off fairly rapidly.

An examination of the *B*-*H* and μ_r-*B* curves shows how the properties of various magnetic materials differ. The effect on machine design can also easily be seen — lower working *B* values necessitate larger section and greater mass to obtain a required flux value. The effect of high permeability materials is also

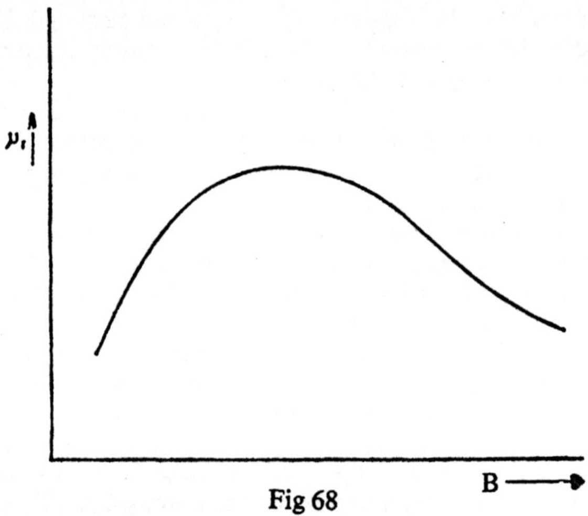

Fig 68

apparent and the shape of the *B-H* curve with the saturation effect, shows the limitation of machine field systems.

The magnetic properties are dependent upon the actual composition of the substance. Thus manganese-steel is practically non-magnetic, but small quantities of carbon or silicon when added to steel vary the shape of the *B-H* curve and sheets of commercial steel marketed under trade names, like Stalloy or Lohys, are available to suit different design requirements.

RELUCTANCE (SYMBOL *S*). This term has been mentioned earlier. It has been likened to the resistance of an electrical circuit. Since flux is proportional to the m.m.f. and is restricted by the reluctance, further investigation will show that reluctance in turn is proportional to the length l of the magnetic circuit and is inversely proportional to its area A. Furthermore it must be inversely proportional to the permeability, since the greater the permeability of the material the greater the flux and hence the smaller the reluctance. We can therefore write, $S = \frac{l}{\mu A}$ and point out here that absolute permeability is being used, it being remembered that $\mu = \mu_o\mu_r$. Thus $S = \frac{l}{\mu_o\mu_r A}$

Calculations on magnetic circuits using magnetic materials are now possible, but it should be noted that, unlike electrical circuit calculations which use $I = \frac{V}{R}$, it is not always necessary to use the comparable relationship of $\Phi = \frac{F}{S}$. The solution of most problems associated with the magnetic circuit can be made without always determining the reluctance, and experience will show the best method of solution. The following examples indicate the alternative way of treating typical simple problems.

Example 45. A solenoid is made up from a coil of 2000 turns, carries a current of 0.25A and is 1m long. An iron rod of diameter 20mm, forms the core for the solenoid and is also 1.0m long. Calculate the total flux produced if the iron has a relative permeability of 1000.

M.M.F. of coil is given by $F = Hl = \frac{IN}{l} \times l = IN$

$$= 0.25 \times 2000 = 500\text{At}$$

$$\text{Area of iron} = \frac{\pi d^2}{4} = \frac{3.14 \times 400 \times 10^{-6}}{4}$$

$$= 3.14 \times 10^{-4} \text{ square metre}$$

Reluctance of iron $= \dfrac{l}{\mu A} = \dfrac{l}{\mu_0\mu_r A}$

or $S = \dfrac{1}{4 \times \pi \times 10^{-7} \times 1000 \times 3.14 \times 10^{-4}}$
$= 2.533 \times 10^6$ At/Wb

Flux $\Phi = \dfrac{F}{S} = \dfrac{500}{2.533 \times 10^6}$
$= 197.3 \times 10^{-6} = 197.3\ \mu$Wb.

ALTERNATIVE SOLUTION

M.M.F. of coil. $F = Hl$ so magnetising force $H = \dfrac{F}{l}$

Thus $H = \dfrac{IN}{l} = \dfrac{0.25 \times 2000}{1} = 500$At/m
Also $B = \mu H = \mu_0\mu_r H = 4 \times \pi \times 10^{-7} \times 1000 \times 500$
$= 0.628$T
Total $\Phi = BA$
$= 0.628 \times 3.14 \times 10^{-4}$ webers
$= 197.3 \times 10^{-6} = 197.3\mu$Wb.

Example 46. A cast-steel ring has a cross-section of 400mm² and a mean diameter of 240mm. It is wound with a coil having 200 turns. What current is required to produce a flux of 400μWb, if the relative permeability of the steel is 1000?

Area of steel $= 400 \times 10^{-6}$ square metre

$\therefore B = \dfrac{\Phi}{A} = \dfrac{400 \times 10^{-6}}{400 \times 10^{-6}} = 1$ tesla

Also $B = \mu_0\mu_r H = 4 \times \pi \times 10^{-7} \times 1000 \times \text{H}$
$B = 4 \times \pi \times 10^{-4} \times H$

So $H = \dfrac{B}{4 \times \pi \times 10^{-4}} = \dfrac{1 \times 10^4}{4 \times \pi}$At/m

M.M.F. of ring F $= Hl = \dfrac{10^4}{4 \times \pi} \times \pi \times 240 \times 10^{-3}$
or F = 600At

$F = IN \qquad \therefore I = \dfrac{F}{N} = \dfrac{600}{200} = 3A.$

ALTERNATIVE SOLUTION

Reluctance of ring $S = \dfrac{l}{\mu_0\mu_r A}$

or $S = \dfrac{\pi \times 240 \times 10^{-3}}{4 \times \pi \times 10^{-7} \times 1000 \times 400 \times 10^{-6}}$
$= 1.5 \times 10^6$ At/Wb

then required m.m.f. $= \Phi S = 400 \times 10^{-6} \times 1.5 \times 10^{6}$
$= 600\text{At}$

Required current $I = \dfrac{600}{200} = 3\text{A}.$

THE COMPOSITE MAGNETIC RING

THE SERIES ARRANGEMENT

Consider a magnetic circuit built up as shown in the diagram (Fig 69). It is obvious that the sections are in series and that the same flux passes through them.

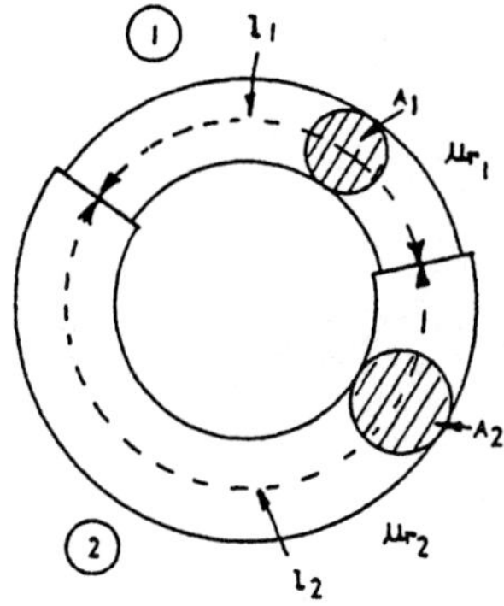

Fig 69

Then total m.m.f. = m.m.f. across section 1 + m.m.f. across section 2.

If the flux is Φ then $\Phi S = \Phi S_1 + \Phi S_2$
where S is reluctance of the composite circuit. Thus:

$$S = S_1 + S_2.$$

Summarising: Total reluctance = the sum of the individual reluctances of the sections for a series arrangement.

THE PARALLEL ARRANGEMENT

Such a magnetic circuit is not frequently encountered but is considered here, being complementary to the series circuit. The arrangement is shown in the diagram (Fig 70).

If the different paths of the magnetic circuit are in parallel, then the necessary m.m.f. is that which will produce the required flux in each part of the circuit considered by itself. Let F = the m.m.f. required to produce fluxes Φ_1, Φ_2, Φ_3, etc. F also produced total flux Φ.

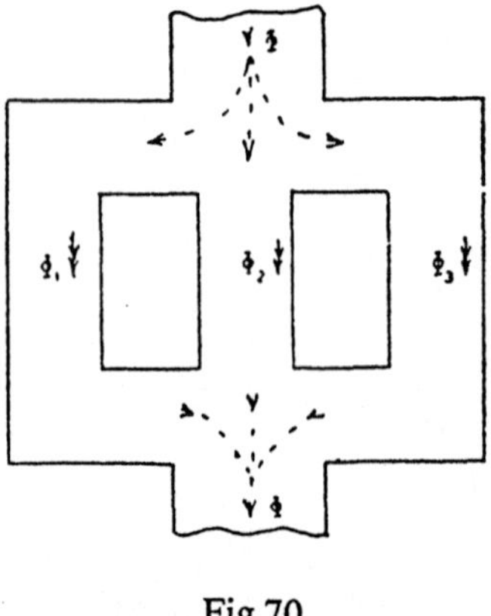

Fig 70

$$So\ \Phi = \frac{F}{\text{Total Reluctance of circuit}} = \frac{F}{S}$$

$$\text{But } \Phi_1 = \frac{F_1}{S_1} \qquad \Phi_2 = \frac{F_2}{S_2} \text{ etc}$$

$$\text{and since } \Phi = \Phi_1 + \Phi_2 + \Phi_3$$

$$\therefore \Phi = \frac{F_1}{S_1} + \frac{F_2}{S_2} + \frac{F_3}{S_3}$$

But F_1, F_2, F_3 are the m.m.f. across the same points of the magnetic circuit and are equal to F.

$$\therefore \frac{F}{S} = F\left(\frac{1}{S_1} + \frac{1}{S_2} + \frac{1}{S_3}\right) or\ \frac{1}{S} = \frac{1}{S_1} + \frac{1}{S_2} + \frac{1}{S_3}$$

$\frac{1}{S}$ is referred to as the *permeance* of a magnetic circuit and the above can be summarised by saying that; the reluctance of a divided magnetic circuit (sections in parallel), can be found by knowing that its permeance is equal to the sum of the permeances of the individual circuits.

Example 47. An iron ring has a mean diameter of 200mm and a cross-section of 300mm². An air gap of 0.4mm is made by a radial saw-cut across the ring. Assuming a relative permeability of 3000 for the iron, find the current required to produce a flux of 250μWb, if the energising coil is wound with 600 turns.

$$\text{Reluctance of iron } S_1 = \frac{(\pi \times 200 \times 10^{-3}) - (0.4 \times 10^{-3})}{4 \times \pi \times 10^{-7} \times 3000 \times 3 \times 10^{-4}}$$

$$= \frac{10^{-3}\,(628 - 0.4)}{4 \times \pi \times 9 \times 10^{-8}}$$

$$= \frac{62.76}{36\pi \times 10^{-6}}$$

$$= 555.2 \times 10^3 \text{ ampere-turns/weber}$$

$$\text{Reluctance of air gap } S_A = \frac{0.4 \times 10^{-3}}{4 \times \pi \times 10^{-7} \times 3 \times 10^{-4}}$$

$$= 1061.5 \times 10^3 \text{ ampere-turns/weber}$$

$$\text{Total Reluctance } S = S_1 + S_A = (555.2 + 1061.5) \times 10^3$$

$$= 1616.7 \times 10^3 \text{ At/Wb}$$

$$\text{Total m.m.f.} = \Phi S = 2.5 \times 10^{-4} \times 1616.7 \times 10^3$$

$$= 404.15\text{At}$$

$$\text{Current} = \frac{404.19}{600} = 0.674\text{A.}$$

ALTERNATIVE SOLUTION

Since $\Phi = 250 \times 10^{-6}$ weber then $B = \dfrac{2.5 \times 10^{-4}}{3 \times 10^{-4}}$

$$= 0.833\text{T}$$

Now H for air is given by:

$$H_A = \frac{0.833}{\mu_0} = \frac{0.833}{4 \times \pi \times 10^{-7}}$$

$$= 663.2 \times 10^3 \text{ At/m}$$

$$\text{Length of air gap} = 0.4 \times 10^{-3} \text{ metre}$$

$$\text{Ampere-turns for air} = 663.2 \times 10^3 \times 0.4 \times 10^{-3}$$

$$= 265.28\text{At}$$

H for iron is given by:

$$H_1 = \frac{0.833}{\mu_0 \mu_r} = \frac{0.833}{4 \times \pi \times 10^{-7} \times 3 \times 10^3} \text{At/m}$$

$$= 221.066\text{At/m}$$

$$\text{Now length of iron path} = (628 - 0.4)\ 10^{-3}$$

$$= 627.6 \times 10^{-3} \text{ metre}$$

$$\text{Ampere-turns for iron} = 221.066 \times 0.6276 = 138.74\text{At}$$

$$\text{Total ampere-turns} = 265.28 + 138.74 = 404.02\text{At}$$

$$\text{Current} = \frac{404.02}{600} = 0.673\text{A.}$$

For the three previous examples, alternative solutions have been given in which the reluctances, as such, for the various sections of the magnetic circuit being considered, have not been found. This alternative method of solution is especially of value when the relevant B and H data for the magnetic material is given in tabular or graphical form. The relative permeability is not given as a specific value and would have to be found before the reluctance could be calculated. Obviously any solution along these lines would be tedious and the following example is recommended to the reader as an instruction on how to solve the type of problem being discussed.

Example 48. An iron ring of square cross-section has an external diameter of 140mm, and an internal diameter of 100mm. A radial saw-cut through the cross-section of the ring forms an air gap of 1mm. If the ring is uniformly wound with 500 turns of wire, calculate the current required to produce a flux of 0.35mWb in the gap.

Magnetic data of the material of the ring are given below and shown in the diagram (Fig 71). Take μ_0 as $4\pi \times 10^{-7}$ H/m.

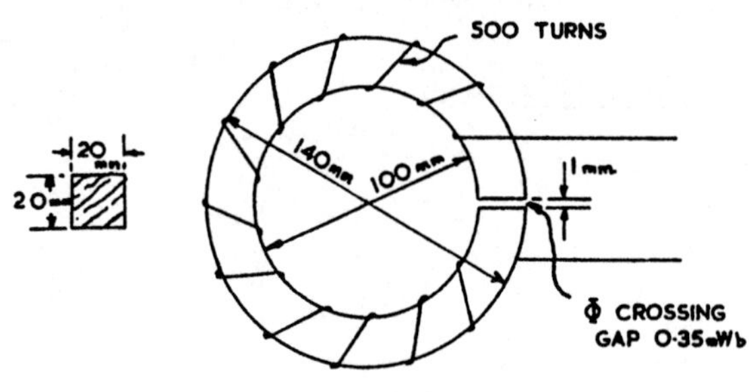

Fig 71

Flux density (T)	0.65	0.89	1.06	1.18
Magnetising Force (At/m)	200	300	400	500

Solution uses the graph (Fig 72) obtained from the above data.

$$\text{Area of iron and air gap} = 20 \times 20 \times 10^{-6}$$
$$= 4 \times 10^{-4} \text{ square metre}$$

$$\text{Length of iron} = \pi \times (\text{mean diameter}) - \text{air gap}$$
$$= (\pi \times 120 \times 10^{-3}) - (1 \times 10^{-3}) \text{ metre}$$
$$= 375.8 \times 10^{-3} \text{ metre}$$

$$\text{Length of air gap} = 1 \times 10^{-3} \text{ metre}$$

$$\text{Flux density for iron and air} = \frac{\Phi}{A} = \frac{0.35 \times 10^{-3}}{4 \times 10^{-4}}$$
$$\therefore B = 0.875\text{T}$$

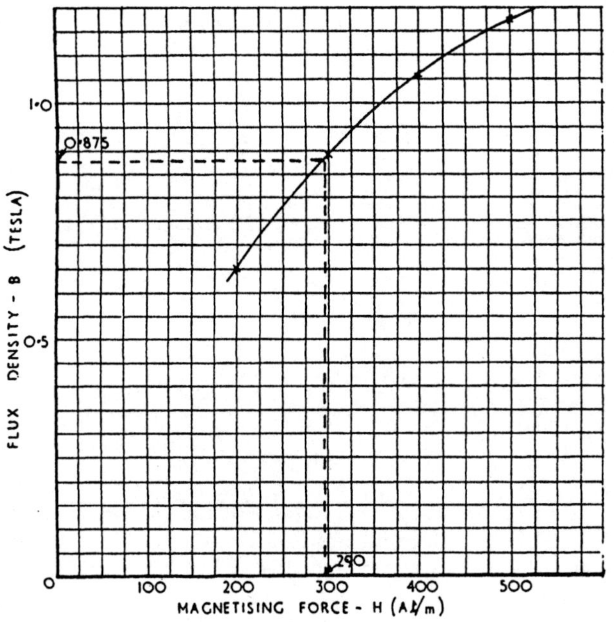

Fig 72

From graph for the iron $H = 290\text{At/m}$ when $B = 0.875\text{T}$

$$\text{But } H = \frac{IN}{l} = \frac{F}{l}$$

$$\therefore F = Hl$$

$$\text{So for iron, m.m.f. } F_1 = 290 \times 375.8 \times 10^{-3}$$
$$= 108.88\text{At}$$

$$\text{For air, since } H = \frac{B}{\mu_0}$$
$$= \frac{0.875}{4 \times \pi \times 10^{-7}}\ \text{At/m}$$

$$\text{and for air, m.m.f. } F_A = \frac{0.875}{4\pi \times 10^{-7}} \times 1 \times 10^{-3}$$
$$= 696.7\text{At}$$

$$\text{Total m.m.f. } F_1 + F_A = 108.9 + 696.7 = 805.6\text{At}$$

$$\text{Current is deducted from} = \frac{805.6}{500} = 1.6112\text{A}$$

Energising current = 1.6A (approx.).

MAGNETIC FRINGING. The diagram (Fig 73) illustrates the manner in which flux is known to bridge an air gap, especially if the latter is comparatively large.

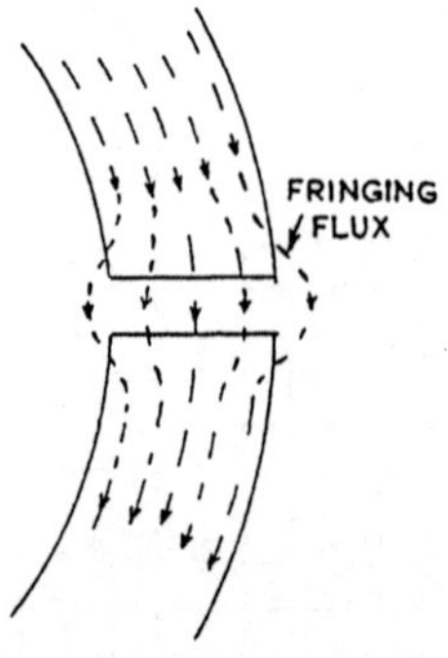

Fig 73

It will be seen that the flux in air tends to occupy a larger area than that of the iron, and the flux density is thus reduced. An allowance can be made for this effect in problems when required, but unless a directive to this effect is given, the area of the air gap can be taken as the area of the iron.

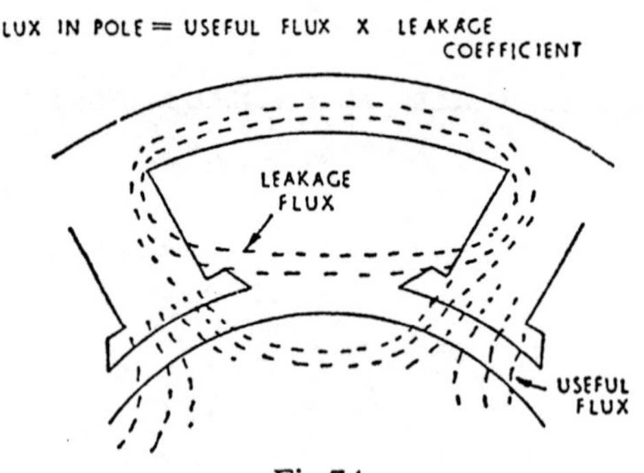

Fig 74

MAGNETIC LEAKAGE. For some magnetic circuits, due to the shape of the iron core and the positioning of the energising coil, a small amount of flux leakage occurs as shown by the diagram (Fig 74). Some lines of flux are not confined to the iron and complete their paths through air. For practical purposes, a factor known as the *leakage coefficient* may be given which, when used, increases the required working flux value by an amount sufficient to allow for this leakage.

Thus: the required total flux = the useful or working flux × leakage coefficient.

The leakage coefficient may be between 1.1 and 1.3.

Example 49. (a) A magnetic circuit has an iron path of length 500mm and an air gap of length 0.5mm, the whole being of uniform square cross-section, 1000mm² in area. Calculate the number of ampere-turns necessary to produce a total flux of 1mWb in the air gap. Ignore fringing, and assume a leakage coefficient of 1.3. The *B-H* curve for the iron is given by the following table:

H (At/m)	100	200	300	400	600	800	1000	1200
B (T)	0.42	0.8	0.98	1.08	1.22	1.3	1.36	1.4

(b) A conductor is passed through the air gap at a speed of 100m/s. If the length of the conductor is greater than the length of the side of the gap, calculate the e.m.f. induced.

The solution uses the graph, obtained from the above data, as shown in diagram (Fig 75).

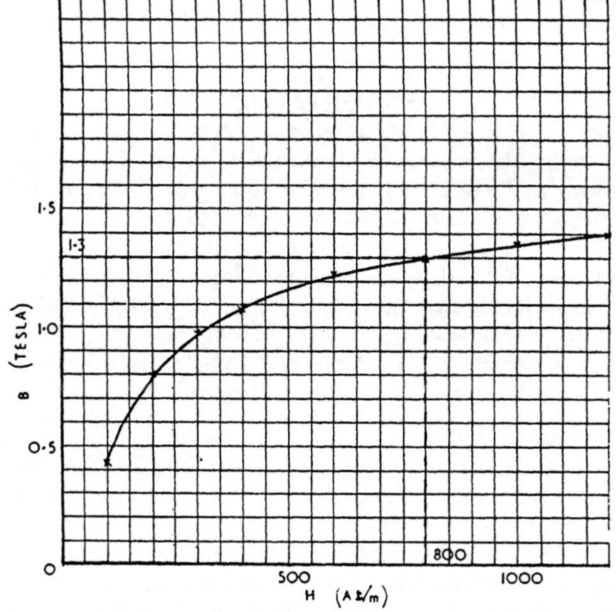

Fig 75

Area of iron and air gap $= 1000 \times 10^{-6}$
$= 10^{-3}$ square metre

Length of iron $= 500 \times 10^{-3} = 0.5$m

Flux density (B) for air $= \dfrac{1 \times 10^{-3}}{10^{-3}} = 1$ tesla.

Flux density (B) in iron $= \dfrac{1 \times 10^{-3} \times 1.3}{10^{-3}} = 1.3$T

For air, since $H = \dfrac{B}{\mu_0} = \dfrac{1}{4 \times \pi \times 10^{-7}}$ At/m

and m.m.f., $F_A = \dfrac{1}{4\pi \times 10^{-7}} \times 0.5 \times 10^{-3}$
$= 398.8$At

From graph for the iron, $H = 800$At/m for a B value of 1.3T
and m.m.f., $F_1 = 800 \times 0.5 = 400$At

Total ampere-turns required $= 398.8 + 400 = 798.8$At

(b) Induced e.m.f. given by $E = Blv$ volts

Since the area of the air gap is that of a square, the side of the square is $\sqrt{1000 \times 10^{-6}} = \sqrt{10 \times 10^{-4}}$
$= 3.162 \times 10^{-2}$ metre

Thus $E = 1 \times 3.162 \times 10^{-2} \times 10^{2}$
$= 3.162$ volts

Note. In the above $B = 1$ tesla. $l = 3.162 \times 10^{-2}$ metre and $v = 100$m/s.

IRON LOSSES

The efficiency of electrical machines and transformers is lowered by the losses which occur in them. Apart from the *Mechanical Losses* such as *Friction* and *Windage* and the *Copper Losses,* due to the resistance of the conductors, an additional loss occurs when a magnetic material is taken through a cyclic variation of magnetisation. This loss is termed the *Iron Loss* and is itself made up of two component losses which are (1) the *Hysteresis Loss* and (2) the *Eddy Current Loss.* The cause of these two sources of power loss will next be considered.

THE HYSTERESIS LOOP

If the magnetising force applied to an iron sample is increased from zero to some maximum value, in exactly the same way as when making the test for a *B-H* curve, and is then reduced to zero, it will be found that the new *B-H* curve, for decreasing values of *H,* will lie above the original ascending curve, and that when *H* is zero *B* is left at some value. The effect of the descending curve being above the ascending one, is said to be due to 'hysteresis', since the *B* values lag behind those for the corresponding *H* strength when increasing. The word hysteresis comes from the Greek meaning 'to lag'. The diagram (Fig 76) shows the effects being discussed.

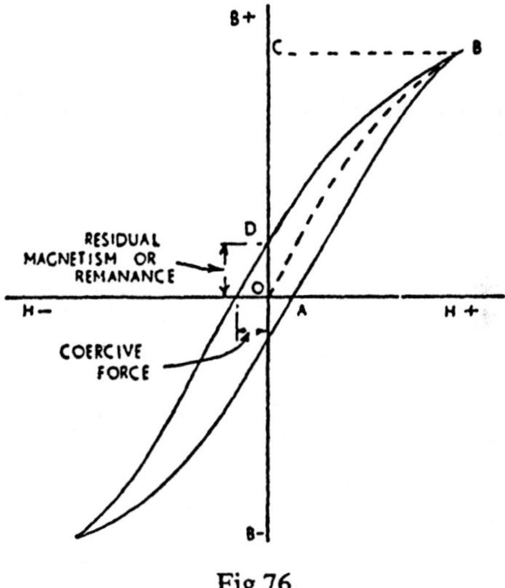

Fig 76

The value of *B* when *H* is zero is the *'remanence'* and is a measure of the residual magnetism. In order to demagnetise the iron it is necessary to apply a negative magnetising force, known as the *'coercive force'*. If now, *H* is increased in the negative direction to its previous maximum value, the curve will reach a value equal to the previous maximum *B* and if *H* is next gradually reduced to zero, reversed and increased to its original maximum, a closed loop will be traced. This is a hysteresis loop and is a measure of part of the iron loss.

To take the iron through the various stages represented by the loop, an alternating magnetising force has to be applied. One method of achieving this is by connecting the energising coil to an a.c. supply, when the iron will continue to go through the same series of changes or magnetic cycles. To confirm that energy is being expended, it will be found that the iron core will register a temperature rise. Although it is not proposed, at this stage, to prove the fact that the area of the loop is a measure of the power loss due to hysteresis, the loop can be regarded as an indicator diagram. More advanced studies will show that the energy absorbed per cubic metre per cycle, due to hysteresis, is given in joules by the area of the loop, provided the scales used for the graph are in the appropriate SI units. During the development of the proof, it would be stated that the energy stored in the magnetic field is represented by the area OABCDO (Fig 76). When the field collapses, energy is returned to the supply which is represented by the area DBCB. The area of the loop OABDO represents the energy lost as heat through hysteresis and is obviously the difference between the energy put into the magnetic circuit when setting up the field and that recovered when the field decays.

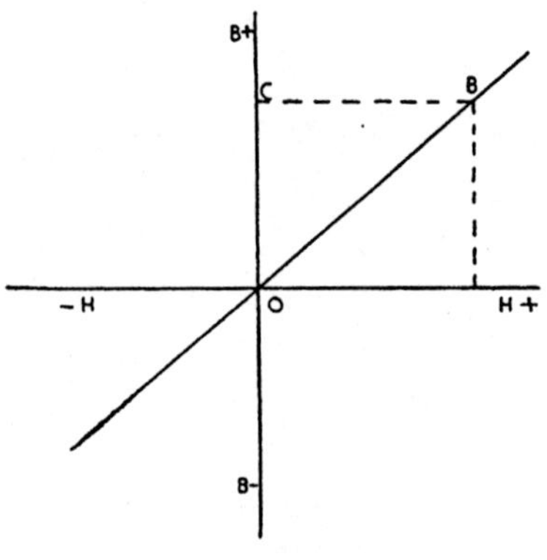

Fig 77

If the iron sample was non-magnetic, *ie* air, then the *B-H* curve would be a straight line, as is shown in the diagram (Fig 77), and the energy stored in the field when it is set up, is represented by the area of the triangle OBC. This energy is recovered when the field collapses. No energy is wasted when air is the medium of a pulsing magnetic field.

For air, area of triangle OBC $= \frac{1}{2}$ OC × CB

$$= \frac{1}{2} B_m \times H$$

where B_m is the maximum flux density value, which has been attained for the H value which was impressed.

But $B = \mu_o H$. ∴ Area of triangle $= \frac{1}{2} \times B_m \times \frac{B_m}{\mu_o} = \frac{B_m^2}{2\mu_o}$

Since the area of the triangle represents the energy stored in air per cubic metre (in joules), it follows that:

For air. Energy stored per cubic metre $= \frac{B_m^2}{2\mu_o}$ joules.

The types of hysteresis loop, as obtained from various magnetic materials, can be grouped into one of three classes as shown by the diagram (Fig 78).

Loop 1 is for hard steel. The large value of the coercive force indicates that the material is suitable for permanent magnets. The area however is large, showing that hard steel is not suitable for rapid reversals of magnetism. Loop 2 rises sharply showing a high μ and a good retentivity (large intercept on B axis). The loop is typical of cast-steel and wrought iron, which are suitable materials for cores of electromagnets and yokes of electrical machines. Loop 3 has a small area and a high μ. The material (mainly alloyed sheet-steels) is suitable for rapid reversals of magnetism and is used for machine armatures, transformer-cores, etc.

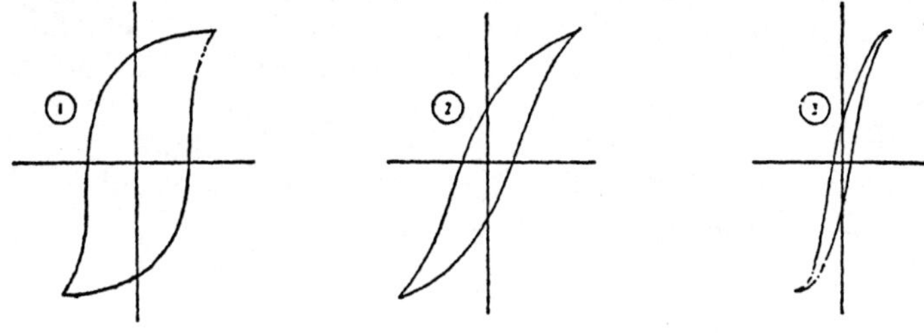

Fig 78

1. HYSTERESIS LOSS. Since this is a function of loop area, the effect of varying B on the area requires consideration. When the value of H is increased *eg* doubled, the value of B is not doubled and consequently the ratio of the loop area is not quadrupled. It is found to increase about 3.1 times and if we assume Area of Loop to be proportional to $B_m{}^x$ then x lies between 1 and 2 — here B_m is the maximum value to which the flux density has been taken. Steinmetz originally found x to be practically constant for all specimens of iron and steel and to equal 1.6 (approximately).

This figure is called the Steinmetz Index and we can now write: power loss per cubic metre per cycle as $P_H \propto B_m{}^{1.6}$

or P_H = constant $\times B_m{}^{1.6} \times f$ joules per second.

Thus $P_H = K_H \times B_m{}^{1.6} \times f$ watts per cubic metre, where K_H is a *hysteresis coefficient,* depending on the material being considered, and f is the cyclic frequency of magnetic alternation. Summarising:

Hysteresis loss $P_H = K_H B_m{}^{1.6} f$ watts per cubic metre — an empirical formula.

2. EDDY-CURRENT LOSS. When an armature rotates in a magnetic field, an e.m.f. is induced in the conductors. Since the conductors are let into slots, it is obvious that the armature teeth can also be looked upon as conductors and that e.m.fs. will be induced across them. Moreover, since the electrical circuit is complete for these e.m.f., currents will flow from end of a tooth through the armature end-plate, along the spider or shaft and back to the other end of the tooth through the opposite armature end-plate. These currents, known as 'eddy currents,' produce a power loss, due to the resistance of the iron circuit, which is $\propto I^2R$ or $\frac{E^2}{R}$. This eddy-current loss is also dependent on a number of other factors — which are considered below. Every attempt is made by machine designers to keep such loss to a minimum. The principal methods by which this is achieved are (a) by laminating the iron circuit and insulating the laminations from each other by varnish, cellulose or paper (b) by using iron with a high specific resistance and (c) by keeping the frequency of the magnetic alternations or cycles to a minimum.

Since a generated voltage is proportional to flux and speed, then $E \propto \Phi N$ or $E \propto B_m f$ where B_m is the maximum flux density and f is the frequency of alternation.

Again since power loss $\propto \frac{E^2}{R}$ we can write:

Power loss $\propto B_m^2 f^2$ or $P_E = K_E B_m^2 f^2$ watts per cubic metre, where K_E is an *eddy-current coefficient* which is dependent upon the type of material being used, its thickness and other dimensions.

PULL OF AN ELECTROMAGNET

It has been seen that the energy stored in a magnetic field in air is given by $\frac{B^2}{2\mu_0}$ joules per cubic metre, where B is in teslas. Consider two poles arranged as shown in the diagram (Fig 79).

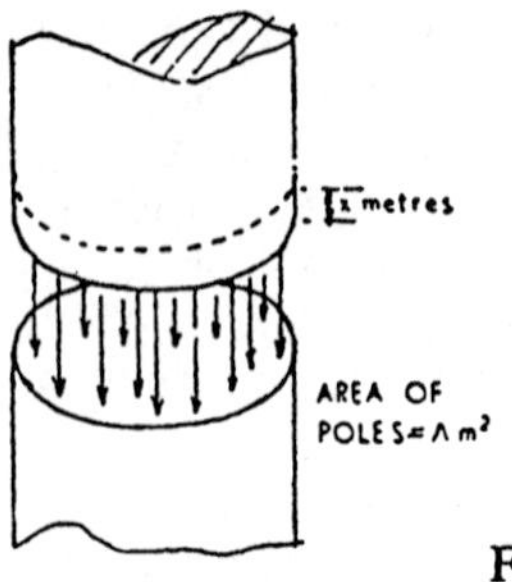

Fig 79

Each has an area A square metres and let F be the force of attraction (in newtons) between the poles.

Let one pole be moved a small distance x (metres) against the force F. Then the work done is Fx newton metres or joules.

The volume of the magnetic field has been increased by Ax cubic metres and therefore the energy stored in the field is increased by $\frac{B^2}{2\mu_0} \times Ax$ joules. This is obviously equal to the work done in separating the poles so that $Fx = \frac{B^2}{2\mu_0} Ax$

or $F = \frac{B^2 A}{2\mu_0}$ newtons, where A is in square metres and B in teslas.

Example 50. An electromagnet is wound with 500 turns. The air gap has a length of 2mm and a cross-sectional area of 1000mm². Assuming the reluctance of the iron to be negligible compared with that of the air gap, and neglecting magnetic leakage and fringing, calculate the magnetic pull when the current is 3 amperes.

M.M.F. of coil = $500 \times 3 = 1500$At

This m.m.f. is used to pass the flux through the air gap, since the reluctance of the iron is negligible.

The magnetising force for the air, is given by 'the ampere-turns per metre' or

$$H = \frac{1500}{2 \times 10^{-3}}$$

Also the flux density in air is B where:

$$B = \mu_0 H$$

$$\therefore B = \frac{4 \times \pi \times 10^{-7} \times 1500}{2 \times 10^{-3}} \text{ tesla}$$

$$= 0.942\text{T}$$

$$\text{Now pull } F = \frac{B^2 A}{2\mu_0} = \frac{0.942^2 \times 1000 \times 10^{-6}}{2 \times 4 \times \pi \times 10^{-7}} \text{ newtons}$$

$$\text{Thus } F = 353.3\text{N.}$$

Example 51. A four-pole d.c. generator has a cast-steel yoke and poles and has a laminated steel armature. The dimensions of the component parts of the magnetic circuit are as follows:

Yoke. Total mean circumference 3.04m — CSA 0.04m²

Pole. Total mean length 0.24m — CSA 0.065m²

Air gap. Total mean length 2mm — CSA 0.065m²

Armature. Total mean path between poles 0.4m — CSA 0.025m²

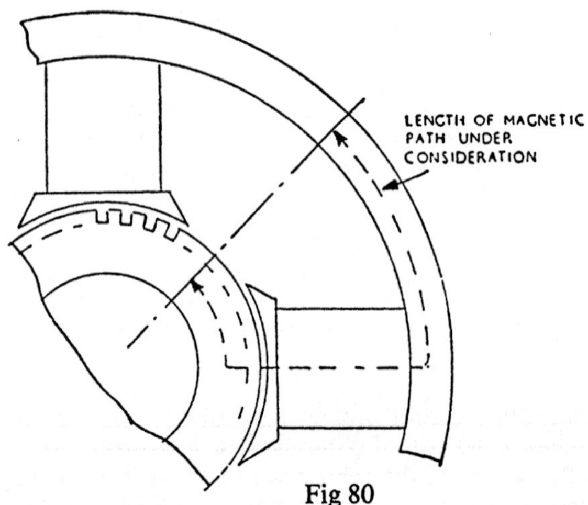

Fig 80

The magnetisation curves are:

	H(At/m)	400	800	1200	1600	2000	2400
Cast Steel	B(T)	0.45	1	1.2	1.3	1.37	1.43
Laminated Steel	B(T)	1	1.34	1.48	1.55	1.6	1.63

Calculate the ampere-turns per pole, for a flux per pole of 0.08Wb in the air gap. The diagram (Fig 80) illustrates the problem and the appropriate magnetic characteristics are shown by the graphs of the diagram (Fig 81).

AIR GAP

Length 2×10^{-3} metre

Area $0.065 = 6.5 \times 10^{-2}$ square metre

$$\Phi = 0.08\text{Wb} \quad B_A = \frac{0.08}{6.5 \times 10^{-2}} = \frac{8}{6.5} = 1.23\text{T}$$

$$H_A = \frac{1.23}{\mu_o} = \frac{1.23}{4 \times \pi \times 10^{-7}} \text{ ampere-turns/metre}$$

Total ampere-turns or M.M.F. for air gap is given by:

$$F_A = \frac{1.23}{4 \times 3.14 \times 10^{-7}} \times 2 \times 10^{-3} \text{ ampere-turns}$$

$$F_A = 1955\text{At}$$

POLE (Cast Steel)

Length 24×10^{-2} metre

Area $0.065 = 6.5 \times 10^{-2}$ square metre

$$\Phi = 0.08\text{Wb} \quad B_P = \frac{\Phi}{A} = \frac{0.08}{6.5 \times 10^{-2}} = 1.23\text{T}$$

From graph, $H_P = 1370$At/m or

Total F_P for pole $= 1370 \times 24 \times 10^{-2} = 330$At

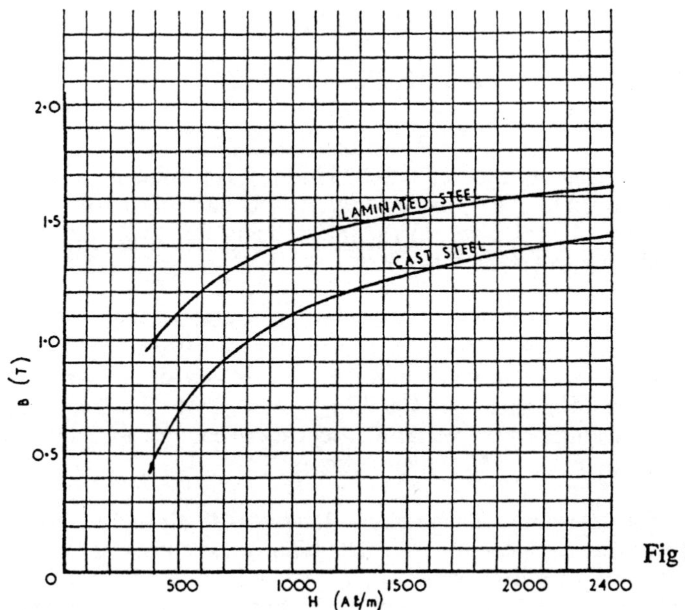

Fig 81

YOKE (Cast Steel)

Length $\frac{3.04}{4} = 0.76 = 76 \times 10^{-2}$ metre (between poles)

or 0.38 magnetic length

Area $= 2 \times 0.04 = 0.08$ square metre. (Note the doubling of the area since full pole area has been taken for the flux)

$= 8 \times 10^{-2} m^2$

$\Phi = 0.08 Wb \quad B_Y = \frac{0.08}{8 \times 10^{-2}} = 1$ tesla

From graph $H_Y = 800 At/m$

Total F_Y for yoke $800 \times 38 \times 10^{-2} = 304 At$

ARMATURE (Laminates)

Length $\frac{0.4}{2} = 20 \times 10^{-2}$ metre (magnetic length)

Area $= 2 \times 0.025 = 0.05 = 5 \times 10^{-2} m^2$

$\Phi = 0.08 Wb \quad B_L = \frac{0.08}{5 \times 10^{-2}} = 1.6T$

From graph $H_L = 2000 At/m$

Total F_L for armature $= 2000 \times 20 \times 10^{-2}$

$= 400 At$

Total M.M.F. per pole $= 1955 + 330 + 304 + 400$

$= 2989 At.$

CHAPTER 6

PRACTICE EXAMPLES

1. A brass rod of cross-section 1000mm² is formed into a closed ring of mean diameter 300mm. It is wound uniformly with a coil of 500 turns. If a magnetising current of 5A flows in the coil, calculate (a) the magnetising force (b) the flux density and (c) the total flux.

2. An electromagnetic contactor has a magnetic circuit of length 250mm and a uniform cross-sectional area of 400mm². Calculate the number of ampere-turns required to produce a flux of 500μWb, given that the relative permeability of the material under these conditions is 2500. Also $\mu_0 = 4\pi \times 10^{-7}$ henry/metre.

3. In a certain magnetic circuit having a length of 1m and a uniform cross-section of 500mm², a magnetising force of 500 ampere-turns produces a magnetic flux of 400μWb. Calculate (a) the relative permeability of the material and (b) the reluctance of the magnetic circuit. $\mu_0 = 4\pi \times 10^{-7}$ H/m.

4. An iron ring having a mean circumference of 1.25m and a cross-sectional area of 1500mm², is wound with 400 turns of wire. An exciting current of 2.5A produces a flux of 0.75mWb in the iron ring. Calculate (a) the permeability (relative) of the iron (b) the reluctance of the iron (c) the m.m.f. of the exciting winding.

5. A U-shaped electromagnet has an armature separated from each pole by an air gap of 2mm. The cross-sectional area of both the electromagnet and the armature is 1200mm² and the total length of the iron path is 0.6m. Determine the ampere-turns necesary to produce a total flux in each air gap of 1.13mWb neglecting magnetic leakage and fringing.
   The magnetisation curve for the iron is given by:

B(T)	0.5	0.6	0.7	0.8	0.9	1.0	1.1
H(At/m)	520	585	660	740	820	910	1030

6. A circular ring of iron of mean diameter 0.2m and cross-sectional area 600mm² has a radial air gap of 2mm. It is magnetised by a coil having 500 turns of wire. Neglecting magnetic leakage and fringing, estimate the flux density in

the air gap, when a current of 3A flows through the coil. Use the magnetic characteristics as given by the graph of Q5.

7. A built-up magnetic circuit without an air gap, consists of two cores and two yokes. Each core is cylindrical, 50mm diameter and 160mm long. Each yoke is of square cross-section 47 × 47mm and is 180mm long. The distance between the centres of the cores is 130mm. Calculate the ampere-turns necessary to obtain a flux density of 1.2T in the cores. Neglect magnetic leakage. The magnetic characteristics of the material are:

B(T)	0.9	1.0	1.05	1.1	1.15	1.2
H(At/m)	200	260	310	380	470	650

8. An iron rod 15mm diameter, is bent into a semi-circle of 50mm inside radius and is wound uniformly with 480 turns of wire so as to form a horse-shoe electromagnet. The poles are faced so as to make good magnetic contact with an iron armature 15 × 15mm cross-section and 130mm long. (a) Calculate the current required to produce a pull of 196.2N between the armature and pole faces. Neglect magnetic leakage. (b) If the armature is fixed so as to leave uniform air gaps 0.5mm wide at each pole face, calculate the ampere-turns necessary to obtain a flux density of 1.15T in the air gap. Neglect leakage and fringing. Use the magnetic characteristics as given by the graph of Q7.

9. Two coaxial magnetic poles each 100mm diameter are separated by an air gap of 2.5mm and the flux crossing the air gap is 0.004Wb. Neglecting fringing calculate (a) the energy in joules stored in the air gap (b) the pull in newtons between the poles.

10. Calculate the ampere-turns per field coil required for the air gap, the armature teeth and the pole, of a d.c. machine working with a useful flux of 0.05Wb/pole, having given: Effective area of air gap 60 000mm^2. Mean length of air gap 5mm. Effective area of pole 40 000mm^2. Mean length of pole 250mm. Effective area of teeth 25 000mm^2. Mean length of teeth 45mm.

Magnetic leakage coefficient = 1.2. Magnetic characteristics of the materials are:

B(T)	1.3	1.4	1.5	1.6	1.8	2.0
H(At/m)	1200	1500	2000	3000	8500	24 000

CHAPTER 7

ELECTROMAGNETIC INDUCTION

Up to now, the only method considered for producing an electromotive force has been that due to chemical action. In the progression of electrical knowledge, history shows that this also was the pattern followed and that electrochemistry was the first branch of the science which was developed and was to play its full part in subsequent electrical investigations of the early nineteenth century. At that time electricity was a subject of interest to the scientist only and had not as yet been accepted as a medium which could be put to use for engineering processes. The chemical cells as were then known, could not produce sufficient energy for practical purposes, nor had any electromagnetic devices been evolved which could be put to engineering applications. As was mentioned in Chapter 5, it was only after the relation between current and magnetism was discovered that attention was turned to various associated findings. These inevitably disclosed the related phenomena of electromagnetic induction and led to the development of machines which were subsequently to interest the engineer as means of producing either electrical or mechanical energy.

The initial electromagnetic *induction* investigations are attributed to Michael Faraday, who in 1821 showed that when the magnetic flux linked or associated with an electrical circuit is changing, an e.m.f. is induced in the circuit. This e.m.f. is shown to last only whilst the change of flux is taking place and the faster the change the greater the e.m.f.

The flux linked with a circuit, which invariably consists of a coil of insulated wire, may be changed in different ways. Thus:

(a) A magnet could be moved in the vicinity of a coil of wire. This principle is used for the a.c. Generator or Alternator.

(b) A coil of wire could be moved in the vicinity of a magnet. This principle is used for the d.c. Dynamo or Generator — the more accepted modern term.

(c) The flux could be changed by varying the current in the energising coil of wire. The ampere-turns are thus varied and the flux produced varies accordingly. This principle is made use of in the operation of the Transformer or the Spark-coil of a petrol-engine ignition system.

For the above three ways of e.m.f. generation, it will be seen that cases (a) and (b) involve relative physical movement between the magnet and the coil. Case (c) however, involves no such movement and the generated e.m.f. is achieved in a stationary coil with which the associated flux only is changing. Thus there are two distinct forms of e.m.f. generation or induction — to use the alternative term. These forms are referred to under two basic headings: (1) Dynamic Induction, (2) Static Induction.

Before these two methods are considered in detail it would be well to give some attention to the meaning of an expression which will be used frequently, namely *'flux-linkages'*.

FLUX-LINKAGES. Earlier studies on magnetism showed that the field of a magnet can be represented by lines of flux emanating from the poles. The strength of the flux can be represented by the number of lines and is measured in webers, while the flux density is measured in teslas. The flux lines make complete loops and the associated conductor or coil of wire in which the e.m.f. is induced can also be considered to consist of a number of turns. Since the number of lines of flux associated with the turns are referred to as flux-linkages, then a magnet with poles of flux strength 3.4μWb associated or linking with a coil of 500 turns is said to result in a condition of $3.4 \times 10^{-6} \times 500 = 0.0017$ weber-turns. The diagram (Fig 82) shows the basic idea of the explanation.

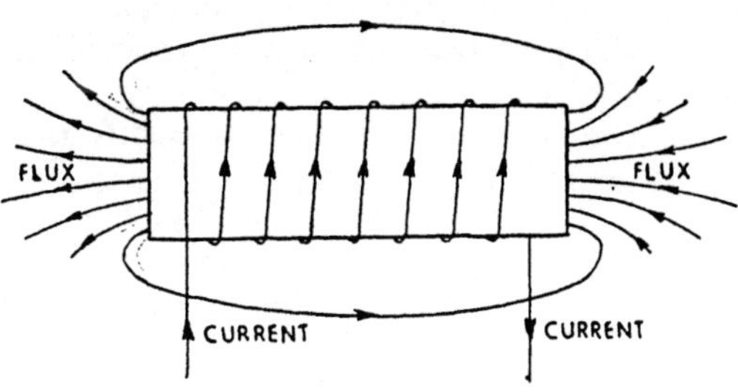

Fig 82

FARADAY'S LAW OF ELECTROMAGNETIC INDUCTION

This summarises the known relationship deduced for the generation of e.m.f. by electromagnetic induction and can be stated as: the magnitude of the e.m.f. produced, whenever there is a change of flux linked with a circuit, is proportional to the rate of change of flux-linkages.

LENZ'S LAW

This identifies a phenomenon always noted for the e.m.f. produced by induction. The law can be stated as: the direction of the current due to the induced e.m.f. will always set up an effect tending to oppose the change which is causing it.

This can be further explained in terms of the magnetic field which would be set up by the current caused by the induced e.m.f. Thus if the flux-linkages tend to increase, then the field produced by the induced current resulting from the induced e.m.f., will tend to oppose this effect, *ie* it would tend to oppose the flux-linkage build up. Similarly, if the flux-linkages tended to be reduced, as when the current in a coil is switched off, then the induced e.m.f. will induce a current which, if allowed to flow, would tend to keep up the flux-linkages to their original value. It is stressed that the action of the induced current does not succeed in preventing the change, but would try to do so during the period when the change is occurring.

Faraday's law is capable of being expressed in mathematical form and thus formulae can be deduced for both static and dynamic electromagnetic induction. These will be considered under each appropriate heading.

STATIC INDUCTION

Consider a coil connected to a d.c. supply. At the instant of switch-on the current produces flux which grows from the centre of the coil outwards, but this flux 'cuts' the coil turns and induces e.m.f. which opposes the current growth (Lenz's Law). Such a circuit in which a change of current causes a change of flux and therefore produces an induced e.m.f. is said to be inductive or to possess *self inductance.*

A circuit has an inductance of 1 Henry if an e.m.f. of 1 volt is induced into the circuit when the current changes at the rate of 1 amp/second.

SELF INDUCTANCE

From the definition it can be seen that if a circuit has an inductance of L Henry and the current changes from I_1 to I_2 amp

amp in t seconds, as shown in Fig 83, then the average induced e.m.f.:

$$Eav = \frac{-L(I_2 - I_1)}{t} \text{ volts}$$

or

$$Eav = -L \times \text{rate of increase of } I = -L\frac{dI}{dt}$$

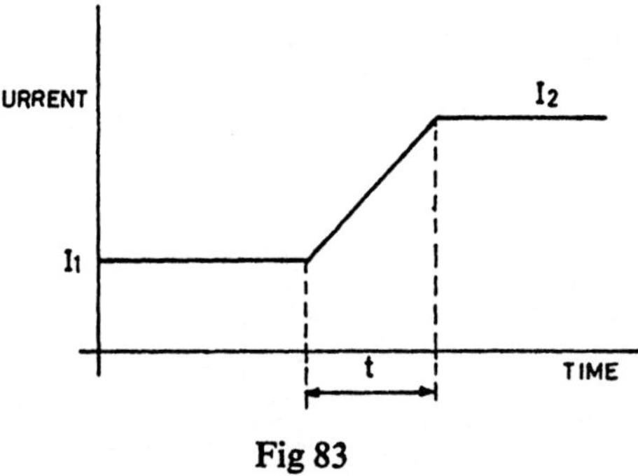

Fig 83

Note: The negative sign indicates that the direction of induced e.m.f. opposes that of the current increase.

Example 52. The current through a coil having an inductance of 0.5 H is reduced from 5A to 2A in 0.05s. Calculate the average e.m.f. induced in the coil.

$$Eav = \frac{-L(I_2 - I_1)}{t} \text{ volts}$$

$$= \frac{-0.5\,(2-5)}{0.05} \text{ volts}$$

$$\therefore Eav = +30 \text{ volts.}$$

N.B. The positive sign indicates that the induced e.m.f. is tending to maintain the current flow.

Since a change of current produces a change of flux linkages, the e.m.f. induced can be expressed in terms of the number of coil turns and the rate of change of flux.

$$Eav = \frac{-N(\Phi_2 - \Phi_1)}{t} \text{ volts} = -N\frac{d\Phi}{dt}$$

Example 53. When a d.c. current passes through an iron cored coil of 2000 turns a magnetic flux of 30mWb is produced. The supply switch is opened and the current falls to zero amperes in

0.12s leaving a residual flux of 2mWb. Find the average value of induced e.m.f.

$$Eav = \frac{-N(\Phi_2 - \Phi_1)}{t}$$

$$= \frac{-2000\ (2 - 30) \times 10^{-3}}{0.12}$$

$$Eav = +466.6 \text{ volts}$$

Note. Once again the +ve sign indicates that the induced e.m.f. is tending to maintain the current flow.

From the foregoing equations:

$$Eav = \frac{-L\ (I_2 - I_1)}{t}$$

$$\text{and} \quad Eav = \frac{-N\ (\Phi_2 - \Phi_1)}{t}$$

$$\therefore \frac{-L\ (I_2 - I_1)}{t} = \frac{-N\ (\Phi_2 - \Phi_1)}{t}$$

$$\therefore \text{Inductance } L = \frac{N\ (\Phi_2 - \Phi_1)}{(I_2 - I_1)} = \frac{Nd\Phi}{dI}$$

$$\text{or Inductance } L = \frac{N\Phi}{I}$$

Hence Inductance L = Flux linkages/amp.

In Chapter 6 flux was expresed in terms of m.m.f. (F) and reluctance

$$F = \Phi S \text{ or } IN = \Phi S \text{ and } \Phi = \frac{IN}{S}$$

$$\textit{Thus } L = \frac{N}{I} \times \frac{IN}{S}$$

$$\therefore \text{Inductance } L = \frac{N^2}{S} \text{ Henrys}$$

E.M.F. DUE TO STATIC INDUCTION

Consider the diagram (Fig 84) which shows two coils A and B of insulated wire. Coil A can be connected to a battery through a switch, whereas B is wound over or placed adjacent to coil A and is connected to a sensitive centre-zero type voltmeter. This type of instrument is used because, as the pointer is positioned at the centre of the scale, a deflection to the left or right depending on the polarity of the supply can be registered.

At the instant of switching on the current in coil A, the flux can be imagined to grow outwards and to cut the turns of coil B. The initial growing is shown by the dotted flux lines becoming fuller until the final condition (full lines) is reached. The cutting of coil B by the flux of A, results in an e.m.f. being induced, its magnitude and direction being governed by Faraday's and Lenz's laws. The flux-linkages *ie* flux linking with the turns (N_B) of coil B increase and if the linking flux grows to a value of Φ webers from its original zero value, then the rate of change of flux-linkages will be the flux-linkages divided by the time (t_1) taken for them to grow, *ie* the time taken for the current to reach its final value.

Thus e.m.f. induced in coil B $= \frac{\text{flux-linkages}}{\text{time}} = \frac{-N_B\Phi}{t_1}$ volts.

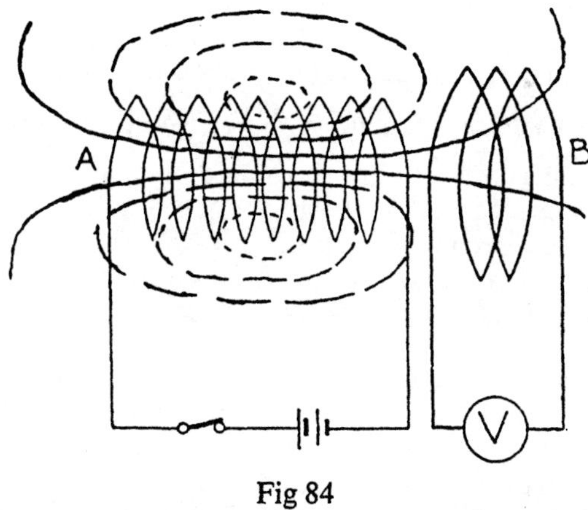

Fig 84

For this equation N_B = turns of coil B, Φ is the flux in webers linking with it and t_1 is the time taken for the energising current to reach its final value *I*. It could be assumed that value *I* is reached immediately the switch is closed, because the flow of electricity is considered to be instantaneous, but here we have a condition where the current takes an appreciable time to reach its full value — due to the inductance of the arrangement. This action will be considered in detail later, but to revert to the initial effects being observed; it will be seen that when the switch for the *primary* coil A is closed, the voltmeter pointer gives a 'kick', say to the left, showing an e.m.f. to be induced in coil B — the *secondary* circuit. The value of e.m.f. $E_B = \frac{-N_B\Phi}{t_1}$ and the voltmeter will show the polarity of coil B to be such that the current, which flowed through the instrument, was in such a direction through the coil as to set up a secondary flux, which would be opposite to the original flux Φ and would try to stop it growing. A further point of importance to note would be that, although a kick of the voltmeter pointer would be seen, yet it would return to the zero position even though current in coil A was allowed to flow indefinitely. Thus an e.m.f. is induced only during the time when the flux-linkages were changing. Further experiments with coil B would show that if the number of turns of wire were doubled, then the induced e.m.f. would be twice as large, even though the flux Φ of coil A was the same. The flux-linkages have been increased and the induced e.m.f. rises in proportion.

Consider next the instant of switching off the current in coil A. The voltmeter will again be seen to kick — to the right this time, showing an induced e.m.f. of reversed polarity. The flow of current in coil B is such as to try and maintain the flux to its original value Φ and again $E_B = \dfrac{-N_B\Phi}{t_2}$ where t_2 is the time taken for 'switching off'. It can be noted here that t_2 need not necessarily equal t_1. If the switch is opened quickly, the current of A will be interrupted very quickly and E_B can be larger at switching off than at switching on; when the rate of growth of the flux is controlled by the inductance and resistance of the circuit.

Up to now we have only considered the induced e.m.f. in coil B and this is said to be due to *Mutual Induction, ie* the mutual action of coil A on B. We now turn our attention to *Self Induction, ie* the conditions appertaining to coil A itself. At the instant of switching on, the flux grows outwards and in so doing, cuts the turns of coil A — the primary circuit. An e.m.f. is thus induced given by $E_A = \dfrac{-N_A\Phi}{t_2}$. Here N_A is the turns of coil A, Φ is the linked flux and t_1 the time taken for the current to reach its full value. As before, the direction of the self-induced e.m.f. E_A will be such as to cause a current to flow in the opposite direction through the battery and will produce a flux which will try to oppose the build-up of flux Φ. We can now see the reason for the opposition to the growth of current in coil A at the instant of switching on and why the current I takes some little time to reach its full value. As before when the switch is opened, the flux collapses and in doing so, again cuts the turns of coil A, inducing a voltage of reversed polarity, which tries to keep the current flowing. Appreciable arcing will be observed at the switch contacts, but if the latter is operated quickly, the circuit will be interrupted quickly in spite of this and E_A will be ineffective. It is stressed however, that this self-induced e.m.f. at 'switching off', can be extremely large in some instances where a large number of turns of an energising winding are associated with a strong magnetic flux. An example would be the opening of the field circuit of a large alternator or d.c. generator and special arrangements are necessary in order to limit the e.m.f. to a safe value and prevent 'break-down' of the insulation by this large induced voltage.

MUTUAL INDUCTANCE

Two coils have a mutual inductance of 1 Henry when a change of current at the rate of 1 ampere/second in one coil produces an e.m.f. of 1 volt in the other.

Consider once again Fig 84. If coils A and B have a mutual inductance of M Henry, and the current in coil A increases from I_1 to I_2 amperes in t seconds then

$$\text{Average e.m.f. induced in B} = \frac{-M\,(I_2 - I_1)}{t}$$

$$= -M \times \text{Rate of increase of current in coil A}$$

The e.m.f. induced into coil B can also be expressed in terms of flux linkages in the same way as was applied to Self Inductance

Let the flux change from Φ_1 to Φ_2 Webers in t seconds due to a change of current from I_1 to I_2 amperes and let coil B have N_B turns.

Thus Average e.m.f. induced in

$$B = \frac{-N_B\,(\Phi_2 - \Phi_1)}{t}$$

$$\text{Hence } \frac{-M\,(I_2 - I_1)}{t} = \frac{-N_B\,(\Phi_2 - \Phi_1)}{t}$$

$$\therefore M = \frac{N_B\,(\Phi_2 - \Phi_1)}{(I_2 - I_1)}$$

$$\text{So Mutual Inductance} = \frac{\text{Change in flux linkages on secondary}}{\text{Change in current in primary}}$$

Example 54. (Self-induction). A coil of 800 turns is wound on a wooden former and a current of 5A is passed through it to produce a magnetic flux of 200 micro-webers. Calculate the average value of e.m.f. induced in the coil when the current is (a) switched off in 0.08 seconds (b) reversed in 0.2 seconds.

(a) $$Eav = \frac{-N(\Phi_2 - \Phi_1)}{t} = \frac{-800 \times (0 - 200 \times 10^{-6})}{0.08}$$

$$Eav = 2 \text{ volts}$$

(b) $\text{Eav} = \frac{-N(\Phi_2 - \Phi_1)}{t}$ here $\Phi_2 = \Phi_1$ numerically but is in the reverse direction, or $\Phi_2 = -\Phi_1$

Thus $\text{Eav} = \frac{2N\Phi_1}{t} = \frac{2 \times 800 \times 200 \times 10^{-6}}{0.2}$

$\text{Eav} = 1.6$ volts

Example 55. (Mutual-induction). If the coil of the above example has a secondary coil of 2000 turns wound onto it, find the e.m.f. induced in this second coil when the current of 5A is switched off in 0.08 seconds. (It can be assumed that all the flux of 200μWb created by the 5A current in the primary links with the secondary coil.)

$$\text{Eav} = \frac{-2000\ (0 - 200 \times 10^{-6})}{0.08} = 5 \text{ volts}$$

Note how the e.m.f. of the secondary is $\frac{5}{2} = 2.5$ times the induced e.m.f. value in the primary. It is in direct proportion to the turns ratio $\frac{2000}{800} = 2.5$. This is the basic principle of the transformer and ignition system spark-coil. It shows how a large voltage can be induced in a secondary coil by the flux associated with a low voltage primary coil. For a petrol-engine ignition system, the e.m.f. in the secondary may be in the region of 8000 volts compared with the 12V applied to the primary. This is achieved by using the appropriate turns ratio for the primary and secondary coils, by providing an iron magnetic circuit to concentrate the flux for maximum linkage and by interrupting the primary circuit quickly by an engine-driven cam-operated switch.

Example 56. The ignition coil of a petrol engine has an inductance of 4.5H and carries a current of 4 amperes. If, when the distributor points close, the circuit current collapses uniformly to zero in 2ms, find the average e.m.f. induced in the coil.

$$\text{Eav} = \frac{-L\ (I_2 - I_1)}{t}$$

$$= \frac{-4.5\ (0 - 4)}{2 \times 10^{-3}}$$

$$\text{Eav} = 9000 \text{ volts} = 9\text{kV}$$

COUPLING FACTOR

There is a relationship between the mutual inductance of two coupled coils and their individual self inductances, depending upon the amount of magnetic coupling between them. The mutual inductance of the two coils can be expressed as the change in flux linkages of one coil to the change in current in the other.

$$M = N_B \frac{d\Phi}{dI_A} \text{ or } M = N_A \frac{d\Phi}{dI_B}$$

$$M^2 = N_B \frac{d\Phi}{dI_A} \times N_A \frac{d\Phi}{dI_B}$$

$$M^2 = N_A \frac{d\Phi}{dI_A} \times N_B \frac{d\Phi}{dI_B}$$

But $$L_A = N_B \frac{d\Phi}{dI_A} \text{ and } L_B = N_B \frac{d\Phi}{dI_B}$$

$$\text{Thus } M^2 = L_A \times L_B$$

$$M = \sqrt{L_A\, L_B}$$

This is the maximum value of mutual inductance available between the coils, but this is very difficult to achieve because of magnetic leakage and fringing effects. The magnetic coupling between the coils is also affected by the separation of the coils and the angular displacement between them. Therefore, in general, mutual inductance is given by:

$M = k\sqrt{L_A L_B}$ where k is the coupling factor.

If perfect coupling can be assumed then $k = 1$.

INDUCTANCE OF TWO COILS IN SERIES

In our study of mutual inductance, we have only considered the effect of two magnetically coupled but electrically separated coils. However the effects of mutual and self inductance can be similarly applied to electrically connected coils.

Consider first the effects of two coils carrying the same current, and wound such that their magnetic fields assist one another (Fig 85a).

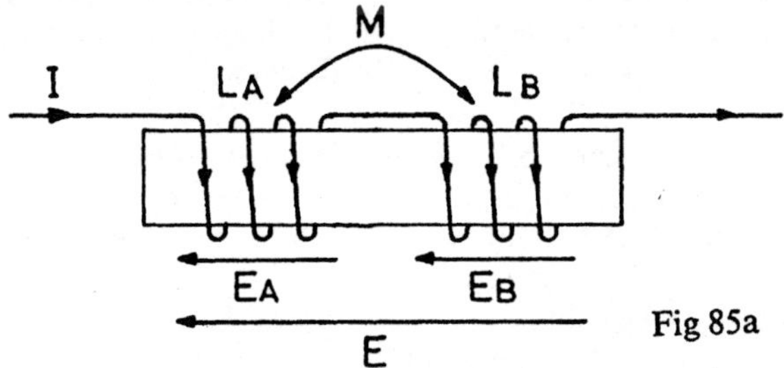

Fig 85a

From our previous work on mutually coupled coils, we have seen that a coil will have an e.m.f. induced into it due to (a) the self inductance of the coil and (b) the mutual inductance to the other coil. Refer to Fig 85a

$$ie \quad E_A = \left(-L_A\frac{dI}{dt}\right) + \left(-M\frac{dI}{dt}\right) = -(L_A + M)\frac{dI}{dt}$$

$$\text{and } E_B = \left(-L_B\frac{dI}{dt}\right) + \left(-M\frac{dI}{dt}\right) = -(L_B + M)\frac{dI}{dt}$$

But E_A and E_B are in series assisting

$$\therefore \quad E = E_A + E_B$$

$$E = \left[-(L_A + M)\frac{dI}{dt}\right] + \left[-(L_B + M)\frac{dI}{dt}\right]$$

$$E = -(L_A + L_B + 2M)\frac{dI}{dt}$$

But $E = -L\frac{dI}{dt}$ Where L is the total inductance.

$\therefore \quad L = L_A + L_B + 2M$ Henrys (for coils assisting)

Similarly we can consider two coils wound in opposition Fig 85b.

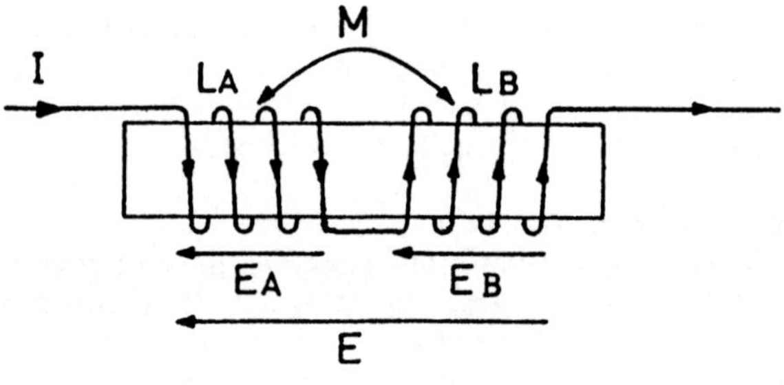

Fig 85b

Note in this case the induced e.m.f. due to mutual inductance opposes that due to self inductance. Hence, by similar proof to that shown for coils assisting, it can be shown that when the coils are opposing $L = L_A + L_B - 2M$ Henrys.

Example 57. Two coils of inductances 10mH and 15mH respectively have a coupling factor of 0.8 between them. What is their combined inductance when they are connected in series (a) assisting and (b) opposing?

$$M = k\sqrt{L_A L_B} = 0.8\sqrt{10 \times 15} \text{ mH}$$
$$= 9.8 \text{ mH}$$

$$L = L_A + L_B + 2M$$
$$= 10 + 15 + (2 \times 9.8) \text{ mH}$$
$$L = 44.6 \text{ mH}$$

(b) $$L = L_A + L_B - 2M$$
$$= 10 + 15 - (2 \times 9.8) \text{ mH}$$
$$L = 5.4 \text{ mH}$$

DYNAMIC INDUCTION

As was mentioned earlier, this condition covers the cases where there is relative movement between a magnetic field and a conductor. Obviously this concerns either a stationary conductor and a moving field or a stationary field and a moving conductor. To avoid repetition of basic theory, the immediate explanations and diagrams will refer to a fixed field and moving conductor.

The diagrams (Fig 86) show a field as produced by two permanent magnets and a conductor which is moved so as to cut the field, thus altering the flux-linkages. The reasoning can be seen as in interpretation of Faraday's and Lenz's laws and three cases are shown.

For case (a) there is seen to be no change of flux-linkages, *ie* no cutting of the field. The conductor is merely moved at a velocity of v metres/second in the same direction of the lines of flux and no e.m.f. is recorded on the voltmeter. For case (b) the conductor is moved at right angles to the field of flux-density B teslas and the voltmeter shows a constant deflection. The flux-linkages can be considered to be changing since the flux lines are cut or can be imagined to stretch and snap as the conductor passes through, to reform again behind the conductor. If the conductor is moved from left to right, a polarity is noted, which reverses if the conductor is moved from right to left. Alternatively, if the field is reversed so that the flux lines are considered to pass from a bottom N pole to a S pole at the top of the diagram, and the conductor is moved from left to right, then a reversed polarity will again be indicated. The investigation will show further deductions. Thus:

The magnitude of the induced e.m.f. varies with the speed of cutting the field or rate of change of flux-linkages. Hence $E \propto v$. Again, if the field being cut is varied by altering the density of the flux, then the e.m.f. will vary as B or $E \propto B$. Obviously also, the longer the conductor cutting a field, the greater will be the magnitude of the e.m.f. and $E \propto l$. Summarising these three conditions we see that $E \propto Blv$. Here l is the length of the conductor in metres.

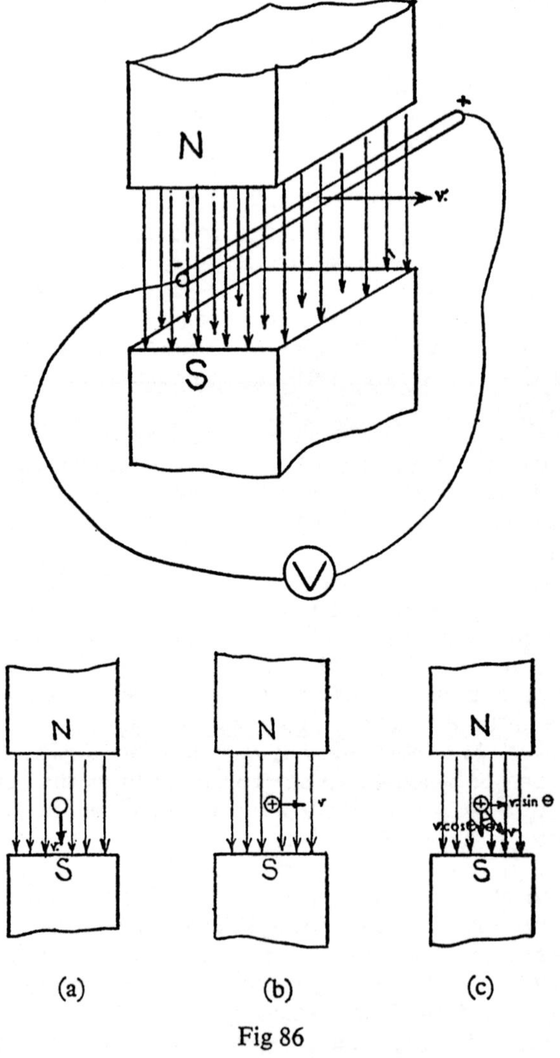

Fig 86

Case (c) of the diagram shows the conductor cutting the field at an angle θ. It is an intermediate condition between cases (a) and (b) and is best treated by resolving v into two component velocities at right angles to each other. Consider $v \cos \theta$ to be the component velocity in the direction of the flux lines, then $v \sin \theta$ will be the other component velocity at right angles to the field.

In accordance with the reasoning for cases (a) and (b) we see that velocity $v \cos \theta$ will be responsible for no induced e.m.f. whereas velocity $v \sin \theta$ will be responsible for such an e.m.f. and $E \propto v \sin \theta$.

$E \propto Blv \sin \theta$ will be a more general expression than those already deduced since it will cover all conditions.

For instance for the condition of case (a) $\theta = 0°$ and since $\sin 0° = 0$ $\therefore$ $Blv \sin 0° = 0$ or $E = 0$ as already stated.
Again for case (b). If $\theta = 90°$ then $\sin 90° = 1$ and $Blv \sin 90° = Blv$ giving $E \propto Blv$.

E.M.F. DUE TO DYNAMIC INDUCTION

As explained above, the induced e.m.f. is found to be proportional to B, l, v and the sine of the angle made by the direction of cutting, and the direction of the field. The actual magnitude of such an e.m.f. can however be deduced in more definite terms thus:

Consider case (b) of Fig 86. In 1 second, the area cut by a conductor of length l metres and moving at a velocity of v metres/second is lv square metres. If the flux density in this area is B teslas, then the flux cut per second by the conductor $= Blv$ webers. Using Faraday's law, we see that Blv can be used as a measure of the magnitude of the induced e.m.f. in volts or

Induced e.m.f. $E = Blv$ volts.

If case (c) is considered the flux cut is proportional to the component of the velocity perpendicular to the field

or induced e.m.f. $E = Blv \sin \theta$ volts.

The above formula can also be deduced as follows; this approach may appear to the reader to be more satisfactory as a proof. The diagram (Fig 87) shows a conductor Q, carrying a

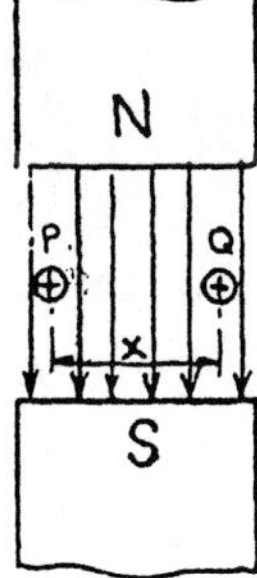

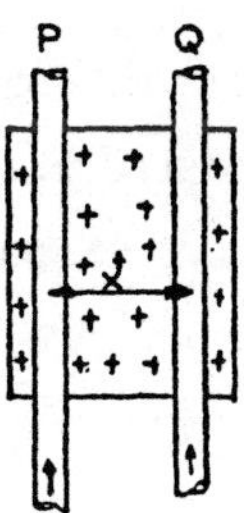

Fig 87

current of I amperes in the direction shown. As before the flux density of the field is taken as B teslas and the length of the conductor as l metres. Using fundamentals already set out in the chapter on electromagnetism, it is known that a force is exerted on a current-carrying conductor in a magnetic field. Thus the conductor in the diagram experiences a force BIl newtons urging it to the left. Accordingly a force of BIl newtons must be applied in the opposite direction to oppose movement of the conductor.

Consider the conductor to move from position Q to position P spaced x metres away. Then work done by the conductor in moving from Q to P = Force × distance = $BIlx$ newton metres or joules.

Let E volts be the e.m.f. induced in the conductor as a result of cutting the field, and t seconds the time taken to complete the operation.

Then Mechanical Power expended $= \dfrac{BIlx}{t}$ watts and if this appeared as electrical power, it would be EI watts

or $EI = \dfrac{BIlx}{t}$ and $E = \dfrac{Blx}{t}$ *volts*

Obviously $\dfrac{x}{t}$ = velocity of cutting v

or as before, $E = Blv$ volts.

From the above, the following deduction can also be made. Blx is (the flux-density × area) or the flux Φ cut by the conductor in moving from position Q to P in time t seconds and since

$$E = \frac{Blv}{t} \text{ then } E = \frac{B \times \text{area}}{t} = \frac{\Phi}{t} \text{ or } E = \frac{\Phi}{t}$$

Thus E (volts) $= \dfrac{\Phi \text{ (webers)}}{t \text{ (seconds)}}$ and we have an alternative formula for the e.m.f. generated in a conductor cutting a magnetic field. In this form, it is similar to that deduced for the statically induced e.m.f. namely $E = \dfrac{N\Phi}{t}$ where N is the number of turns of the coil, and Φ is the change of flux.

The formula can thus be $E = \dfrac{\text{flux-linkages}}{\text{time}}$ or $E = \dfrac{\Phi}{t}$ since the flux-linkages are numerically equal to Φ, there being only one conductor.

Important Note. $\dfrac{\Phi}{t}$ is $\dfrac{\text{Flux cut}}{\text{time}}$ and thus we have an alterna-

tive way of stating Faraday's law, which can now be expressed as:
'The e.m.f. generated in a conductor is proportional to the rate of cutting lines of flux or is proportional to the flux cut/second.'

The above form of Faraday's law is more applicable to dynamic induction and will be used repeatedly in connection with the Generator, Motor and Alternator.

Example 58. A conductor is moved to cut a magnetic field at right angles. Find the e.m.f. induced in it, if the average density of the field is 0.45 teslas, the length of conductor is 80mm and the speed of cutting is 8.88 metres/second.

In the Formula $E = Blv$ we have

$$E = 0.45 \times 0.08 \times 8.88 = 0.32 \text{ volts.}$$

An alternate solution could be:

Area swept by the conductor/second = 0.08×8.88 square metres. The flux in this area would be $\Phi = 0.45 \times 0.08 \times 8.88$ webers.

$$\text{and e.m.f.} = \text{flux cut per second}$$

$$\text{or } E = \frac{0.45 \times 0.08 \times 0.88}{1} = 0.32 \text{ volts.}$$

Example 59. A four-pole generator has a flux of 12mWb/pole. Calculate the value of e.m.f. generated in one of the armature conductors, if the armature is driven at 900 rev/min.

In 1 revolution a conductor cuts $4 \times 12 \times 10^{-3} = 0.048$Wb

$$\text{Time of 1 revolution of the armature} = \frac{1}{900} \text{ minutes}$$

$$= \frac{60}{900} \text{ or } \frac{1}{15} \text{ seconds}$$

$$\therefore \text{Rate of cutting flux} = \frac{\text{Flux cut per revolution}}{\text{time taken to complete a revolution}}$$

$$\text{Thus } E = \frac{0.048}{\frac{1}{15}} = 0.048 \times 15 = 0.72 \text{ volts/conductor.}$$

THE WEBER. Although this unit has been introduced, it has not as yet been satisfactorily defined, since the accepted definition is based on the principle of electromagnetic induction. Thus we have:

'The weber is that magnetic flux which, when cut by a conductor in one second, generates in the conductor an e.m.f. of value equal to one volt.'

Alternative ways of defining the weber or SI unit of flux are: 'An e.m.f. of one volt is generated when a conductor cuts flux at the rate of one weber/second', or 'an e.m.f. of one volt is generated when the flux linked with one turn changes at the rate of one weber /second.'

DIRECTION OF INDUCED E.M.F. (HAND RULES)

The direction of the induced e.m.f. can be deduced from first principles, using Lenz's law or by the application of a rule which was first enunciated by Professor T. A. Fleming and is now commonly known as FLEMING'S RIGHT-HAND RULE. It should be noted that there is also a Left-Hand Rule and to avoid later confusion in the students' mind, the point is made at this stage and the following aid is suggested for memorising the appropriate rule to suit the circumstances. Thus the Generator is studied *before* the Motor and for the average person, the use of the Right-hand is preferred *before* that of the left. Therefore use the Right-hand rule for the Generator and the Left-hand rule for the Motor. The Generator is a machine concerned with the generated or induced e.m.f. in armature conductors and thus for conditions of electrodynamic induction, we use the right-hand rule which can now be explained.

RIGHT-HAND RULE (Fleming's). Consider a conductor in a magnetic field as shown in the diagram (Fig 88a).

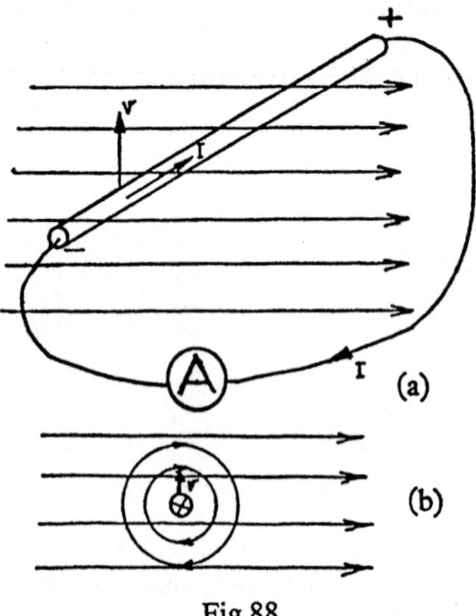

Fig 88

Imagine the magnetic field to be in the direction from left to right and the conductor to be moved at right angles and upwards with a velocity of v metres/second. The e.m.f. across the ends of the conductor is assumed as shown, *ie* the polarity is such that, if the ends of the conductor are joined externally through an ammeter, current will flow as indicated. Its direction in the conductor is seen and if attention is given to Fig 88b, it will be deduced that the field due to the conductor current is clockwise, to strengthen the field at the top and weaken the field at the bottom. Thus according to Lenz's law, opposition is offered to the motion of the conductor as one imagines the field lines concentrating or massing before the conductor, stretching and then snapping. A force of opposition to the direction of movement is apparent and the assumed polarity must be correct to confirm the action which takes place in practice. If a reversed polarity was assumed, the current would be in the opposite direction and field weakening would occur above the conductor and strengthening below. This would result in a driving force behind the conductor which would be a motoring rather than a generating condition. There is no opposition to moving the conductor and since such a condition is not possible, this alternative e.m.f. polarity assumption must be incorrect.

Since the original assumption actually is confirmed by practical conditions, the right hand can be drawn and used to find the direction of the induced current and therefore the induced polarity. This is shown in the diagram (Fig 89).

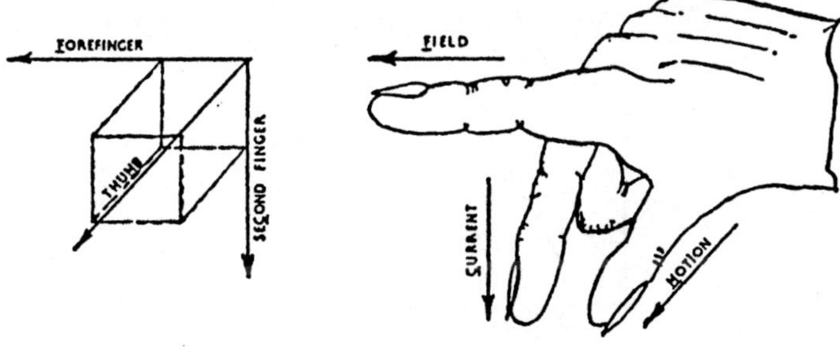

Fig 89

To use the rule, place the thumb, index finger and second finger of the right hand at right angles to each other. Point the index finger in the direction of the flux lines and the thumb in the direction of moving the conductor. The current in the conductor, due to the induced e.m.f. would be in the direction indicated by the second finger. For the example being considered (Fig 88), current would be into the paper as deduced from first principles.

THE SIMPLE MAGNETO-DYNAMO

Once the principles of electromagnetic induction were discovered, it soon became evident that the way was open to constructing a machine, in the true sense of the word, which could convert mechanical energy into electrical energy and thus generate electricity as a result of being driven by a prime-mover, such as a steam engine or water turbine. The idea of making insulated conductors move through a stationary magnetic field presented no difficulties for a small machine and so the basic construction of such a magneto-dynamo followed fundamental requirements. A typical machine is therefore, illustrated in the diagram (Fig 90), and consists of permanent magnets to provide the field and a simple coil which is mounted on but insulated from a shaft which can be rotated. In order to allow contact to be made with the moving conductors, they are connected to slip-rings which are mounted on but insulated from the shaft. Fixed *brushes'* in turn, contact the slip-rings to make sliding connections and allow an external circuit to be energised.

It will be seen that the coil consists of two 'active' conductors which have been designated AB and CD. These are connected in

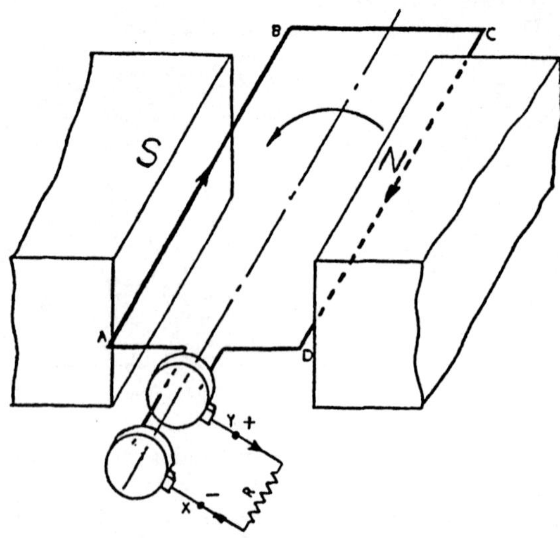

Fig 90

series by the connection BC which, together with the front connections to the slip-rings, plays no part in the generation of e.m.f. but merely serves as a means of carrying current to the external circuit. The load resistance of the external circuit has been shown as concentrated in R and is connected to the terminals X, Y of the machine.

Consider the operation of the machine as follows:
As one conductor AB moves down through the field, the other CD moves up and the induced e.m.fs. will be such that A is +ve relative to B and C is +ve relative to D. The induced current, if allowed to flow, would be as shown by the arrows and, since it is from terminal Y to terminal X through the external circuit, Y is +ve with respect to X. It would be well for the student at this point to try the right-hand rule for himself and satisfy himself as to the polarity of the terminals for the half-revolution being considered. It should be noted that the right-hand rule as described, can be applied here to conductor AB, the condition

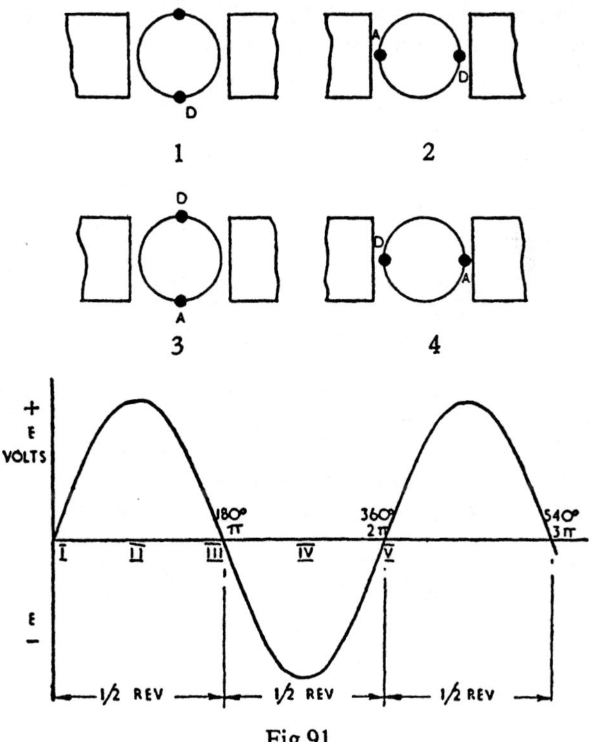

Fig 91

being that AB is moving from the top vertical position round past the centre of the magnet pole and then onto the bottom vertical position. The position where it moves past the pole at right angles is of particular importance, being a condition of maximum e.m.f.

After the coil has rotated a half revolution, conductor DC begins to move downwards and AB upwards. The polarity induced is now in reverse to that for the 1st half revolution, D being +ve relative to C and B is +ve relative to A. Terminal X is now the +ve terminal and Y is the negative. An alternating e.m.f. is generated, as shown in the diagram (Fig 91), which also illustrates four positions of the coil viewed from the slip-ring end. For position 1, A and D are moving horizontally along the field and no e.m.f. is being generated. A similar condition exists for position 3, but for positions 2 and 4 maximum e.m.f. is being generated, since field-cutting at right-angles is taking place. For intermediate positions, the general conditions of e.m.f. generation, as represented by $E \propto Blv \sin \theta$, is followed, since the conductors are cutting a uniform field at an angle but are moving at a uniform velocity. Thus the e.m.f. generated at any instant is not constant but varies and it is customary to use a small letter for what is termed the *instantaneous* value.

The expression $e = Blv \sin \theta$ volts gives the magnitude of the voltage being generated, provided the correct units are used when substituting. If the voltage is plotted to a base of revolutions, degrees or radians, a waveform such as that illustrated will be obtained.

THE SIMPLE D.C. GENERATOR

The simple magneto-dynamo machine, as described above (Fig 90), or electrical generator, as it is now called — to use the modern term; is seen to have a uniform field arranged to be cut by conductors as shown. It will provide an e.m.f. whose magnitude varies sinusoidally, that is the e.m.f. polarity and value follows a sine waveform. A sinusoidal waveform is desirable for a.c. working but for the d.c. generator, modifications are necessary to achieve a substantially constant uni-directional voltage magnitude and polarity. It is now apparent that a distinction is being made between the generation of direct current and alternating current and from here on the division between the two methods of generating, transmitting and utilising electrical energy will become marked. In this book, it is hoped that the study of both d.c. and a.c. theory will be given equal attention.

It is not proposed to enter, at this stage, into a discussion as to the relative advantages of alternating current over direct current or vice-versa; but it is stressed that the major portion of electrical theory is concerned with a.c. circuits and machines and that if later study difficulties are to be minimised, then full attention must be given to a.c. theory right from the beginning.

The first of the modifications referred to above for the d.c. machine, involves the introduction of an iron or magnetic material into the armature or moving-coil part of the assembly. The coil made up from insulated wire or strip, is wound onto an iron armature which is carried on the shaft. The magnet system is also provided with iron pole-shoes or extensions as shown in the diagram (Fig 93). Since the length of the flux path through air is now reduced to two small air-gaps, the remainder being through the iron of the armature and field system; the flux density or *B* value in the air-gaps is increased and the conductors will therefore cut a stronger field. Again as the air-gaps are now small and of constant width, the flux lines will cross them as shown and the field will be uniform over the pole-faces. The moving conductors thus pass from a small arc with substantially no magnetic flux into a large arc of constant flux density. The flux lines are seen to be radial in the gaps and are cut at right angles for most of the distance under the pole. The e.m.f. waveform is now as shown in the diagram (Fig 92), *ie* it is proportional to the flux density through which it passes.

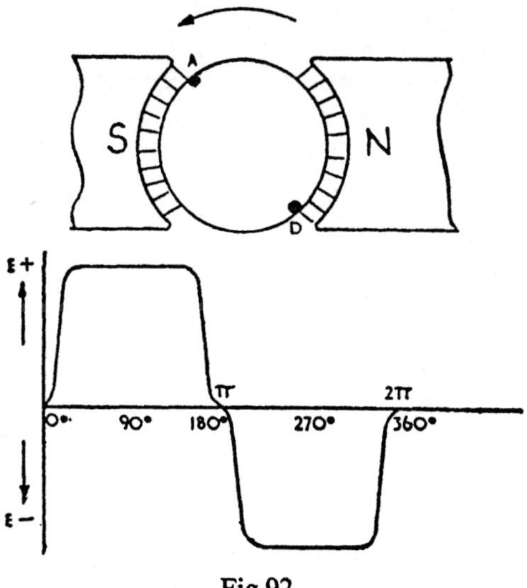

Fig 92

COMMUTATION. To obtain a constant unidirectional e.m.f. or to produce the true d.c. generator as is used in practice, the next step in modification is, to fit a form of automatic reversing switch or *commutator* which, even though the moving coil continues to generate an alternating e.m.f., ensures that a unidirectional or 'rectified' e.m.f. appears at the terminals of the machine. The diagram (Fig 93) shows how a commutator is

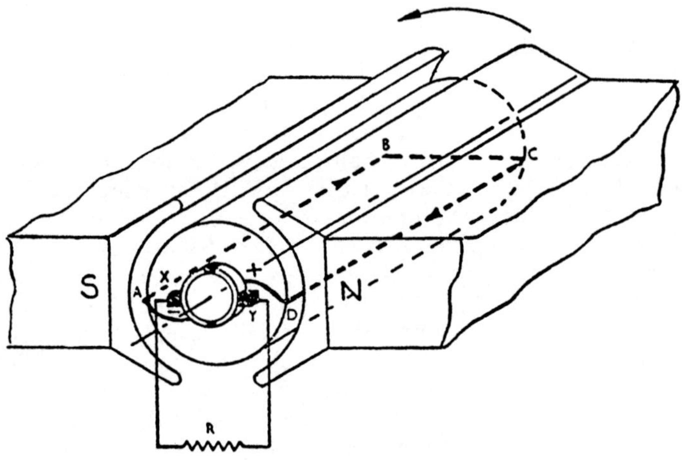

Fig 93

fitted. It consists of a metal slip-ring which is split into two parts, each insulated from the other and from the shaft. The ends of the coil are connected to each half or segment of the commutator. The stationary brushes are so adjusted that they bridge the gap in the slip-rings at the instant when the e.m.f. induced in the coil has zero value and is due to reverse.

The diagram (Fig 94) shows the side view of the commutator and the reversing action of the switching arrangement can be more clearly seen. The diagrams can be considered to be complementary to those of Fig 91 although only conditions for positions 2 and 4 are shown. It is apparent that the obvious position for the brushes is on the 'magnetic neutral axis' and that the brush Y will always be the –ve and brush X the –ve terminal. The new shape of the waveform is also shown.

For position 2 it will be seen that –ve end D of conductor CD is connected to the +ve brush Y, whereas the +ve end A of AB is connected to the –ve brush X. For position 4 when the e.m.fs have reversed in the conductors of the coil, end D which is now

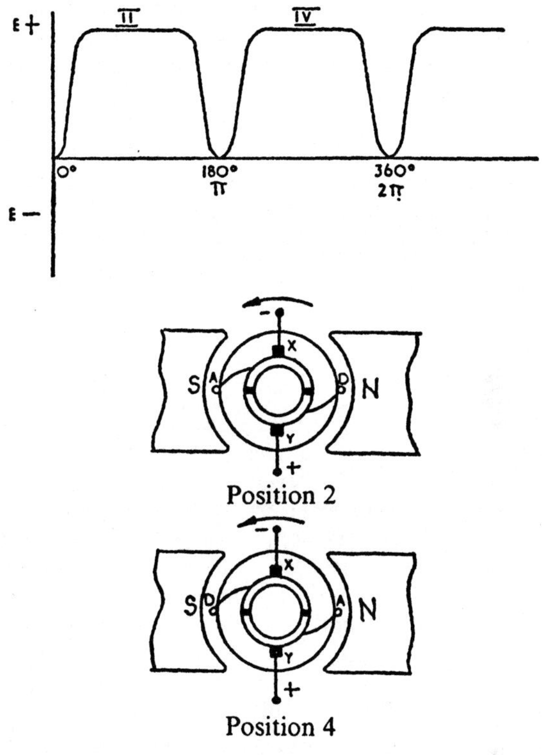

Position 2

Position 4

Fig 94

+ve is connected to the −ve brush X and end A of AB is now −ve and connected to the +ve brush Y. It is well to remind the reader here, that the brush polarity is decided by the direction of current flow in the external circuit. Thus current flows from Y (the +ve brush) to X (the −ve brush) and then to A onto B, etc. The apparent anomaly of +ve end A being connected to a −ve brush and so on is thus explained. The resulting effect of the commutating action is to produce a pulsating but unidirectional e.m.f. at the terminals of the generator.

PRACTICAL REQUIREMENTS. To obtain a more uniform e.m.f., the two-part commutator and single coil can be repeated to give an arrangement employing a greater number of segments and a larger number of coils. Each coil can consist of a number of turns to give a larger output voltage.

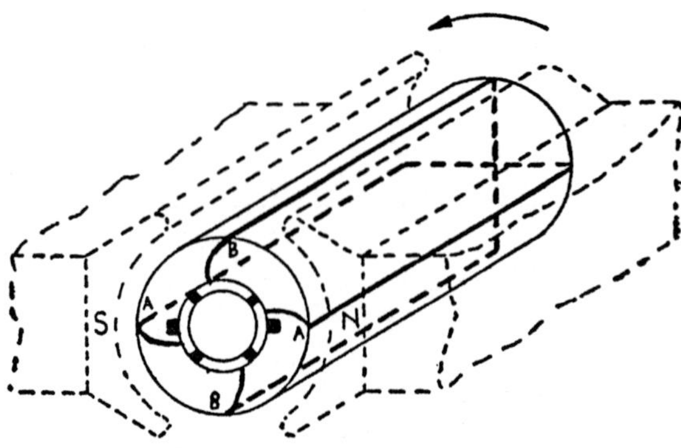

Fig 95

The example shown in the diagram (Fig 95) is an armature with two coils at right angles. It follows that for this arrangement when coil A develops maximum e.m.f., coil B generates no e.m.f. and when the armature rotates through a quarter of a revolution, the conditions would be vice-versa. The accompanying diagram (Fig 96) shows the waveforms of the generated e.m.fs. The generator terminal voltage never falls to zero but it is obvious that two distinct disadvantages are still evident. Firstly, all the conductors are not used to maximum advantage since only one coil at a time is being employed for supplying the external circuit. Secondly, but of prime importance is the new condition of commutation. Since the brushes must be placed in a position to contact the coil in which e.m.f. is being generated, it follows that if the generator is on load, *ie* supplying current, then at the instant when the connected segments leave the brush, since an e.m.f. still exists and current is flowing, arcing will take place at the brushes. If coil A is being considered and Figs 95 and 96 are noted, it will be seen that at the instant when the gap between segments is being bridged by the brushes, coil A is still cutting the field and coil B has only just entered the field. Thus coil A tends to be short-circuited by a coil in which the e.m.f. may not have risen to the required value and current will flow in the coil B. This current is diverted from the load current and also adversely affecting the commutation. If the number of coils is increased, the tendency would be to give a smoother output voltage but continued arcing at the brushes would persist. This arrangement is obviously not satisfactory.

In the early period of development of a satisfactory arrangement for the armature conductors many ideas were introduced.

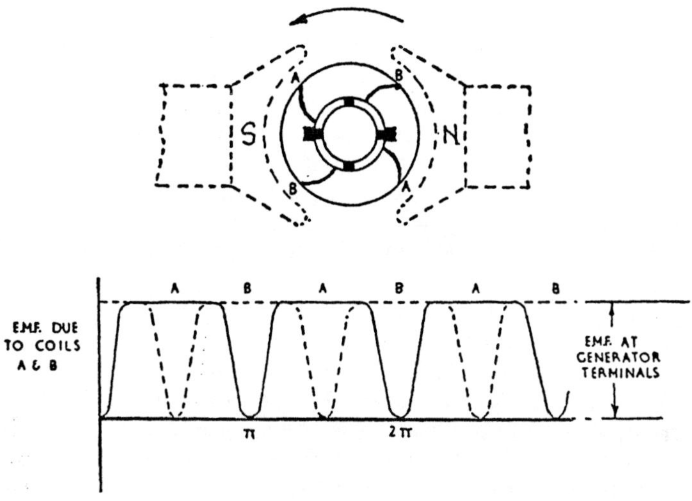

Fig 96

One such arrangement was incorporated in the Gramme-Ring Armature, which involved a special construction, in that the armature iron circuit was built up as a ring and the conductors were connected in series, with tappings being brought out to the commutator segments. The conductors were thus part of a continuous winding, but it will be noted that only the conductors on the outside of the ring are active, whereas that part of the winding which is on the inside and at the ends of the ring, cuts no flux, and is thus responsible for no e.m.f. This was a most uneconomical arrangement since it wasted conductor material. Although the Gramme-Ring Armature is shown (Fig 97), it is stressed that this never was a commercial proposition. The only reason why it is being described here is that it shows the student

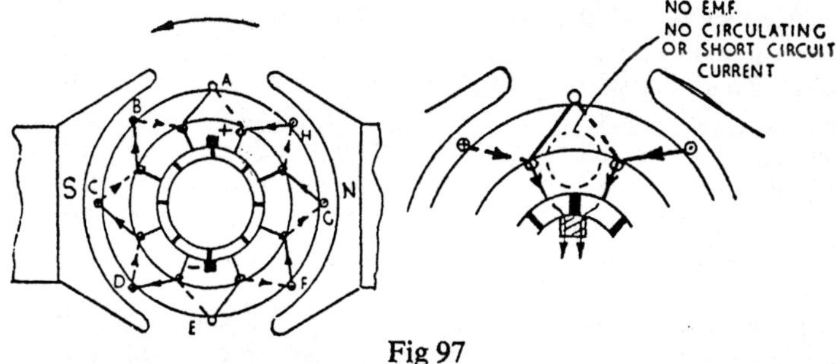

Fig 97

how a continuous closed winding can be made to operate as a suitable source of generated e.m.f. Furthermore satisfactory commutation is possible and all 'active' conductors in the air-gaps are used to advantage. It is not intended that the reader should give the arrangement undue attention nor should he consider the actual details worth remembering. All his attention should be given to the 'drum' type armature winding which is described next and dealt with more fully in books on the practical aspect of modern machine construction and operation.

The ring arrangement shows how a continuous winding is possible, in which the induced e.m.f. automatically divides the armature into two parallel paths, the flow of current being as shown. The coils in which no e.m.f. is being generated are short-circuited as they pass under the brushes and since there is no current, no sparking occurs and commutation is correct. All other active conductors are however used to advantage and are connected in series to supply the e.m.f. at the terminals. Even though the armature is rotating, the disposition of the active conductors in space can be considered to be stationary. As any one coil passes from one side of a brush to the other, it leaves the series circuit of one parallel branch of the armature winding to be replaced by a similar coil, which has just been commutated or short-circuited by passing under a brush. Thus one coil enters into the series circuit of one of the parallel paths as one coil leaves and the total e.m.f. being generated is substantially constant.

It should be noted that for a 2-pole machine the armature winding is automatically divided into two parallel paths. For a 4-pole machine the winding will be divided into four parallel paths and so on. The armature can now be looked upon as a battery made up from a number of similar cells. An armature with one hundred conductors, being wound for a 4-pole machine, will mean twenty-five conductors are in series for one parallel path and there are four such parallel paths. If 1V is generated in one conductor, the e.m.f. of one parallel path is 25V. This is also the e.m.f. of the machine, since as for a battery, the e.m.f. of the arrangement, is the e.m.f. of one parallel path. Again if the size of the conductor used for the winding is suitable for carrying 10A, then the current carried by one parallel path is 10A and the total current that can be expected from the armature is $4 \times 10 = 40$A.

THE DRUM WINDING. The diagram (Fig 99) shows the basic arrangement. This armature winding arrangement is accepted as the only modern method of connecting the active conductors together. The amount of 'copper' on the armature is used to maximum advantage since, except for the overhang at the back of a coil and the front connections to the commutator segments, the winding consists of lengths of copper conductors which are so placed to cut magnetic flux and thus generate e.m.f. The basic winding uses a number of coils in series between brushes, these coils being arranged at constant angles to each other. The resultant e.m.f. is thus more uniform and larger since many coils in series are employed. The connecting up of such coils presents difficulties; but any requirement can be realised by proper choice of coil numbers, the span of a coil, number of parallel paths, etc. An example of a simple drum winding can now be considered.

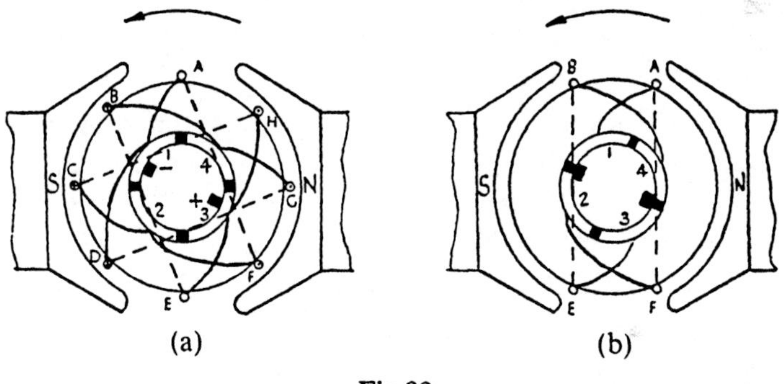

Fig 99

A, B, C, D, etc are insulated conductors fitted into slots cut into the iron of the armature. There are also four commutator segments Nos 1, 2, 3 and 4. The conductors may be connected to each other and to the segments in a variety of ways and one possible arrangement is shown in the diagram.

With rotation as shown, the e.m.f. in B, C and D are from front to back, while for F, G and H the e.m.f. are from back to front. The full lines show connection to segments and the dotted lines connections, which constitute the overhang of the coils, at the back.

Starting at the −ve brush on segment 1, current enter the armature from the external circuit and divides into two parallel paths. One path passes to A and flowing down this comes at the back to F, hence up across to segment 2 and then onto and down C to H, from where it rises up and goes to the brush on segment 3. This would be the +ve brush. The other current path is from segment 1 to conductors D, G, B, E and onto segment 3 or the +ve brush. There are thus two circuits in parallel and it will also be noticed that, as a brush passes from one segment to the next, one coil is short-circuited and the brush must be located so as to short the coil at the instant when its e.m.f. and resultant current is zero. Such an instant is shown in Fig 99b giving correct commutation conditions for the short-circuited coils between segments 1 and 2 and between segments 3 and 4.

Example 60. A slow-speed d.c. generator has an armature of diameter 3.0m and active conductors of length 510mm. The average strength of the field in the air-gap is 0.8T and the armature speed is 200 rev/min. If the armature has 144 conductors arranged in 8 parallel paths, find the e.m.f. generated at the machine terminals.

Using formula $E = Blv$ volts

then $B = 0.08$ teslas. $l = 510 \times 10^{-3}$m and v is obtained thus:

In 1 second the armature revolves $\frac{200}{60}$ or $\frac{10}{3}$ times. Also in 1 revolution one conductor travels $\pi d = 3.14 \times 3 = 9.42$m

So in 1 second the conductor travels $9.42 \times \frac{10}{3} = 31.4$m

E.M.F. generated per conductor $= 0.8 \times 510 \times 10^{-3} \times 31.4$

$= 12.8$ volts

Now there are 144 conductors in 8 parallel paths

So the conductors in series in each parallel path $= \frac{144}{8} = 18$

Thus the e.m.f. generated in 1 parallel path $= 18 \times 12.8$

$= 230.4$V.

CHAPTER 7

PRACTICE EXAMPLES

1 Calculate the e.m.f. generated in the axles of a railway train when travelling at 100km/h. The axles are 1.4m in length and the component of the earth's magnetic field density is 40μT

2. Find the generated e.m.f./conductor of a 6-pole d.c. generator having a magnetic flux/pole of 64mWb and a speed of 1000 rev/min. If there are 468 conductors, connected in six parallel circuits, calculate the total generated e.m.f. of the machine. Find also the total power developed by the armature when the current in each conductor is 50A.

3. An iron-cored coil of 2000 turns produced a magnetic flux of 30mWb when a current of 10A is flowing from the d.c. supply. Find the average value of induced e.m.f. if the time of opening the supply switch is 0.12 second. The residual flux of the iron is 2mWb.

4. A one-turn armature coil has an axial length of 0.4m and a diameter of 0.2m. It is rotated at a speed of 500 rev/min in a field of uniform flux density of 1.2T. Calculate the magnitude of the e.m.f. induced in the coil.

5. When driven at 1000 rev/min with a flux pole of 20mWb, a d.c. generator has an e.m.f. of 200V. If the speed is increased to 1100 rev/min and at the same time the flux/pole is reduced to 19mWb, what is then the induced e.m.f.?

6. A coil of 200 turns is rotated at 1200 rev/min between the poles of an electromagnet. The flux density of the field is 0.02T and the axis of rotation is at right angles to the direction of the field. The effective length of the coil is 0.3m and the mean width 0.2m. Assuming that the e.m.f. produced is sinusoidal, calculate (a) the maximum value of e.m.f. (b) the frequency.

7. A coil of 1200 turns is wound on an iron core and with a certain value of current flowing in the circuit, a flux of

4mWb is produced. When the circuit is opened, the flux falls to its residual value of 1.5mWb in 40ms. Calculate the average value of the induced e.m.f.

8. The armature of a four-pole generator rotates at 600 rev/min. The area of each pole-face is $0.09m^2$ and the flux density in the air-gap is 0.92T. Find the average e.m.f. induced in each conductor. If the armature winding is made up of 210 single-turn coils connected so as to provide four parallel paths between the brushes, find the generator terminal voltage.

9. A solenoid 1.5m long is wound uniformly with 400 turns and a small 50 turns coil of 10mm diameter is placed inside and at the centre of the solenoid. The axes of the solenoid and the coil are coincident. Calculate (a) the flux linked with the small coil when the solenoid carries a current of 6A and (b) the average e.m.f. induced in the small coil when the current in the solenoid is reduced from 6A to zero in 50ms.

10. Two coils A and B having 1000 and 500 turns respectively are magnetically coupled. When a current of 2A is flowing in coil A it produced a flux of 18mWb, of which 80 per cent is linked with coil B. If the current of 2A is reversed uniformly in 0.1s, what will be the average e.m.f. in each coil?

CHAPTER 8

ELECTROSTATICS AND CAPACITANCE

ELECTRIC FIELD

This term has already been used in association with the potential difference required to cause electron movement in a circuit. It will be dealt with in more detail under the heading of Electrostatics, but must be given some attention here, since it is directly associated with electron or current flow.

Elemental bodies such as those made of metal, have the same electrical properties of their atoms. If therefore, the atoms are charged; in that they have become ions, either through the removal of electrons making them +vely charged or through an over abundance of electrons, making them −vely charged, then the main body of which they are composed, is +vely or −vely charged. Since atoms endeavour to remain neutral, ions will always tend to acquire or discard electrons by interchange with neighbouring atoms. The same property will exist for charged bodies and if a +vely charged body (deficient in electrons) is placed in contact with a −vely charged body (excess electrons), then electron flow will occur from the second to the first or a current will flow from the first body to the second, until both bodies have the same degree of charge. Before the bodies were placed in contact with each other, a force would be detected between them and the adjacent space would show signs of such a force. The space within which this force could be detected would be an *electric field.* Thus a potential difference is said to exist between the charged bodies, which results in an electric field. When the bodies are placed in contact, an equalising of charges takes place or current flows and we have the basic requirement that a p.d. must exist between two points before a current will flow. It is not proposed here to consider how such a potential difference is created, but earlier studies have shown that an appropriate electrical device, such as a battery or generator, functions by developing an electromotive force (e.m.f.), resulting in a p.d. between its terminals and thus between the two bodies being considered. If such e.m.f. or p.d. is maintained continually by the action of the battery or generator, then a continuous current will flow. Our related studies are concerned with the broad field of electrical engineering which deals with *dynamic electricity* and its effects, and forms the main subject of this book. If however, an e.m.f. or p.d. is only maintained until the current commences to flow and then falls to zero, in accordance with the dying away of current as the charges on both bodies equalise, then our studies are concerned with the subject of *static electricity* or *electrostatics,* which forms the main subject matter for this chapter.

Before commencing with a study of electrostatics, the student is reminded that, although he has been introduced to the electron, which is a unit of charge, this is much too small for practical purposes. Experiment shows the −ve charge of an electron to equal 1.6×10^{-19} coulombs. The coulomb is the practical unit of quantity of charge and thus:

$$1 \text{ coulomb} = 6.3 \times 10^{18} \text{ electron charges.}$$

Again, passage of charges constitutes a current and the practical unit of current is the ampere, which can be defined in terms of the coulomb and the time taken for this to pass. Thus if a charge of 1 coulomb takes 1 second to pass through a point in a circuit, then the rate of flow of electricity is 1 coulomb per second and the current is 1 ampere.

Thus 1 ampere (1A) = 1 coulomb/second.

Both the coulomb and the ampere are used as units of practical engineering and have been defined at other appropriate stages in this book. The properties of conductors and insulators are also described elsewhere but can be explained in terms of the electron theory thus:

A *conductor* is a material, like metal, carbon and certain liquids, which contains mobile electrons that move under the influence of an applied potential difference and so allow the free passage of current.

An *insulator* is a material which has very few free electrons. Examples are materials, like rubber, glass, mica and most oils, in which the electrons are bound strongly to the nucleus. Since little movement of electrons occurs, current flow is negligible.

ELECTROSTATICS

Mention has been made in earlier chapters of the existence of static charges but, for the reader who is consolidating his studies at this point, we can repeat the introduction made before proceeding with further work. From earliest times the presence of electricity was known because of its simple attraction and repulsion effects. The Greeks for instance, knew that a piece of amber, when rubbed, attracted light bodies such as pieces of cork and fibrous material. The amber was said to be charged with electricity and the phenomena to be discussed here, are concerned with the presence of electric charges at rest, *ie* electrostatics.

Experiments show that the simplest method of generating static electricity is by rubbing or friction. Thus a glass rod when rubbed with silk is electrified and can attract pieces of paper, but if a similarly treated glass rod is suspended by a thread, and brought near the original charged glass rod, then a repulsion effect would be noted. An ebonite rod rubbed with fur would also be found to be charged and, if brought near to the suspended charged glass rod, attraction would be noted.

Summarising, we can say that the glass and ebonite acquire charges which can be of two types, termed positive (+ve) and negative (−ve), and that like charges repel whereas unlike charges attract. The allocation of the type, +ve charge to the glass rod and −ve charge to ebonite, is purely arbitrary, but the general theory is that all uncharged bodies consist of +ve and −ve charges which neutralise each other. If these charges are separated by some applied effort, then their presence becomes detectable and if they are caused to move from one body to another then their movement is explained by the passage of current. It will be noted that these assumptions line up with the electron theory already studied and it is apparent that a negatively charged body has an excess of electrons and a positively charged body is deficient in electrons. In the uncharged state, the atoms of the material are neutral, *ie* the charges due to the electrons and protons exactly balance.

THE ELECTROSCOPE. The deductions made from experiments in electrostatics are fundamental to theory, but to assist in the demonstrations a simple detector of charges is necessary. Such a detector, termed a gold-leaf electroscope is generally used for the investigations. This consists of two leaves of gold foil attached to a metal rod, which in turn is contained in a glass jar from which it is insulated as shown in the diagram (Fig 100). A metal disc may be fitted to the rod and the container may be a metal box-like frame with glass sides. The leaves of the electroscope are arranged to hang downwards when no charges are involved, but if a charge is imparted to the instrument the leaves are seen to diverge as shown.

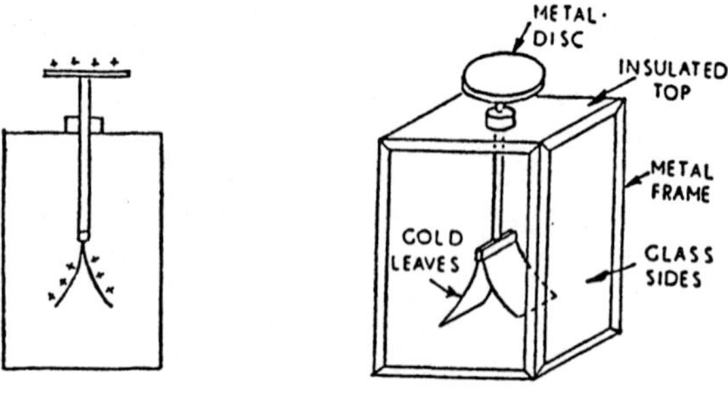

Fig. 100

The action of the electroscope is apparent when the instrument is studied. Assume a +ve charge to be given to the electroscope by stroking the disc with a glass rod which has been charged by rubbing with silk. The +ve charge imparted to the disc spreads all over the insulated metal and the leaves, having the same charges, are repelled and diverge. The condition is as shown in the diagram (Fig 100). If an ebonite rod, −vely charged by rubbing it with fur, is brought near the electroscope, the leaves will be seen to converge. The explanation is that the +ve charges on the electroscope are attracted by the −vely charged rod and rise up to concentrate in the area of the disc. The charge on the leaves thus diminishes and the repulsion force between them falls. In the same way, a +vely charged glass rod brought near the instrument will produce a further divergence of the leaves, since the +ve charges already present are repelled down towards the leaves. The density of charge in this region thus increases and increased divergence is indicated. The instrument, though seldom seen in use outside the laboratory, is useful for assisting with simple demonstrations and will be referred to frequently.

POTENTIAL DIFFERENCE

If, when two bodies are charged, as described earlier, and are brought into contact, a minute current is noted to flow between them while the charges equalise, then in the isolated and electrified state, a potential difference will exist between them. For the bodies to maintain their charges they must be insulated from earth, *ie* mounted on insulating rods. In the case of the electroscope, the charge given to it is with respect to earth; the gold leaves, rod and disc of the instrument being one body and the mass of earth being the other. Thus the leaves are charged

+ve to earth if a +ve charge is given to the electroscope. Similarly a −ve charge given to the instrument means that the leaves are −ve with respect to earth. It can thus be seen that if two bodies are charged +ve and −ve, they are at a potential to each other, *ie* a potential difference exists between them and they are also at a potential to earth. One body is +ve to earth and the other −ve to earth, the earth mass itself being considered to be at zero potential.

ELECTROSTATIC CHARGING

The effect of friction has already been mentioned and in practical engineering, it is the most important cause of electric charging. The build-up of charges can be quite considerable and precautions must be taken in the artificial silk, paper, rubber, cable-making and associated industries to discharge the bales or coils of material after they have been processed. Such processing involves, kneading, rolling, drawing, etc and the friction effects can cause large voltages to be generated, which would be uncomfortable and even dangerous to persons handling the material. The electrostatic charging of aircraft and motor-vehicles is a well-known hazard and in the case of the former, because of the large voltages possible, a means of earthing is necessary before persons can alight from the aeroplane. Electrically conducting rubber tyres have been developed to this end. For motor-cars the problem is not as important, since the charges that result are small. Nevertheless precautions are necessary for special load-carrying road vehicles such as petrol-tankers, which should be 'earthed' before unloading or loading of fuel is commenced. The action of the 'lightning-conductor' will be mentioned shortly, but its use is concerned with the +vely and −vely charging of clouds which results from atmospheric activity.

CHARGING BY INDUCTION. Imagine the electroscope to be uncharged and a +vely charged body is brought near to the disc. Then mobile electrons of the metal are attracted up into the disc and +ve charged atoms or ions are displaced to the leaves as shown in the diagram (Fig 101a). If now the disc is touched with the finger (Fig 101b), electrons from the main earth mass flow up and neutralise the ions or +ve charges and as a result the leaves collapse or converge. If next, the finger is taken off and then the adjacent +vely charged body is removed, the leaves would be seen to diverge again slightly (Fig 101c). The −ve charges, previously held by the adjacent +vely charged body, are allowed to spread all over the electroscope and charges of the same polarity spread down to the leaves.

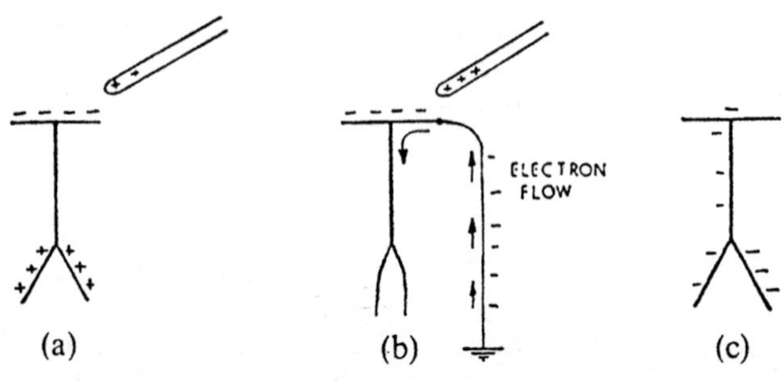

Fig. 101

It will be noted that charging by induction results in a charge of opposite polarity being produced. For the example, the inducing charge was +ve and a −ve charge resulted on the instrument. If a −ve inducing charge had been used, a +ve charge would result on the electroscope.

DISTRIBUTION CHARGE

The statements set out below are the results of experiments with a charged electroscope and a proof plane. The latter is a small metal disc fitted with an insulated handle. The proof plane is placed in contact with the body being investigated and is electrified to the same polarity. If brought near a charged electroscope, movement of the leaves with the appropriate interpretation will enable conclusions to be made. The following are some deductions made as a result of such investigations as illustrated by the diagram (Fig 102).

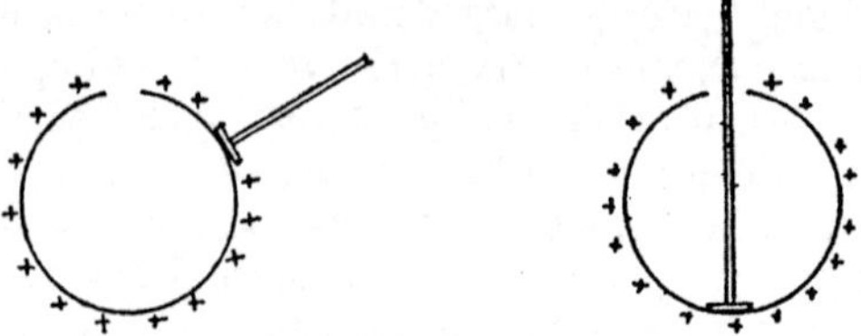

Fig. 102

1. A hollow body such as that shown, is only charged on the outside. Thus a proof plane contacted with the outer surface and presented to a charged electroscope, will show a deflection. If contacted with the inside, it will show no deflection.

2. If a sphere is charged, the charge is spread uniformly over its surface and the *surface density* is uniform. If a charged body is non-spherical, the charge concentration is found to be greatest in the region where the radius of curvature is the smallest. The diagram (Fig 103) depicts what is stated, the charge distribution or surface density being represented by the dotted envelope.

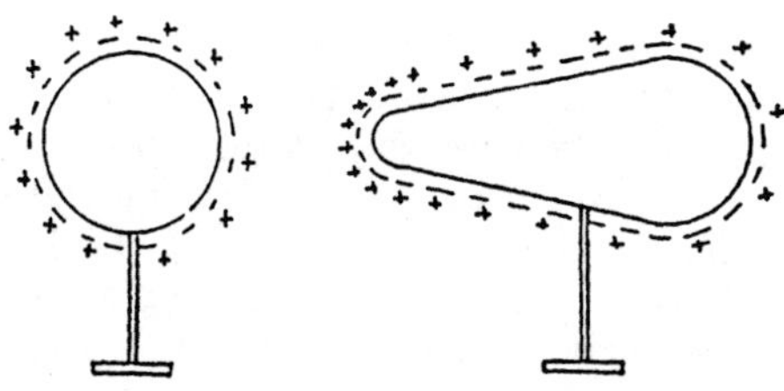

Fig. 103

3. If a charged body is shaped to have a sharp point, then the charge concentrates at the point and the surface density may be so great that dust or particles in the air, coming in contact with the body, are charged and repelled. On moving away, each particle removes a small portion of the original charge and the effect of a point on a charged body is to discharge the body. The action of the lightning conductor can now be explained.

Consider a +vely charged cloud in the vicinity of a high building which is fitted with such a lightning conductor, consisting of a copper rod and conductor which is well earthed. The building, and thus the rod, acquires a −ve charge by induction, with the air particles becoming negatively charged by contact. These particles move into space towards the cloud tending to neutralise it. Alternatively, the space between the cloud and building being charged by the −ve particles, has its insulating effect lowered until a break-down occurs and a spark (lightning discharge) passes between cloud and earth. A current will flow during the discharge as electrons pass from earth to cloud which is thus discharged with safety. The current is conducted along a suitable path, having a low resistance and damage is avoided, since minimum energy dissipation results.

ELECTROSTATIC FIELDS OF FORCE

When two bodies are charged, a force of attraction or repulsion is produced depending on the polarity of the charges. The magnitude of this force relative to the charges can be investigated, especially as a conception of the conditions appertaining is possible, if the existence of *'lines of flux'* is assumed. Since a force would be exerted on a small +vely charged body, placed adjacent to a large +vely charged body, it can be stated that an electric field of force exists in the space around the large charged body. If the small body is considered to be small enough to constitute a +ve charge then, if free to move, it would travel in a definite direction in the electric field and the path traced by it would represent a line of flux. Thus the large +vely charged body can be considered to have many lines of flux passing out from it, The similarity with the representation used for the magnetic field will be noted. In considering the electrostatic fields shown below in the diagram (Fig 104), one important fact should be observed. Each line of flux terminates at the surface of the charged body and does not pass through the body itself to form a closed path, as in the case of a magnetic line of flux. The medium through which the lines of electric flux pass is called the *'dielectric'* and the lines are shown to terminate at the surface of another body where balancing charges of opposite polarity appear. Thus Fig 104a shows a +vely charged metal sphere in the centre of a room. Lines leave the surface perpendicularly, in all directions and planes, to terminate on the earthed walls, floor and ceiling, the earth mass being negative to the charged body.

Fig 104b, shows the field arrangement associated with two bodies when charged in opposition, and Fig 104c shows the field with like charges on the bodies. Fig 104d, shows the

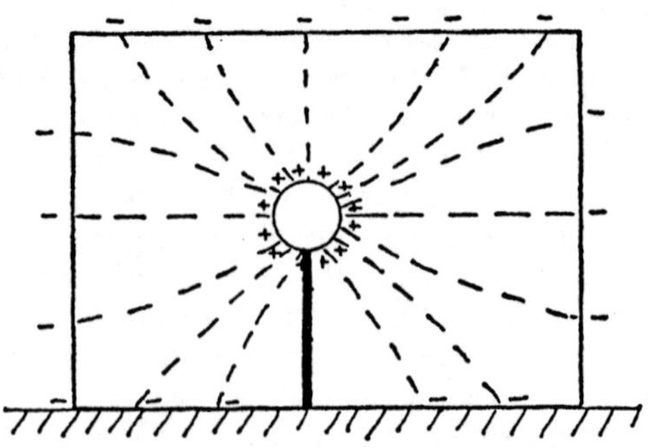

Fig. 104(a)

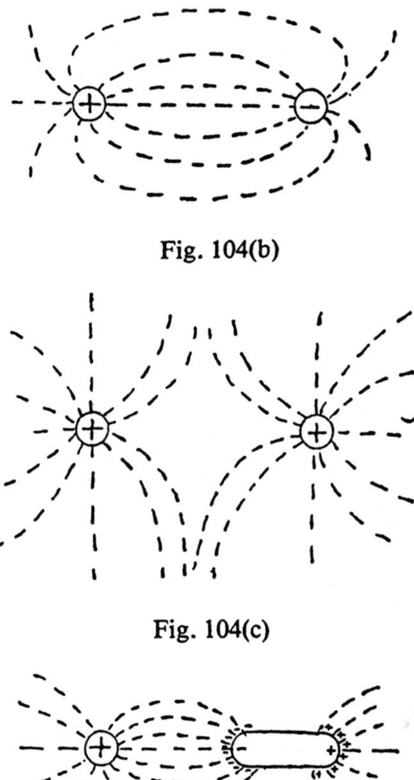

Fig. 104(b)

Fig. 104(c)

Fig. 104(d)

arrangement when an uncharged body is placed in the field and how induced polarities result. As for the magnetic field, the lines of flux can be imagined to be elastic threads which tend to contract if allowed to do so.

ELECTROSTATIC FLUX

As for the magnetic field so for the electric field, we can introduce the term flux. The symbol Ψ (the Greek letter psi) and the number of electrostatic lines of flux, passing through a particular medium, is called the flux. Since the practical unit of charge is the coulomb, then in order to establish an electrostatic unit we consider one line of flux to emanate from 1 coulomb. Thus a charge of 10 coulombs will have in association a flux of 10 lines, or Ψ will equal 10 coulombs. Since most practical electrostatic work is concerned with *Capacitors* which are made up from flat plate-like conductors, placed adjacent and parallel to each other and as the medium between the plates or dielectric carries the flux, it is convenient to introduce the term *'electric flux density'* — symbol D. Thus $D = \frac{\Psi}{A}$ where A is the area of the dielectric in square metres. Then $D = \frac{\Psi}{A}$ or $\frac{Q}{A}$ coulomb per square metre.

ELECTRIC POTENTIAL

The basic idea of electric potential has already been introduced when it was seen that, if two bodies are charged and connected together, then as a current flowed while the charges equalise, there must have been a difference of potential between them. Again, since the earth mass can be taken as being at zero potential then, if a body charged with Q coulombs of electricity is connected to earth, a current will flow. The current will be from the body to earth if it is +vely charged, and from earth to the body if −vely charged. The direction of the current is not important but work is done during this period. If we consider that 1 joule of work is done whilst 1 coulomb is passed, then the potential difference of electric potential of the body must have been 1 volt. Voltage is therefore used as a measure of the potential difference between the body and earth.

The reasoning already made can also be applied to the practical capacitor, when the plates are charged with respect to each other. The condition of a charged capacitor can usefully be considered to enable further deductions. If a +ve unit charge is placed in the field between the charged capacitor plates, a force will urge it towards the −ve plate. The force is taken as a measure of the *intensity* or *strength* of the field. The symbol for electric field strength or field intensity is E and the force is measured in newtons. Thus a charge of Q coulombs placed in an electric field of intensity E, will experience a force of $F = QE$ newtons or $E = \frac{F}{Q}$ newtons per coulomb.

An alternative method of considering the strength of the field is to take it as an electric force or electrical potential gradient, when it can be measured in volts per metre. This is the more usual unit.

Thus, $E = \frac{V}{d}$ where d is the distance between the plates in metres.

Both expressions for electric field intensity and potential gradient or electric force are numerically the same. If practical units are substituted, the same amount of work would be done by 1 newton of force acting through 1 metre distance between the plates, as would be done by 1 coulomb being conveyed by a pressure of 1 volt.

Thus, since $E = \frac{F}{Q}$ or $\frac{V}{d}$ then $\frac{F}{Q} = \frac{V}{d}$ or $Fd = VQ$

Note.
1 joule = 1 volt × 1 coulomb, or 1 joule = 1 newton × 1 metre.

THE CAPACITOR

Numerous references have already been made to the electrical capacitor or condenser, as it used to be called originally. In its simplest form, it consists of two metal plates separated from each other by a film of air. As will be seen later, the area of the plates, distance of separation and nature of the dielectric (insulating medium), all influence the performance of a capacitor, but the basic fact learned from any experiments made with such an arrangement is that it can store electricity. Thus if the plates are connected to a source of supply through a sensitive milliammeter, then a current will be seen to pass at the instant of closing the switch. The current will quickly fall to zero, because the potential difference between the plates will be seen to rise, as indicated by an appropriate voltmeter. The form of current fall-off or decrement, forms a separate item of investigation, but the capacitor can now be assumed to have attained a 'charged' state. If the supply is disconnected and the plates are shorted together, a discharge current will be found to flow which will be opposite in direction, and although initially large, will soon decay away to zero. The initial voltage, although showing the value of the charging supply, will also fall away to zero.

Experiments with a simple capacitor will soon establish the basic relationship between quantity of electricity that can be stored and the charging voltage. The former will be found to be proportional to the latter or $Q \propto V$. Since this is direct proportionality, a constant can be introduced to give the expression:

$$Q = CV$$

C is termed the capacitance and a unit is possible by simply defining it in terms of unit quantity and unit voltage. Thus the unit of capacitance is the farad and a capacitor is said to have a capacitance of 1 farad if 1 coulomb of electricity is stored when 1 volt is applied across the plates.

Example 61. Find how many electrons are displaced when a potential difference of 500V is established between the plates of a 4μF capacitor.

Since $Q = CV$

then $Q = 4 \times 10^{-6} \times 500 = 2 \times 10^{-3}$ coulombs

But 1 coulomb $= 6.3 \times 10^{18}$ electrons

$\therefore$ No of electrons $= 12.6 \times 10^{15}$.

CAPACITOR SYSTEMS

Such units can be connected in series or parallel and the student should compare the expressions giving equivalent capacitance values with those giving equivalent resistance values, for comparable arrangements.

SERIES CONNECTION. The arrangement is shown in the diagram (Fig 105).

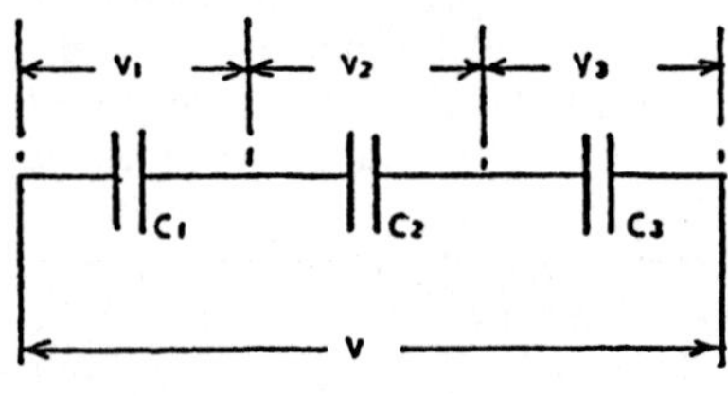

Fig. 105

Let the capacitors have values of C_1, C_2 and C_3 farads respectively, and let the applied voltage V be dropped as shown.

Then since $V = V_1 + V_2 + V_3$

and since $V_1 = \dfrac{Q_1}{C_1}$ $V_2 = \dfrac{Q_2}{C_2}$ and $V_3 = \dfrac{Q_3}{C_3}$

we can write:

$$V = \frac{Q_1}{C_1} + \frac{Q_2}{C_2} + \frac{Q_3}{C_3}$$

If C is taken to be the equivalent capacitance of the arrangement then

$$V = \frac{Q}{C}$$

or $$\frac{Q}{C} = \frac{Q_1}{C_1} + \frac{Q_2}{C_2} + \frac{Q_3}{C_3}$$

but the same current flows through each capacitor for the same time. $\therefore Q = Q_1 = Q_2 = Q_3$ and the above can be simplified to:

$$\frac{1}{C} = \frac{1}{C_1} + \frac{1}{C_2} + \frac{1}{C_3} \text{ etc.}$$

PARALLEL CONNECTION. The arrangement is shown in the diagram (Fig 106), from which it will be seen that the same voltage is applied to each capacitor.

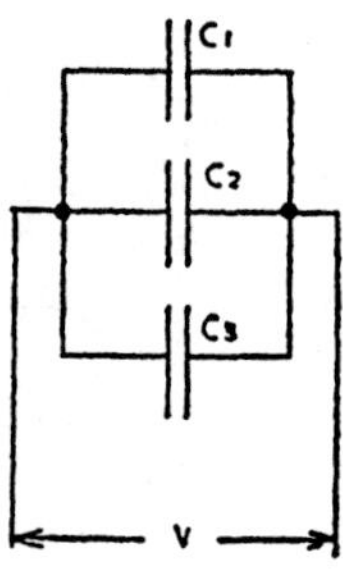

Fig 106

Then for each capacitor $Q_1 = C_1V \quad Q_2 = C_2V \quad Q_3 = C_3V$

If the total quantity of charge $= Q$ then obviously

$$Q = C_1V + C_2V + C_3V = V(C_1 + C_2 + C_3)$$

$$\text{or } \frac{Q}{V} = C_1 + C_2 + C_3$$

If C is taken as the equivalent capacitance of the arrangement, then $Q = CV$ or $CV = V(C_1 + C_2 + C_3)$

when $C = C_1 + C_2 + C_3$.

Example 62. If two capacitors of values 100μF and 50μF respectively, are connected (a) in series (b) in parallel, across a steady applied voltage of 1000V, calculate the joint capacitance

(a) Series. Joint capacitance C is given by:

$$\frac{1}{C} = \frac{1}{100} + \frac{1}{50}$$

$$C = \frac{100}{3} = 33.33\mu\text{F}.$$

(b) Parallel. Joint capacitance is given by $C = 100 + 50$ or $C = 150\mu\text{F}$.

CAPACITOR CURRENT. From the relation $Q = CV$, the following can also be deduced.

Since $Q = It$ then $It = CV$

$$\text{or } I = C\frac{V}{t}$$

The expression shows that current only flows when the voltage across a capacitor is changing, since $\frac{V}{t}$ represents a rate of change of voltage. The current at any instant can also be found, if the rate of change of the voltage is known at that instant. If however, the rate of change is uniform for a period of time, then a constant current will flow. This is illustrated by the example.

Example 63. The p.d. across the plates of a $50\mu F$ capacitor varies thus:

From time $t = 0$ to $t = 1$ millisecond, V rises uniformly from 0 to 200 volts

From time $t = 1$ to $t = 3$ millisecond, V is constant at 200 volts

From time $t = 3$ to $t = 5$ millisecond, V falls uniformly from 200 volts to 0

Illustrate the voltage variations on a graph and deduce the shape of the current wave during the period of 5 milliseconds.

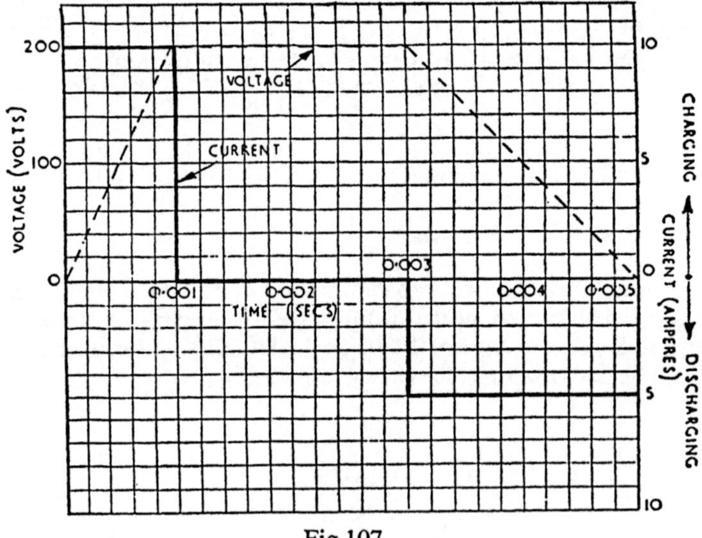

Fig 107

$$\text{Since } Q = CV = It \quad \text{then } It = CV \quad \text{or } I = C\frac{V}{t}$$

(a) $I = 50 \times 10^{-6} \times \dfrac{(200 - 0)}{0.001} = 10\text{A}$ (charging)

(b) $I = 50 \times 10^{-6} \times \dfrac{(200 - 200)}{0.002} = 0$ amperes

(c) $I = 50 \times 10^{-6} \times \dfrac{(0 - 200)}{0.002} = -5\text{A}$ (discharging)

The required graphs are shown on the attached diagram (Fig 107).

ENERGY STORED IN AN ELECTRIC FIELD OR DIELECTRIC

Consider the voltage to rise uniformly across the plates of a capacitor, to a value of V volts, in a time of t seconds. The average value of the p.d. would then be $\frac{V}{2}$ volts and the charging current would be constant, of value equal to I amperes. The average power supplied during the charging period would be $\frac{V}{2} \times I$ watts and the energy fed in would be $\frac{V}{2} \times I \times t$ joules.

This energy is not converted into heat, since a capacitor has no resistance, but does work in establishing the electric field.

It is this energy which is stored, and can be recovered when the field collapses as the capacitor discharges. Thus:

$$\text{Energy stored} = \frac{V}{2} It \text{ joules or} = \frac{V}{2} Q \text{ joules}$$

$$\text{or alternatively, } W = \tfrac{1}{2}CV^2 \text{ joules.}$$

Example 64. Consider the capacitor arrangement of Example 62 and calculate the total energy stored for a steady applied voltage of 1000V, for both the series and parallel connection.

(a) Series.

$$\begin{aligned} \text{Energy stored is given by } W &= \tfrac{1}{2}CV^2 \text{ joules} \\ &= \tfrac{1}{2} \times 33.33 \times 10^{-6} \times 1000^2 \\ &= 16.67\text{J} \end{aligned}$$

(b) Parallel.

$$\begin{aligned} \text{Energy stored or } W &= \tfrac{1}{2} \times 150 \times 10^{-6} \times 1000^2 \\ &= 75\text{J.} \end{aligned}$$

It will be noted that the equivalent capacitance value for each arrangement has been used for C in the energy expression.

RELATIVE PERMITTIVITY

If a parallel-plate capacitor is made up so that one plate is connected to earth and the other to an electroscope, the effect of altering the nature of the dielectric can be observed. With air as the insulating medium between the plates, the capacitor can be charged to a given value as would be indicated by the amount of divergence of the leaves of the instrument. If now a sheet of insulating material, such as a slab of paraffin-wax, was interposed into the air-gap, a converging effect of the leaves is produced, indicating that the charge appears to have been reduced. This can be shown to be an incorrect supposition since, if the insulation is removed the leaves again diverge to the original extent. An alternative assumption to be made is that the capacitance of the arrangement has been increased, *ie* the capacitor is now capable of accepting a greater amount of charge for the same divergence of the leaves. This is confirmed if the capacitor is charged to an amount giving the original extent of divergence of the leaves, and the wax insulation is then removed. The leaves would be seen to diverge to a far greater extent, showing a larger charge to have been imparted to the arrangement. The experiment shows that capacitance can be varied by the nature of dielectric or by a property which is termed its *permittivity.* This property can be likened to the permeability of a magnetic substance, and at this stage is considered as relative only. It can thus be defined quite easily, as the ratio of the capacitance of a capacitor, with the material being considered as the dielectric, to the capacitance of the same capacitor with air — or more truly a vacuum, as the dielectric. Another term for relative permittivity is the *dielectric constant.* The symbol used is ϵ_r, (the Greek letter — small Epsilon).

Typical values for relative permittivity are Air 1.0006, Paraffin Wax 2.2, Mica 4.5 to 8, Glass 4 to 10.

ABSOLUTE PERMITTIVITY

It is not proposed to do much extensive work on dielectrics as used for capacitors or cables, but the absolute value of permittivity needs to be considered, since ϵ_r as introduced is only relative to air. As for the magnetic circuit, we can write:

Absolute permittivity = relative permittivity × permittivity of free space

$$\text{Thus } \epsilon = \epsilon_r \times \epsilon_o$$

PERMITTIVITY OF FREE SPACE

Some comparisons have already been made with the magnetic circuit and permittivity can best be understood by comparing it with permeability — its magnetic equivalent.

Since permeability was defined as the ratio $\dfrac{\text{Flux density}}{\text{Magnetising force}}$

$$\text{or } \mu = \frac{B}{H}$$

Then similarly; Permittivity $= \dfrac{\text{Electric flux density}}{\text{Electric force}}$ or $\epsilon = \dfrac{D}{E}$

For air, permittivity is measured to be

$$\frac{1}{4\pi \times 9 \times 10^9} \text{ farads per metre}$$

Thus $\epsilon_o = 8.85 \times 10^{-12}$ farads per metre.

Although other alternatives can be derived to allow estimation of the value of ϵ_o, it will be seen that the following deduction will give the required answer, if unity is taken for the dimensions of area A and spacing d between the plates.

$$\text{Then since } \epsilon_0 = \frac{D}{E} = \frac{Q/A}{V/d} = \frac{Q}{V} \times \frac{d}{A}$$

$$\text{or } \epsilon_0 = \frac{CV}{V} \times \frac{d}{A} = \frac{Cd}{A}$$

When d and A are made equal to 1, ϵ_0 = the capacitance value of the arrangement. For a vacuum, the capacitance value of the standard capacitor, using unity for A and d, will be measured to be 8.85×10^{-12} SI units.

$$\text{or } \epsilon_0 = \frac{1}{4\mu \times 9 \times 10^9} \text{ which is also expressed as}$$

8.85×10^{-12} farads per metre.

Note. Although a vacuum has been mentioned for the above capacitor arrangement, air can be taken as the dielectric, since the effect of the variation is small enough to be neglected.

CAPACITANCE OF PARALLEL-PLATE CAPACITOR

Consider the area of the plates to be A square metres and their spacing to be d metres, *ie* this would be the thickness of the dielectric. The applied voltage can be taken as V volts, resulting in a charge of Q coulombs. Since the charge Q is assumed to be uniformly distributed over the whole area of the plates, the electric flux density D would be $\frac{Q}{A}$.

The electric force or potential gradient E in the dielectric would be $\frac{V}{d}$ volts per metre and permittivity ϵ would be, by definition, $= \frac{D}{E}$.

$$\text{Thus } \epsilon = \frac{D}{E} \text{ or } \epsilon = \frac{Q/A}{V/d} = \frac{Qd}{VA}.$$

$$\text{Whence } \epsilon = \frac{CVd}{VA} = \frac{Cd}{A} \text{ or } C = \frac{\epsilon A}{d} \text{ but } \epsilon = \epsilon_0 \epsilon_r$$

$$\text{So } C = \frac{\epsilon_0 \epsilon_r A}{d} \text{ farads.}$$

Example 65. A capacitor consists of two parallel metal plates, each 300mm by 300mm, separated by a sheet of polythene 2.5mm thick, having a relative permittivity of 2.3. Calculate the energy stored in the capacitor when connected to a d.c. supply of 150V.

$$C = \frac{8.85 \times 10^{-12} \times 2.3 \times (300 \times 10^{-3})^2}{2.5 \times 10^{-3}} \text{ farads}$$

$C = 733 \times 10^{-12}$ F
$C = 733$ pF

$$\text{Energy stored} = \tfrac{1}{2}CV^2 \text{ joules}$$
$$= \tfrac{1}{2} \times 733 \times 10^{-12} \times 150^2$$
$$= 8.25 \times 10^{-6} \text{ joules}$$
$$= 8.25\ \mu\text{J}$$

Example 66. A capacitor of 5μF charged to a p.d. of 100V, is connected in parallel with a similar uncharged capacitor. What quantity of electricity would flow into the second capacitor and to what voltage would it be charged?

Consider the first capacitor designated by the letter A, then since $Q = C_A V$, $Q = 5 \times 10^{-6} \times 100 = 5 \times 10^{-4}$ coulombs.

When the capacitor B is connected across A, charge will pass from A to B until the potential of each is the same. The arrangement of capacitors can now be considered as a parallel connection or the joint capacitance is the same as that of 1 unit of 10μF.

Applying the formula $Q = CV$

$$\text{Then } V = \frac{Q}{C} = \frac{5 \times 10^{-4}}{10 \times 10^{-6}}$$
$$\text{or } V = 50 \text{ volts}$$

The final voltage will be 50V.

This could have been deduced more simply from the fact that the capacitors are similar and charge will pass from A to B until the potential of each is the same.

TRANSIENT EFFECTS IN D.C. CIRCUITS

In this chapter we have considered the factors affecting capacitance, and charge on capacitor plates, without reference to their effects in practical circuits. In Chapter 10 we shall return to the effects of capacitance in a.c. circuits but, at this stage it is appropriate to consider the application of a d.c. voltage to a capacitive circuit.

In a resistive d.c. circuit the current rises to its final steady state and the p.d. across it settles to a steady value almost instantaneously. Similarly on opening the circuit, current and p.d. fall instantaneously to zero. However, when a capacitor and resistor, or an inductor and resistor, are connected in series to a d.c. supply, it is found that the final steady state of current and p.d. is not achieved immediately. The change takes a short period of time depending upon the circuit component values. These changes of current and p.d. up to the final steady state conditions are said to be transient values. In volume 7 a more detailed explanation will be made of growth and decay of voltages and currents in d.c. circuits but an initial explanation is made here to enable the introduction to exponential curves.

CAPACITOR IN A D.C. CIRCUIT

Consider Fig 108 *C* is an unchanged capacitor in series with a resistor *R* connected to a d.c. supply *V*.

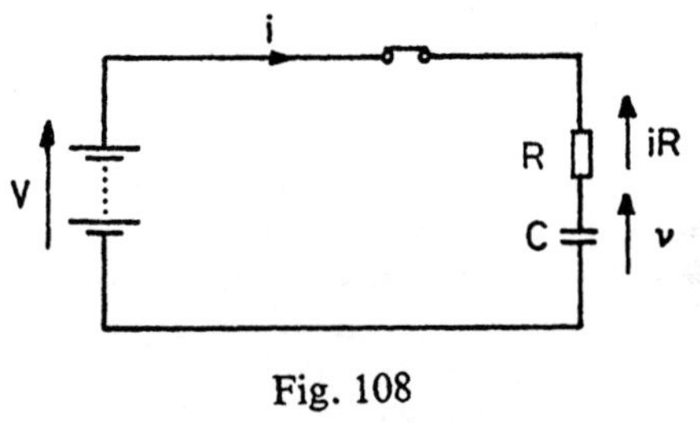

Fig. 108

$$V = p.d. \text{ across } R + p.d. \text{ across } C$$
$$V = iR + v$$
$$\therefore i = \frac{V - v}{R}$$

At the instant of switch on, the instantaneous value of voltage across the capacitor (v) is zero, and the current (i) is maximum, being only limited by the resistance *R*.

Hence on switch on $i = \frac{V}{R}$, but this current is the charging current of the capacitor and, as the capacitor charges, v increases which in turn decreases i. The rate of charge therefore decreases until, when the capacitor is fully charged, the current is zero. However zero current is achieved at infinite time but, for practical purposes, it can be assumed that this occurs in a time equal to five times the initial rate of charge. If the initial rate of charge had been maintained, the capacitor would have been fully charged in a definite time depending upon the circuit components. This time is referred to as the Time Constant and is given the symbol τ (Greek letter TAU)

$$\text{Thus } \tau = CR \text{ seconds}$$

Since the rate of charge decreases a graph of Capacitor Voltage against time is non-linear, but it does follow an exponential law as shown in Fig 109.

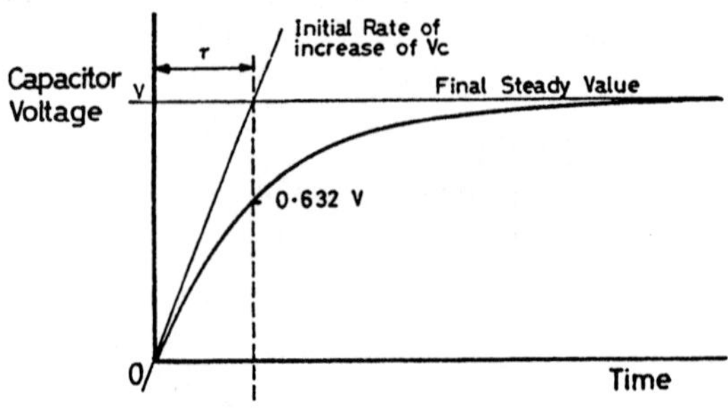

Fig. 109

The equation for this line is:

$$v = V(1 - e^{-t/\tau})\text{ volts}$$

Where v = Instantaneous p.d. across capacitor
V = Applied circuit voltage
t = Time from switch on
τ = CR = Time constant

It should also be noted that the capacitor voltage will be only 0.632 of the final steady value after a time equal to one time constant. Similarly during a second time constant, the voltage would only increase by 63.2 per cent of the remaining voltage, and each subsequent time interval thereafter voltage would similarly increase. Since current i is a maximum value when capacitor voltage v is zero, and the voltage increases exponentially; it follows that the current will decrease exponentially, as shown in Fig 110.

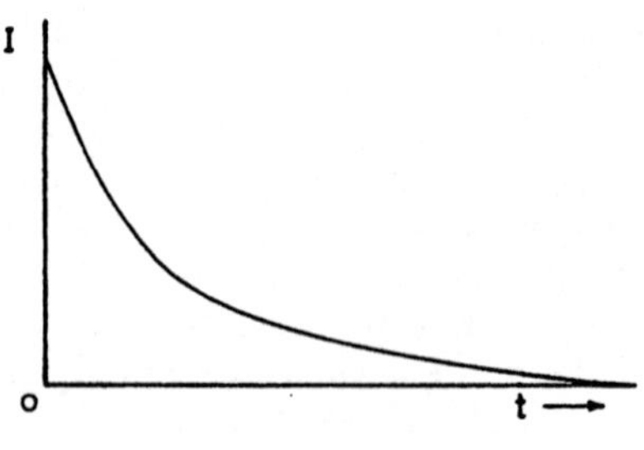

Fig. 110

The equation for this curve is:

$$i = Ie^{-t/\tau} \quad \text{or} \quad i = \frac{V}{R} e^{-t/\tau}$$

Consider now the fully charged capacitor C in Fig 111 being discharged through resistor R.

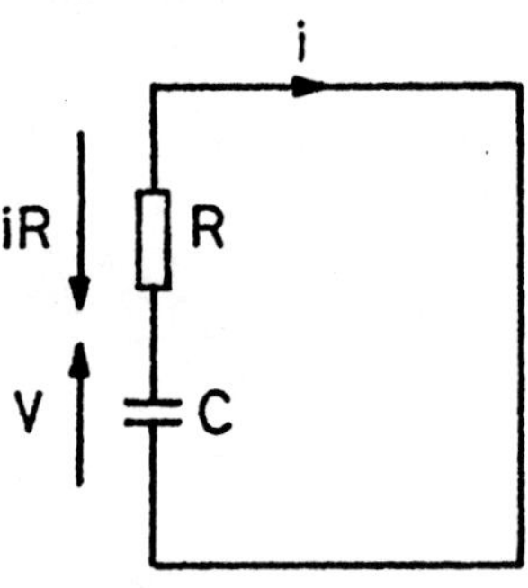

Fig. 111

Capacitor C acts as a source of supply and, as it discharges, the p.d. across it falls and once again the curve is exponential.

Since the current is limited by the resistance $\left(i = \frac{v}{R}\right)$, at the start of discharge the current is maximum $\left(i = \frac{V}{R}\right)$, thus the current curve also follows a falling exponential curve.

Hence on discharge:

$$v = Ve^{-t/\tau}$$

$$\text{and } i = Ie^{-t/\tau}$$

Example 67. A 40 kΩ resistor and a 20 μF capacitor are connected in series to a 200V d.c. supply. Find the circuit current and the p.d. across the capacitor after 0.2 seconds from switch on.

$$\text{Time Constant } \tau = CR = 20 \times 10^{-6} \times 40 \times 10^{3}$$
$$= 0.8 \text{ s.}$$

$$i = \frac{V}{R} e^{-t/\tau} = \frac{200}{40 \times 10^{3}} e^{-0.2/0.8}$$
$$= 5 \times 10^{-3} e^{-0.25}$$

$\therefore$ After 0.2 s $\quad i = 3.89$ mA

$$v = V(1 - e^{-t/\tau}) = 200\,(1 - e^{-0.2/0.8})$$

$\therefore$ After 0.2 s $\quad v = 44.24$ volts

Example 68. A 20 μF capacitor, fully charged to a voltage of 300V, is discharged through a 1 MΩ resistor. Find the time taken for the capacitor voltage to fall to 60 volts.

$$\tau = CR = 20 \times 10^{-6} \times 1 \times 10^{6}$$
$$= 20 \text{ s}$$
$$v = Ve^{-t/\tau}$$
$$60 = 300e^{-t/20}$$
$$0.2 = e^{-t/20}$$
$$\frac{-t}{20} = -1.61$$
$$t = 32.2 \text{ s.}$$

CHAPTER 8

PRACTICE EXAMPLES

1. Two capacitors of 0.02μF and 0.04μF are connected in series across a 100V d.c. supply. Find the voltage drop across each unit.

2. For the circuit shown, calculate the effective capacitance between A and B. The capacitance values shown, are in microfads.

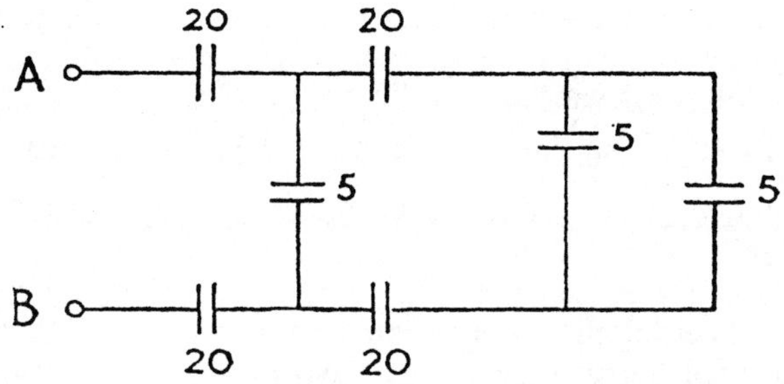

3. A variable capacitor having a capacitance of 1000μF is charged to a p.d. of 100V. The plates of the capacitor are then separated by means of an insulated rod, so that the capacitance is reduced to 300μF. Find, by calculation, whether the p.d. across the capacitor will have changed.

4. A plate capacitor consists of a total of 19 metal-foil plates each 2580mm^2 and separated by mica 0.1mm thick. Find the capacitance of the assembly if the relative permittivity of mica is 7.

5. A p.d. of 10kV is applied to the terminals of a capacitor consisting of two circular plates, each having an area of 10 000mm^2, separated by a dielectric 1mm thick. If the capacitance is $3 \times 10^{-4}\mu$F, calculate the electric flux density and the permittivity of the dielectric.

6. A capacitor consists of two parallel metal plates, each 200mm by 300mm, separated by a sheet of polythene 3.5mm thick, having a relative permittivity of 3.0. Calculate the energy stored in the capacitor when connected to a d.c. supply of 300V.

7. Calculate the capacitance value of a capacitor which has 10 parallel plates separated by insulating material 0.3mm thick. The area of one side of each plate is $1500mm^2$ and the relative permittivity of the dielectric is 4.

8. Two capacitors A and B having capacitances of $20\mu F$ and $30\mu F$ respectively are connected in series to a 600V d.c. supply. Determine the p.d. across each capacitor. If a third capacitor C is connected in parallel with A and it is then found that the p.d. across B is 400V, calculate the value of C and the energy stored in it.

9. A d.c. voltage of 500V is applied to a $40\mu F$ capacitor. Find the value of the charging current at the instants when the voltage is varying as follows:

Time $\left(\frac{1}{1000}\text{ sec.}\right)$	0–1	1–2	2–3	3–4	4–5
Voltage values	0–100	100–150	160 const	150–50	50–0

10. A single-phase concentric cable takes a charging current of 1A per kilometre, when connected to 11kV, 50Hz mains. The diameter of the inner core is 12 mm and the insulation has a radial thickness of 10mm. Calculate the permittivity of the insulating material.

CHAPTER 9

BASIC A.C. THEORY

Introduction is made by quickly revising the relevant fundamentals of Chapter 7. The diagrams (Fig 112a and b) show an elementary form of a.c. generator in which a coil is rotated in a uniform magnetic field. The sides of the coil, *ie* conductors, cut the magnetic flux and thus, an e.m.f. is induced which, from first principles is $e = Blv$ volts. The letter e, for the value of induced e.m.f., has been introduced here because this value is

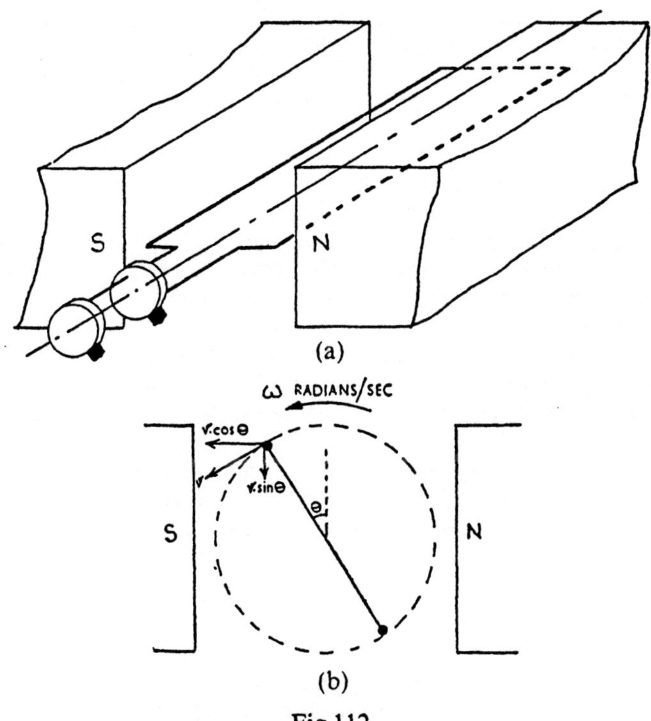

Fig 112

not constant but, as will be seen, varies from instant to instant. Thus even though the coil is rotated at a uniform velocity v, the rate of cutting is not constant, but depends upon the angle at which the conductors cut flux. The velocity can be resolved into a cutting component ($v \sin \theta$) and a non-cutting component ($v \cos \theta$). The cutting velocity component only is responsible for e.m.f. and can be used to give a general expression which gives the e.m.f. at any instant, as $e = Blv \sin \theta$ volts.

THE A.C. WAVEFORM

In the expression $e = Blv \sin \theta$, as for any alternator, B, l and v can be assumed to be constant and made equal to K. The expression now becomes $e = K \sin \theta$ and a further value for K can be obtained if we consider the instant when the coil sides are cutting the field at right angles. Velocity component $v \sin \theta$ generates a maximum e.m.f. which can be designated as E_m. At this instance $e = E_m$ and we can write $E_m = K \sin \theta$. But $\sin \theta = \sin 90 = 1$. $\therefore E_m = K$.

Substituting back in the expression, we have:

$$e = E_m \sin \theta$$

The above is an important equation which shows that the generated e.m.f. varies sinusoidally. e is termed the *instantaneous* value and E_m the *maximum* value.

If attention is now turned to a waveform plotted to a time or angle base, it will be remembered, from work already done elsewhere on vectors, that a sine wave can be deduced from the vertical component of a rotating vector—for electrical work, such a rotating vector is called a *phasor*. If the length of the phasor is made to represent E_m, then for any angle θ, the instantaneous value is the vertical projection and this also can be used as an ordinate for the waveform, when plotted to an angle or time base. The diagram (Fig 113) illustrates the procedure for deducing a waveform and the method is summarised thus:
Draw a circle of radius made equal to the maximum value of the wave. Starting from the horizontal, move the phasor through a known angle and project the vertical value onto an angle or time scale. Choose suitable scales so as not to distort the sinusoidal shape of the wave. The following reminder shows the connection between the construction and the representation of a sinusoidally induced e.m.f. Since, from the triangle illustrated,

$$\frac{e}{E_m} = \sin \theta \text{ so } e = E_m \sin \theta.$$

This expression is in the form deduced previously but can be further modified to suit the representation by a phasor. In

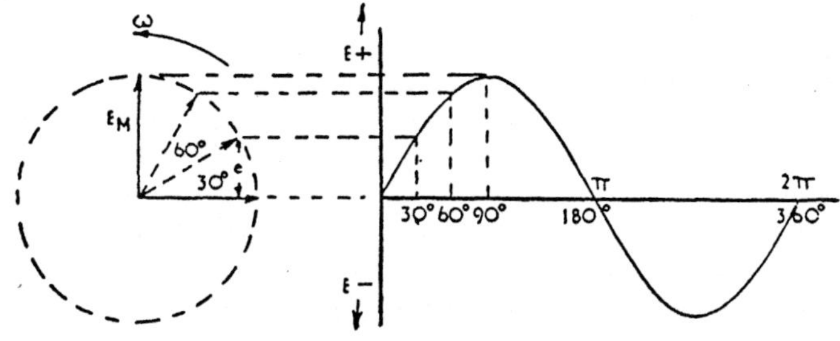

Fig 113

accordance with accepted procedure, assume that the phasor E_m rotates from the zero or horizontal position in an anticlockwise direction with an angular velocity of ω radians/second (ω is the Greek letter—small omega). Then $\theta = \omega t$ where t is the time in seconds and the equation can now be written as $e = E_m \sin \omega t$.

The diagram (Fig 114) following below, shows some of the terms used in connection with a.c. theory. *Periodic time* = the time for 1 cycle. The *frequency f* of the wave = the number of complete cycles in the interval of 1 second. In accordance with SI recommendations, the name *hertz* (Hz) is now being adopted for frequency measurement. This replaces the older term of cycles per second (c/s). Present marine practice uses either 50Hz or 60Hz a.c. systems. The maximum value reached by the wave is also called its *peak* value, or its *amplitude* and, as mentioned earlier, the value at any instant is termed its instantaneous value and is denoted by a small letter such as *e*. In passing, we can observe that sinusoidal current conditions can also occur and that the expression for a current following a sine-wave law can similarly be written as $i = I_m \sin \omega$t.

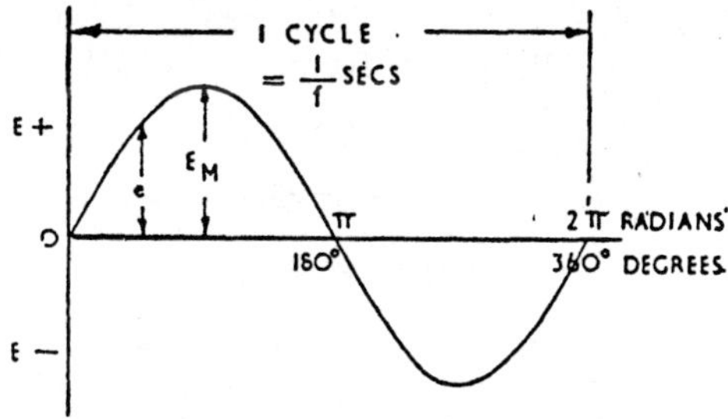

Fig 114

Following the introduction of the sine wave, as derived from a phasor, and the generation of a sinusoidal e.m.f. by a rotating coil, the treatment can be combined still further by the following deduction. The phasor is assumed to be rotating at a constant angular velocity of ω radians/second and the waveform, if considered to have a frequency of f hertz, will stretch out to cover in 1 second, an angle of $360f$ degrees or $2\pi f$ radians. The phasor meanwhile will have passed through ω radians in 1 second and it follows that ω can be equated to $2\pi f$ ° or $360f$ °. The earlier deduced expressions can now be written in their most useful form, namely:

$$e = E_m \sin 2\pi ft.$$

The above is the first fundamental formula of a.c. theory and should be given full attention. It is important to note that if $\frac{22}{7}$ or 3.14 is substituted for π, then the angle will be in radians and can be converted into degrees by multiplying by 57.3. The simpler method is to substitute 180° for π, thus converting into degrees directly.

Example 69. Find the instantaneous value of a 50Hz sinusoidal e.m.f. wave of maximum value 100V, at an instant of time 0.003 seconds after the zero value.

Substituting in $e = E_m \sin 2\pi ft$ we have

$e = 100 \sin (2 \times 180 \times 50 \times 0.003)$

or $e = 100 \sin 54°$

and $e = 80.9$V.

Important Note. A catch problem can occur when the instantaneous value is given and the time is required. Attention should be given to the following example, which illustrates the point being made.

Example 70. Find the first time after zero, when the instantaneous value of a sinusoidal current wave is 6.8A. The maxiınum value is 12A and the frequency is 50Hz. Find also the second time after zero.

Here $i = I_m \sin 2\pi ft$

or $6.8 = 12 \sin (2 \times 180 \times 50 \times t)$

Thus $\frac{6.8}{12} = \sin (180 \times 100 \times t)$

and $0.566 = \sin (18 \times 10^3 \times t)$. Further solution of this equation can only be made by use of electronic calculator or sine tables, from which an angle can be found whose sine equals 0.566.

Thus let θ = this angle, then $\sin\theta = 0.566$.
θ is seen to be 34° 30′ or 34.5°

$\therefore\ \sin\theta = \sin(18 \times 10^3 \times t)$ or $18 \times 10^3 \times t = 34.5$

and $t = \dfrac{34.5}{18 \times 10^3} = 1.9\text{ms}$

The second time value required, is obtained by finding the time for a $\frac{1}{2}$ cycle and then subtracting the interval, from a zero value, necessary for the instantaneous value to attain a height of 6.8A.

Thus time for $\frac{1}{2}$ cycle = $\frac{1}{100}$ seconds = 0.01s.
So second time required = 0.01 − 0.0019
= 0.0081s or 8.1ms.

REPRESENTATION OF SINUSOIDAL ALTERNATING QUANTITIES

Earlier it has been shown that an alternating voltage or current can be represented by an expression such as $e = E_m \sin 2\pi ft$ or $i = I_m \sin 2\pi ft$ and that this method of notation conveys all that is required to be known about the quantity, *ie* the fact that it follows a waveform whose amplitude, frequency and instantaneous value, at any particular time, can be found. This method of notation is called trigonometrical representation.

TRIGONOMETRICAL REPRESENTATION. This is useful for two quantities which are alternating, but not necessarily in the same simultaneous manner. Thus an alternating voltage of 50Hz can cause an alternating current in a circuit which will alternate at 50Hz. The current need not however, be in phase with the voltage, which latter may reach its maximum value a little time before the current reaches its maximum value. The voltage is said to *lead* the current or the current to *lag* the voltage. There is a *phase difference* between the two quantities or between their waveforms and such a phase difference is shown by the inclusion of the phase angle (in radians). Thus if two current waveforms are represented by

$i_1 = I_{1m} \sin 2\pi ft$ and $i_2 = I_{2m} \sin(2\pi ft + \frac{\pi}{3})$, it means that the second waveform leads the first by an angle of $\frac{\pi}{3}$ radians or $\frac{180}{3} = 60°$. A third waveform written as $i_3 = I_{3m} \sin(2\pi ft - \frac{\pi}{6})$ is seen to lag the first or reference waveform by $\frac{180}{6} = 30°$.

The trigonometrical form of representation, being a mathematical expression, can be used for the usual trigonometrical operations, such as multiplication, division, expansion, etc and such applications will be well illustrated in the course of a.c. theory as this is developed.

PHASOR REPRESENTATION. This, commonly used for a.c. quantities such as current, voltage, flux, etc, has already been introduced in terms of vectors. In Volume II dealing with Mechanics a vector was introduced and defined. Since voltages or currents are quantities whose magnitudes and directions are known, they can be described by rotating vectors but since 'phase' is usually involved, it is now customary to represent these by phasors. Thus a voltage phasor can be drawn to scale, its length representing the magnitude of the voltage and the direction in which it acts can be shown by an arrow. This technique has already been introduced earlier in this chapter, and we can now proceed to consider the accepted methods of phasor operation.

The relation between and the graphical deduction of, a wave-form from a phasor has already been covered. Since for most common and practical a.c. work, waveforms and instantaneous values are of comparatively little importance compared with magnitude and phase, the use of a phasor alone, as shown in the diagram (Fig 115), makes representation much simpler. Even further simplification and understanding can result from the correct use of *phasor diagrams*. Such diagrams are used to illustrate a.c. circuit relationships and are particularly useful if more than one current and/or voltage is being considered at the same time.

Note. For most practical work *r.m.s.* (root mean square) values are used in electrical engineering. The full meaning of the term will be dealt with later in this chapter, but it is mentioned here, because it is more convenient to make a phasor equal to

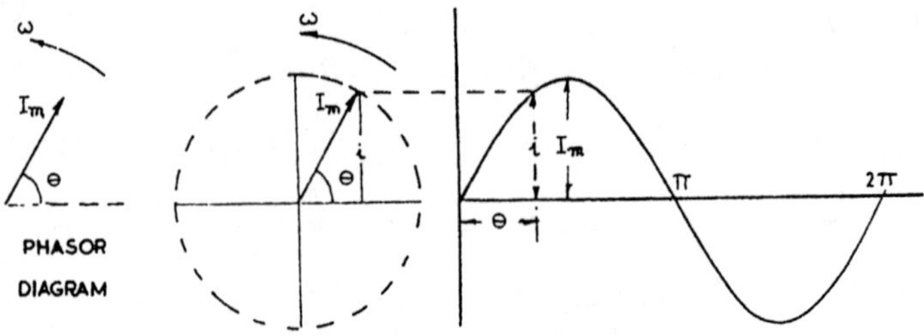

Fig 115

this value rather than the maximum value—since the representation for a maximum value holds good for an r.m.s. value. This modification will however only be introduced at a later stage.

Phase difference can also be shown by phasors. Consider two 50 hertz sinusoidal voltages represented by E_{1m} and E_{2m}. The phase angle ϕ is known, the voltages being of the same frequency but out of phase. The voltages can be written as $e_1 = E_{1m} \sin \omega t$ and $e_2 = E_{2m} \sin (\omega - \phi)$ where the angle ϕ can be assumed to be say 60° or $\frac{\pi}{3}$ radians and $e_2 = E_{2m} \sin (\omega - \frac{\pi}{3})$. The wave-forms can be drawn graphically as described earlier and since the first leads the second then, if the instant when the first is going through its zero value is considered as the start of the angle or time scale, the first wave can be considered as the reference and the phasor diagram can be drawn as shown in the diagram (Fig 116).

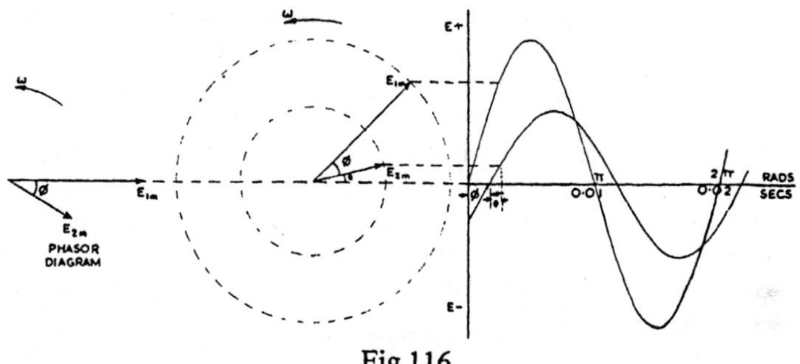

Fig 116

It will be seen that for the phasor diagram, we merely depict the two phasors and their relation to each other. The first phasor has been taken as the reference and the second is seen to lag it by

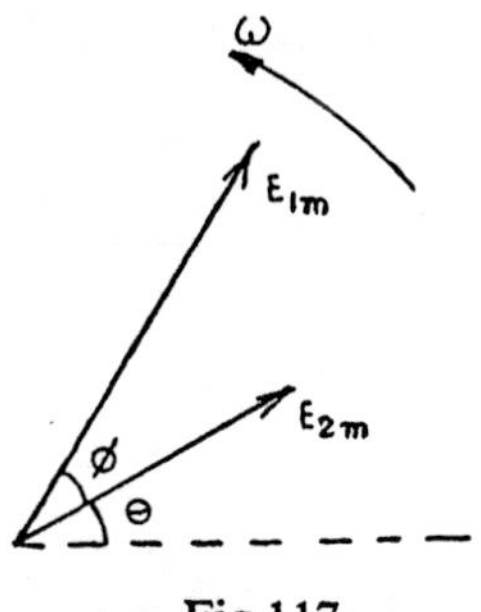

Fig 117

an angle ϕ. Direction of rotation is anti-clockwise so E_{2m} is behind E_{1m} by the angle ϕ. If an instant θ degrees later in time has to be considered then the diagram can be drawn as shown (Fig 117), the horizontal being taken as the zero time or reference axis.

ADDITION AND SUBTRACTION OF ALTERNATING QUANTITIES

When two or more sinusoidal voltages or currents act in a circuit the resultant can be obtained in either of the following ways (1) By Trigonometrical Methods (2) By Phasor Methods.

(1) TRIGONOMETRICAL METHODS. These methods require a good knowledge of trigonometrical identities and follow recognised procedures. Examples of their uses will occur in later studies.

(2) PHASOR METHODS. The resultant of two or more phasors may be obtained (a) *Graphically* or (b) *Mathematically*.

(a) The *Graphical Method* is performed by setting out the phasors to scale at the given angles, completing the parallelogram or polygon and measuring the resultant. The diagram (Fig 118) shows the method employed. Phasor addition is shown. To subtract a phasor, reverse its direction and proceed as before.

In the diagram, phasor addition is shown: (I) by completing the parallelogram to obtain the resultant of two phasors and then using this resultant with a third phasor to obtain the final resultant, (II) by completing the polygon as shown. Both methods are cumbersome and have the disadvantage that errors are cumulative.

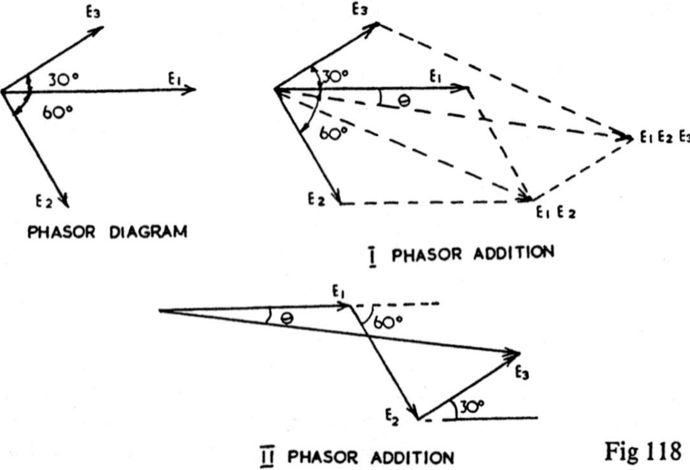

Fig 118

If the resultant of two individual waveforms is required, then either of two procedures can be followed.

The first procedure uses the known fact that the sum of any two sine waves of the same frequency is itself a sine wave. Thus any instantaneous value on the resultant wave is the sum of the individual instantaneous values taken from the other waves. Each waveform is drawn graphically in accordance with the method already outlined, care being taken to displace one from the other by the given phase angle. By adding instantaneous values, as shown in the diagram (Fig 119b), the resultant instantaneous value is obtained to give a point on the resultant wave. In the example $e = e_1 + e_2$. Other points are obtained in a similar manner and a smooth curve drawn by joining the points.

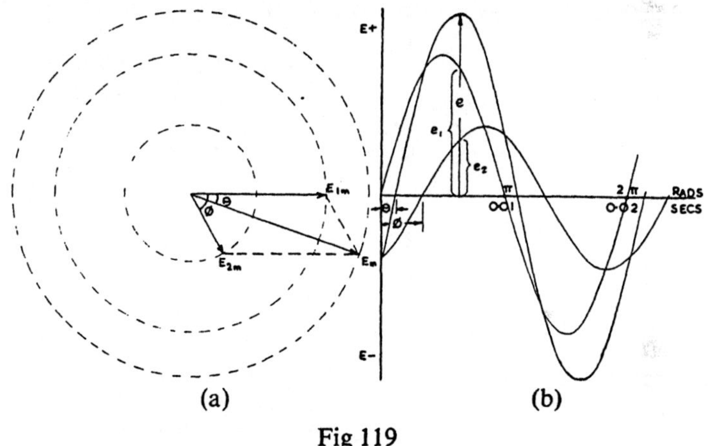

Fig 119

Note. One waveform should always be used as the reference, the other and the resultant being drawn to its base and zero value.

The alternative procedure for obtaining the resultant waveform is as follows. Since the individual voltages or currents can be considered as phasors, the resultant of any two or more values can be obtained in the same manner, as already described above. If the parallelogram is completed, as in Fig 119a, the resultant E_m will give the maximum value of the wave of the resultant. Using E_m as the radius of the largest circle, the resultant waveform can be deduced as before. As an example, if E_{1m} = 10 volts, E_{2m} = 6 volts and the phase difference is 60°, then the resultant E_m would be found to be 14 volts, E_m would also be found to lag 21° 45′ behind E_{1m}. The same procedure as to magnitude and phase angle would give the resultant r.m.s. value, if r.m.s. values were used for the component values instead of maximum values.

The second method, of obtaining the resultant waveform as described above, is obviously the quicker method if the component waveforms are not required. The resultant phasor may be obained graphically or by one of the mathematical methods as detailed next.

(b) The *Mathematical Method* can be performed in one of two ways: (i) by using the Cosine Rule, (ii) by resolving into horizontal and vertical components.

(i) The Cosine Rule or a modification of the same, can be used to advantage if the resultant of only two phasors is required. Consider the diagram (Fig 120). Then the resultant E can be obtained from $E = \sqrt{E_1^2 + E_2^2 + 2E_1 E_2 \cos \phi}$ where E_1 and E_2 are the given phasors and ϕ the angle between them. The phase angle θ of the resultant can be obtained from the Sine Rule.

$$\text{Thus } \frac{E}{\sin (180 - \phi)} = \frac{E_2}{\sin \theta}$$

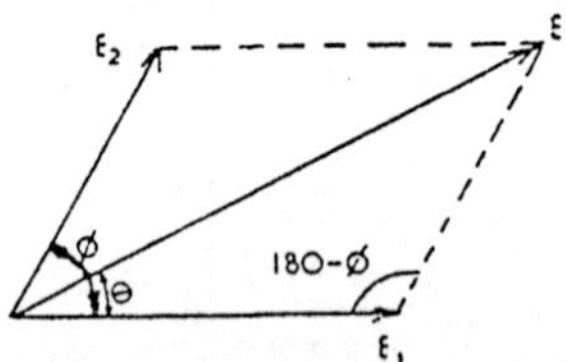

Fig 120

For more than two phasors, the resultant is used with a third phasor and so on. The method, next to be described, is advocated for the summation of more than two phasors, since it is quicker and is to be encouraged, being the method on which the treatment of series and parallel a.c. circuits is based.

(ii) Horizontal and Vertical Components. Any phasor can be split into two components, which are at right angles to each other and together produce the same effect as the original phasor. Thus in the diagram (Fig 121) the e.m.f. phasor E can be split into a horizontal and a vertical component. If E lies at an angle ϕ to the horizontal, then the horizontal component will be $E \cos \phi$ and the vertical component will be $E \cos (90 - \phi)$ or $E \sin \phi$.

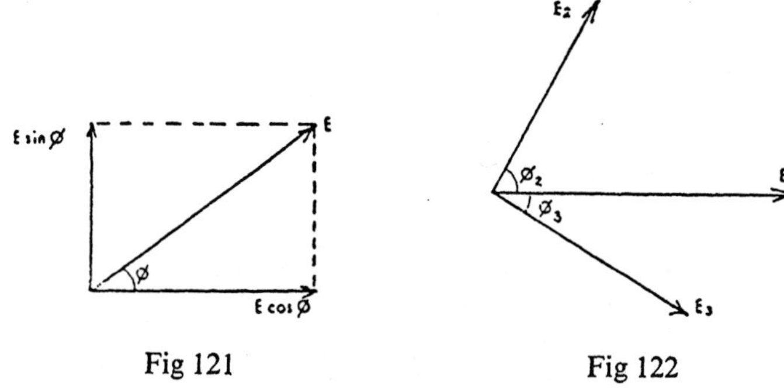

Fig 121 Fig 122

If all the phasors as shown in the diagram (Fig 122) are to be added and the resultant obtained then:

The sum of the horizontal components would be

$$E_H = E_1 \cos \phi_1 + E_2 \cos \phi_2 + E_3 \cos \phi_3.$$

Similarly the sum of the vertical components would be

$$E_V = E_1 \sin \phi_1 + E_2 \text{Sin}\, \phi_2 - E_3 \sin \phi_3.$$

Note. Due allowance must be made for the signs. Thus if the vertical components are considered to be +ve when acting upwards, then $E_3 \sin \phi_3$ must be subtracted from the sum since it acts in the downwards or −ve direction.

The resultant E is obtained from $E = \sqrt{E_H^2 + E_V^2}$ and ϕ, the angle at which it acts, can be found from the sine, cosine or tangent values. Thus $\cos \phi = \dfrac{E_H}{E}$

The method is illustrated by the following example.

Example 71. Find the resultant of the following currents

$$i_1 = 5 \sin \omega t$$

$$i_2 = 4 \sin \left(\omega t + \frac{\pi}{3}\right)$$

$$i_3 = 3 \sin \left(\omega t - \frac{\pi}{6}\right)$$

Express the resultant in a trigonometrical form *ie* in the same form as the individual quantities.

The diagram of Fig 123 should be considered with this solution.

$$\text{Then } I_H = 5 \cos 0 + 4 \cos 60 + 3 \cos 30$$
$$= (5 \times 1) + (4 \times \tfrac{1}{2}) + (3 + 0.866)$$
$$= 9.598\text{A}$$

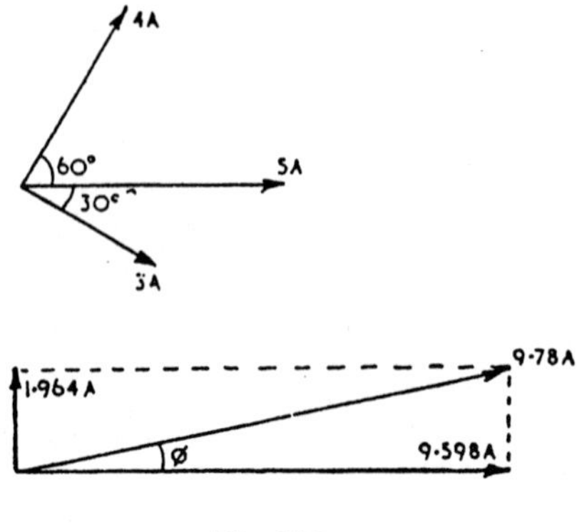

Fig 123

$$\text{And } I_V = 5 \sin 0 + 4 \sin 60 - 3 \sin 30$$
$$= (5 \times 0) + (4 \times 0.866) - (3 \times \tfrac{1}{2})$$
$$= 1.964\text{A}$$

$$\text{From which } I = \sqrt{I_H^2 + I_V^2} = \sqrt{9.598^2 + 1.964^2}$$
$$= 9.78\text{A}$$

$$\cos \phi = \frac{I_H}{I} = \frac{9.598}{9.78} = 0.98$$

$$\phi = 12° \text{ (approx)} = \frac{\pi}{15} \text{ radians}$$

$$\text{So } i = 9.78 \sin\left(\omega t + \frac{\pi}{15}\right).$$

Note. The following points are of interest in this example (a) In line with mathematical practice, phasors drawn to the right and those drawn upwards are given +ve signs, whereas those drawn to the left and those drawn downwards are assumed −ve. Thus in the example all the I_H components are +ve. The phasor diagram should be considered. For the I_V, it will be seen that 3 sin 30° is considered to act downwards and is therefore subtracted from 4 sin 60° which acts upwards. (b) The resulting sign of I_V also indicates whether the resultant I is in the 1st or 4th quadrant, *ie* whether it lags or leads the reference which in this case is the horizontal. In the solution ϕ is found to approximate to 12° which is $\left(\frac{180}{12} = 15\right)$ equal to $\frac{\pi}{15}$ radians. The resultant can thus be written as shown.

It will be noted that in the treatment introducing the Mathematical Method and in the above examples (Figs 120 to 123), the suffix m has been omitted from the e.m.f. symbol E. This is intended to illustrate that, as stated earlier, the method is equally applicable to maximum values and to r.m.s. values. The meaning of r.m.s. values will next be considered but as these are the most commonly used in a.c. work, it is important to appreciate that phasor representations, applications and solutions will be used without further introduction.

ROOT MEAN SQUARE AND AVERAGE VALUES

R.M.S. OR EFFECTIVE VALUE

The magnitude of an alternating current varies from instant to instant and the power dissipated in a resistance varies accordingly. The energy given out over a period of time manifests itself as heat. A resulting temperature rise is attained, which is steady and can be considered to be due to a constant power dissipation, *ie* due to the passage of a constant current which gives the same heating effect in the same time. Thus from a heating aspect, any value of alternating current can be assumed to have an equivalent value of direct current. When it is remembered that the heating effect is proportional to 'current squared'— since $P = I^2R$, then the magnitude of this equivalent value can be deduced as follows.

Let I amperes be the equivalent direct current which has the same average heating effect as the alternating current of varying instantaneous value i amperes.

For the d.c. condition

Energy expended = (current^2 × resistance) × time
$= I^2Rt$

For the a.c. condition

Energy expended = (mean or average of i^2R) × time
= (mean or average i^2) × R × t

By assumption since both energy conditions are considered to be equal then I^2Rt = (mean or average of i^2)Rt

or $I = \sqrt{\text{mean or average of } i^2}$

Thus the effective or r.m.s. value of an alternating current is the square root of the mean or average of the squares of the instantaneous values. This is true for the shape of any half cycle.

Note. The above deduction shows where the term 'root mean square' comes from, as the alternative to 'effective'. In practical electrical engineering, the term r.m.s. is the most commonly used, but a further older and little used alternative is the virtual value. Thus r.m.s. value, effective value and virtual value all mean the same.

The following definition will be found useful.

'The r.m.s. or effective value of an alternating current or voltage is that value of direct current or voltage which, when passed through or applied to a given resistor for a time of 1 cycle, produces the same amount of heat as the alternating current or voltage.'

It will be noted that in the definition, voltage has been mentioned although the r.m.s. value of a voltage wave was not specifically mentioned earlier. It will be seen that the heating effect on a resistor of value R ohms, has been taken as the basis of discussion, but an alternative to $P = I^2R$ is $P = \frac{V^2}{R}$, so a r.m.s. value of an alternating voltage wave of instantaneous value v, could have been deduced in a manner similar to that set out earlier.

The r.m.s. value of a waveform can be obtained graphically shown in the diagram (Fig 124). To do this, the instantaneous values of current or voltage should be plotted to a time or angle base, suitable scales having been chosen. Sub-divide the base of one half cycle into equal divisions and erect the mid-ordinates i_1, i_2, i_3, etc up to i_n. Measure these to scale and substitute in the expression.

$$I = \sqrt{\frac{i_1^2 + i_2^2 + i_3^2 + \ldots i_n^2}{n}}$$

It will be seen that only a half cycle has been considered, because the next half cycle is similar to the first, even though it is considered −ve. As the square of the current ordinates is

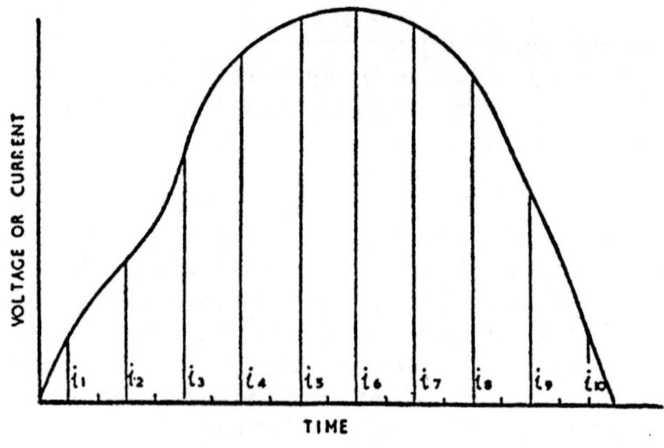

Fig 124

required, +ve values will result and the r.m.s. average, if taken over a complete cycle will be the same as for a half cycle. It should be noted that the Mid-ordinate Rule has been applied to the ordinates squared and not to the ordinates directly. If the latter had to be done, the average value would be obtained. This will be described later in the chapter.

Example 72. The following are the results of measurements taken at intervals over a half cycle of alternating voltage:

Time (t milliseconds)	0	0.45	0.95	1.5	2.1	2.5	3.1	3.9	4.5	5.0
Voltage (V volts)	0	20	36	40	37.5	33	32	31	20	0

Calculate the r.m.s. value and frequency of the wave.

The solution is shown in the diagram (Fig 125), from which it can be seen that, if the waveform is plotted to suitable scales, then dividing the base into ten equal parts, mid-ordinates can be drawn and measured as indicated.

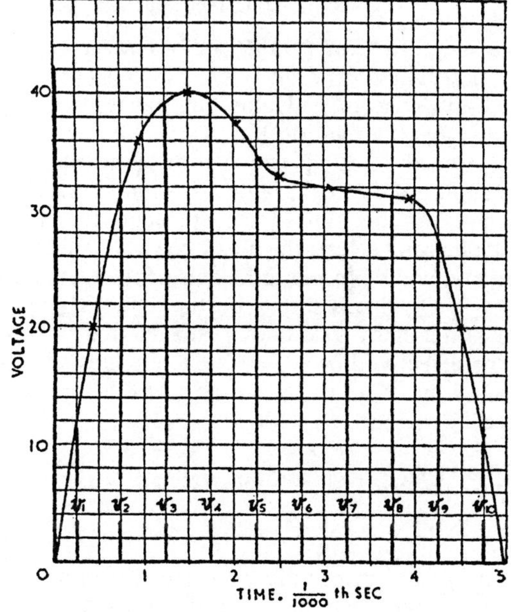

Fig 125

Then $v_1 = 12$	and $v_1^2 = 144$
$v_2 = 32$	$v_2^2 = 1024$
$v_3 = 39.5$	$v_3^2 = 1560$

$v_4 = 39.5$	$v_4^2 = 1560$
$v_5 = 35$	$v_5^2 = 1225$
$v_6 = 32.5$	$v_6^2 = 1056$
$v_7 = 31.5$	$v_7^2 = 992$
$v_8 = 31$	$v_8^2 = 961$
$v_9 = 28$	$v_9^2 = 784$
$v_{10} = 11$	$v_{10}^2 = 121$
	Total 9427

Mean or average of $v^2 = \frac{9427}{10} = 942.7$

$$\text{R.M.S. value} = \sqrt{942.7} = 30.7 \text{ volts}$$

Time for a half cycle $= \frac{5}{1000}$th seconds, so time for a whole cycle $= \frac{2 \times 5}{1000} = \frac{1}{100}$ second and frequency = 100 hertz.

For a sine wave, the r.m.s. value can be shown mathematically to be 0.707 times the maximum value. The most direct approach involves a knowledge of Calculus but the following graphical method illustrates the relationship.

Consider an alternating current of sinusoidal waveform having a maximum or peak value of 4 amperes. The current-squared curve can be plotted to represent the instantaneous values of the heating effect; the mean height of this curve can be determined and will be found to be half of the maximum height, *ie* half of 16 = 8. This is shown on the diagram (Fig 126). The number of amperes of continuous current that would give the same heating effect will be $\sqrt{8} = 2.828$ amperes and thus is the square-root of the mean of the squares of the current, *ie* a true r.m.s. value.

The ratio of r.m.s. to maximum value $= \frac{2.828}{4}$

or $\frac{\text{R.M.S. value}}{\text{Maximum value}} = \frac{2.828}{4} = \frac{0.707}{1}$ and this ratio is true for any alternating current or voltage which is sinusoidal.

As mentioned earlier, the r.m.s. value is always used and assumed in practical electrical engineering. Sine-wave working is also assumed and any departure from this will always be clearly stated. Thus if a supply voltage is given as 220V, this would be the r.m.s. value and actually the voltage varies over a cycle between zero and $\frac{220}{0.707} = 311.2$ volts.

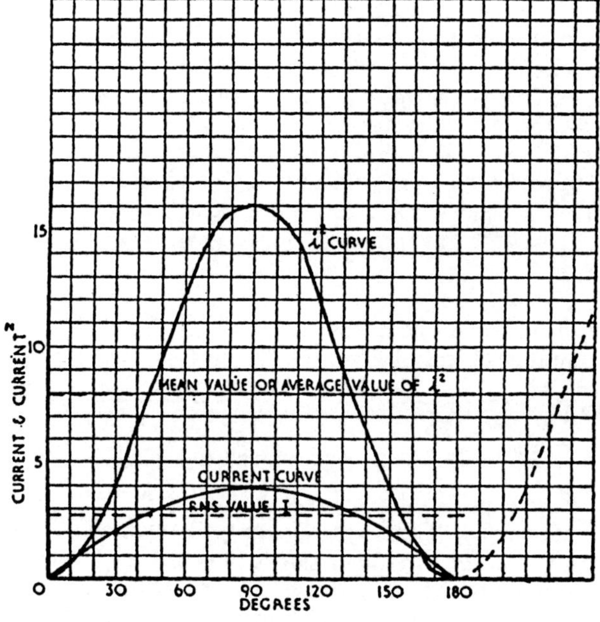

Fig. 126

AVERAGE VALUE

This is the true average value—as is understood mathematically, of a half cycle since that of a full cycle is zero. The average value of a waveform is of particular interest for devices which do not depend on the effect of current squared. An example, of such a device, would be a rectifier. Rectifiers have been developed to convert alternating current into direct current, without the need of rotating machinery, and are available in various forms. In the domestic type of 'mains' radio set, gas or vapour-filled valves are provided which convert the a.c. mains voltage to a direct voltage for operating the other valves in the set. Alternatively a 'metal' or semiconductor rectifier may be used, since these are being developed to suit all requirements. For the marine industry gas or vapour type rectifiers, the most common of which is the 'mercury-arc' rectifier, have not been used to any extent for shipboard work, although they may be seen in shipbuilding and dockyards for providing the 'shore supply' to d.c. ships. Metal and semiconductor rectifiers are however, being used in various forms for a.c. ships, in sizes both large enough to supply direct current for the 'de-gaussing' gear and small enough to build into moving-coil indicating instruments, to enable these to be used for a.c. circuits.

The d.c current or voltage has a value equal to the average value of the a.c. waveform being rectified and for a non-sinusoidal wave it can be obtained graphically, as follows: Referring to the diagram (Fig 124) of the waveform considered earlier, let I_{av} equal the average value, then

$$I_{av} = \frac{i_1 + i_2 + i_3 \ldots i_n}{n}.$$

Example 73. Consider the same waveform and ordinates which are the subject of Example 72. Find the average value.

$$\text{Here } I_{av} = \frac{12 + 32 + 39.5 + 39.5 + 35 + 32.5 + 31.5 + 31 + 28 + 11}{10}$$

$$= \frac{292}{10} = 29.2\text{A or Average value} = 29.2\text{A}.$$

For a sine wave the average value is 0.6365 times the maximum value. This can be proved mathematically or graphically. As for the r.m.s. value, the most direct method of calculation involves a knowledge of Calculus but the graphical method can be checked by plotting a sine wave, to a time or angle base. Sub-divide the base into equal divisions, erect the mid-ordinates and obtain the average value by using the mid-ordinate rule or substituting in the expression

$$I_{av} = \frac{i_1 + i_2 + i_3 + \ldots i_n}{n}$$

As stated earlier, the ratio of average to maximum value is $\frac{2}{\pi}$ or $\frac{\text{Average value}}{\text{Maximum value}} = \frac{2}{3.14} = 0.6365.$
This ratio is true for any alternating current or voltage which is sinusoidal.

FORM FACTOR

This factor, when given a numerical value, states how near a waveform approaches the theoretical ideal sine wave. For any waveform, it can be defined as the ratio of the r.m.s. to the average value.

$$\text{Thus Form Factor} = \frac{\text{R.M.S. value}}{\text{Average value}}$$

For a sine wave, the Form Factor is 1.11. This can be obtained from:

$$\text{Form Factor} = \frac{0.707 \text{ Maximum value}}{0.6365 \text{ Maximum value}} = \frac{0.707}{0.6365} = 1.11.$$

Example 74. For the problem already considered, in obtaining the r.m.s. and average values, the form factor would be $\frac{30.7}{29.2} = 1.05$.

PEAK FACTOR

The term 'peak factor' may occasionally be encountered when dealing with a.c. waveforms. This can be defined as the ratio of the maximum value to the r.m.s. value. Thus:

$$\text{Peak Factor} = \frac{\text{Maximum value}}{\text{R.M.S. value}}$$

For a sine wave the Peak Factor would be

$$\frac{1}{0.707} = 1.41.$$

CHAPTER 9

PRACTICE EXAMPLES

1. Three circuits carrying currents of I_1, I_2 and I_3 are joined in parallel. $I_1 = 4A$, $I_2 = 6A$, lagging I_1 by 30° and $I_3 = 2A$, leading I_1 by 90°. Find, by a phasor construction drawn to scale, the resultant current and its phase angle with reference to current I_1.

2. A sinusoidal, 25Hz a.c. voltage has a maximum value of 282.8V. Find the time interval, after the zero value, when the voltage wave reaches (a) its first, (b) its second instantaneous value of 200V.

3 A sinusoidal e.m.f. of 100V maximum value is connected in series with an e.m.f. of 80V maximum value, lagging 60° behind the 100V e.m.f. Determine the maximum value of the resultant voltage and its phase angle with respect to the 100V e.m.f.

4. The following figures give the instantaneous values of an alternating current, which varies smoothly over one half cycle.

Time (milli-seconds)	0	1	2	3	4	5	6	7
Current (amperes)	0	0.4	0.75	1.1	1.4	1.7	1.9	2.0

Time (milli-seconds)	8	9	10
Current (amperes)	1.8	1.3	0

Plot the curve of current and find its r.m.s. value. Calculate the power dissipated when the above current flows through a resistance of 8Ω.

5. Three currents of peak values 10A, 17.32A and 20A respectively meet in a common conductor. The 17.32A current lags the 10A current by 90 electrical degrees and leads the 20A current by 60 electrical degrees. Draw a phasor diagram and find the value of the resultant current, giving its phase relation with respect to the 10A current.

6. An alternating voltage, in the form of a sine wave, having a peak value of 340V, is applied to the ends of a 24Ω resistor. Calculate the r.m.s. value of the current in the resistor.

7. Represent by phasors, the following e.m.f.s.,
$e_1 = 100 \sin \omega t$,
$e_2 = 50 \cos \omega t$, $e_3 = 75 \sin\left(\omega t + \frac{\pi}{3}\right)$,
$e_4 = 125 \cos\left(\omega t - \frac{2\pi}{3}\right)$.

Determine by calculation the values of E and θ if
$$e_1 + e_2 + e_3 + e_4 = E \sin(\omega t + \theta).$$

8. Two alternators are arranged to be coupled to the same prime-mover in a manner which allows the phase-angle, between their generated e.m.f.s., to be varied. If the machines are connected in series and generate 100V and 200V respectively, find the total output voltage when the phase difference is; (a) zero, (b) 60°, (c) 90°, (d) 120°, and (e) 180°.

9. A stepped alternating current wave has the following values over equal intervals of time.

Value (amperes)	4	6	6	4	2	0	0	−2	−4
Time Interval (seconds)	0–1	1–2	2–3	3–4	4–5	5–6	6–7	7–8	8–9 etc.

Plot the waveform and find what value of direct current would give the same heating effect.

10. The 50Hz e.m.f.s., induced in four separate coils of an a.c. generator, are each of maximum value 4V and are successively 10 degrees out of phase. If these coils are connected in series, find by calculation and phasor construction, the resultant maximum value, expressing this in the form $e = E_m \sin(\omega t + \theta)$, where θ is the angle of phase difference with respect to the first coil.

CHAPTER 10

THE A.C. CIRCUIT

The method of introduction to the a.c. circuit chosen here, is again not the recognised approach as used in most text books. Experience in teaching the subject has convinced the author that the essential fundamentals should be introduced as soon as possible and the details should be considered as experience is acquired. Consequently at this stage, a number of new terms are mentioned and relationships are stated. These should be memorised as soon as possible.

IMPEDANCE

For the a.c. circuit, conditions are followed which are somewhat comparable with those for Ohm's law, as it has been propounded for the d.c. circuit. Thus we can consider a modification of Ohm's law as applicable to the a.c. circuit; where the current is found to be directly proportional to the applied voltage and also to be inversely proportional to the opposition of the circuit to the flow of current. This opposition is called the circuit *Impedance* (Symbol—Z. Unit—the ohm) and is due to more than the mere ohmic resistance R of the circuit. The difference between Z and R is now considered further.

For the a.c. circuit, the current flowing is given by

$$\frac{\text{Applied voltage}}{\text{Impedance}}$$

$$\text{Thus Current} = \frac{\text{Voltage}}{\text{Impedance}} \text{ or } I\,(\text{amperes}) = \frac{V\,(\text{volts})}{Z\,(\text{ohms})}$$

Note the variations of the relationship. Thus:

$$I = \frac{V}{Z} \text{ or } V = IZ \text{ or } Z = \frac{V}{I}$$

For the d.c. circuit, it is known that $I = \frac{V}{R}$ where R is the ohmic resistance of the circuit. If a wire, of resistance R ohms, is laid out as a straight conductor and connected to a d.c. supply of V volts, the current I when measured, would be given by

$\frac{V}{R}$ amperes. If the same wire is next connected to an a.c. supply of volts, the current *I,* when measured, would be given by $\frac{V}{Z}$ amperes and would be found to be of the same magnitude as for the d.c. test. For this case *Z* and *R* would be equal or the impedance would be made up of resistance only. The circuit is then said to be 'resistive' or 'non-inductive'.

If the wire is next wound into a coil or solenoid and the same voltage *V* is applied, the current will be smaller, *ie* the new impedance would be greater than the ohmic resistance. Again, if an iron core is inserted into the solenoid, the impedance would increase further and it would be apparent that the impedance is made up of ohmic resistance together with some extra current-limiting quantity. This extra quantity is termed *'Reactance'* (Symbol—*X*. Unit—the ohm) and it should be noted here that the impedance is *not* given by a straight-forward arithmetical summation of resistance *R* and reactance *X* but by a right-angled relationship where:

$$\text{Impedance} = \sqrt{\text{Resistance}^2 + \text{Reactance}^2}$$

$$\text{or } Z \text{ (ohms)} = \sqrt{R^2 \text{ (ohms)} + X^2 \text{ (ohms)}}$$

The relationship between *R, X* and *Z* can be represented by the 'Impedance Triangle' as shown by the diagram (Fig 127).

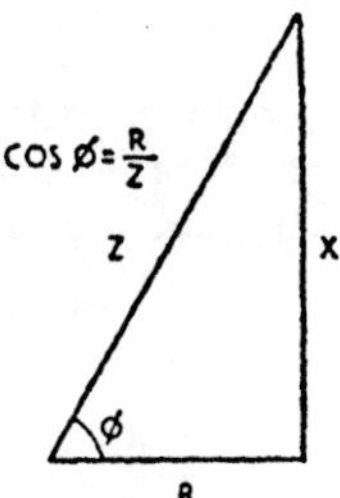

Fig. 127

The angle ϕ is called the 'Phase-angle' and cos ϕ is a measure of the *'Power Factor'* of the circuit. It is apparent that much explanation is now necessary to assist with the understanding of the terms introduced and the relationships stated. The reactance as mentioned, is furthermore of only one particular form, namely *inductive reactance.* If a coil with its associated magnetic field is concerned with the a.c. circuit, then its inductive reactance must be known, which in turn requires a knowledge of the *Inductance* and the supply frequency. The term 'inductive reactance' is thus associated with an *inductor* coil — usually iron-cored. An alternative to the term inductor is *reactor* — usually used for a large coil built to pass heavy currents and strengthened to withstand the great associated electromagnetic forces.

In contrast to inductive reactance there is also *capacitive reactance,* a term associated with a *condenser* or *capacitor* — to use the modern and advocated name. The capacitor and capacitive reactance will be considered in due course, but immediate attention is now given to inductance and inductive reactance.

Example 75. An inductor (sometimes called a choke-coil) has an ohmic resistance of 3Ω and a reactance of 4Ω. If it is connected to a 20V a.c. supply, find the current which flows and the power factor at which the coil operates. Note the diagram (Fig 128).

Here $R = 3\Omega$ and $X = 4\Omega$. Also since $Z = \sqrt{R^2 + X^2}$

Then $Z = \sqrt{3^2 + 4^2} = 5\Omega$

The current $I = \frac{V}{Z} = \frac{20}{5} = 4A$

The circuit power factor is given by $\cos\phi = \frac{R}{Z}$ (From the impedance triangle). Thus $\cos\phi = \frac{3}{5} = 0.6$ (lagging).

The term 'lagging' is associated with an a.c. circuit which contains inductive reactance. A more complete explanation will follow as the theory is developed.

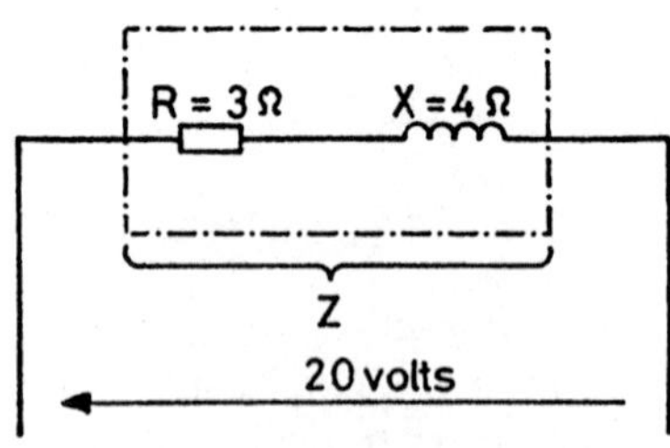

Fig. 128

The diagram shows how the circuit is represented. It will be seen that all the ohmic resistance is considered to be concentrated in a resistor *R* and the reactance in an inductor *X,* even though they together constitute the impedance *Z* of the choke-coil. The dotted rectangle represents the choke-coil and will generally be omitted in future diagrams.

INDUCTANCE

Basic electromagnetic induction theory, as covered in Chapter 7, has already shown that whenever the magnetic flux linked with a circuit is changing, an e.m.f. is induced in the circuit. Faraday's law shows that the value of the induced e.m.f. is proportional to the rate of change of flux-linkages and this e.m.f. only exists whilst a change of flux-linkages is taking place. Since, for an a.c. circuit, the current is changing continually, then the associated flux of a coil carrying such a current is also changing continually. This means that the flux-linkages are changing and an induced e.m.f. is thus being continually generated. By Lenz's law this would be a 'back e.m.f.', tending to oppose the change which is causing it.

When the current in an inductive circuit is made to change, it is apparent that, due to the property of inductance being present, the current value will, at the instant of change, be controlled by more than merely the voltage and resistance. During the changing or *transient* conditions, a back e.m.f. is being generated and new conditions of voltage balance occur. For the a.c. circuit, since current is varying sinusoidally and is thus changing constantly, inductance will have an important and continuous effect. This is considered next.

INDUCTIVE REACTANCE. Imagine a choke-coil having no resistance and only inductance of value L henrys, An alternating voltage of V volts is applied giving a current of I amperes. The diagram (Fig 129) represents the current and voltage conditions.

Assume the current of I amperes (r.m.s. value) to be sinusoidal. Since the induced e.m.f. $= \frac{LI}{t}$ or $= L \times$ *(rate of change of current)*, then at point 'a' the current value is zero, but it is increasing at its maximum rate, since the slope of the waveform is steepest at this point. Maximum induced e.m.f. will therefore occur at this instant 'a' and, since this is a 'back e.m.f.', by Lenz's law it must oppose the supply voltage. The voltage waveform conditions will be as illustrated and the corresponding phasor diagram is also shown with I used as the reference phasor. At point 'b' on the current wave, no e.m.f. is induced since current is maximum and not changing at this instant. Thus there is a 90° phase difference between the current and the induced voltage (E_X) and a further 90° difference between the current and the applied voltage (V_X). *Note.* This condition applies to a circuit with Inductance only.

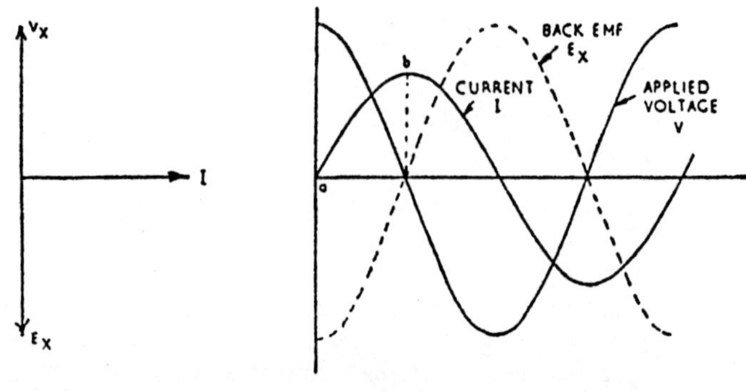

Fig. 129

Again referring to Fig 129, it is seen that, as the current rises to its maximum value I_m in the first quarter cycle, flux-linkages LI_m are set up, $\left(\text{since } L = \frac{N\Phi}{I} \text{ or } N\Phi = LI\right)$. As current falls to zero in the second quarter cycle, the linkages are destroyed. For the next half cycle the same number of linkages are set up and destroyed.

The change of flux-linkages in 1 cycle $= 4LI_m$ and the change of flux-linkages in 1 second $= 4fLI_m$ (f is the frequency). Also, since the average value of induced e.m.f. = rate of change of flux-linkages.

$$\text{Then average e.m.f.} = \frac{\text{Flux-linkages}}{\text{time}} = \frac{4fLI_m}{1} \text{ volts}$$

Thus back e.m.f. $E_{Xav} = 4fLI_m$ volts.

The supply voltage is equal and opposite, its value being V_X (r.m.s.) or V_{Xav} (average). Since r.m.s. values are preferred the following conversion is necessary.

$$\text{Since } V_{Xav} = \frac{2}{\pi} V_{Xm} \text{ and } V_{Xav} = 4fLI_m$$

$$\text{then } \frac{2}{\pi} V_{Xm} = 4fLI_m \quad \text{or} \quad V_{Xm} = 2\pi fLI_m$$

giving $0.707 V_X\text{m} = 2\pi fL \times 0.707 I_m$ or $V_X = 2\pi fLI$.

Thus the voltage drop in an inductor or reactor is seen to be proportional to the current and a constant which involves the circuit inductance and the supply frequency. This constant is given the name 'reactance' and since it is for an inductive circuit, we represent it by the symbol X with the suffix L.

Thus $X_L = 2\pi fL$ ohms and for a purely inductive circuit

$$V_X = IX_L \text{ where } X_L = 2\pi fL.$$

Inductive reactance is measured in ohms and is seen to be proportional to both frequency and inductance. Since resistance has been neglected, the phase relationship between the reactive voltage drop IX_L and current is seen to be fixed at 90° or these quantities are in quadrature with respect to each other. This condition will be deduced again as the various conditions of the a.c. circuit are considered.

1. CIRCUIT WITH PURE RESISTANCE

The circuit conditions are illustrated by the diagram (Fig 130).

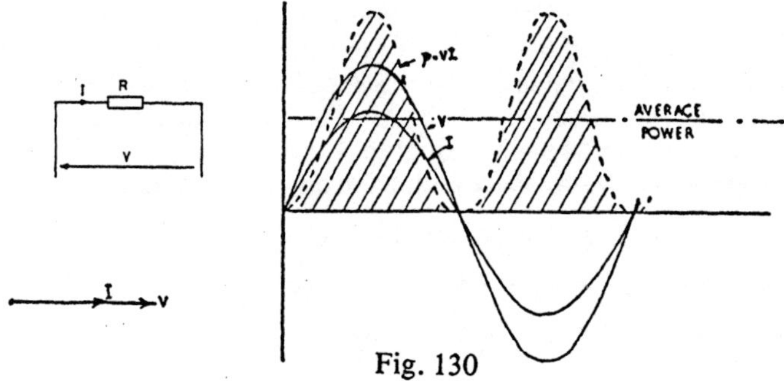

Fig. 130

Assume a sinusoidal voltage of value $v = V_m \sin \omega t$ to be applied to a purely non-inductive resistor of value R ohms. The applied voltage has to overcome the ohmic voltage drop at every instant or we can write $i = \frac{v}{R}$ and maximum current will occur when the voltage is a maximum since $I_m = \frac{V_m}{R}$.

Again since $\frac{v}{R} = \frac{V_m}{R} \sin \omega t \quad \therefore i = I_m \sin \omega t.$

Thus we see that the circuit current is also sinusoidal and is in phase with the applied voltage. The phasor diagram can also be drawn as shown.

R.M.S. values can be used since $0.707 I_m = \frac{0.707 V_m}{R}$ or $I = \frac{V}{R}$

The power condition can be deduced thus:

Power at any instant is given by $p = vi$

or $p = V_m \sin \omega t \, I_m \sin \omega t = V_m I_m \sin^2 \omega t$

$$= V_m I_m \frac{(1 - \cos 2\,\omega t)}{2} \ldots *$$

Average power P

$= \text{Average value of } \left\{\frac{V_m I_m}{2} - \frac{V_m I_m \cos 2\omega t}{2}\right\}$

$= \frac{V_m I_m}{2} - 0$. Since the average value of a cosine wave is zero.

Thus $P = \frac{V_m}{\sqrt{2}} \times \frac{I_m}{\sqrt{2}} = VI$

or $P = VI$ (watts).

From the expression marked thus*, it will be seen that the power wave is a periodic quantity, always +ve and of twice supply frequency. These facts would also be confirmed if the power wave was plotted by obtaining values of v and i for corresponding instants in time and multiplying these together to give p, the power value at that instant. The resulting power wave would be seen to be fully displaced above the horizontal and its maximum value would be equal to $V_m I_m$. Being symmetrical, its average value would be obtained from the distance its axis is displaced from the horizontal. This would be $\frac{V_m I_m}{2}$. This value would then be a measure of the average power.

Thus: $P = \frac{V_m I_m}{2} = \frac{V_m}{\sqrt{2}} \times \frac{I_m}{\sqrt{2}}$ or $P = VI$(watts).

For the resistive circuit being discussed, power is found to equal the product of voltage and current, but further considerations will show that this is only true for non-inductive circuits. In general, if an attempt is made to co-relate the power with the voltage and current or volt amperes of the circuit, it will be found that the product of V and I must be multiplied by a factor, which for convenience is called the 'power factor'.

For the condition of a purely resistive circuit, if we write $P = VI \times$ power factor, then it is obvious that the power factor must equal unity. It will again be seen later that the power factor is related to the ratio of the resistance and impedance of a circuit and that it can be obtained from $\frac{R}{Z}$. From the impedance triangle of the general circuit, $\frac{R}{Z}$ is the cosine of the phase-angle ϕ between circuit voltage and current. Thus $\cos \phi = \frac{R}{Z}$ = the power factor.

The assumption already made that $P = VI \times$ power factor can now be written as $P = VI \cos \phi$. Furthermore, since here the circuit is resistive $Z = R$ or $\cos\phi = \frac{R}{Z}$ and $\frac{R}{Z} = 1$, giving the condition of unity power factor already mentioned.

Note. The following deduction is also of value and is seen to be identical to that for the d.c. circuit. Since $P = VI\cos\phi$ we can write $P = VI\frac{R}{Z} = \frac{V}{Z}IR = I \times I \times R$ or $P = I^2R$ (watts).

Example 76. An electric fire rated at 2kW, is connected to a 220V supply. Find the current which will flow and the resistance value of the fire element.

Since an electric fire consists of a heating element which is purely resistive, it can be assumed that the circuit operates at unity power factor. Thus $\cos\phi = 1$ or the general expression $P = VI\cos\phi$ becomes $P = VI$

$$\text{Therefore } I = \frac{P}{220} = \frac{2000}{220} = 9.1\text{A}$$

$$\text{Again } Z = \frac{V}{I} = \frac{220}{9.1} = 24.2\Omega$$

$$\text{Here } Z = R \quad \therefore R = 24.2\Omega$$

$$\text{Alternatively } P = I^2 R \quad \therefore R = \frac{P}{I^2} = \frac{2000}{9.1^2}$$

$$\text{whence } R = \frac{2000}{82.81} = 24.2\Omega.$$

2. CIRCUIT WITH PURE INDUCTANCE

The basics of this condition have already been introduced in this Chapter under Inductance and Reactance, but since further deductions are necessary, the circuit is again illustrated by the diagram (Fig 131) which shows a coil with no resistance, but having inductance of value L henrys.

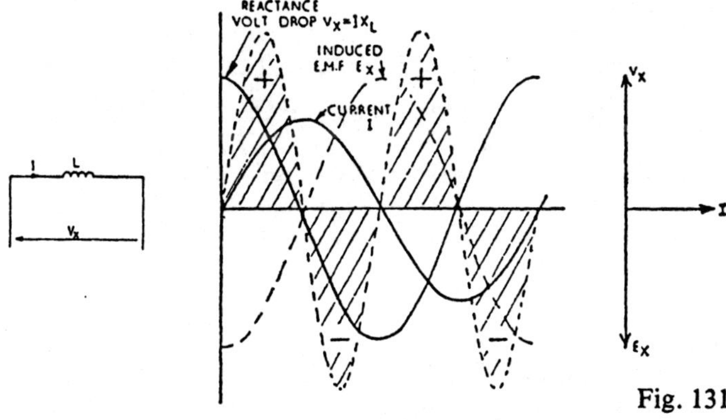

Fig. 131

Assume a sinusoidal current of value given by $i = I_m \sin \omega t$ to be flowing through the coil. Since i is varying sinusoidally, the magnetic field is also varying and a sinusoidal self-induced e.m.f. is set up opposing the applied voltage at every instant. Treatment of the a.c. circuit with inductance only, showed that the e.m.f. of self-inductance or 'back e.m.f.' can be regarded as being equivalent to a voltage drop, which is caused by the current and by a property, with which the circuit is credited. This property is termed Inductive Reactance (Symbol—X_L. Unit—the ohm). Thus we have $E_X = V_X = I \times X_L$. X_L has already been shown to equal $2\pi fL$. The associated phasor diagram can now be considered with the waveform. E_X is the e.m.f. of self-inductance, displaced 90° behind the current I. V_X is the supply voltage and, being 90° ahead of the current, is 180° out of phase with E_X. V_X is thus always equal and opposite to E_X. This is illustrated by the waveforms and can also be shown thus:

It has been seen that the e.m.f. of self-induction can be written mathematically as $e = -L\frac{di}{dt}$. By Lenz's law, since it at all times opposes the supply voltage, we can write $v = L\frac{di}{dt}$.

Thus $e = -L\dfrac{di}{dt} = -\dfrac{L\,d(I_m \sin \omega t)}{dt} = \omega L\, I_m \cos \omega t$ *or*

$e = \omega\, L\, I_m\left(\sin \omega t - \frac{\pi}{2}\right)$. Similarly v can be deduced as

$v = \omega\, L\, I_m\left(\sin \omega t + \frac{\pi}{2}\right)$.

Note v is 180° ahead or anti-phase with e.

Since ωL is the reactance X_L, it being remembered that $\omega = 2\pi f$ we have

$e = E_m \sin\left(\omega t - \frac{\pi}{2}\right)$ and $v = V_m \sin\left(\omega t + \frac{\pi}{2}\right)$ giving the 90° phase displacement between the current and voltage waves as shown on the diagram. *Note.* $V_m = X_L\, I_m$. The relationship $X_L = 2\pi fL$ is a fundamental and has been deduced earlier.

The power condition can be deduced thus:

Power at any instant is given by $p = vi$

$$\text{or } p = V_m \sin\left(\omega t + \frac{\pi}{2}\right) I_m \sin \omega t$$
$$= V_m\, I_m \sin \omega t \cos \omega t$$
$$= V_m\, I_m \frac{\sin 2\,\omega t}{2}$$

Thus the instantaneous power $p = \frac{V_m}{\sqrt{2}} \times \frac{I_m}{\sqrt{2}} \sin 2\,\omega t$

$= VI \sin 2\,\omega t \ldots$ *

Average power P = Average of value of $VI \sin 2\,\omega t$

= 0. Since the average value of a sine wave is zero.

From the expression marked thus*, it will be seen that the power wave is a sine wave of double frequency. It will be symmetrically disposed about the horizontal and the average value will be zero. This would also be confirmed, if the power wave was plotted from values obtained from the voltage and current waves. Since the axis of the power wave is not displaced above but lies along the horizontal, the value of the average power utilised must be zero because the +ve halves of the power wave are exactly equal to the −ve halves. This indicates that when power is taken from the supply, to establish the magnetic field associated with the coil, it is returned to the supply when the magnetic field collapses.

If the general expression $P = VI \times$ power factor is adopted for this circuit, then the power factor must equal zero; since $P = 0$. If used in the form $P = VI \cos\phi$, then $\cos\phi = 0$. This can be further deduced from $\cos\phi = \frac{R}{Z}$ as stated earlier. Here $R = 0$ therefore $\cos\phi = \frac{0}{Z} = 0$.

Summarising we can say that a circuit with inductance only and no resistance would be purely imaginary but would give rise to a zero power-factor working condition.

Example 77. A 220V, 50Hz supply is applied to a choke-coil of negligible resistance and the circuit current is measured to be 2.5A. Find the inductance of the coil and the power dissipated.

Since $Z = \frac{V}{I}$ then $Z = \frac{220}{2.5} = 88\Omega$

Now $R = 0 \;\therefore X_L = Z$ or $X_L = 88\Omega$

also $X_L = 2\pi fL$

so $L = \frac{X_L}{2\pi f} = \frac{88}{2 \times 3.14 \times 50}$

or $L = 0.28$H

Also as

$R = 0 \quad \cos\phi = \frac{0}{88} = 0$

$\therefore P = 220 \times 2.5 \times 0 = 0$

Alternatively since $P = I^2 R$

then $P = 2.5^2 \times 0 = 0$

3. CIRCUIT WITH RESISTANCE AND INDUCTANCE IN SERIES

Consider a pure resistance and a pure inductance in series as shown in the diagram (Fig 132). It should be noted that for the practical choke-coil or reactor, resistance and inductance are physically inseparable, but for diagrammatic purposes they can be shown as two individual components R and L.

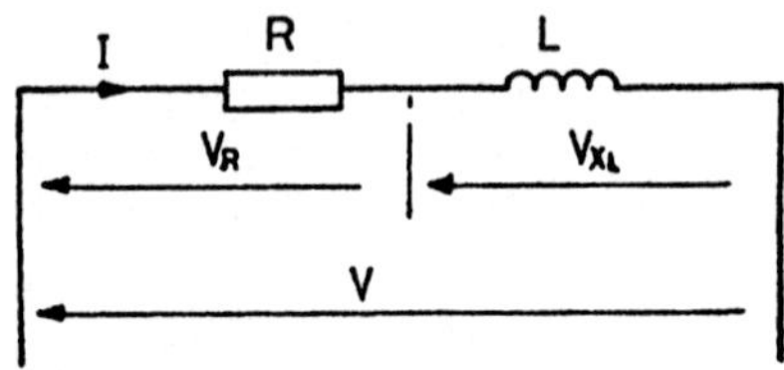

Fig. 132

Since the circuit conditions for both resistance and inductance have already been dealt with then, from the deductions already made, it can be considered that if a current I is assumed to flow, two voltage drops $V_R = IR$ and $V_{XL} = IX_L$ will exist. These must together form the applied circuit voltage and for greater convenience we can assume that the applied voltage V consists of two components. One such component V_R would be the voltage necessary to overcome the resistance voltage drop of the circuit and the other component V_{XL} would be the voltage necessary to overcome the reactance voltage drop or would oppose the induced back e.m.f. of self-induction. Since these two components are at right-angles to each other, as shown by consideration of Circuit Conditions 1 and 2, it follows that the applied voltage is the resultant of the two components. The relationships being discussed can be illustrated by the diagram (Fig 133) which shows the relevant waveforms and also the appropriate phasors.

For the phasor diagram, the current is common to both components, it being a series circuit, and is used as the reference phasor. The resistance voltage drop $V_R = IR$ is in phase with current and is drawn horizontally. The reactance voltage drop $V_{XL} = IX_L$ is at right-angles to the current and has been drawn vertically. The e.m.f. of self-inductance has also been shown but will be omitted from now on since it serves no useful purpose on the phasor diagram.

The resultant of V_R and V_{XL} is obviously V, the applied voltage and the current is thus seen to lag V by an angle ϕ which is the phase-angle of the circuit.

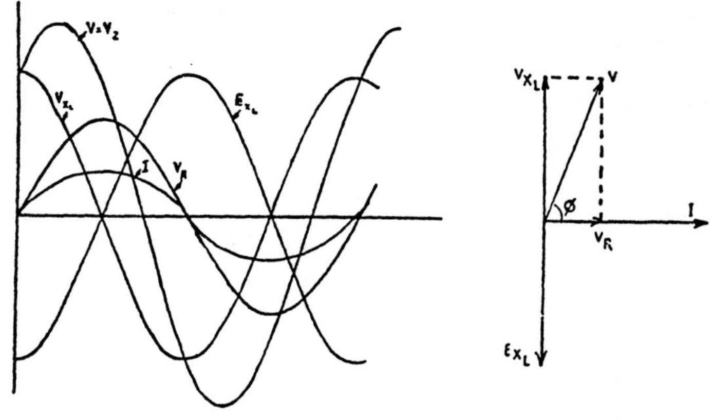

Fig. 133

Simplification of the phasor diagram can be useful in that it allows the appropriate 'voltage triangle' to be extracted, which in turn can be modified to give the 'impedance triangle'. These are shown in the diagram (Fig 134) and are used to prove the assumptions already made at the start of the Chapter.

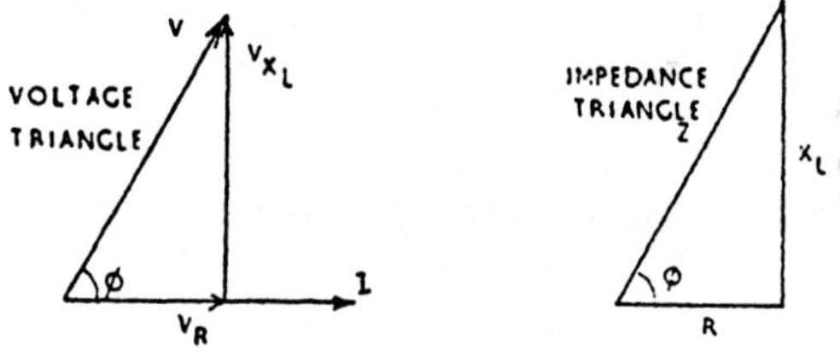

Fig. 134

Thus from the voltage triangle we have $V = \sqrt{V_R^2 + V_{XL}^2}$ and $\cos\phi = \dfrac{V_R}{V}$

also since $V_R = IR$ and $V_{XL} = IX_L$ the above can be written as

$$V = \sqrt{(IR)^2 + (IX_L)^2} = I\sqrt{R^2 + X_L{}^2}$$

If Z is taken to be the equivalent impedance of the circuit then $V = IZ$ or $Z = \dfrac{V}{I}$. Thus $IZ = I\sqrt{R^2 + X_L{}^2}$ whence we have the impedance triangle relationship of:

$$Z = \sqrt{R^2 + X_L{}^2} \text{ and } \cos\phi = \frac{R}{Z}$$

The power condition for the *R, L* series arrangement can now be deduced as follows:
The diagram (Fig 135) shows the basic waveforms of *v* and *i*, redrawn to allow the power wave to be deduced.

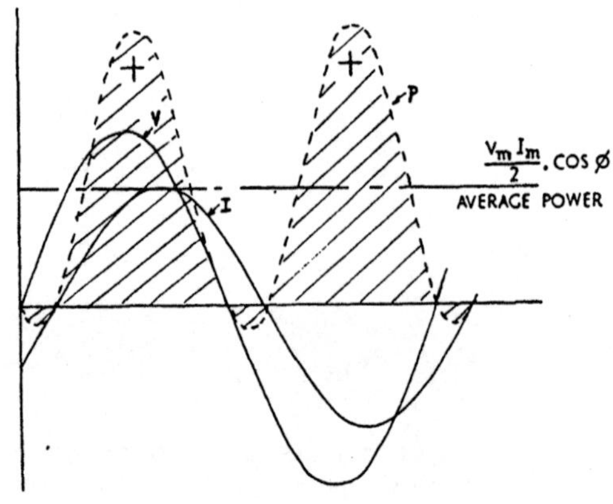

Fig 135

Let $v = V_m \sin \omega t$ be the applied voltage and
$i = I_m \sin (\omega t - \phi)$ be the circuit current lagging the voltage by the angle ϕ.

Then the instantaneous power $p = vi$
$= V_m \sin \omega t \times I_m \sin (\omega t - \phi)$

or $p = V_m I_m \sin \omega t \sin (\omega t - \phi)$

$$= V_m I_m \left\{\frac{\cos \phi - \cos (2\omega t - \phi)}{2}\right\}$$

$$= \frac{V_m}{\sqrt{2}} \times \frac{I_m}{\sqrt{2}} \left\{\cos \phi - \cos (2\omega t - \phi)\right\} \ldots *$$

Thus $p = VI \cos \phi - VI \cos (2\omega t - \phi)$.

The Average power P = Average of $VI \cos \phi$ − Average of $VI \cos (2\omega t - \phi)$

Whence $P = VI \cos \phi - 0$. Since the average of a cosine wave is 0.

Note. $VI \cos \phi$ is a constant quantity, thus its average value is apparent.

In the expression, as finally deduced, $P = VI \cos \phi$, we recognise the term power factor as it has been already introduced. We now see how it can vary between the limits of 1 and 0, to enable the extreme conditions of pure *R* or pure *L* to be satisfied. If we examine the power wave we also see how power factor or $\cos \phi$ is involved with the actual displacement of the power wave axis above the horizontal. The expression marked thus * shows that, here again, the power wave is periodic and of twice supply frequency. It consists of +ve and −ve areas, showing that some power can be returned to the supply, the amount depending on the power factor at which the circuit operates. The greater +ve net result of the power wave area, the greater the power consumption of a circuit and the nearer unity the power-factor condition.

It should be noted that the only component responsible for power consumption is the resistance and that the deduction made previously may well be repeated here to advantage.

Thus since $P = I^2 R$ and $I = \frac{V}{Z}$ we can write:

$$P = I \times I \times R = \frac{V}{Z} IR \text{ or } P = VI\frac{R}{Z} = VI \cos \phi.$$

Power factor will also be the subject of further treatment later, but it may be useful to point out, at this stage, that the product VI is frequently referred to as the 'volt amperes' of the circuit and may suggest the 'apparent power'. P, we know, is the 'true or active component of power', so we have the relation.

$$\text{True power} = \text{Apparent power} \times \cos \phi$$

and hence the name power factor for cos ϕ.

Example 78. A circuit has a resistance value of 25Ω and an inductance value of 0.3H. If it is connected to a 230V, 50Hz supply, find the circuit current, the power factor and the power dissipation.

$$X_L = 2\pi fL = 2 \times \pi \times 50 \times 0.3 = 94.2\Omega$$

$$Z = \sqrt{25^2 + 94.2^2} = 97.5\Omega$$

$$I = \frac{230}{97.5} = 2.36\text{A}$$

$$\text{Power factor} = \cos \phi = \frac{R}{Z} = \frac{25}{97.5} = 0.256 \text{ (lagging)}$$

$$P = VI \cos \phi = 230 \times 2.36 \times 0.256 = 139\text{W}$$

or $P = I^2 R = 2.36^2 \times 25 = 139$W.

It will be noted that the word 'lagging' has been introduced after the power-factor figure. This is usual to indicate whether the current lags or leads the voltage. The latter is possible for certain circuit conditions which will be considered as study progresses, but it can be assumed that inductive circuits always operate with a lagging power factor. The term is used for the circuit current with respect to the applied voltage, *ie* the current lags the voltage.

Example 79. A coil of wire dissipates 256W when a direct current of 8A is flowing. If the coil is connected to an alternating applied voltage of 120V, the same current flows. Find the resistance, and impedance of the coil and the power dissipated on a.c.

D.C. condition. Since $I = 8$A and $P = 256$W, then the applied voltage must be $\frac{256}{8} = 32$V. The coil resistance must also be $\frac{32}{8} = 4\Omega$

A.C. condition. Since $I = 8$A and the applied voltage is 120V then the coil impedance must be $\frac{120}{8} = 15\Omega$

The resistance is, as for the d.c. case $= 4\Omega$

The power dissipated $= I^2 R$ or $8^2 \times 4 = 256$W, as for the d.c. condition.

The latter part of the problem could have been solved by $P = VI \cos \phi = 120 \times 8 \times \cos \phi$ and $\cos \phi$ obtained from $\frac{R}{Z}$

$$\text{Thus } \cos \phi = \frac{4}{15} = 0.266 \text{ (lagging)}$$

$$\text{and } P = 960 \times 0.266 = 255.4\text{W}.$$

CAPACITANCE

The property of capacitance is attributed to an electrical condenser or capacitor, to use the modern term. A more complete treatment of the capacitor was made in Chapter 8, but since the component is mainly associated with a.c. circuits, it is necessary at this stage to revise the arrangement and action briefly, before proceeding with a.c. circuit theory.

If two conductors, arranged as plates, are separated by insulation, such as air or mica and are connected to a d.c. voltage, then at the instant of making connection, a current will flow. This current is of maximum value at the instant of switching-on but gradually dies away to zero value. This is termed a 'charging' current and the action is explained by considering the insulation to be put into a state of electrical stress. The effect of this stress can be likened to a 'back e.m.f.' which gradually builds up in the capacitor to oppose the supply voltage. Once the capacitor has been charged and this voltage has built up, its presence becomes apparent from the fact that, if the supply voltage is lowered, this back e.m.f. asserts itself, and causes a current to flow in the reverse direction, *ie* it causes a 'discharging' current.

Since an alternating voltage is varying all the time, either rising or falling, it follows that, if it is applied to a capacitor or a circuit with capacitance, then a 'to and fro' or a.c. current will flow. This is explained by the fact that, as the voltage across the capacitor plates rises, a charging current results and as the voltage falls, a discharging current results, due to the stress e.m.f. already mentioned. The magnitude of the current depends on the Capacitive Reactance (Symbol—X_C, Unit—the ohm) of the circuit. This term is introduced to correspond with inductive reactance and an expression for it will require to be developed. A unit for capacitance must however, be introduced first. It can be shown that the quantity of electricity or charge which passes into a capacitor is dependent on the applied voltage. Therefore $Q \propto V$ or $Q = CV$ where C is a constant depending on the construction of the capacitor, *ie* upon its dimensions, nature of the insulation used between the plates — called the dielectric, upon the spacing between the plates, etc. A measure of this constant, termed its capacitance, can now be given as a definition in units already known to us. Thus:

A capacitor is said to have a unit capacitance of 1 farad if 1 volt charges it with a quantity of electricity of 1 coulomb. It should be remembered that a coulomb results from an average current of 1 ampere passing for 1 second. It is also pointed out that a farad is a very large unit and the micro-farad (μF) is the more practical unit. $1\mu F = 1 \text{ farad} \times 10^{-6}$.

It should be noted that current only flows if the voltage across the plates is changing. In the above expression V is the change in p.d. across the plates from 0 to V volts, when the switch is closed. If the p.d. increases by v volts in t seconds and i is the average charging current, we can then write.

$$q = Cv \text{ or } it = Cv \text{ or } i = \frac{Cv}{t}$$

For very short instants in time, the above will become:

$$i = \frac{Cdv}{dt}$$

CAPACITIVE REACTANCE. The action of a capacitor when an alternating voltage is applied across its plates can now be considered in detail and is illustrated by the diagram (Fig 136). As the voltage rises from 0 to V_m in a quarter cycle, the charge on the plates rises to CV_m, since $Q = CV$. During the next quarter cycle, the charge falls to zero. For the next quarter cycle, the charge rises to $-CV_m$ and again finally falls to zero for the last quarter of a cycle. The total change of charge for a complete

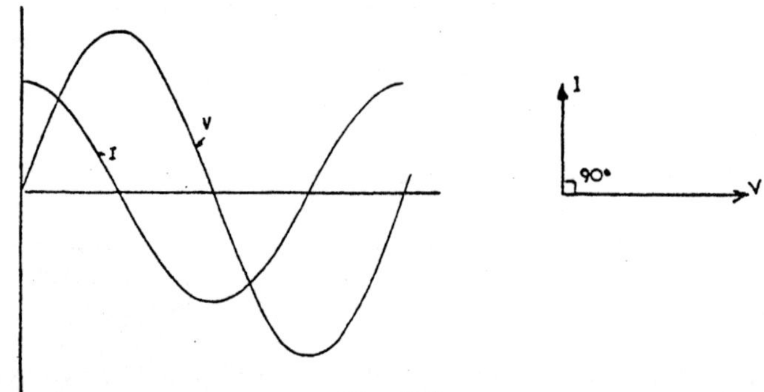

Fig. 136

cycle is thus $4CV_m$ and this occurs f times a second. The average current during this time is $\frac{Q}{t}$

or $I_{av} = \frac{4fCV_m}{t} = 4fCV_m$ since t is 1 second.

For a sine wave $I_m = \frac{\pi}{2} \times I_{av}$. Thus $I_m = \frac{\pi}{2} \times 4fCV_m$ $= 2\pi fCV_m$ or in r.m.s. values $0.707\ I_m = 2\pi fC \times 0.707V_m$ and $I = 2\pi fCV$.

Whence $\frac{V}{I} = \frac{1}{2\pi fC} = X_C$. Here X_C is called the capacitive reactance and the expression is in line with that developed for inductive reactance.

Thus $X_c = \frac{1}{2\pi fC}$ megohms or $\frac{10^6}{2\pi fC}$ ohms. Here C is in microfarads.

X_C itself is usually measured in ohms. It should be noted that the current wave is seen to lead the voltage wave by 90° since maximum current occurs at the instant of maximum rate of charge of voltage. If a phasor diagram is drawn, then it follows that the current phasor I is 90° ahead of the applied voltage phasor V.

The above conclusions can also be shown thus:
Let $v = V_m \sin \omega t$ be the sinusoidal voltage applied across the plates of the capacitor. Since $i = C\frac{dv}{dt}$ then $i = \frac{Cd\,(V_m \sin \omega t)}{dt}$

or $i = C\omega V_m \cos \omega t$

$= C\omega V_m \sin\left(\omega t + \frac{\pi}{2}\right)$. The capacitor current is in quadrature with the voltage. It is also sinusoidal. If $\frac{1}{\omega C}$ is called the capacitance reactance and is made equal to X_C

then $i = \frac{V_m}{X_C} \sin\left(\omega t + \frac{\pi}{2}\right)$ and i becomes a maximum when the wave becomes a maximum or $\sin\left(\omega t + \frac{\pi}{2}\right) = 1$

Thus $I_m = \frac{V_m}{X_C}$ or $0.707 I_m = \frac{0.0707 V_m}{X_C}$ giving $I = \frac{V}{X_C}$

Summarising here $V = IX_C$ as for the inductive circuit, except that here $X_C = \frac{1}{\omega C} = \frac{1}{2\pi f C}$.

The current leads the voltage by 90° and unlike the inductive circuit, where it was said that a choke-coil without resistance was not possible, here it is pointed out that a capacitor has negligible resistance and thus Circuit Condition 4, as set out below, can exist and is practical.

4. CIRCUIT WITH PURE CAPACITANCE

This condition has already been considered above in detail. The circuit diagram however is shown (Fig 137) and the waveforms and phasor diagram repeated, since the power condition has yet to be considered.

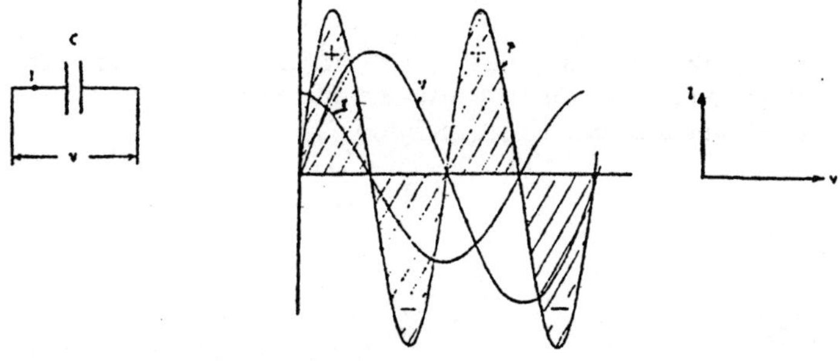

Fig. 137

The power at any instant $p = vi$

$$\text{or } p = V_m \sin \omega t \times I_m \sin\left(\omega t + \frac{\pi}{2}\right)$$

$$= V_m I_m \sin \omega t \cos \omega t$$

$$= V_m I_m \frac{\sin 2\,\omega t}{2} = \frac{V_m}{\sqrt{2}} \times \frac{I_m}{\sqrt{2}} \sin 2\,\omega t$$

$$= VI \sin 2\,\omega t \ldots *.$$

Thus average power $P = 0$ since the average of a sine wave is zero. The expression marked thus * shows the power wave to be periodic and of double frequency. The diagram also shows power to be +ve at the times when the voltage is increasing and energy is being put into the electrostatic field of the capacitor. When the voltage decreases the power is shown as −ve, *ie* energy is being recovered from the field as the capacitor discharges and thus no power is wasted. The power wave is seen to be symmetrical about the axis and the circuit power factor is zero. Thus if the expression $P = VI \cos \phi$ or $P = VI \times$ power factor is applied to this condition, it follows that $\cos \phi = 0$, as is borne out from $\cos \phi = \frac{R}{Z} = \frac{0}{Z} = 0$, there being no resistance.

Example 80. A capacitor of value 200μF is connected across a 220V, 50Hz supply mains. Find the current which would be recorded and the circuit impedance.

Here $X_C = \frac{1}{2\pi fC} = \frac{1}{2\pi \times 50 \times 200 \times 10^{-6}}\ \Omega$

$X_C = 15.92\ \Omega$

The current is given by $\frac{V}{X_C}$ or $I = \frac{220}{15.92} = 13.8\text{A}$

Since there is no circuit resistance then the impedance is made up of reactance only or $Z = X_C = 15.92\Omega$.

5. CIRCUIT WITH RESISTANCE AND CAPACITANCE

The diagram (Fig 138) illustrates the circuit conditions and the technique employed for its consideration is similar to that used for the inductive circuit of Condition 3.

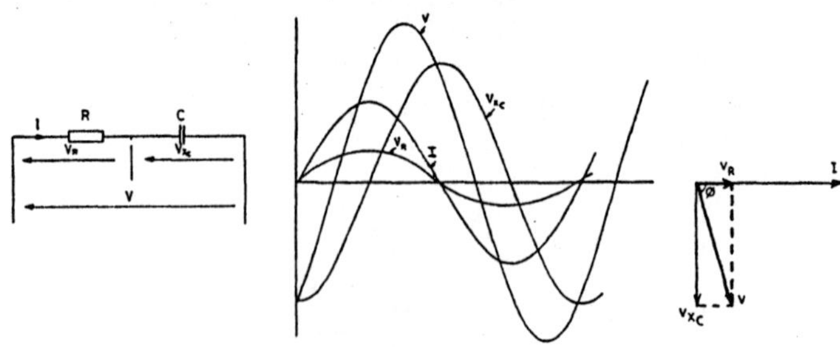

Fig. 138

The applied voltage V is considered to be resolved into two components V_R and V_{XC}. One component V_R overcomes the resistance voltage drop due to the passage of current I, and the other component V_{XC} maintains the charging current of the capacitor and is at all times equalled and sustained by the internal stress voltage. As seen from Condition 4, there is a 90° phase displacement between V_{XC} and I. If current is used as the reference for the waveform and phasor diagram, since it is common to R and C (this being a series circuit), then the conditions shown can be deduced. If the voltage triangle (shown heavy) is extracted, the impedance triangle and relationships can be found thus:

$$V = \sqrt{V_R^2 + V_{Xc}^2} = \sqrt{(IR)^2 + (IX_C)^2} = I\sqrt{R^2 + X_C^2}$$

If Z is the circuit impedance then $\frac{V}{I} = Z = \sqrt{R^2 + X_C^2}$

or $Z = \sqrt{R^2 + \left(\frac{1}{2\pi fC}\right)^2}$ As before $\cos\phi = \frac{R}{Z}$

The power relation follows the form already used several times.

Thus power at instant $p = vi$

where $i = I_m \sin \omega t$ *and* $v = V_m \sin(\omega t - \phi)$

$$\text{then } p = V_m I_m \sin \omega t \sin(\omega t - \phi)$$

$$= V_m I_m \left\{\frac{\cos\phi - \cos(2\omega t - \phi)}{2}\right\}$$

$$\text{or } p = \frac{V_m}{\sqrt{2}} \times \frac{I_m}{\sqrt{2}} \left\{\cos\phi - \cos(2\omega t - \phi)\right\}$$

$$= VI\cos\phi - VI\cos(2\omega t - \phi).$$

Average power P = Average of $VI\cos\phi$ − Average of $VI\cos(2\omega t - \phi)$ or $P = VI\cos\phi - 0$. Since the average of a cosine wave is zero.

Thus $P = VI\cos\phi$ — the form of expression already encountered several times. If the power wave is plotted as before it will be as shown in the diagram (Fig 139).

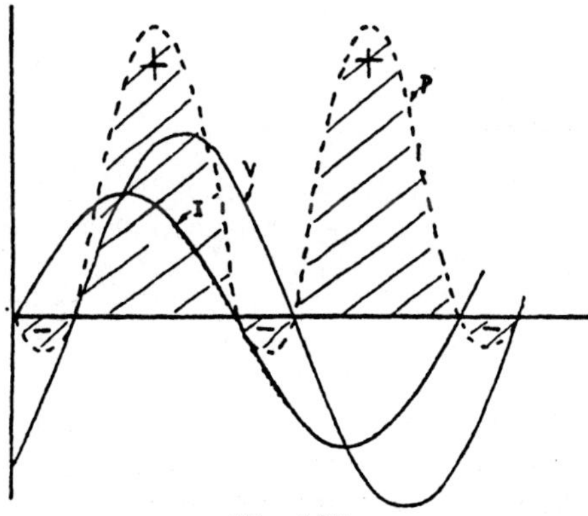

Fig. 139

The power wave is seen, as before to be periodic, of double frequency and consisting of +ve and −ve sections. The average value can be found from the amount by which the axis is displaced above the horizontal and this displacement varies with the power factor of the circuit. Thus Conditions 1 and 4 are also covered. If $X_C = 0$ and the circuit is purely resistive then $\cos \phi = 1$ and the wave will be fully displaced above the horizontal. If $R = 0$ and the circuit is purely capacitive then $\cos \phi = 0$ and the wave will be symmetrical about the horizontal giving $P = 0$. As for the inductive circuit, the only component responsible for the dissipation of power is resistance.

As before $P = I^2 R$ or $P = I \times I \times R = VI \dfrac{R}{Z}$

and $P = VI \dfrac{R}{Z}$ or $P = VI \cos \phi$ as already deduced.

Example 81. A 500W, 100V bulb is to be connected across 250V, 50Hz mains. Find the value of the capacitor required to be connected in series.

Current taken by bulb is $\dfrac{500}{100} = 5\text{A}$

Resistance of lamp $= \dfrac{100}{5} = 20\Omega$

On 250V, impedance of the circuit is to be $\dfrac{250}{5} = 50\Omega$

Thus $X_C = \sqrt{50^2 - 20^2}$
$= 45.8\Omega$

Again $X_C = \dfrac{1}{2\pi f C}$

or $C = \dfrac{1}{2\pi f X_C}$

$\therefore C = \dfrac{1}{2\pi \times 50 \times 45.8} F$

giving $C = 69.5\mu\text{F}$.

THE SERIES CIRCUIT

From the work which has now been completed and the various circuit conditions considered under the Cases 1 to 5, we can see the general form of technique employed for a series circuit. The phasor diagram is easily drawn with current being used for the reference phasor. From this diagram is deduced the circuit relationships and expressions. This method is employed below.

INDUCTIVE IMPEDANCES IN SERIES

The diagram (Fig 140) shows the circuit arrangement and the deduced phasor diagram.

Impedances A and B, consisting of resistances and reactances of values R_A, R_B, X_A and X_B ohms respectively, are connected in series. From the phasor diagram we can deduce an expression for the total circuit impedance Z, noting that it is *not* equal to $Z_A + Z_B$.

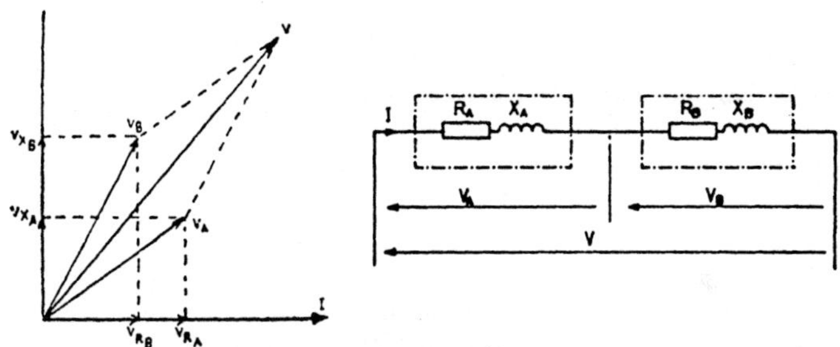

Fig. 140

Using the diagram we have:

$$V = \sqrt{V_R^2 + V_X^2} = \sqrt{(V_{RA} + V_{RB})^2 + (V_{XA} + V_{XB})^2}$$

$$= \sqrt{(IR_A + IR_B)^2 + (IX_A + IX_B)^2}$$

$$\text{or } V = I\sqrt{(R_A + R_b)^2 + (X_A + X_B)^2}$$

If Z is the equivalent circuit impedance then:

$$Z = \frac{V}{I} = \sqrt{(R_A + R_B)^2 + (X_A + X_B)^2}$$

or summarising, for more than two inductive impedances,

$$Z = \sqrt{(R_A + R_B + R_C \ldots)^2 + (X_A + X_B + X_C \ldots)^2}$$

Also the power factor is given by:

$$\cos\phi = \frac{V_R}{V} = \frac{IR}{IZ} = \frac{R}{Z} = \frac{(R_A + R_B + \ldots)}{Z}$$

The example given below shows how simply the above expression can be adapted for practical use.

Example 82. Two coils A and B are connected in series to 50Hz mains. The current is 1A and the voltage across each coil is measured to be 45V and 70V respectively. When the coils are connected to a d.c. supply, the current is also 1A, but the voltages across the coils are now 20V and 40V respectively. Find the impedance, reactance and resistance of each coil, the total circuit impedance, the applied a.c. voltage and the power factor of the complete circuit.

On D.C.	On A.C.
$R_A = \frac{20}{1} = 20\Omega$	$Z_A = \frac{45}{1} = 45\Omega$

$R_B = \frac{40}{1} = 40\Omega$ $\qquad Z_B = \frac{70}{1} = 70\Omega$

Then $X_A = \sqrt{45^2 - 20^2}$
$= 40.3\Omega$
Also $X_B = \sqrt{70^2 - 40^2}$
$= 57.4\Omega$
Total $R = 20 + 40 = 60\Omega$
Total $X = 40.3 + 57.4 = 97.7\Omega$
Total impedance $Z = \sqrt{60^2 + 97.7^2}$
$= 114\Omega$
Applied voltage $= 114 \times 1 = 114\text{V}$
Circuit power factor $= \frac{R}{Z} = \frac{60}{114} = 0.53$ (lagging).

INDUCTIVE AND CAPACITIVE IMPEDANCES IN SERIES

The diagram Fig 141) shows the arrangement

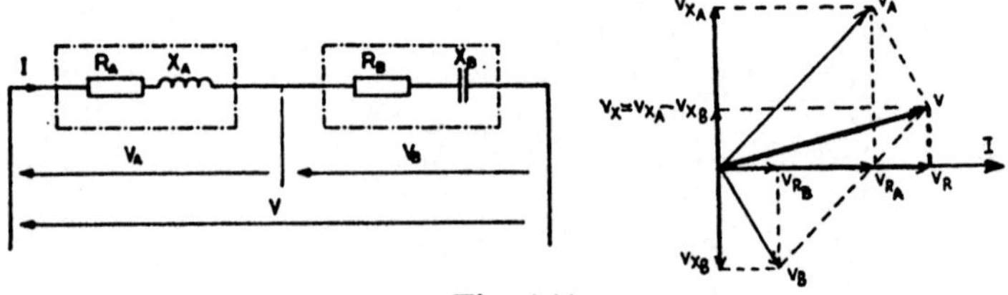

Fig. 141

From the phasor diagram we can deduce the expression for the total circuit impedance Z. It will be noted that although V_{RA} and V_{RB} are in phase and can be added, V_{XA} and V_{XB} are anti-phase and the resultant of the vertical phasors must be obtained by subtraction. Thus from the resulting final diagram (shown heavy):

$$V = \sqrt{V_R^2 + V_X^2} = \sqrt{(V_{RA} + V_{RB})^2 + (V_{XA} - V_{XB})^2}$$
$$= \sqrt{(IR_A + IR_B)^2 + (IX_A - IX_B)^2}$$
$$= I\sqrt{R_A + R_B)^2 + (X_A - X_B^2)}$$

If Z is the equivalent circuit impedance then,

$$Z = \frac{V}{I} = \sqrt{(R_A + R_B)^2 + (X_A - X_B)^2}$$

Summarising $Z = \sqrt{R_A + R_B)^2 + (X_A - X_B)^2}$

Also for the circuit, the power factor

$$\cos\phi = \frac{V_R}{V} = \frac{IR}{IZ} = \frac{R}{Z} \text{ or, } \cos\phi = \frac{R_A + R_B}{Z}$$

The above two circuit conditions give rise to the general series circuit, which is set out below.

THE GENERAL SERIES CIRCUIT

From work already done on circuit theory, a fundamental expression can be deduced from the phasor diagram (Fig 142).

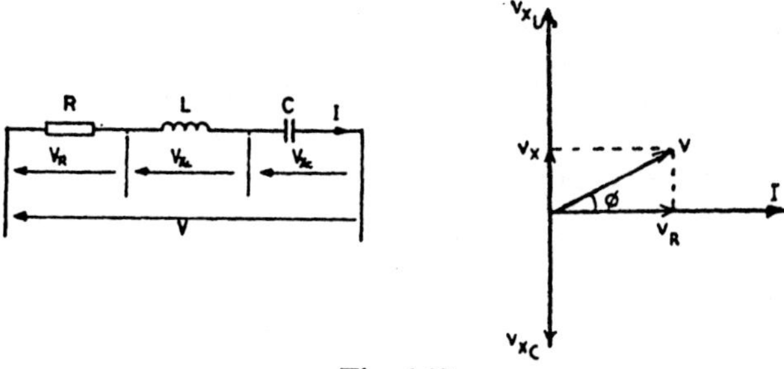

Fig. 142

Since it is a series circuit, current is common and can also be used as the reference phasor. It will be noted that the condition being considered is similar to that already considered for Inductive and Capacitive Impedances in series, except that all resistance of the circuit is taken as being contained in one resistor R. Then for the phasor diagram.

$V_R = IR$ and is in phase with the current
$V_{XL} = IX_L$ and is 90° ahead of I
$V_{XC} = IX_C$ and is 90° behind I

V_{XL} and V_{Xc} are 180° out of phase or anti-phase and a phasor difference can be obtained where $V_X = V_{XL} - V_{Xc}$. Here V_{XL} is assumed greater than V_{Xc}.

Further deduction from the diagram is possible, thus:

$$V = \sqrt{V_R^2 + V_X^2} = \sqrt{V_R^2 + (V_{XL} - V_{Xc})^2}$$
$$= \sqrt{(IR)^2 + (IX_L - IX_C)^2}$$
$$= I\sqrt{R^2 + (X_L - X_C)^2} \text{ or } \frac{V}{I} = \sqrt{R^2 + (X_L - X_C)^2}$$

If Z is taken as the equivalent impedance of the circuit, then

$$Z = \frac{V}{I}$$

$$\textit{or } \frac{V}{I} = Z = \sqrt{R^2 + (X_L - X_C)^2}$$

$$\text{Thus } Z = \sqrt{R^2 + \left(2\pi fL - \frac{1}{2\pi fC}\right)^2}$$

Example 83. A series circuit is made up of a choke-coil of resistance 20Ω and inductance 0.08H, connected in series with a 100μF capacitor. If the circuit is conneced across 200V, 50Hz mains, find (a) the circuit current and (b) its power factor.

Here $X_L = 2\pi fL = 2\pi \times 50 \times 0.08 = 25.2\Omega$

$$X_C = \frac{1}{2\pi fC} = \frac{1}{2\pi \times 50 \times 100 \times 10^{-6}} = 31.75\Omega$$

Resultant reactance $= X = X_L - X_C = 25.2 - 31.75$
$= -6.55\Omega$

The $-$ve sign denotes that the capacitive reactance predominates and that the phasor diagram will be as shown (Fig 143a).

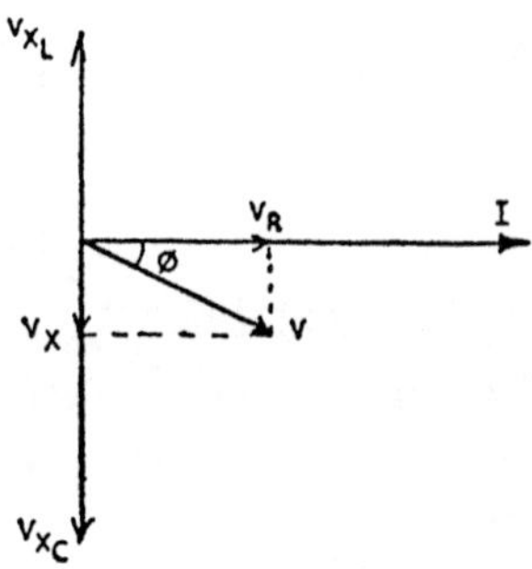

Fig 143a

From the diagram as before $V = \sqrt{V_R^2 + V_X^2}$
or $Z = \sqrt{R^2 + X^2}$
$\therefore Z = \sqrt{20^2 + 6.55^2}$
$= 21\Omega$

The circuit current $= \dfrac{200}{21} = 9.5\text{A}$

The power factor is given by $\cos \phi = \dfrac{R}{Z} = \dfrac{20}{21} = 0.95$ (leading), *ie* the current leads the voltage since the capacitive reactance of the circuit predominates.

SERIES RESONANCE. From the phasor diagram (Fig. 143b) it can be seen that a state can occur when V_{XL} and V_{XC} are equal, and also in this condition V_R is equal to the supply voltage V. An examination of the general series circuit expression

$$Z = \sqrt{R^2 + \left\{2\pi fL - \frac{1}{2\pi fC}\right\}^2}$$

will show that a condition can arise when $2\pi fL = \dfrac{1}{2\pi fC}$ in magnitude, *ie,* the capacitive reactance equals the inductive reactance. Under this condition $Z = R$ and the circuit is said to be in a 'state of resonance'. The current passed will be limited by

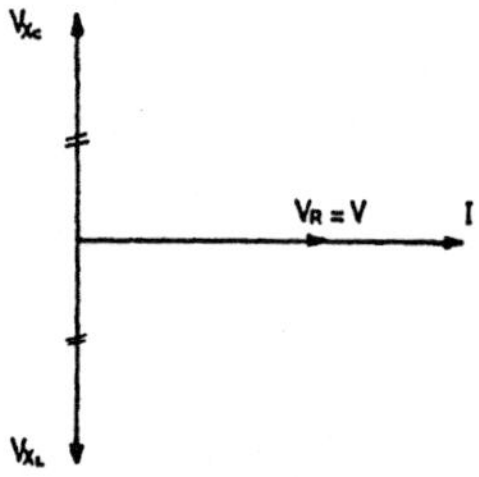

Fig. 143b

the value of R only and although large voltages can be present across components L and C, their effect on the supply voltage V is not evident. *Series resonance* as this condition is called, is used to advantage for some practical purposes, especially radio work.

Since inductive and capacitive reactances vary with frequency, a point will exist where the reactances are equal at a particular frequency. This is referred to as *Resonant Frequency.*

Since $$X_L = X_c$$

$$2\pi fL = \frac{1}{2\pi fC}$$

$$\therefore \text{ Resonant Frequency } f = \frac{1}{2\pi\sqrt{LC}}$$

Example 84. A 4μF capacitor is connected in series with a coil of inductance 39.6mH and resistance 40Ω to a 200V a.c. supply. Calculate (a) the frequency when the current is a maximum value and (b) the p.d. across the capacitor at this frequency.

When current is maximum Z is minimum.

ie, $Z = R$ and circuit is in resonant condition

$$\therefore \text{ Maximum current } I = \frac{V}{R} = \frac{200}{40}$$

$$I = 5\text{A at resonance}$$

$$\text{(a) Resonant frequency } f = \frac{1}{2\sqrt{LC}}$$

$$= \frac{1}{2\pi\sqrt{39.6\times10^{-3}\times4\times10^{-6}}}$$

$$F = 399.89 \text{ Hz (say 400 Hz)}$$

$$X_C = \frac{1}{2\pi fC} = \frac{10^6}{2\pi\times400\times4}$$

$$X_C = 99.5\Omega$$

$$\text{(b) p.d. across Capacitor } V_{XC} = IX_C = 5\times99.5$$

$$= 497.5 \text{ volts}$$

Example 85. A coil of unknown inductance and resistance is connected in series with a 25Ω, non-inductive resistor across 250V, 50Hz mains. The p.d. across the resistor is found to be 150V and across the coil 180V. Calculate the resistance and inductance of the coil and also find its power factor.

The circuit diagram and phasor diagram (Fig 144) are shown.

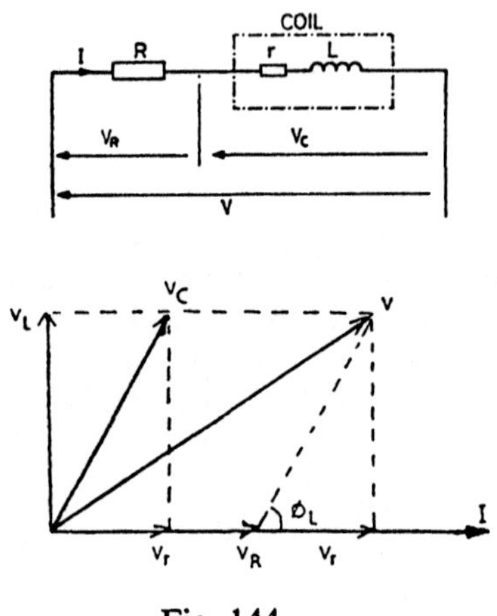

Fig. 144

This example is important in that it involves basic fundamentals and yet has a simple solution.

The phasor diagram is first explained with the various voltage drops considered in detail. V_R is the voltage drop across resistor $R = IR$. V_C is the voltage drop across the coil and is the resultant of two voltage drops, V_r across the resistance of the coil $= Ir$ and V_L across the reactance of the coil $= IX_L$. V_r is in phase with current and V_L is 90° ahead of the current. From the phasor diagram it is apparent that V is the resultant of V_C and V_R and that the expression given for simple phasor summation can be applied here.

Thus $V = \sqrt{V_R^2 + V_C^2 + 2V_RV_C \cos\phi_L}$

or $250^2 = 150^2 + 180^2 + 2 \times 150 \times 180 \times \cos\phi_L$

$\therefore$ $62\,500 = 22\,500 + 32\,400 + 54\,000 \cos\phi_L$

or $54\,000 \cos\phi_L = 62\,500 - 54\,900$

$$\cos\phi_L = \frac{7600}{54\,000} = 0.141 \text{ (lagging)}$$

The current flowing $= \dfrac{150}{25} = 6\text{A}$

The impedance of the coil $= \frac{180}{6} = 30\Omega$

Resistance of coil $= Z \cos \phi_L = 30 \times 0.141 = 4.23\Omega$

.Reactance of coil $= \sqrt{30^2 - 4.23^2} = 29.7\Omega$

Inductance of coil $= \frac{29.7}{2 \times \pi \times 50} = 0.0945\text{H}$

Power factor of coil $= \cos \phi_L = 0.141$ (lagging).

Example 86. A moving-iron voltmeter with a resistance of 1732Ω and an inductance of 0.625H registers 110V with maximum deflection on a 50Hz, a.c. circuit. It is required to be placed in a 230V, 50Hz a.c. circuit in series with a non-inductive resistor. Find the value of R, the required resistor.

X of meter $= 2 \times \pi \times 50 \times 0.625$
$= 196.25\Omega$

Z of meter $= \sqrt{1732^2 + 196.25^2}$
$= 1744\Omega$

Current for full-scale deflection $= \frac{110}{1744} = 0.063\text{A}$

On 230V. New circuit impedance must be

$$\frac{230}{0.063} = 3650\Omega$$

and $3650^2 = (1732 + R)^2 + 196.25^2$

$13\,322\,500 = (1732 + R)^2 + 38\,433.7$

$(1732 + R)^2 = 13\,284\,066$

or $1732 + R = 3640$

$R = 3640 - 1732 = 1908\Omega$.

Example 87. A coil of resistance 10Ω and inductance 0.1H is connected in series with a capacitor of capacitance 150μF, across a 200V, 50Hz supply. Calculate (a) the inductive reactance, (b) the capacitive reactance (c) the circuit impedance (d) the circuit current (e) the circuit power factor (f) the voltage drop across the coil (g) the voltage drop across the capacitor.

(a) Inductive reactance $= 2\pi fL = 2 \times \pi \times 50 \times 0.1$
$= 31.4\Omega$

(b) Capacitive reactance $= \frac{1}{2\pi fC} = \frac{10^6}{2 \times \pi \times 50 \times 150}$
$= 21.2\Omega$

(c) Resultant reactance $= 31.4 - 21.2 = 10.2\Omega$ (inductive)

Impedance $= \sqrt{R^2 + X^2} = \sqrt{10^2 + 10.2^2}$
$= 14.28\Omega$

(d) Circuit current $= \frac{200}{14.28} = 14\text{A}$

(e) Power factor $= \frac{10}{14.28} = 0.7$ (lagging)—Since the circuit reactance is net inductive

(f) Impedance of coil $= \sqrt{10^2 + 31.4^2}$
$= 33\Omega$

Voltage drop across coil $= 14 \times 33 = 462$V

(g) Voltage drop across capacitor $= 14 \times 21.2 = 296.8$
$= 297$V.

Note. The point already made in connection with resonance. Although resonance is not occurring here, the condition is working towards this and large voltages can be built up across components. Thus the fact that the voltages across the coil and capacitor are larger than the supply voltage is in accordance with theory and all the values are in order.

CHAPTER 10

PRACTICE EXAMPLES

1. A circuit has a resistance of 3Ω and an inductance of 0.01H. The voltage across its ends is 60V and the frequency is 50Hz. Calculate (a) the impedance (b) the power factor (c) the power absorbed.

2. A 100W lamp for a 100V supply, is placed across a 220V supply. What value of resistance must be placed in series with it so that it will work under its proper conditions? If a coil is used instead of the resistor and if the resistance of the coil is small compared to its reactance, what is the inductance of the coil? The frequency is 50Hz. What is the total power absorbed in each case?

3. An inductive load takes a current of 15A from a 240V, 50Hz supply and the power absorbed is 2.5kW. Calculate (a) the power factor of the load (b) the resistance, reactance and impedance of the load. Draw a phasor diagram showing the voltage drops and the current components.

4. Two inductive circuits A and B are connected in series across 230V, 50Hz mains. The resistance values are A 120Ω; B 100Ω. The inductance values are A 250mH; B 400mH. Calculate (a) the current (b) the phase difference between the supply voltage and current (c) the voltages across A and B (d) the phase difference between these voltages.

5. Two coils are connected in series. When 2A d.c. is passed through the circuit, the voltage drop across the coils is 20V and 30V respectively. When passing 2A a.c. at 40Hz, the voltage drop across the coils is 140V and 100V respectively. If the two coils in series are connected to a 230V, 50Hz supply, find the current flowing.

6. A simple transmission line has a resistance of 1Ω and a reactance at normal frequency of 2.5Ω. It supplies a factory with 750kW, 0.8pf (lagging) at a voltage of 3.3kV. Determine the voltage at the generator and its power factor. Find also the output of the generator and draw the phasor diagram.

7. A non-inductive resistor of 8Ω is connected in series with an inductive load and the combination placed across a 100V supply. A voltmeter (taking negligible current) is connected across the load and then across the resistor and indicates 48V and 64V respectively. Calculate (a) the power absorbed by the load (b) the power absorbed by the resistor (c) the total power taken from the supply (d) the power factors of the load and whole circuit.

8. A circuit, consisting of a resistor and a capacitor connected in series across a 200V, 40Hz supply, takes a current of 6.66A. When the frequency is increased to 50Hz and the voltage maintained at 200V, the current becomes 8A. Calculate the value of resistance and capacitance and sketch a phasor diagram (not to scale) for either frequency.

9. A coil, having an inductance of 0.5H and a resistance of 60Ω, is connected in series with a 10μF capacitor. The combination so formed is now connected across a sinusoidal supply and it is found that, at resonance, the p.d. across the capacitor is 100V. Calculate the current flowing in the circuit under this condition. Sketch the phasor diagram (not to scale).

10. A certain coil has a resistance of 400Ω and, when connected to a 60Hz supply, an impedance of 438Ω. If the coil is connected in series with a 40μF capacitor and a p.d. of 200V, 50Hz is applied to the circuit, find the current and the p.d. across the coil and the capacitor.

CHAPTER 11

A.C. CIRCUITS (continued) AND SYSTEMS

POWER IN THE A.C. CIRCUIT

From the various circuit conditions considered in Chapter 10, *ie* resistance in series with inductive reactance, resistance in series with capacitive reactance and resistance in series with both inductive and capacitive reactance; it was seen that, the current flowing was sinusoidal and displaced from the applied sinusoidal voltage by an angle ϕ, termed the phase angle of the circuit. The general expressions were:

For voltage $v = V_m \sin \omega t$ and for current $i = I_m \sin (\omega t - \phi)$ a lagging phase angle being assumed for convenience.

The instantaneous power $p = vi = V_m I_m \sin \omega t \sin (\omega t - \phi)$

$$\text{or } p = V_m I_m \left\{ \frac{\cos \phi - \cos (2\omega t - \phi)}{2} \right\}$$

$$= VI \cos \phi - VI \cos (2\omega t - \phi)$$

and average power $P = VI \cos \phi - 0$

$$\text{or } P = VI \cos \phi$$

VI is frequently called the '*apparent power*' of the circuit and *P* is referred to as the '*active power*'.

Then active power = apparent power × power factor.

The reason for calling $\cos \phi$ the 'power factor' can now be readily seen. It is the factor by which the apparent power must be multiplied to obtain the active value of power expended in a circuit.

So power factor $= \dfrac{\text{active power}}{\text{apparent power}}$ or $\cos \phi = \dfrac{P}{VI} = \dfrac{I^2R}{IZI}$

$= \dfrac{R}{Z}$ as deduced earlier.

The following is also of interest.

Since active power = apparent power × power factor then P (watts) = VI (volt amperes) × $\cos \phi$

Thus wattage is given by the volt amperes multiplied by the power factor. This can be expressed by $W = VA \cos \phi$ or $kW = kVA \cos \phi$.

Note. The term kVA is an accepted method of giving the rating of an a.c. generator, motor or transformer and it must be remembered that it *does not* indicate the power rating. More information is required before the latter can be deduced, and the power factor is usually specified at the same time. The volt amperes or VA of a circuit is a term in itself but more has yet to be said about its usage. It has been retained from the early days of electrical engineering, before standardising of terms and symbols was recognised as being beneficial and VA or kVA, as a rating, is now used internationally for a.c. circuits and machines.

ACTIVE AND REACTIVE COMPONENTS

These terms are usually used in connection with current but under certain conditions can be applied to voltage and power. Consider the phasor diagram (Fig 145), for a simple a.c. circuit with current lagging the voltage by an angle ϕ. If the current I is considered to be split into its two quadrature components I_a and I_r as shown then $I_a = I \cos \phi$ and $I_r = I \sin \phi$.

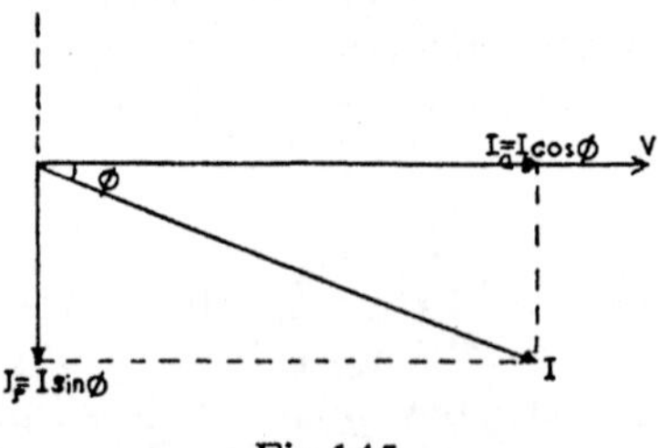

Fig 145

Since $I \cos \phi$ is a current, in phase with the voltage V and we know $VI \cos \phi$ is the measure of the power expended in a circuit, then it appears that $I \cos \phi$ is the component of current which is responsible for power dissipation. Thus $I \cos \phi$ is called the *active* power, wattful or working component of current. Similarly $I_r = I \sin \phi$, being always at right angles to voltage, is responsible for no power and is called the *reactive*, wattless or idle component of current. The example further illustrates these terms.

Example 88. A single-phase a.c. motor of 15kW and 90 per cent efficiency is run from a single-phase supply of 400V. Find the current taken from the mains, if the motor operates at 0.8 power factor (lagging). What is the value of the active current.

the reactive current and the motor rating in volt amperes.

Motor power output = 15kW = 15 × 1000 watts

Motor power intput = $\frac{15 \times 1000}{90} \times 100$ watts

= 16 667W or 16.7kW

The volt ampere rating = $\frac{16.67}{0.8}$ = 20.84kVA

The line current is obtained by dividing the volt-ampere value by the supply voltage. Thus $I = \frac{20.84 \times 1000}{400} = 52.1$A

Active component of current $I_a = 52.1 \times 0.8 = 41.7$A.

Reactive component of current = $I_r = 52.1 \times 0.6 = 31.3\,A$.

Note. If 16 667 was divided by 400, then I_a would have been obtained directly.

Thus $I_a = \frac{16\,667}{400} = 41.7$A. I could then be obtained by $\frac{41.7}{0.8} = 52.1$A and I_r as before, by $I \sin \phi = 52.1 \times 0.6$.

It is well to point out the simple relation for sin ϕ being 0.6 when cos ϕ is 0.8. This is obviously referring to a right-angled triangle of sides 10, 8 and 6. Similarly for examples, cos ϕ is frequently given as 0.707 or sometmes 0.7. This is referring to a right-angled isosceles triangle and sin ϕ in this case is also 0.707 or 0.7 (approx).

THE PARALLEL CIRCUIT

The parallel circuit is being treated under a separate heading, to remind the student that procedure is different to that for the series circuit. Nevertheless it will be seen that the method employed follows the familiar technique of phasor summation, *ie* that of resolving into the horizontal and vertical components or, to be more in line with the latest terms introduced in this chapter, into active and reactive components. The branches of the parallel circuit are made up of simple R, X_L or X_C values in series, and all work done in this connection will be in no way altered. For a parallel circuit it is pointed out that the same voltage is applied to all branches and it is usual to work with V as the reference for the phasor diagram. The current condition is often written as $\bar{I} = \bar{I}_1 + \bar{I}_2 + \bar{I}_3$ etc. The dash above the I is to remind one that, this is a phasor summation and not an arithmetical one. Thus all correct operations for a phasor summation must be performed.

INDUCTIVE IMPEDANCE IN PARALLEL

Assume two inductive impedances to be connected in parallel as shown in the diagram (Fig 146). Impedance Z_1 is made up of a resistance R_1 and inductive reactance X_1 whereas Z_2 is made up of resistance R_2 and inductive reactance X_2. The phasor diagram and circuit relationships are also shown. Since V is common to both branches it is used as the reference phasor. The problem is to find I where $\bar{I} = \bar{I}_1 + \bar{I}_2$.

Here $I_1 = \dfrac{V}{Z_1}$ and $I_2 = \dfrac{V}{Z_2}$

Resolving into active and reactive components, using arbitrary signs, we have $I_a = I_1 \cos \phi_1 + I_2 \cos \phi_2$

and $I_r = -I_1 \sin \phi_1 - I_2 \cos \phi_2$. It should be remembered that these phasors are vertically downwards.

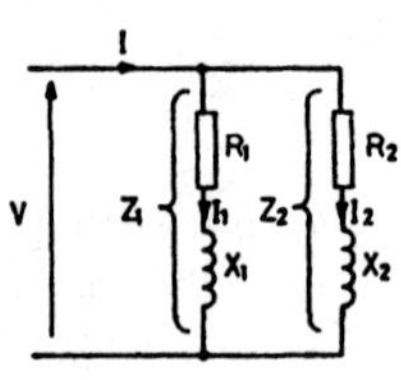

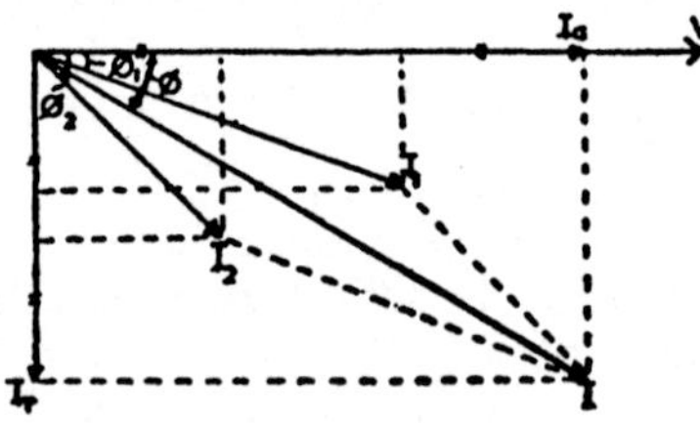

Fig 146

Then $I = \sqrt{I_a^2 + I_r^2}$ and $\cos \phi = \dfrac{I_a}{I}$. Here $\cos \phi$ is the power factor of the whole circuit.

Example 89. In the circuit shown above, let $R_1 = 3\Omega$ and $X_1 = 4\Omega$ while $R_2 = 8\Omega$ and $X_2 = 6\Omega$. If the applied voltage is $20V$, find the total current supplied and the power factor of the complete circuit. Find also the total power expended.

$$Z_1 = \sqrt{R_1^2 + X_1^2} = \sqrt{3^2 + 4^2} = 5\Omega$$

Then $I_1 = \dfrac{20}{5} = 4\text{A}$

$$Z_2 = \sqrt{R_2^2 + X_2^2} = \sqrt{8^2 + 6^2} = 10\Omega$$

and $I_2 = \dfrac{20}{10} = 2\text{A}$

$\cos \phi_1 = \dfrac{3}{5} = 0.6$ (lagging) $\sin \phi_1 = \dfrac{4}{5} = 0.8$

$\cos \phi_2 = \dfrac{8}{10} = 0.8$ (lagging) $\sin \phi_2 = \dfrac{6}{10} = 0.6$

Also $I_a = (4 \times 0.6) + (2 \times 0.8) = 4A$

$I_r = -(4 \times 0.8) - (2 \times 0.6) = -4.4A$

Whence $I = \sqrt{4^2 + 4.4^2} = 5.95A$

Circuit power factor $\cos \phi = \dfrac{4}{5.95} = 0.67$ (lagging)

Power expended $= 20 \times 5.95 \times 0.67 = 80W$

The above can be checked thus:

Power in branch 1 $= I_1^2R_1 = 4^2 \times 3 = 48W$

Power in branch 2 $= I_2^2R_2 = 2^2 \times 8 = 32W$

Total 80W.

INDUCTIVE AND CAPACITIVE IMPEDANCES IN PARALLEL

The procedure for solving problems, associated with this type of circuit, follows that outlined above, except that due allowance is made for the directions and signs when adding the reactive components. Thus in the diagram (Fig 147), impedance Z_2 is made up of resistance R_2 and capacitive reactance X_2 in series. The phasor for the reactive component of current is drawn vertically upwards and is allocated a +ve sign, whereas the reactive component of current for branch 1 is allocated a −ve sign. The total of the reactive components is thus a difference, as will be noted. The voltage is again used as the reference for the phasor diagram.

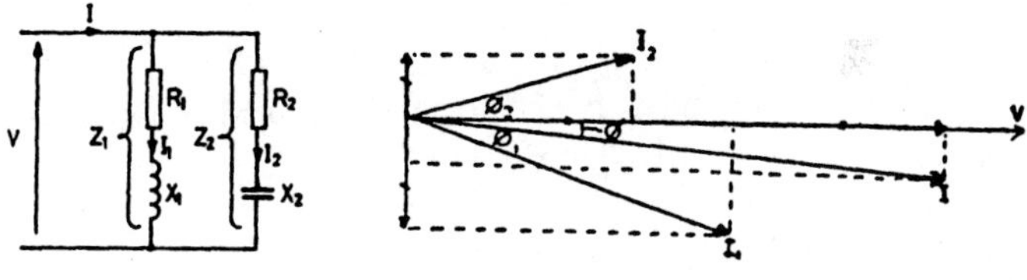

Fig 147

As before $I_a = I_1 \cos \phi_2 + I_2 \cos \phi_2$

and $I_r = - I_1 \sin \phi_1 + I_2 \sin \phi_2$. I_r will carry either a +ve or −ve sign, decided by the relative values of $I_1 \sin \phi_1$ and $I_2 \sin \phi_2$. Thus the resulting reactive component will act either upwards or downwards and the resultant circuit current may be lagging or leading as shown by the example. As before $I = \sqrt{I_a^2 + I_r^2}$ and $\cos \phi = \frac{I_a}{I}$. The qualifying term lagging or leading is decided by the sign of I_r.

Example 90. A circuit consists of two branches in parallel. Branch A consists of a 20Ω resistor in series with a 0.07H inductor, while branch B consists of a 60μF capacitor in series with a 50Ω resistor. Calculate the mains current and the circuit power factor, if the voltage is 200V at 50Hz.

Branch A. $X_A = 2\pi fL = 2 \times 3.14 \times 50 \times 0.07$

$= 22\Omega$

$R_A = 20\Omega$

$\therefore Z_A = \sqrt{20^2 + 22^2} = 29.7\Omega$

Thus $I_A = \dfrac{200}{29.7} = 6.74\text{A}$ and $\cos\phi_A = \dfrac{R_A}{Z_A} = \dfrac{20}{29.7}$

$= 0.674$ (lagging)

$\sin\phi_A = \dfrac{X_A}{Z_A} = \dfrac{22}{29.7} = 0.74$

Branch B. $X_B = \dfrac{1}{2\pi fC} = \dfrac{10^6}{2 \times 3.14 \times 50 \times 60}$

$= 53\Omega$

$R_B = 50\Omega$

$\therefore Z_B = \sqrt{50^2 + 53^2} = 72.8\Omega$

Thus $I_B = \dfrac{200}{72.8} = 2.75\text{A}$

$\cos\phi_B = \dfrac{R_B}{Z_B} = \dfrac{50}{72.8} = 0.686$ (leading)

$\sin\phi_B = \dfrac{X_B}{Z_B} = \dfrac{53}{72.8} = 0.728$

Then $I_a = (6.74 \times 0.674) + (2.75 \times 0.68)$

$= 4.55 + 1.885 = 6.43\text{A}$

$I_r = -(6.74 \times 0.74) + (2.75 \times 0.728)$

$= -5 + 2.005 = -2.995\text{A}.$

Note. The mains current will lag, since the effect of the inductive branch predominates.

$I = \sqrt{I_a^2 + I_r^2} = \sqrt{6.43^2 + 2.995^2}$

$I = 7.1\text{A}$

$\cos\phi = \dfrac{I_a}{I} = \dfrac{6.43}{7.1} = 0.902$ (lagging).

The mains current is 7.1 amperes and the circuit operates at a lagging power factor of 0.9.

PARALLEL RESONANCE

Before passing on to the more practical applications of parallel working, it would be well to point out that, a condition of resonance can occur for the parallel circuit. This condition is often termed 'current resonance' to distinguish it from 'voltage resonance' as dealt with for the series circuit. It will be seen, from the example set out above, that a condition can arise when $I_A \sin \phi_A = I_B \sin \phi_B$ and as these are the reactive components of currents in inductive and capacitive branches, then they will oppose each other tending to produce a total reactive component of zero value. The remaining active components will total to give the line current, since $I = \sqrt{I_a^2 + 0} = I_a$ and the combined circuit will operate at unity power factor. This is illustrated by the phasor diagram (Fig 148), from which it is seen that, since the power factors of both branches are low, the phase angles ϕ_A and ϕ_B are large and $I_A \cos \phi_A$ and $I_B \cos \phi_B$ are small compared to the reactive components.

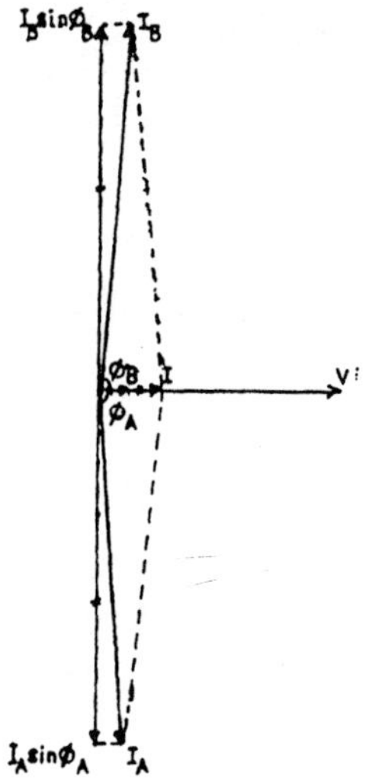

Fig 148

Large currents can flow in the choke-coil and capacitor branches, which are very much greater than the main supply current and these, are therefore, not supplied from the line. On examining the power waves for an inductor and capacitor, it will be seen that they are directly opposite in phase, as are the current waves. It can be assumed that as the capacitor discharges, the power given out is absorbed by the choke in building up its field. When the field collapses, the power released charges the capacitor and there is a current due to oscillation of power between choke and capacitor. Apparatus using such a circuit is an oscillator and has many applications in radio and electrical filter circuits. If no supply is available the current is not maintained, due to energy loss in the circuit resistance which, however small, cannot be neglected. To maintain the oscillatory current, the resistance loss must be supplied at the correct frequency from the external supply force.

POWER-FACTOR IMPROVEMENT

The full meaning and advantages of this technique, which is much used in practical electrical engineering work, is best illustrated by the use of an example, as now considered.

Example 91. (a) Two inductive coils of resistance values 5Ω and 8Ω and inductance values of 0.02H and 0.01H respectively are connected in parallel across a 240V, 50Hz supply. Find the coil currents, the circuit current and its power factor.

The arrangement is shown by the diagram (Fig 149). A phasor diagram is also drawn.

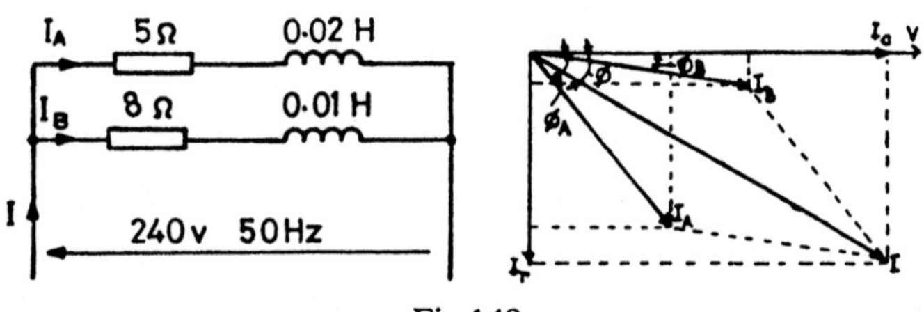

Fig 149

$$\text{Branch A. } X_A = 2\pi fL = 2 \times \pi \times 50 \times 0.02$$
$$= 6.28\Omega$$
$$Z_A = \sqrt{5^2 + 6.28^2}$$
$$= 8.02\Omega$$
$$I_A = \frac{240}{8.02} = 29.8\text{A} \quad \cos\phi_A = \frac{5}{80.2}$$
$$= 0.622 \text{ lagging}$$
$$\sin\phi_A = \frac{6.28}{8.02} = 0.78$$
$$\text{Branch B. } X_B = \tfrac{1}{2} \text{ that of branch A, since } L \text{ is halved}$$
$$= 3.14\Omega$$
$$Z_B = \sqrt{8^2 + 3.14^2}$$
$$= 8.6\Omega$$
$$I_B = \frac{240}{8.6} = 27.9\text{A} \quad \cos\phi_B = \frac{8}{8.6}$$
$$= 0.93 \text{ lagging}$$

$$\sin \phi_B = \frac{3.14}{8.6} = 0.366$$

Then $I_a = (29.8 \times 0.622) + (27.9 \times 0.93)$
$= 44.6A$
$I_r = (29.8 \times 0.78) + (27.9 \times 0.366)$
$= 33.5A$

It will be noted that the arbitrary −ve sign has not been used here since, both branches are inductive and there is no doubt as to, the resultant current being lagging.

Then $I = \sqrt{44.6^2 + 33.5^2}$
$I = 55.6A$

$$\text{p.f.} = \cos \phi = \frac{44.6}{55.6} = 0.801 \text{ (lagging)}.$$

Example (contd) (b) Find the effect on the main circuit current and power factor, if a capacitor of 400μF, was connected across the supply in parallel with the coils.

The phasor diagram (Fig 150), shows the new conditions.

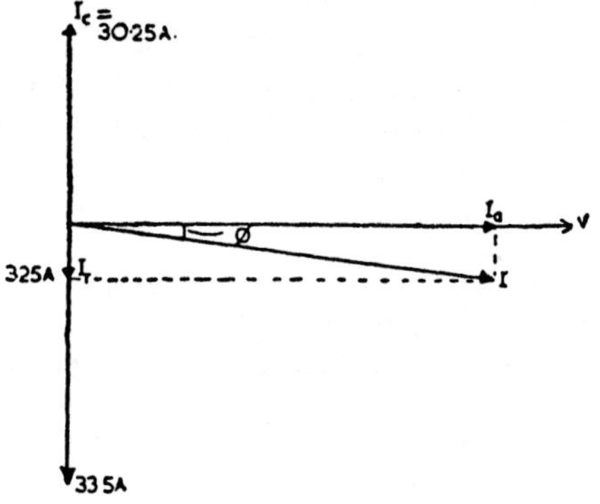

Fig 150

Branch C. Reactance of capacitor $X_C = \frac{1}{2\pi fC}$

$$\text{Thus } X_C = \frac{10^6}{2 \times \pi \times 50 \times 400}$$

$$= 7.95\Omega$$

$$\therefore I_C = \frac{240}{7.95} = 30.25A$$

and there being no resistance in branch C—only capacitive reactance, then $\cos \phi_C = 0$ and $\sin \phi_C = 1$

Again I_C acts at 90° to the voltage and is wholly reactive, there being no active component. Then I_a as before = 44.6A

and $I_r = -23.3 - 10.2 + 30.25$
or $I_r = -33.5 + 30.25 = -3.25\text{A}$

It will be seen that the arbitrary signs have been introduced here, because the reactive current of branch C acts in the opposite direction to that of branches A and B.

The circuit current is now:

$$I = \sqrt{I_a^2 + I_r^2} = \sqrt{44.6^2 + 3.25^2}$$
$$\therefore I = 44.6\text{A}$$

and $\cos\phi = \dfrac{44.6}{44.6} = 1.0$ *ie* unity.

From the above example it is seen that, by connecting a capacitor in parallel with the inductive loads, the total line current is reduced from 55.6 to 44.6 amperes and the overall circuit power factor is improved from 0.8 (lagging) to unity. The resulting advantages of the arrangement are now considered in detail.

ADVANTAGES OF P.F. IMPROVEMENT

For the majority of commercial loads, the current lags behind the voltage, due to the inductance of the apparatus or the operating characteristics of motors and control gear. Typical values of power factor are:

System supplying lighting loads only: power factor (lagging) = 0.95.

System supplying lighting and power loads: power factor (lagging) = 0.75 to 0.85.

System supplying power loads: power factor (lagging) = 0.5 to 0.7.

The lower the power factor, the greater the line current need be for any given load *kW* or output power rating and the attendant disadvantages are:

(a) The transmission losses in the supply cables or power lines are increased: these being given by I^2R, where R is the cable or line resistance.

Thus for a given amount of power transmitted, the current at 0.7 power factor is $\dfrac{1}{0.7} \times$ current at unity power factor $= 1.43 \times$ current at unity power factor. Also the transmission loss at 0.7 power factor is $(1.43)^2 \times$ loss at unity power factor $= 2 \times$ loss at unity power factor.

(b) Because of the larger currents resulting from a low power factor, there will be a greater voltage drop in the supply lines resulting in a lower voltage at the load. Conversely the size of the conductors must be increased to keep the voltage drop figure to an acceptable value.

(c) Again because of the larger current resulting from a low power factor, the size of the current-carrying conductors in transformers, control-gear and alternators must be larger than need be. This means that the physical dimensions of the equipment must be larger and that advantage is not taken of good design. The equipment is also more costly.

(d) The 'Regulation'—a term used for the 'sitting-down' of the voltage of generating and transmitting plant, is adversely affected by a low power factor. The lower the power factor, the greater the internal voltage drop in this equipment *ie* armature reaction and attendant effects are worsened.

Electricity supply authorities encourage good power factor operation by offering advantageous tariffs. Power factor can be improved by adding to existing loads, apparatus which will take sufficient leading current to minimise the lagging current of the load. Static capacitors are frequently used for this purpose.

Example 92. A 40kW load, operating at 0.707 power factor (lagging), is supplied from 500V, 50Hz mains. Calculate (a) the capacitor value required to raise the line power factor to unity (b) the capacitance required to raise the power factor to 0.95 (lagging)

$$\text{(a) Load } kVA = \frac{kW}{\cos\phi} = \frac{40}{0.707}$$

$$\text{Load current} = \frac{40 \times 10^3}{0.707 \times 500} = 113.15\text{A}$$

$$\begin{aligned}\text{Active component of load current } I_1 &= I_1 \cos\phi_1 \\ &= 113.15 \times 0.707 \\ &= 79.997\text{A} = 80\text{A}\end{aligned}$$

$$\begin{aligned}\text{Reactive component of load current } I_1 &= I_1 \sin\phi_1 \\ &= 113.15 \times 0.707 \\ &= 80\text{A}.\end{aligned}$$

To nullify this reactive current, a capacitor can be fitted to operate in parallel with the load. This capacitor must pass a similar value of reactive current as shown by the phasor diagram (Fig 151a).

Thus I_C must be 80A. Reactance X_C of capacitor must be

$$\frac{500}{80} = 6.25\Omega$$

Since $X_C = \dfrac{1}{2\pi fC}$ then $\dfrac{10^6}{2\pi fC} = 6.25$ where C is in microfarads

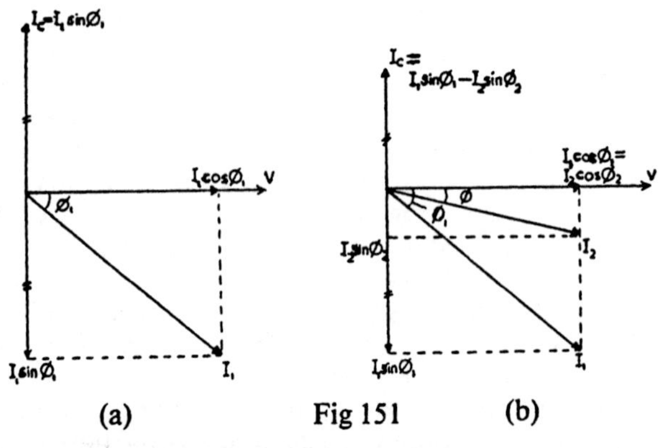

(a) Fig 151 (b)

$$\text{Thus } C = \frac{10^6}{2 \times 3.14 \times 50 \times 6.25}$$

$$C = 510\mu\text{F}.$$

(b) For this part of the problem it will be seen that $I_1 \sin \phi_1$ is not to be cancelled completely since, the line phase angle is only to be reduced from ϕ_1 to ϕ_2 and line current to a new value I_2. This is illustrated by the diagram (Fig 151b). Since the power or active component remains the same, then for this condition $VI_2 \cos \phi_2 = 40\ 000$ as before.

$$\therefore I_2 = \frac{40 \times 10^3}{500 \times 0.95} = 84.2\text{A}$$

Also since $\cos \phi_2 = 0.95$ (lagging) $\sin \phi_2 = 0.312$ and $I_2 \sin \phi_2 = 84.2 \times 0.312 = 26.1$A.

The line reactive current component has now to be reduced from 80A to 26.1A = 53.9A. This then must be the new value of I_C or a capacitor must be used which takes a current of 53.9A.

$$\text{Thus } X_C = \frac{500}{53.9} = \frac{10^6}{2\pi fC} \text{ (}C\text{ being in microfarads)}$$

$$\text{or } C = \frac{10^6 \times 53.9}{500 \times 2 \times 3.14 \times 50}$$

$$C = 343\mu\text{F}.$$

It will be noted that 343μF will bring the line power factor to 0.95 and that a further (510 − 363) = 167μF would be needed to bring the value to unity. Since the cost of a capacitor depends on its capacitance value and little advantage is gained by improving the power factor above 0.95, it is not always necessary to achieve unity power factor working. In this way some saving to the consumer can be effected. It is important to note that, although power factor is improved, an increased power output is not obtained from the load. Students frequently have the erroneous idea that, if for example, the power-factor working of a circuit supplying a 5kW motor is improved, then the motor will then give an output greater than 5kW. This is not the case. All that is achieved is that, by connecting an additional item of static apparatus across the motor, the total line current is reduced, *ie* a condition can be attained when the minimum supply current required for a specified power output is used. This minimum supply current reduces all the disadvantages already enumerated but the motor current itself remains unaltered.

kW, kVA and *kVAr*

As mentioned earlier in the chapter, the above terminology is much used in practical electrical engineering and some revision is necessary, before the full treatment is considered. The diagram (Fig 152a and b) should be considered with the text. For a circuit, where the current and voltage are out of phase, the phasor diagram is as shown. Current I can be resolved into an in-phase or active component $I \cos \phi$ and an out-of-phase or reactive component $I \sin \phi$. $I \cos \phi$ is responsible for all the power dissipated by the circuit, since $P = VI \cos \phi$, and is also called the power or wattful component, whilst $I \sin \phi$ is responsible for no power, being at right angles to the voltage, and is called the wattless or reactive component.

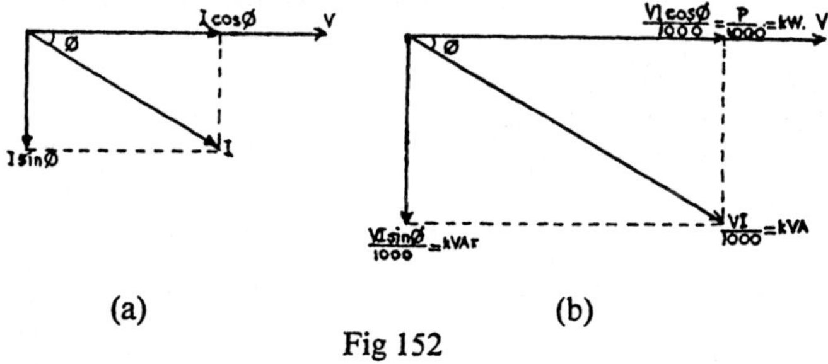

(a) (b)

Fig 152

From the expression $P = VI \cos \phi$ it is seen that P can be the 'active' component of VI (symbol S—see *Note*), and if the term volt amperes or kilovolt amperes is used for VI then the kVA (kilovolt amperes) can be regarded as being resolved into two components, one of which is the power component. The term W or kW (kilowatts) can be used to describe this component and the other component termed the 'volt amperes reactive' or 'reactive kilovolt amperes' can be designated by VAr or $kVAr$. If the current phasors of Fig 152a are multiplied by V, the new condition becomes more apparent and leads to a power diagram. The product VI (S) is shown as the volt amperes (VA) or $\frac{VI}{1000} = kVA$ and is referred to as the 'apparent power'.

Since $VI \cos \phi = P$ then $VA \cos \phi = W$ and $kVA \cos \phi = kW$. kW is a measure of the 'active power', in line with, original definition for power factor. *ie* the ratio of active power to apparent power.

Thus: power factor or $\cos \phi = \dfrac{kW}{kVA}$

Similarly $VI \sin \phi$ (Q—see *Note*), or $kVA \sin \phi$ is the 'reactive power' or volt amperes reactive designated by $kVAr$ and from the power diagram (Fig 152b), we have:

$$\text{Apparent Power} = \sqrt{\text{Active Power}^2 + \text{Reactive Power}^2}$$

$$\text{Summarising } kW = kVA \cos \phi. \quad kVAr = kVA \sin \phi$$

$$kVA = \sqrt{kW^2 + kVAr^2}$$

$$\cos \phi = \frac{kW}{kVA} \qquad \sin \phi = \frac{kVAr}{kVA}$$

Note. The symbols S, P and Q are recommended as substitutes for VI, $VI \cos \phi$ and $VI \sin \phi$ but it is probable that the units: kilovolt amperes, kilowatts and kilovolt amperes reactive will continue to be used and shown on the phasor diagrams, since this is the older, though basically inconsistent, practice of the electrical power engineer. The appropriate alternative has been introduced and shown where it is considered to be appropriate in this chapter.

For summarising we have:

$$P = VI \cos \phi \qquad Q = VI \sin \phi \quad \text{and} \quad S = VI$$

$$\text{Thus } S = \sqrt{P^2 + Q^2} \text{ and } P = S \cos \phi \quad Q = S \sin \phi$$

$$\cos \phi = \frac{P}{S} \text{ and } \sin \phi = \frac{Q}{S}.$$

It must be remembered that the kVA values of various loads are not in phase and therefore cannot be added arithmetically. kW values are all active components, are in phase and can be added. $kVAr$ values are reactive components, they can be in-phase or in anti-phase and can be added, provided due allowance is made for the sign. This is shown by the following examples.

Example 93. Two loads are connected in parallel. Load A is 800kVA at 0.6 (lagging). Load B is 700kVA at 0.8 power factor (lagging). Find the total *kW*, *kVA* and overall power factor of the joint loads.

See the diagram (Fig 153).

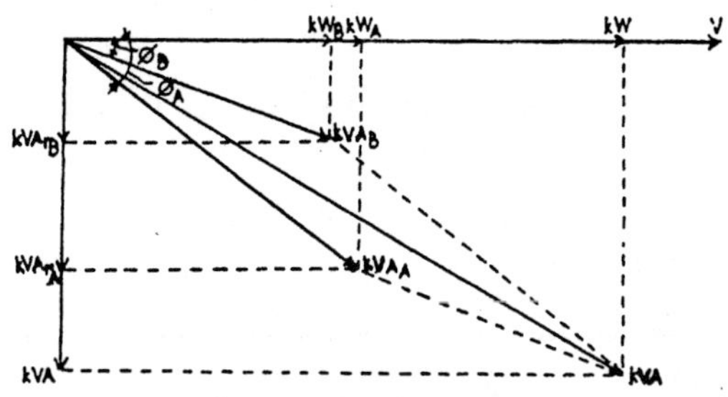

Fig 153

$$\text{For load A. } \cos\phi_A = 0.6,\ \sin\phi_A = 0.8$$

$$\text{Active power, } P_A = VI_A\cos\phi_A = 800 \times 0.6 = 480\text{kW}$$

$$\text{Reactive power, } Q_A = VI_A\sin\phi_A = 800 \times 0.8 = 640\text{kV Ar}$$

$$\text{For load B. } \cos\phi_B = 0.8,\ \sin\phi_B = 0.6$$

$$\text{Active power, } P_B = VI_B\cos\phi_B = 700 \times 0.8 = 560\text{kW}$$

$$\text{Reactive power, } Q_B = VI_B\sin\phi_B = 700 \times 0.6 = 420\text{kV Ar}$$

$$\text{Total active power, } P = 480 + 560 = 1040\text{kW}$$

$$\text{Total reactive power, } Q = 640 + 420 = 1060\text{kVAr}$$

$$\text{Total apparent power, S} = \sqrt{1040^2 + 1060^2}$$

$$= 1485\text{kVA}$$

$$\text{Overall power factor} = \frac{1040}{1485} = 0.7 \text{ (lagging)}$$

Problems involving a number of loads, may be best treated by setting out the power-diagram components in tabular fashion as shown. An arrow has been drawn in to illustrate the phasor direction and to remind the student as to which columns can be added arithmetically.

Example 94. A 220V, single-phase alternator supplies the following loads:

(a) 20kW at unity power factor for lighting and heating.

(b) A 75kW induction motor having an efficiency of 90.5 per cent operating at a power factor of 0.8 (lagging).

(c) A synchronous motor taking 50kV A at a power factor of 0.5 (leading).

Find the total kVA, current and the power factor of the combined load.

Load (a) can be set into the columns directly as shown.

Load (b) Motor power output = rating as given = 75kW

$$\text{Motor input active power} = \frac{75}{0.905} = 82.9\text{kW}$$

$$\text{Apparent power} = \frac{82.9}{0.8} = 103.6\text{kVA}$$

Load (c) can also be set into the columns directly.

Load	kVA (S) ↘	kW or → $kVA\cos\phi$ (P)	$kVAr$ or ↕ $kVA\sin\phi$ (Q)	$\cos\phi$	$\sin\phi$
a	20 →	20 →	0	1	0
b	103.6 ↘	82.9 →	−62.16 ↓	0.8	0.6
c	50 ↗	25 →	43.3 ↑	0.5	0.866
		127.9 →	−18.86 ↓		

$$\text{Total apparent power (S)} = \sqrt{127.9^2 + 18.9^2} = 129\text{kV A}$$

$$\text{Total current} = \frac{129 \times 1000}{220} = 588\text{A}$$

$$\text{Resultant power factor} = \frac{127.9}{129} = 0.99 \text{ (lagging)}$$

Note. reactive component of inductive load predominates, hence the resultant lagging power-factor condition.

POWER-FACTOR IMPROVEMENT (kVA method)

Treatment of problems follows lines, similar to those set out for the 'current method'. The diagram for the load condition is built up by splitting the original load kVA into its kW and $kVAr$ components. Since the kW remains the same, then for a new power-factor condition for the supply, the final $kVAr$ value is obtained by reducing the original $kVAr$ by an amount equal to the $kVAr$ of the apparatus being added. Such apparatus must use no power and the static capacitor is such an item of equipment. The added $kVAr$ being leading, will reduce the lagging *kVAr* of the supply. It should be noted that if a synchronous motor is used to obtain a better overall power factor, then this also contributes output power which must be taken into account. This was illustrated by the previous example.

Example 95. A 400V, 50Hz, 20kW, single-phase induction motor has a full-load efficiency of 91.15 per cent and operates at a power factor of 0.87 (lagging). Find the *kVAr* value of the capacitor to be connected in parallel to improve the circuit power factor to 0.95 (lagging). Find also the capacitance value of this capacitor. The diagram (Fig 154), illustrates the problem and solution.

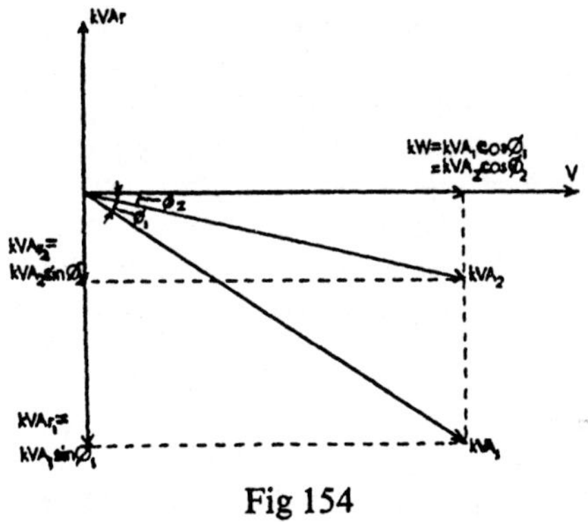

Fig 154

Motor output = motor rating as given = 20kW

$$\text{Motor active input power } P_1 = \frac{20}{0.9115} = 21.94\text{kW}$$

$$\text{Motor apparent power } S_1 = \frac{21.94}{0.87} = 25.22\text{kVA}$$

Also, since $\cos\phi_1 = 0.87$ then, from tables, $\sin\phi_1 = 0.493$. Thus Q_1 or $S_1 \sin\phi_1 = 25.22 \times 0.493 = 12.44$kVAr

Although the power factor of the circuit is to be improved to $\cos\phi_2$, the power of the circuit is not altered $\therefore P_1 = P_2$

or $S_1 \cos\phi_1 = S_2 \cos\phi_2$ whence

$$S_2 = \frac{S_1 \cos\phi_1}{\cos\phi_2}$$

$$\text{or } S_2 = 25.22 \times \frac{0.87}{0.95} = 23.1\text{kVA}$$

Again $\cos\phi_2 = 0.95$ therefore, from tables, $\sin\phi_2 = 0.3123$ and $Q_2 = 23.1 \times 0.3123 = 7.21$kVAr

Required Q value = 12.44 − 7.21 = 5.23kVAr. This therefore must be the rating of the capacitor.

$$\text{Capacitor current } I_C = \frac{5230}{400} = 13.75\text{A}$$

$$\text{Capacitor reactance } X_C = \frac{400}{13.75} = 30.59\Omega$$

$$\text{or } X_C = 30.59 = \frac{10^6}{2 \times \pi \times 50 \times C}$$

where C is the value in microfarads.

$$\text{Hence } C = \frac{10^4}{30.59 \times \pi} = 104\mu\text{F}.$$

POLYPHASE WORKING

The student who intends to have a good practical knowledge of electrotechnology, must make himself thoroughly conversant with the terms, relationships and theory of polyphase working. The importance of the work now to be covered cannot be too strongly stressed. Experience has shown that most students consider this part of theory to be 'that little extra, which breaks the camel's back' and accordingly give it insufficient attention at first. The result is that much hasty revision is necessary when the various a.c. machines are to be studied later. Detailed attention to fundamentals will bring long-term advantages and, although the next book will be devoted to more advanced a.c. technology, the subject matter now to be considered must be treated as basic and essential to such further studies. It must be both understood and memorised.

THREE-PHASE SYSTEMS

Universal practice has established 3-phase systems to be the most advantageous for polyphase working. A single-phase supply, as is usual for small installations, can always be obtained from a 3-phase system and in this way the relative advantage of either system is available. The 2-phase system is only rarely used and its unimportance does not warrant its study at this stage. More than 3-phase arrangements, such as 6-phase, have relatively fewer and even more specialised applications and here we confine our investigation to 3-phase working only.

Consider a 2-pole magnet, as shown in the diagram (Fig 155), to be rotated inside an external stationary armature or stator. Three coils are shown equally displaced, with 'starts' and 'finishes' marked symmetrically to make a regular arrangement. Induced e.m.fs. will result in each coil, which are identical in magnitude but displaced in phase by 120 electrical degrees. A phase sequence R-Y-B is assumed *ie* the rotor turns so that the red-phase voltage reaches its maximum 120° before the yellow-phase voltage reaches its maximum and the latter 120° before the blue-phase voltage, as shown by the waveform diagram. A sinusoidal distribution of rotor flux is assumed and that sine-wave e.m.fs. are induced. The methods by which this is achieved will be considered under the detailed study of the alternator. The three separate coils can be used to supply three independent single-phase loads, but advantages are obtained by interconnecting the coils or phase windings and the two important methods are described as either the STAR or DELTA connection.

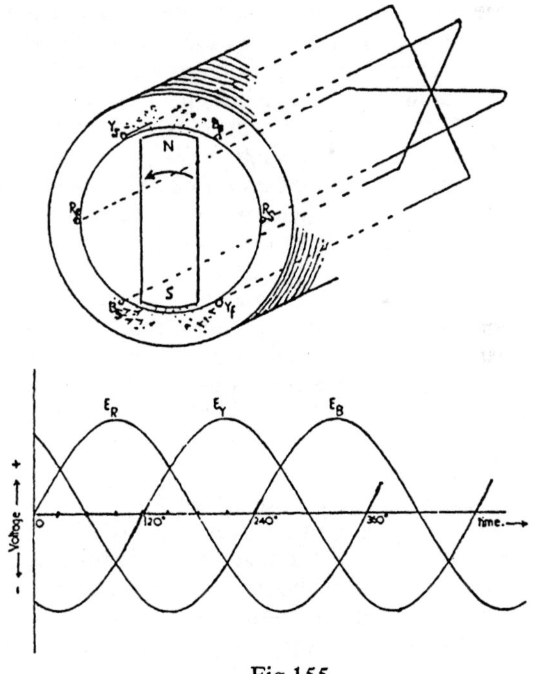

Fig 155

STAR OF Y CONNECTION

The diagram (Fig 156), shows the arrangement and it will be seen that here, the three coils or phase windings are connected so that either all the *starts* or *finishes* are joined together to form the star-point *ie* 'corresponding ends' are connected together.

Similarly the supply lines are connected to the free ends, remote from the star-point.

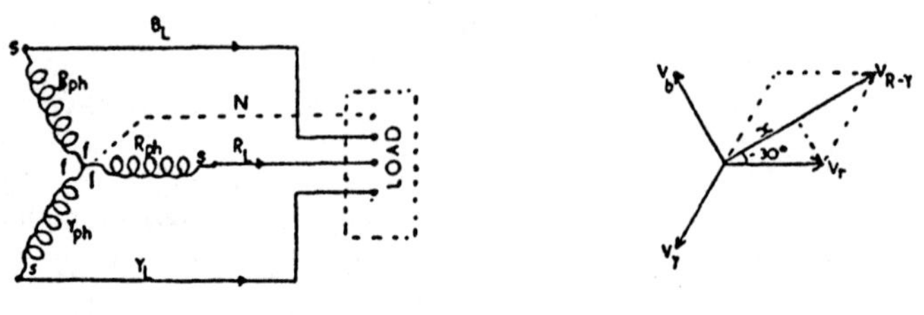

Fig 156

The phasor diagram has been drawn in terms of voltage with the red-phase voltage (V_r) used as the reference. The notation being used from now on should be observed. It will be seen that the small letter suffix denotes the phase value, while the capital letter denotes the line value. The lines have been identified with the colours of the phases to whose 'starts' they have been connected. The double suffix such as V_{R-Y} denotes the voltage between lines, the example being the Red to Yellow line voltage.

Assume the condition when the red-phase voltage wave is positive and that the 'start' of the red-phase winding is +ve with respect to the 'finish' or neutral point. Current will flow through the lines and load as shown. For the example and condition being considered this is possible because, for the yellow phase at the same instant, its start will be −ve with respect to its finish, since the yellow phase waveform is in its −ve half-cycle.

Thus for the phasor diagram, the voltage between the red and yellow lines is obtained by the phasor *difference* of V_r and V_y.

Since a phasor difference is being considered, the resultant is obtained by reversing one phasor with respect to the other and completing the parallelogram. From the deduction set out below, it will be seen that the line voltage is √3 times a phase voltage. This relation also holds for the other lines and the associated phases. A further point of importance for the star connection, is that the line current equals the phase current or $I_L = I_{ph}$.

Consider the phasor diagram. Let the line voltage $V_{R-Y} = 2x$

But $\frac{x}{V_r} = \cos 30°$ $\therefore x = \frac{\sqrt{3}}{2} V_r$ or $2x = \sqrt{3} V_r$

Hence $V_{R-Y} = \sqrt{3} V_r$

or the voltage between lines = √3 × a phase voltage.

Thus $V_L = \sqrt{3} \times V_{ph}$.

For a star connection the following must therefore be remembered.

Line voltage = $\sqrt{3}$ Phase voltage

or $V = \sqrt{3}V_{ph} = 1.732V_{ph}$

Line current = Phase current or $I = I_{ph}$

It will be noted that the subscript L, as in V_L and I_L, is omitted when generalising. This is usual and both V and I can be assumed to be line values. Again the relations deduced have been derived for an alternator or source of supply but they also relate to a star-connected load as the example shows.

Example 96. Three 50Ω resistors are connected in star across 415V, 3-phase mains. Calculate the line and phase currents and the power taken from the supply.

Since the load is balanced, the voltage across each resistance is the correct phase voltage. Thus $V_{ph} = \frac{415}{\sqrt{3}} = 240\text{V}$

Phase current = line current or $I_{ph} = I$

$= \frac{240}{50} = 4.8\text{A}$

Power dissipated by one phase of load

$= I_{ph}^2 R_{ph}$

$= 4.8^2 \times 50\text{W}$

$= 1152\text{W}$ or 1.152kW

and 3-phase power from the supply

$= 3 \times 1.152 = 3.456\text{kW} = 3.5\text{kW}.$

USE OF THE NEUTRAL

One obvious use of the star-connection is for distribution, since two voltages are available to the consumer, one for lighting and the other for power. Either 1-phase or 3-phase loading is also possible and this is shown by the diagram (Fig 157).

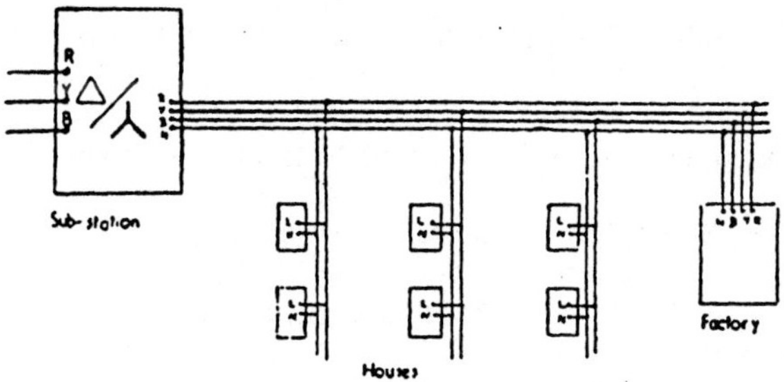

Fig 157

BALANCED LOAD

A 3-phase load is said to be balanced when the currents in all three phases are equal and their phase angles are the same. If an instant in time is considered, as shown, on the diagram (Fig 158), it will be seen that the sum of instantaneous values of the currents $i_r + i_y + i_b = 0$. Since these currents meet at the load neutral point and the resultant flows through the neutral line then the neutral carries no current and need not be used for balanced loading.

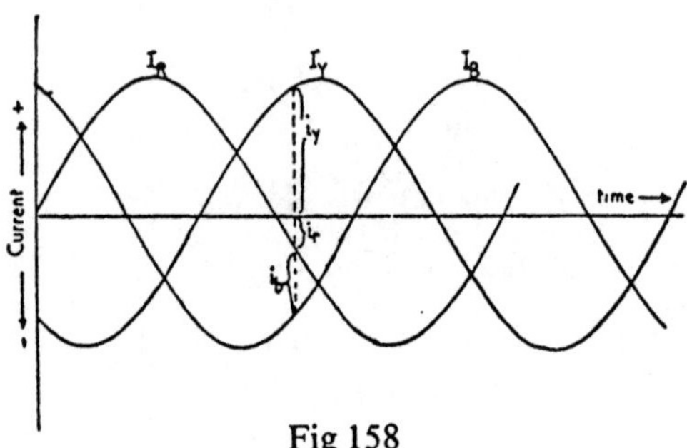

Fig 158

UNBALANCED LOAD

A neutral must be used if the load phase currents are unequal or if their phase angles are different. The neutral line will carry the unbalanced current *ie* the resultant of the three line currents. Since this neutral current is a phasor sum, it can be obtained graphically or mathematically, as shown by the example.

Example 97. The loads of a 4-wire, 3-phase system are:
Red line to neutral current = 50A, power factor of 0.707 (lagging).
Yellow line to neutral current = 40A, power factor = 0.866 (lagging)
Blue line to neutral current = 40A, power factor = 0.707 (leading)

Determine the value of the current in the neutral wire. The solution is worked with reference to the diagram (Fig 159).

I_r = 50A lagging the voltage by 45° since cos 45 = 0.707
I_y = 40A lagging the voltage by 30° since cos 30 = 0.866
I_b = 40A leading the voltage by 45° since cos 45 = 0.707

Resolving into horizontal and vertical components.

$$I_H = (50 \times \cos 45) - (40 \times \cos 30) - (40 \times \cos 15)$$
$$= (50 \times 0.707) - (40 \times 0.866) - (40 \times 0.966)$$
$$= 35.35 - 34.64 - 38.64 = -37.93\text{A}$$

$I_V = -(50 \times \sin 45) - (40 \times \sin 30) + (40 \times \sin 15)$
$= -(50 \times 0.707) - (40 \times 0.5) + (40 \times 0.259)$
$= -35.35 - 20 + 10.36 = -44.99$

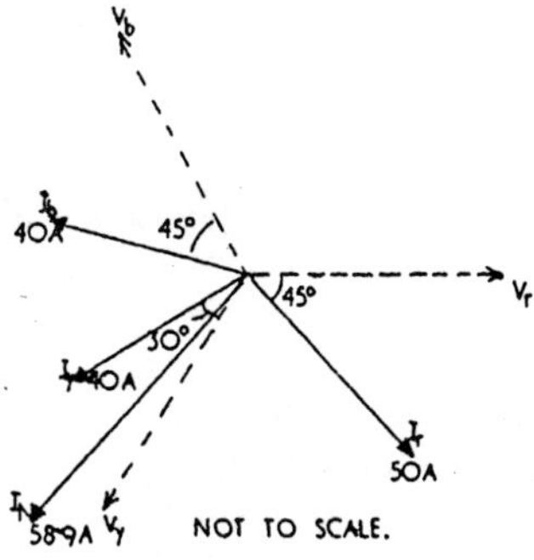

Fig 159

Current in the neutral is the resultant.
or $I_N = \sqrt{37.93^2 + 44.92^2}$
$= 58.9\text{A}$

$$\cos \theta = \frac{-37.93}{58.9} = -0.653$$
$$\theta = 49.5°.$$

Note. The −ve sign gives the quadrant in which I_N lies. This is shown on the diagram.

DELTA OR Δ (MESH) CONNECTION

The arrangement is shown in the diagram (Fig 160). For this connection, the three-phase windings are arranged to form a closed circuit by connecting 'uncorresponding ends' *ie* the *start* of one phase to the *finish* of another phase. Thus R start is connected to B finish, Y start to R finish, etc. The same reasoning as was introduced for Fig 156, is applied here, except that the voltages are considered. Thus if the diagram represented the three equal phase voltages, then it would be seen that, for any instant of time, the sum of the instantaneous values of voltages $v_r + v_y + v_b = 0$. For the mesh or closed winding, since the sum of the instantaneous voltages is zero, no circulating current flows round the mesh. The lines are taken from the junction points and for this connection, it is obvious that the voltage developed across a phase is the voltage provided for the connected lines.

Thus V_L or $V = V_{ph}$

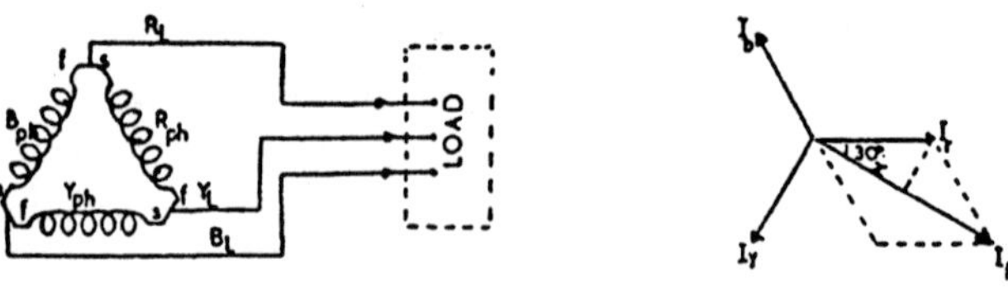

Fig 160

It will be noted that the lines have been identified with the colours of the phases to whose starts they have been connected. Assume the condition when the red-phase voltage is positive *ie* that the start of the red-phase winding is +ve with respect to the finish. Current will flow through from R phase into R line as shown. At this same instant the voltage in B phase is negative *ie* its finish is +ve with respect to its start. Thus it is also correctly connected for feeding current into the R line and a line current is thus obtained by considering the phasor *difference* of two phase currents. The resultant line current is obtained by reversing a phase current (I_b) and combining it with I_r as shown in the diagram (Fig 160).

As before $x = I_r \cos 30° = I_{ph} \cos 30°$

$$\text{or } x = \frac{\sqrt{3}}{2} I_{ph} \text{ and as } I_L = 2x = \frac{2\sqrt{3}I_{ph}}{2}$$

Hence I_L or $I = \sqrt{3}I_{ph}$

Thus for a delta connection, Line voltage = Phase voltage and Line current = $\sqrt{3}$ Phase current.

This relationship can be deduced for any line and the connected phases and will give the same result. As before V and I are used for line values and V_{ph} and I_{ph} for phase values.

Example 98. Three 50Ω resistors are connected in delta across 415V, 3-phase lines. Calculate the line and phase currents and the power taken from the mains.

Voltage across 1-phase resistor = 415V

Current in one phase of load $= \frac{415}{50} = 8.3$A

Since the load is balanced, line current $= \sqrt{3} \times 8.3$ amperes
$= 14.38$A

Power in one phase of load $= 8.3^2 \times 50 = 3445$W

Power in three phases of load $= 3 \times 3.445 = 10.3$kW.

THREE-PHASE POWER

For a *Star*-connected load. $V = \sqrt{3}V_{ph}$ and $I = I_{ph}$

The power expended in one phase $= V_{ph} I_{ph} \cos \phi$

and the power expended in three phases

$= 3V_{ph}\, I_{ph} \cos \phi$

Converting to line values, the above becomes:

$$\text{Three-phase power} = 3\frac{V}{\sqrt{3}} I \cos \phi$$

$$\text{or } P = \sqrt{3}VI \cos \phi$$

For a *Delta*-connected load $V = V_{ph}$ and $I = \sqrt{3}I_{ph}$

The power expended in one phase $= V_{ph}\, I_{ph} \cos \phi$

and the power expended in three phases

$= 3V_{ph}\, I_{ph} \cos \phi$

Converting to line values the above becomes:

$$\text{Three-phase power} = 3V\frac{I}{\sqrt{3}} \cos \phi$$

$$\text{or } P = \sqrt{3}\ VI \cos \phi$$

Thus the general expression holds, irrespective of the type of connection, namely for either a star or delta connection:

Three-phase power is given by $\sqrt{3}VI \cos \phi$.

Example 99. A 75kW, 400V, 3-phase, delta-connected induction motor has a full-load efficiency of 91 per cent and operates at a power factor of 0.9 (lagging). Calculate the line and phase currents at full load.

$$\text{Output power} = 75 \times 10^3 \text{ watts}$$

$$\text{Input power} = \frac{75 \times 10^3 \times 100}{91} \text{ watts}$$

$$\text{also } P = \sqrt{3}VI \cos \phi$$

$$\text{So } \sqrt{3} \times 400 \times I \times 0.9 = \frac{75 \times 10^5}{91}$$

$$\text{and } I = \frac{75 \times 10^4}{1.732 \times 4 \times 9 \times 91} \text{ amperes}$$

$$\text{or } I = 132.2\text{A}$$

$$\text{Motor phase current } \frac{132.\text{A}}{\sqrt{3}} = 76.3\text{A}.$$

THREE-PHASE kVA, kW AND $kVAr$

Since power factor can be defined as the ratio of true power to apparent power, this can be applied to three-phase working. Thus:

$$\text{power factor} = \frac{\text{active power}}{\text{apparent power}}$$

Again from earlier deductions, we see that irrespective of star or delta connection $P = \sqrt{3}VI \cos \phi$. Accordingly:

$$\cos \phi = \frac{P}{\sqrt{3}VI}$$

It follows that for three-phase working, in order that the definition for power factor should apply,

$$\text{apparent power } (S) = \sqrt{3}\ VI$$

Note the introduction of $\sqrt{3}$—distinguishing this condition from single-phase working.

Again it is known that $\cos\phi = \frac{P}{S}$ *or* $\frac{kW}{kVA}$ and, in exactly the same way, it follows that; *Three-phase* $kVA = \frac{\sqrt{3}VI}{1000}$

This final deduction is most important.

Example 100. A 3-phase, 400V motor takes a current of 16.5A when the output is 9kW. Calculate (a) the kVA input, (b) the power factor, if the efficiency at this load is 89 per cent.

(a) kVA input $= \frac{\sqrt{3}\ VI}{1000} = \frac{\sqrt{3} \times 400 \times 16.5}{1000}$

$= 11.43\text{kVA}$

(b) Output power $= 9\text{kW}$

True active power $= \frac{9.0}{0.89} = 10.11\text{kW}$

So power factor $= \frac{\text{active power}}{\text{apparent power}} = \frac{10.11}{11.43}$

$= 0.88$ (lagging).

Example 101. A 3-phase, star-connected alternator supplies a delta-connected induction motor at 600V. The current taken is 40A. Find (a) the phase voltage of the alternator (b) the current in each phase of the motor. Refer to the diagram (Fig 161).

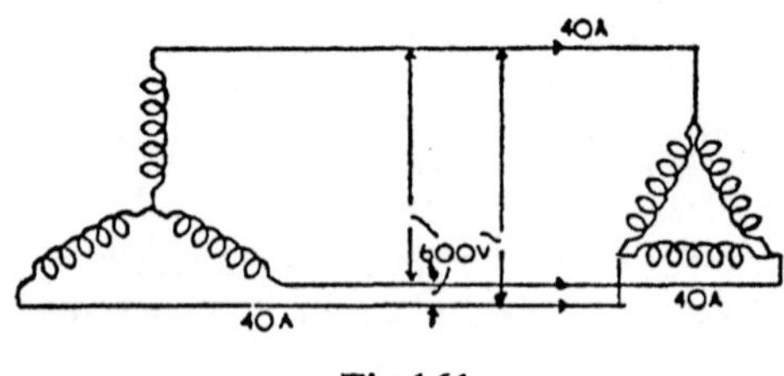

Fig 161

(a) For a star connection $V = \sqrt{3}\ V_{ph}$

$\therefore V_{ph} = \frac{V}{\sqrt{3}} = \frac{600}{\sqrt{3}} = 346\text{V}$. This is the alternator phase voltage.

(b) For a delta connection $I_{ph} = \frac{I}{\sqrt{3}} = \frac{40}{\sqrt{3}} = 23.1\text{A}$

This is the current in each motor phase.

(c) If the motor operates at a power factor of 0.8 (lagging) and an efficiency of 88 per cent, find the *kVA* rating of the alternator and power output of the motor.

$$\text{Apparent power rating of alternator} = \frac{\sqrt{3}VI}{1000} = \frac{\sqrt{3} \times 60 \times 40}{1000} = 41.6\text{kVA}$$

$$\text{Motor apparent input power} = \frac{\sqrt{3}\ VI}{1000} = \frac{1.732 \times 600 \times 40}{1000} = 41.6\text{kVA}$$

$$\text{True active input power} = 41.6 \times 0.8 = 33.28\text{kW}$$

$$\text{Output power} = 33.28 \times \frac{88}{1000} = 29.21\text{kW}.$$

Example 102. A delta-connected load is shown by the diagram (Fig 162). If the supply voltage is 400V, 50Hz, calculate the red-line current.

Assume the currents as shown and maintain the correct phase sequence of R, Y and B. If the red-line current is assumed to feed current into the red phase of the load, the blue-phase current would be in the opposite sense or a phasor difference is involved.

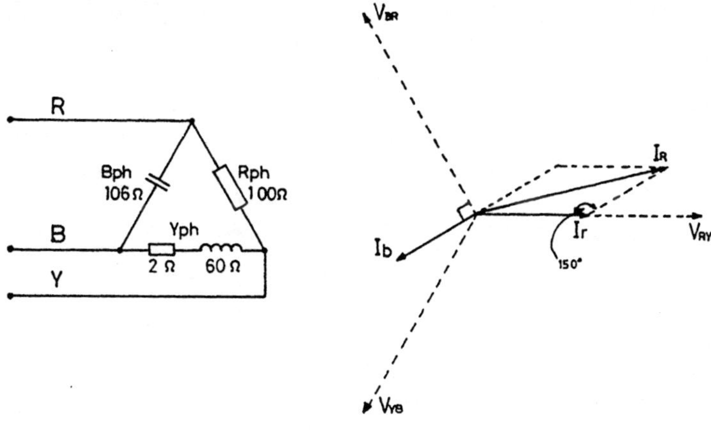

Fig 162

Here $I_r = \frac{400}{100} = 4$A in phase with V_{RY}

Also $I_b = \frac{400}{106} = 3.7$A leading V_{BR} by 90°

Reversing I_b and using the Cosine Rule

$$I_R = \sqrt{I_r^2 + I_b^2 - 2I_rI_b \cos 150°}$$
$$= \sqrt{4^2 + 3.7^2 - 2 \times 4 \times 3.7 \cos 150°}$$

or $I_R = 7.4$A.

CHAPTER 11

PRACTICE EXAMPLES

1. A coil consumes 300W when connected into a d.c. circuit of 60V. It consumes 1200W when connected into an a.c. circuit of 130V. What is the reactance of the coil?

2. A circuit consists of two branches A and B in parallel. Branch A has a resistance of 12Ω and a reactance of 3Ω, whilst the values of branch B are 8Ω and 20Ω respectively. The circuit is supplied at 100V. Calculate the current in each branch and the supply current.

3. An inductive circuit of resistance 50Ω and inductance 0.02H is connected in parallel with a capacitor of value 25μF across a 200V, 50Hz supply. Find the total current taken from the supply and its phase angle.

4. Two coils of resistances 8Ω and 10Ω and inductances 0.02H and 0.05H respectively, are connected in parallel across 100V, 50Hz mains. A capacitor of capacitance value 80μF in series with a resistor of 20Ω is then connected in parallel with the coils. Find the total current taken from the mains and its phase angle with respect to the applied voltage.

5. A single-phase motor has an input of 50.6A at 240V, the power input being 10kW, and the output 9kW. Calculate the value of the apparent power, power factor and of the efficiency.

6. A single-phase motor running from a 230V, 50Hz supply takes a current of 11.6A when giving an output of 1.5kW, the efficiency being 80 per cent. Calculate the capacitance required to bring the power factor of the supply current to 0.95 (lagging). Calculate also the *kVAr* rating of the capacitor.

7. The load taken from a single-phase supply consists of:
   (a) Filament lamp load of 10kW at unity power factor.
   (b) Motor load of 80kV A at 0.8 power factor (lagging).
   (c) Motor load of 40kV A at 0.7 power factor (leading).
   Calculate the total load taken from the supply in *kW*

and in kVA and the power factor of the combined load. Find the 'mains' current if the supply voltage is 250V.

8. Three equal impedances of 10Ω, each with a phase angle of 30° (lagging), constitute a load on a 3-phase alternator, giving 100V per phase. Find the current per line and the total power when connected as follows. (a) Alternator in star, load in star. (b) Alternator in star, load in delta. (c) Alternator in delta, load in delta. (d) Alternator in delta, load in star.

9. A 500V, 3-phase, star-connected alternator supplies a star-connected induction motor which develops 45kW. The efficiency of the motor is 88 per cent and the power factor is 0.9 (lagging). The efficiency of the alternator at this load is 80 per cent. Determine (a) the line current, (b) the power output of the alternator, (c) the output power of the prime-mover.

10. A 400V, 3-phase system takes 40A at a power factor of 0.8 (lagging). An over-excited synchronous motor is connected to raise the power factor of the combination to unity. If the mechanical output of the motor is 12kW and the efficiency is 91 per cent, find the kVA input to the motor and its power factor. Find also the total power taken from the supply mains.

CHAPTER 12

THE D.C. GENERATOR

It is not intended in this book to cover the finer points of d.c. machine construction, operation and maintenance since these matters are better appreciated if dealt with in a practical manner. It is necessary however, to set out here the basic features of the machine as it has been developed to date and also to follow up on the theory already commenced in Chapter 7. Once the general construction of the machine is outlined, full attention can then be given to further theory and it can be considered from its functional point of view, which classifies it under the heading of (1) The Generator (2) The Motor.

D.C. MACHINE CONSTRUCTION

The principal features of the machine will be described under (a) the field system or stator and (b) the armature or rotor. The diagram (Fig 163) shows the usual representation for a d.c. compound wound generator.

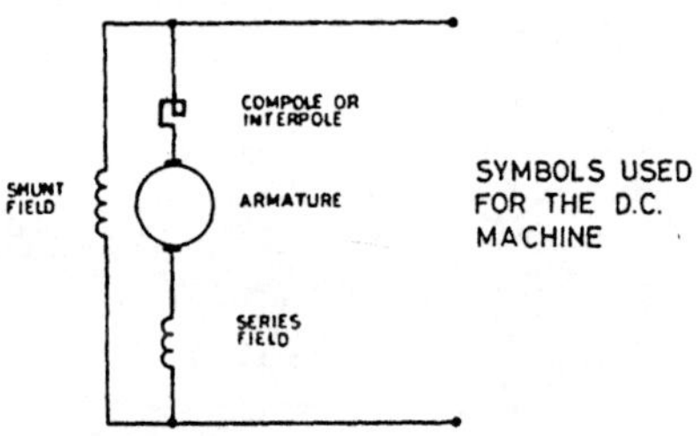

Fig 163

(a) FIELD SYSTEM

This includes the magnet arrangement comprising the poles and yoke, the field coils and interpoles (when fitted). Note that the interpoles are essentially part of the armature electrical circuit and will be mentioned later under this latter heading.

POLES AND YOKE. The former are the cores of the machine electromagnets and are usually fitted with pole-shoes which concentrate the field across the air-gaps in which the conductors move. The yoke is an extension of the magnet system, forms the main frame of the machine and serves to carry flux from and to the poles. The diagram (Fig 164) shows typical ways in which the field system can be constructed.

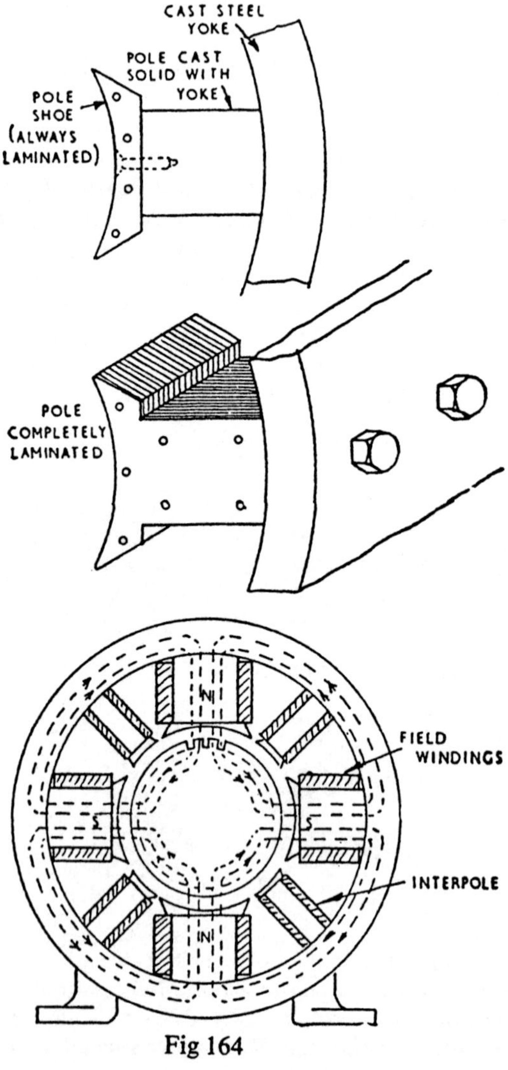

Fig 164

The poles and yoke can be constructed from cast steel or fabricated from mild steel sheet which has been cut and rolled into shape. The poles may be part of the yoke, but for modern machine construction, they are more usually built up from thin laminations, riveted together and shaped to include the pole shoes.

FIELD COILS. The field coils are basically of two types (i) Shunt Coils which consist of a large number of turns of fine wire and (ii) Series Coils which are made from a few turns of thick cable or conductor. The shunt coils are built up on a 'bobbin' or 'former' but the series coils may be self-supporting. The diagram (Fig 165), shows a typical cross-section of construction; the insulation used is decided by the class of machine and its duty.

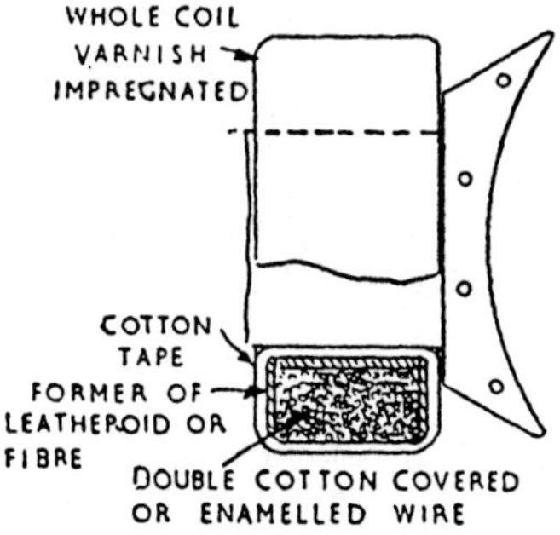

Fig 165

(b) THE ARMATURE

This consists of the armature core, the windings, the shaft and commutator. The brushes, although not part of the armature, are considered here since they work in conjunction with the commutator.

ARMATURE CORE. This is built up from iron laminations which are clamped between two end plates. The laminations are insulated from each other to minimise induced circulating currents, called 'eddy currents', and if clamping bolts are used to pass through the core as shown in the diagram (Fig 166), then they must be insulated. Modern techniques use stamped laminations which are pressed onto and 'keyed' to the shaft, the end plates being screwed onto the shaft for smaller machines. For larger designs a

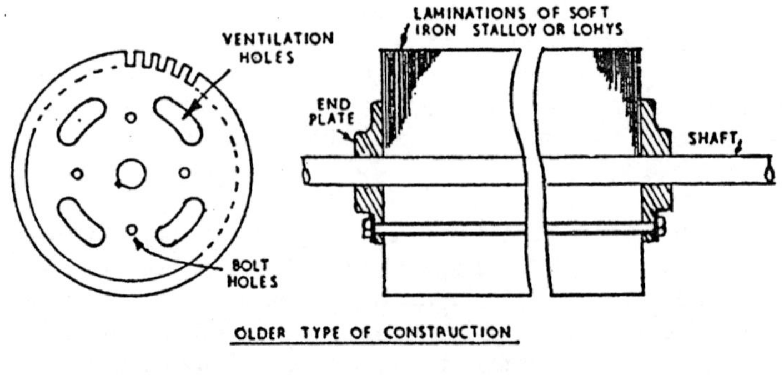

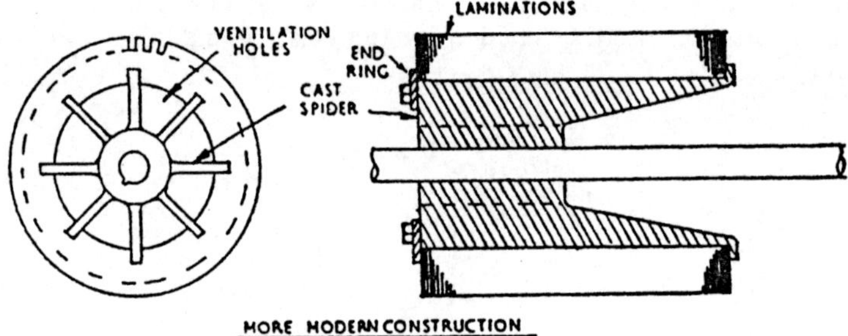

Fig 166

'spider' is employed which allows ventilation and keeps the iron required to a minimum.

WINDINGS. The number of conductors, their size, shape, etc, are decided by the design requirements for the machine. The diagram (Fig 167) shows a typical method of locating and holding the coil sides in position. In the sketch, a wedge made of bakelised paper or bakelised fabric is shown, but open slots with a closing piece of fibre and 'binders', made from high-tensile

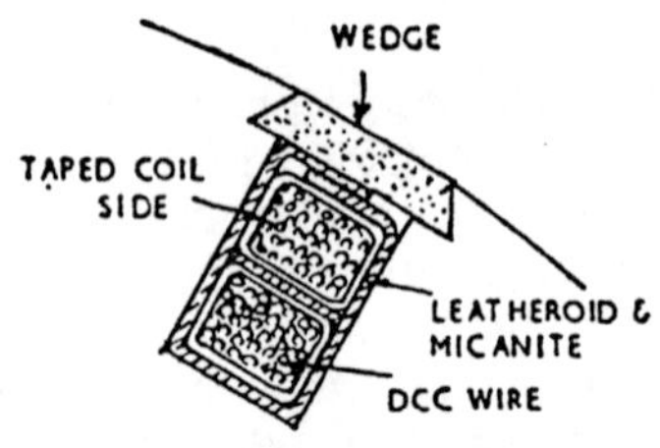

Fig 167

steel wire, are frequently used. For small machines, mainly motors; as for vacuum-cleaners, cabin-fans, etc, the armature windings consist of enamelled or cotton-covered wire placed in position by hand winding. Semi-enclosed slots are used with a fibre insert closing the slots.

The method of arranging the conductors to form a closed winding will be considered in greater detail after the machine construction has been dealt with.

SHAFT. This is made from the best forged mild steel and is designed so that it will not deflect unduly when running up to its maximum speed.

COMMUTATOR. This consists of copper segments, insulated from each other by mica. The width at the top of a segment may be up to 6mm and the segments may be mounted on but insulated from a sleeve, which is secured to the shaft, and are clamped by an end-ring which can be bolted or screwed as shown in the diagram (Fig 168). Insulated cone-shaped rings, made from micanite are used to insulate the segments from the steel clamping assembly. The armature windings are brought out and soldered to the segments. The mica must be undercut between adjacent segments.

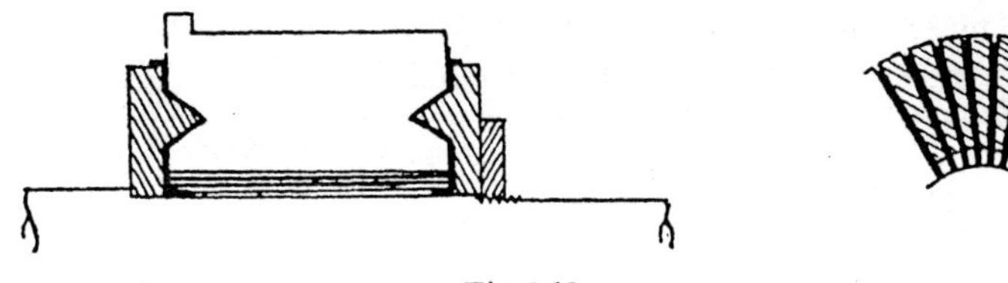

Fig 168

BRUSHES. Any one brush is pressed onto the commutator by means of the pressure arm and is connected to the holder by means of a pig-tail of braided copper wire which is moulded into the brush. One or more brush-holders may be carried on an insulated spindle which is mounted on the brush rocker-ring. The brush rocker-ring is arranged to be clamped firmly once the brush position has been set. Brushes of modern d.c. machines are always of moulded carbon and graphite, the grade of hardness being chosen to suit the running conditions. The diagram (Fig 169) shows a typical arrangement.

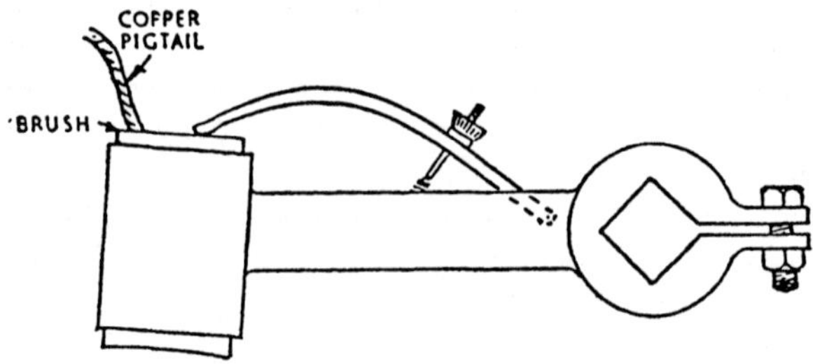

Fig 169

BEARINGS For most industrial d.c. machines the bearings are of the ball or roller type. Advantages are (i) its axial length is shorter than that of the journal type (ii) after initial packing with grease, service for a long period can be obtained. Journal bearings, *ie* sleeve types, give quieter running and are frequently preferred for marine work, since they resist 'transmitted' vibration troubles better. They are usually of the 'ring-oiler' pattern. The steel shaft runs in a brass or cast-iron sleeve lined with white metal. For small and medium size machines, the bearings are carried in the end shields, but for large machines, the bearings are carried in separate pedestals.

D.C. ARMATURE WINDING ARRANGEMENTS

The simplest winding possible would be built up from single-turn coils of span equal to one 'pole pitch', *ie* 180° for a two-pole machine. For a four-pole machine the coil span would still be one pole pitch, but now 90 mechanical degrees. This is illustrated by the diagram (Fig 170).

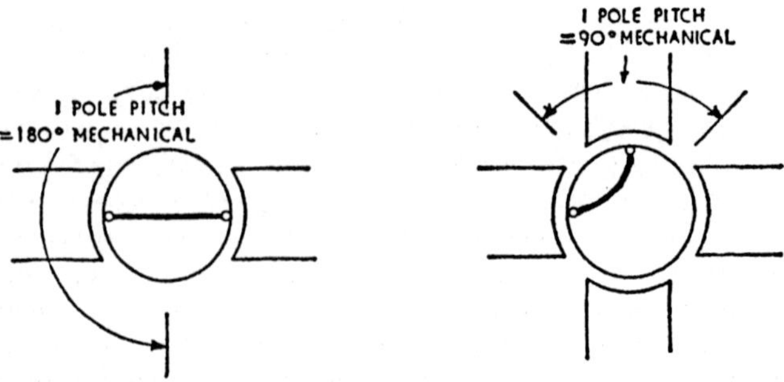

Fig 170

In practice it is not usual to make the span equal to one pole pitch exactly and many small machines have an odd number of slots. Each slot carries two coil sides, *ie* it contains more than one conductor. D.C. windings are usually of the 2-layer type, a coil side lying at the bottom of the slot and another at the top. Sometimes more than 2, such as 4, 6 or even 8 coil sides may be contained in 1 slot since it may not be practicable to have too many slots. There are two basic methods of connecting up the conductors on an armature after they have been formed into either single or multi-turn coils, and the complete winding falls into one of two distinct types namely (a) a wave winding or (b) a lap winding.

(a) The WAVE or two-circuit Winding. This winding results in there always being two paths in parallel irrespective of the number of poles of the machine. Two sets of brushes only are necessary but it is usual to fit as many sets of brushes as the machine has poles. The diagram (Fig 171a) shows the essential layout.

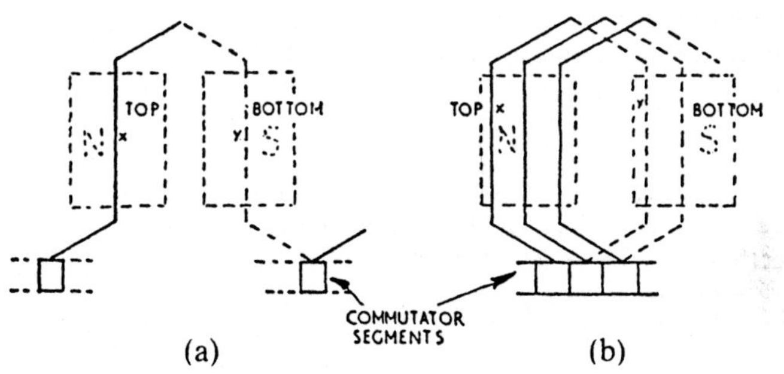

Fig 171

(b) The LAP or multi-circuit Winding. This winding results in as many paths in parallel as the machine has poles. There are as many sets of brushes as the machine has poles. The diagram (Fig 171b) also shows the essential layout.

In building up a winding it is essential to connect coil elements in such a manner that the induced e.m.f. in the conductors add, in much the same way as cells are connected in series so that their e.m.fs. add to give the required battery voltage. Thus conductor X is in series with conductor Y which occupies relatively the same position as X but is under a pole of reversed polarity. The coil element so formed by conductors XY should then be connected in series with a similarly placed coil element under a pair of poles so that the required voltage for a parallel path of the armature can be attained in this manner. For a wave winding the connection can be readily seen from the diagram already introduced and for a lap winding the same rule is followed, except that all the coil elements under a pair of poles are connected in series before the winding progresses to connect up the conductors under the next pair of poles.

The example, which follows, will illustrate both simple lap and wave windings. A small armature is to be designed to have one-turn coils — one turn comprising 2 conductors. There are to be 8 coils. There will be 1 commutator segment to a coil, *ie* 8 commutator segments. If only 2 coil-sides are to be accommodated in a slot then there must be 8 armature slots and if a four-pole system is to be used then there will be 2 slots/pole. This will give a true pole pitch of 2 — the pole pitch being the number of armature slots divided by the number of poles. Since the sides of a coil should be under the influence of the correct field poles, the winding pitch must be as nearly as possible equal to the pole pitch. Thus the winding pitch would also be equal to 2 or a coil should embrace 2 teeth.

The LAP winding is considered first, being suitable for this armature. Now for such a winding, the connecting up of the conductors is such that the winding progresses round the armature by being pitched alternatively forwards and backwards. For our example, if (Fig 172) is considered, it will be seen that conductor No 1 is connected to No 6 which is spaced 2 teeth away. No 6 is then connected to No 3 and so on. The

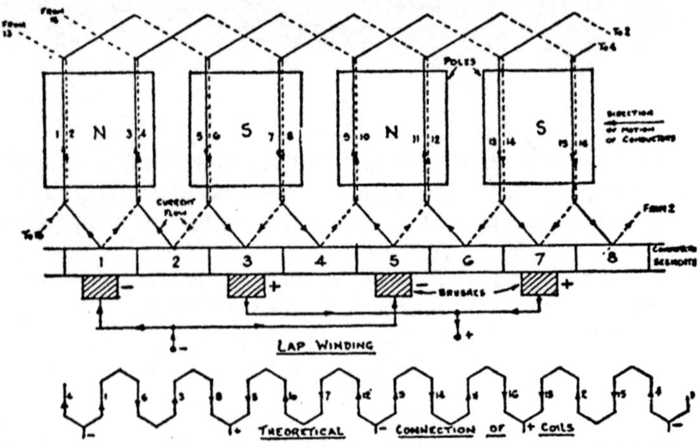

Fig 172

winding thus progresses by 1 slot until it is closed by all the slots having been occupied and conductor No 15 being connected to No 1 through No 4.

If now, for our example, a WAVE winding is required then a preliminary examination would show that this could not be achieved. If the winding started at No 1 proceeded to No 6 and then on through Nos 9 and 14 it would close back onto conductor No 1. It is obvious that an armature with 8 slots would not be suitable for such a wave winding and one of 7 or 9 slots should be considered. A nine-slot armature winding would give a winding pitch of length slightly less than the true pole pitch length and is considered as suitable. Consider now the diagram (Fig 173). Here conductor No 1 is connected to No 6 as before which in turn is connected to Nos 9, 14, 17 and then to No 4, *ie* the winding passes into the slot beyond that at which the start was made. The winding, thus does not close immediately and if the connecting-up proceeds as described, it will be seen that the winding will progress four times round the armature before the close is made at the starting slot by conductor No 11 being joined to No 1 through No 16. This then would be a suitable winding but 9 coils would be used with 9 armature slots and 9 commutator segments.

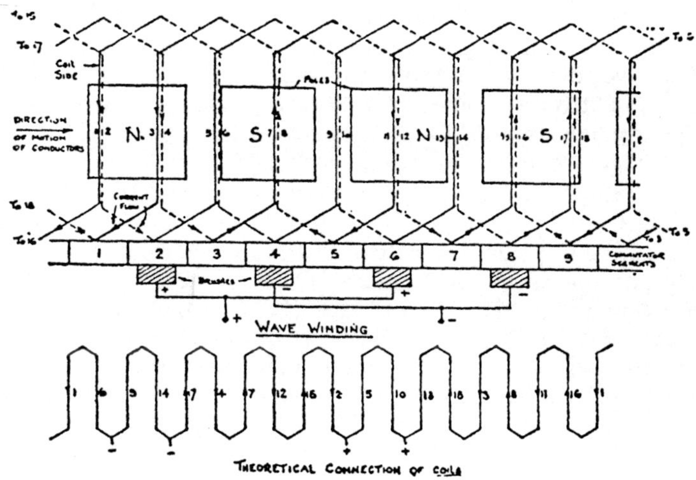

Fig 173

More details on armature windings will be found in a book dealing more fully with the practical subject since machine design and armature winding is specialists' work. However, it is of interest to find where the brushes are to be placed on the commutator and one accepted way is to draw out the 'equivalent ring' winding.

EQUIVALENT RING WINDINGS. The diagrams (Figs 172 and 173) show how the connection of the conductors can be set out. On the armature winding diagram, current flow is assumed in the conductors under the poles, this current flow being due to the induced e.m.fs. Thus if, for the conductors under a *N* pole, the direction of the current is assumed from the bottom of the page to the top, then for those under a *S* pole, it would be from the top of the page to the bottom. If next, the winding is drawn out as a ring winding, as shown at the bottom of each figure and the conductors are marked correctly then:

For the Lap winding No 1 should lead to No 6 and then onto 3 and so on.

For the Wave winding No 1 should lead to No 6 and then onto 9 and so on.

The assumed current directions can be transferred from the armature winding and drawn in to show that:

For the Lap winding the currents meet at the junctions of Nos 8 and 5, 16 and 13. These would be +ve points or the brushes should be placed on segments Nos 3 and 7 — these +ve brushes. The −ve points would be at the junction of Nos 4, 1 and 12, 9 or the −ve brushes should be placed on segment Nos 1 and 5. The ring winding also shows that there are 4 paths in parallel between +ve and −ve terminals of the machine.

For the Wave winding the currents meet at the junction between conductor Nos 6 and 17. It will be noted that no current (induced e.m.f.) has been attributed to conductors which are in the position between the poles. Thus a +ve point is at the junction 6-9 or 14-7, there in effect being no potential drop in loop 9 and 14. A brush could be placed at either commutator segment No 4 or 8 and to maintain uniformity, brushes may be placed at both these points and connected as shown to form the +ve terminal of the machine. Similarly −ve points occur at the junctions of 2 and 13 or the actual joints 2-5 and 10-13. Brushes may be placed at these points, there being no current (induced e.m.f.) in loop 5, 10. As for the +ve terminal, brushes are placed on segment Nos 2 and 6 and joined together to form the −ve terminal. The ring winding also shows that there are 2 paths in parallel for a wave winding.

THE D.C. GENERATOR

Theory of the d.c. armature and commutator has shown that commercially, direct current is best obtained by using an armature wound with a number of coils so connected, that all the coils, except those being short-circuited by the brushes, are in the circuit. The armature being a continuous closed winding splits itself electrically into a number of parallel paths. It has also been shown that there are two fundamental ways of winding an armature (a) with a lap winding or (b) with a wave winding. An important rule can now be stressed which is that:

For a lap winding the number of parallel paths in the armature is always equal to the number of poles.

For a wave winding the number of parallel paths is always two, irrespective of the number of poles provided for the machine.

Interpoles or commutating-poles, sometimes termed compoles, do not perform any function of the main poles and are to be disregarded for the rule just enunciated.

THE E.M.F. EQUATION

Consider the diagram (Fig 174) and the factors for a machine as given below. A simple expression for the composite armature is now deduced and it is stressed that this is of the utmost importance. It must be memorised and the student should be capable of proving it from first principles.

Let N = the speed of the machine in rev/min, P = the number of poles. Φ = the flux/pole (webers). Z = the number of armature conductors. A = the number of parallel paths of the armature winding.

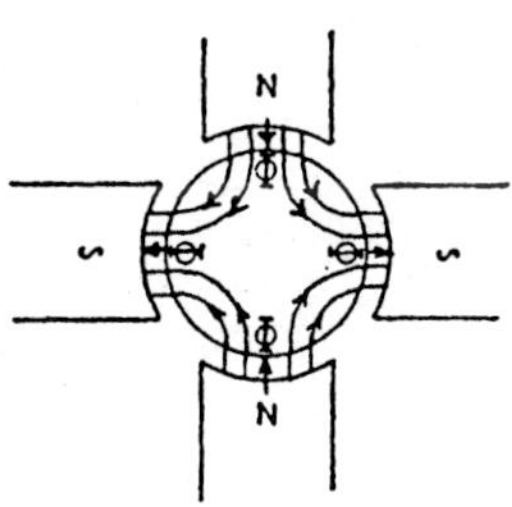

Fig 174

In one second the armature revolves $\frac{N}{60}$ times

and in 1 revolution, one conductor cuts a flux of $P \times \Phi$ webers

$\therefore$ In 1 second one conductor cuts $P\Phi \frac{N}{60}$ webers.

From Faraday's law the magnitude of the e.m.f. generated in volts is given by the flux cut/second

so the e.m.f. generated in 1 conductor $= \frac{P\Phi N}{60}$ volts.

If the armature winding is divided into A parallel paths then the e.m.f. of one parallel path is also the e.m.f. of the machine. Now in a parallel path there are $\frac{Z}{A}$ conductors in series,

so e.m.f. of 1 parallel path = e.m.f. of the machine $= \frac{P\Phi N}{60} \times \frac{Z}{A}$

Thus $E = \frac{Z\Phi N}{60} \times \frac{P}{A}$ volts, where E = the generated voltage.

Example 103. The armature of a four-pole, shunt generator is lap wound and generates 216 volts when running at 600 rev/min. The armature has 144 slots with six conductors/slot. If this armature is rewound and wave connected, determine the e.m.f. generated at the same speed and flux/pole.

From the e.m.f. equation $216 = \frac{(6 \times 144) \times \Phi \times 600 \times 4}{60 \times 4}$

or $\Phi = \frac{216}{60 \times 144}$ webers

Note. This is a Lap-wound armature so $A = P = 4$.
For a Wave-wound armature $A = 2$

$$\therefore E = \frac{6 \times 144}{60} \times \frac{216}{60 \times 144} \times \frac{600}{2} \times 4 = 216 \times 2$$
$$= 432\text{V}.$$

CHARACTERISTICS

By characteristics are meant graphs which show the behaviour of any type of machine when under investigation. As an example, consider the e.m.f. equation. It is noted that, for any given machine, all the factors except Φ and N are constant. The equation can be written $E = k\Phi N$ where $k = \frac{ZP}{60A}$

Thus $E \propto \Phi$, if N is kept constant and $E \propto N$, if Φ is kept constant.

If Φ and N are both varied then E will vary accordingly. Thus we see that the voltage generated can be controlled by varying the speed or the flux of the machine. This can be shown by deducing the 'no-load' characteristics.

ASSOCIATED MAGNETIC CIRCUIT EFFECTS

In Chapter 6 consideration was made of the iron or steel cored electromagnet and since the magnetic circuit forms an essential part of the d.c. machine, it is necessary here to consider two effects which influence generator characteristics.

The first noticeable effect is that of *residual magnetism.* Experiments with a piece of iron show that when such a specimen is made the core of an electromagnet and it is magnetised by passing a current through the energising coil then, when the current is switched off and the magnetising m.m.f. has been removed, the magnetism or magnetic flux will not completely disappear. In other words, some magnetism remains when, in theory, it should be non-existent, and it is important for the reader to appreciate that the effect does occur.

The second factor which is of importance, is the *saturation effect* of an iron sample when it is being subjected to a magnetomotive force. If a magnetic circuit uses iron as the medium for conveying the flux then, as the magnetomotive force is increased, the flux increases in proportion up to a stage, when the straight-line relation between the m.m.f. and the flux Φ or the consequent flux density B, is no longer followed. Thus if, for an iron sample, the B value is plotted against H — the magnetising force/metre length, a graph called 'the B-H curve' is obtained. The method of testing a sample and a detailed explanation of the results obtained was also discussed in full in Chapter 6, but the point being stressed here is that the resulting graph is a straight line for only a short section of its length. It will be of the form shown in the diagram (Fig 175) which indicates that the iron sample appears to saturate, *ie* no matter how much the m.m.f. is increased, once the curve bends over and flattens out then it would be apparent that no increased flux Φ or consequent B value would be produced, irrespective of the strength of the magnetising force.

The flattening out or saturation effect is considered to be due to all the molecular magnets having orientated themselves in the direction of the magnetic field and thus, it can be expected that, the saturating effect will make itself apparent when investigating the relation of the generated voltage E to the flux Φ in the magnetic circuit of the d.c. machine.

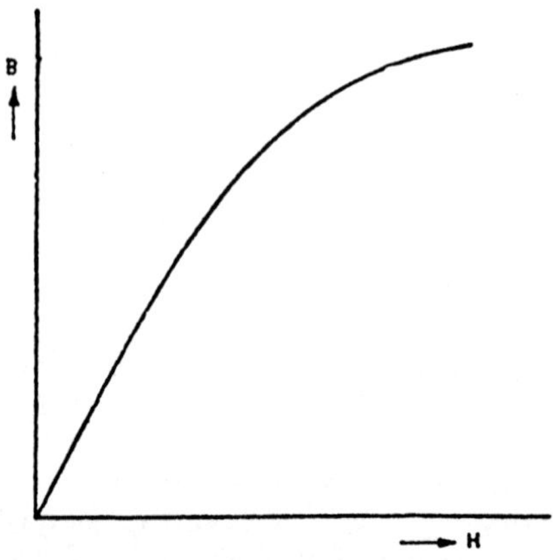

Fig 175

THE NO-LOAD CHARACTERISTIC

Theory of the machine has so far shown that the generated voltage is dependent on the machine flux and speed and the no-load characteristics, already mentioned, can now be considered as detailed under the following (a) and (b) headings.

(a) VARIATION OF E WITH N (Flux Φ constant). A permanent-magnet generator is seldom used for practical applications, but the investigation to be considered, can be made by controlling the current of separately-energised field electromagnets. This current, referred to as the field current I_f when flowing through the field coils, gives effect to a m.m.f. which results in the flux in the air-gaps. If this current is kept at a constant value I_{f1}, then the flux will be constant, and tests can be made by varying the speed at which the machine is driven and by noting values together with the corresponding voltage being generated.

Since flux Φ is constant and as $E \propto N$, a straight-line graph as shown by (1) of the diagram (Fig 176), will result. If the field current is next adjusted to a smaller value I_{f2} and is kept constant, then when the test is repeated, a straight-line graph such as (2) will result and the deduction assumed, namely that E varies directly with N, will be proved.

(b) VARIATION OF E WITH Φ (Speed N constant). As explained above, variation of flux can most readily be effected by controlling the energising current I_f in the field coils or the 'exciting current' as it is often called. If no residual magnetism is present

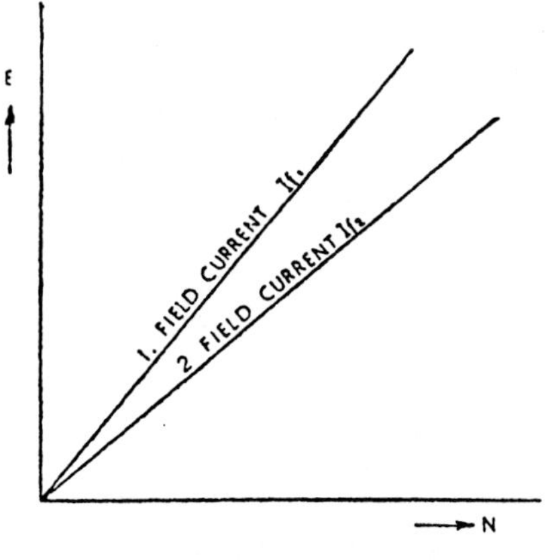

Fig 176

in the field system, then, if I_f is increased, the m.m.f. is increased and the flux in the air-gaps increases. The generated e.m.f. increases accordingly and a *B-H* type of curve (1) as shown in the diagram (Fig 177), is obtained if E is plotted to a base of I_f. *Note.* Φ cannot be readily measured but its effects can be gauged by knowing the appropriate values of the exciting current.

Curve (1) at first increases as a straight line, flattening out to a horizontal as the magnet system saturates. When saturation has occurred, if the field current is reduced, Curve (2) will result. This curve will be found to lie slightly above the original curve (1) and it is seen that for decreasing values of I_f, the values of E are above those obtained for the ascending curve (1). The cause of the difference between curves (1) and (2) is Magnetic Hysteresis. When the field current is eventually reduced to zero, some generated e.m.f. is found to be present while the machine is run at the constant speed N. This e.m.f. is due to the residual magnetism; which will be shown later to be essential, if a generator is required to be self-exciting. The e.m.f. due to residual magnetism can only be removed by demagnetising the field system. If the value of I_f is increased again, Curve (3) will be followed which closes up on curve (1). The diagram has been drawn to accentuate the difference between curves (1) and (2). In the modern machine this difference is not appreciable and if a mean curve is drawn, shown dotted, this is known as the *'magnetisation'* or *'open-circuit characteristic'* (O.C.C.) curve.

The Magnetisation Curve or O.C.C. is of sufficient importance to require the direction of students to this effect. It will be mentioned repeatedly during theory relating to generators (alternating current and direct current) and motors and many problems will require reference to it before they can be solved. It is plotted in a standard fashion as Generated Voltage or E.M.F. to a base of Field Current. This point should also be noted.

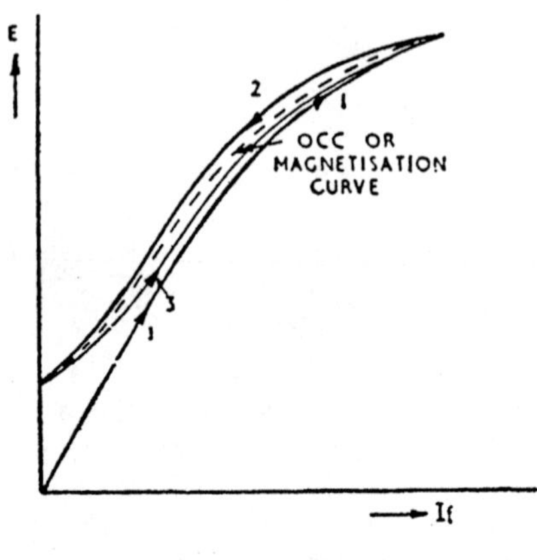

Fig 177

TYPES OF D.C. GENERATOR

Under this general heading, the machine can be classified in different ways. Since the types of generator are altered by variations of the magnet system, in that either the magnetic material or the connection of the field energising coils can differ; in this book the machines are described in the following order. (a) The Permanent-magnet type of Generator. (b) The Separately-excited type of Generator. (c) The Self-exciting type of Generator, which may be further subdivided under the practical headings of (c i) Shunt-connected, (c ii) Series-connected, (c iii) Compound-connected.

(a) THE PERMANENT-MAGNET TYPE OF GENERATOR

This type of generator is not used to any great extent because of the difficulty of making large permanent magnets and of varying the magnetic field so as to control the generator output. The most common use is for specialised applications such as for electrical tachometers (speed indicators), hand-operated insulation testers (the Megger is an example) and primary exciters for large alternators.

THE LOAD CHARACTERISTIC. Since Φ is constant, the load characteristic will be almost identical to the no-load characteristic which has already been considered. A tachometer arrangement is considered in the diagram (Fig 178).

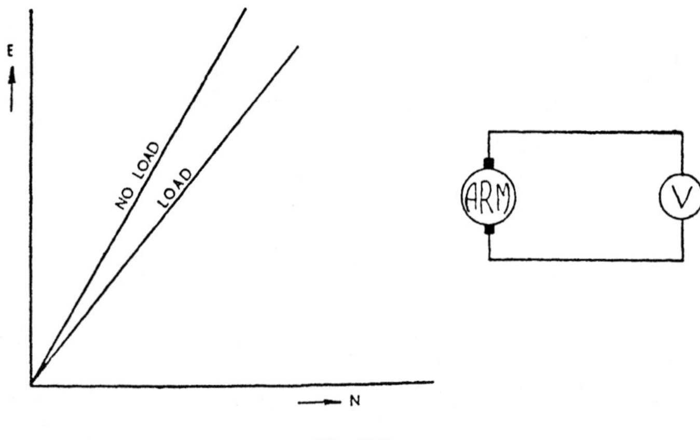

Fig 178

If the voltmeter used is sensitive, *ie* it requires very little current, the generator output current will be small so that armature voltage drop (I_aR_a) will be negligible. Here R_a is the ohmic resistance value of the armature and I_a the armature current. Thus the load terminal voltage V is approximately equal to the generated e.m.f. E and the voltmeter can be calibrated in revolutions per minute.

(b) THE SEPARATELY-EXCITED TYPE OF GENERATOR

Knowing that $E \propto N$ if Φ is constant, then the no-load characteristic will be a straight line as has been considered earlier. If however N is constant and Φ is varied the characteristic will vary as the B-H curve and an open-circuit characteristic (O.C.C.) as described earlier will result. The two variations of the characteristic are shown in Figs 176 and 177.

THE LOAD CHARACTERISTIC. This characteristic is obtained by setting the field current at a value which gives the normal rated voltage at the correct speed, and by applying load in stages so as to take currents of value between zero and some 25 per cent overload.

For a small generator such loading can best be applied by switching in banks of similar wattage lamps which are thus connected in parallel. If the terminal voltage V is plotted against the load current I_L, the External Load Characteristic would be obtained as is shown in the diagram (Fig 179).

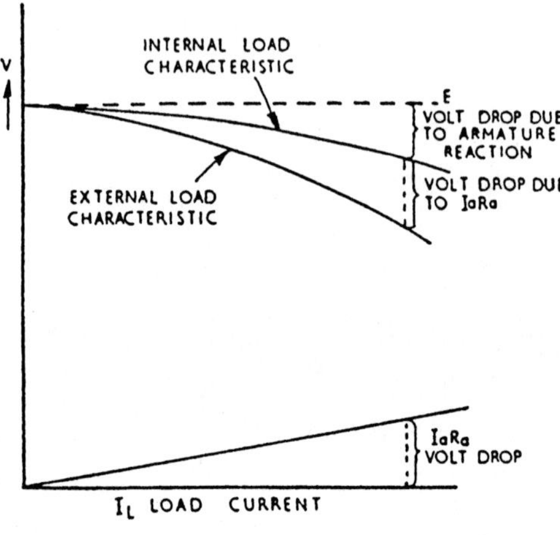

Fig 179

If the machine is stopped and the armature resistance R_a is measured by the ammeter/voltmeter method and a separate low-voltage supply, then the I_aR_a voltage-drop line can be plotted as shown. If various I_aR_a voltage-drop values are added to the external characteristic the Internal Load Characteristic can be obtained by construction. The difference between this line and the horizontal line of the theoretical generated e.m.f. E will illustrate the voltage drop due to *armature reaction* effects. Armature reaction will be explained very much later in detail (Book 7), but can be described here briefly, by saying that the passage of current through the armature sets up a magnetic field which interacts with the main field, tending to weaken and distort the latter. Thus the magnitude of the generated e.m.f. is reduced and commutation is effected adversely.

The load characteristics have been introduced to illustrate the effects which are responsible for a voltage drop inside the generator itself, when the machine is on load. In problems the armature reaction effect is seldom mentioned but the armature is usually credited with a resistance value greater than its ohmic value in order to allow for a total internal voltage drop. The voltage equation would be:

$$E = V + I_aR_a$$

The separately-excited generator in its d.c. form is only used for specialised applications such as, machines used for supplying current to electroplating vats. Under these conditions some 6000 to 10 000 amperes may be required at voltages of 6 to 12 volts and the output can best be controlled by varying a separately-excited field.

(c i) THE SHUNT-CONNECTED GENERATOR

The diagram (Fig 180) shows the typical diagram of connections for this machine. Here the armature current is fed to both the load circuit and the parallel field circuit which, although it takes a very small current in comparison with the load, may require to be considered — especially for problems.

Thus: $I_a = I_f + I_L$.

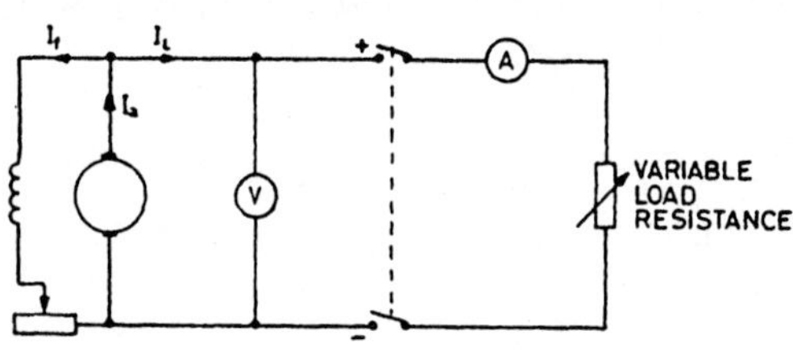

Fig 180

The shunt field of the machine is connected across the terminals of the generator. The field coils form a high-resistance circuit, they are wound with many turns of fine wire, *ie* the ampere-turns are produced by a small current value and a large number of turns.

As before: $E = V + I_a R_a$

Example 104. A four-pole, wave-wound generator delivers 40kW at 200V. Its armature has 181 turns and a resistance of 0.01Ω. The air-gap flux/pole is 0.02Wb. Calculate the speed of the machine, neglecting any voltage drop at the brushes and taking the shunt-field resistance as 50Ω.

$$\text{Load current} = \frac{\text{kilowatts}}{\text{voltage}} \text{ or } I_L = \frac{40\,000}{200} = 200\text{A}$$

$$\text{Shunt-field current } I_f = \frac{V}{R_f} = \frac{200}{50} = 4\text{A}$$

$$\text{So armature current } I_a = 200 + 4 = 204\text{A}$$

$$\text{Voltage drop in armature} = 204 \times 0.01 = 2.04\text{V}$$

Generated voltage = terminal voltage + voltage drop in armature

$$\text{or } E = V + I_aR_a = 200 + 2.04 = 202.04\text{V}$$

$$\text{Also } E = \frac{Z\Phi N}{60} \times \frac{P}{A} \text{ and } N = \frac{60A}{Z\Phi} \times \frac{E}{P} = \frac{60 \times 2 \times 202.04}{362 \times 0.02 \times 4}$$

$$\text{Thus } N = 840 \text{ rev/min.}$$

Note. Here the armature has been assumed to be wound with single-turn coils, *ie* 2 conductors/turn, since no extra data is given to assume otherwise.

THEORY OF SELF-EXCITATION. Since the shunt-connected generator utilises the principle of self-excitation, it is necessary at this stage to explain the theory involved. If the field system has residual magnetism, then rotation of the armature will generate some small e.m.f. This e.m.f. will cause a field current which will produce more flux, which in turn causes more e.m.f., hence more field current, more flux and e.m.f., to give a continual building-up condition. The voltage continues to rise and only steadies when the voltage drop across the field equals the terminal voltage.

Note. The field current must be in the correct direction through the field coils, to assist the build-up of the original residual flux.

Summarising, the conditions necessary for self-excitation are:

i. There must be residual magnetism — sufficient to generate a small e.m.f. when the armature is rotated at the correct speed.

ii. The shunt-field circuit must be continuous and so connected that current flow will cause a flux to built up, to assist the original residual flux.

iii. The shunt-field circuit resistance must be less than the *critical resistance* as determined from the open-circuit characteristic (O.C.C.) when the machine is running at a particular speed.

Critical resistance will be explained and defined below to assist the understanding of the conditions for satisfactory self-excitation to occur. It must be emphasised that the subject matter, already dealt with under the heading of self-excitation and that to follow, is of the utmost importance and forms the subject of many examination questions and problems.

THE MAGNETISATION CURVE OR O.C.C. APPLIED TO SELF-EXCITATION, CRITICAL RESISTANCE

The diagram (Fig 181) illustrates, as a means of revision, the circuit and characteristic which is obtained by making tests with a generator being separately excited. The initial part of the O.C.C. graph is somewhat complex, in that the effect of residual magnetism is apparent, but if it is assumed that the graph started from zero then it is usually straight, since the magnetic circuit of a machine involves air-gaps and saturation conditions are reached only gradually. Full saturation conditions are seldom attained. It is again stressed that the O.C.C. is dependent on speed.

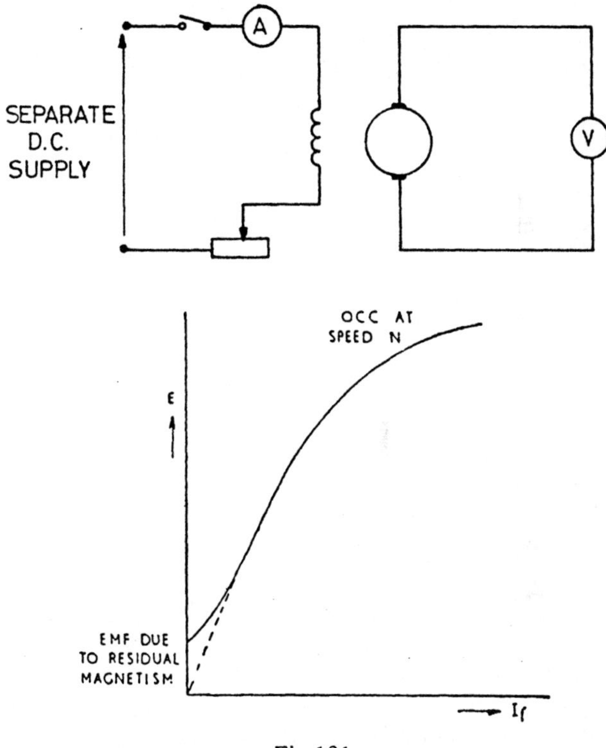

Fig 181

When the field is shunt-connected, provided the conditions set out above are fulfilled, the generator will self-excite and an open-circuit voltage value is attained, where the voltage drop in the shunt field is equal to the generated terminal voltage. This condition is illustrated in the diagram (Fig 182) and is best understood by considering the associated O.C.C. and field voltage-drop line.

The magnetisation curve for any particular speed N is drawn from test results obtained by separate excitation. Imagine the shunt field and regulator to have a resistance of R_f ohms. Then

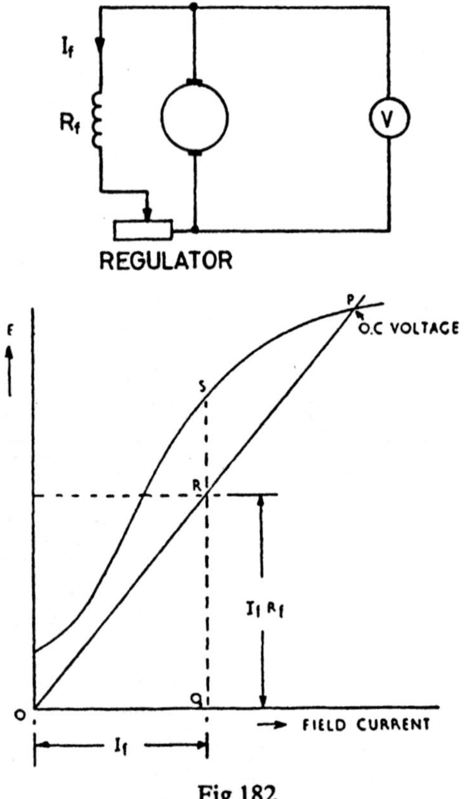

Fig 182

assuming a current of value I_f amps to flow, the field voltage drop will be I_fR_f volts. Plot this value (example point R) and extend the straight line through R from zero to cut the O.C.C. at point P. For this point of intersection (P), the voltage drop across the field equals the applied terminal voltage and conditions are balanced. Consider also the I_f condition shown, where the generated voltage SQ is greater than the field voltage drop RQ by SR volts. More current will flow in the field circuit because of this voltage difference and both graphs will rise until a point of intersection is reached. The reasoning and procedure for estimating the O.C. voltage is further illustrated by the following example.

Example 105. A d.c. generator when separately excited and driven at 1000 rev/min gave the following test values on open circuit.

Field current (A)	0	0.16	0.48	0.66	0.8	1.0	1.29
O.C. e.m.f. (V)	6.25	50	150	200	225	250	275

The field windings are then shunt connected. Find (i) the voltage to which the machine will self-excite on open circuit when driven at 1000 rev/min and the resistance of the shunt-field circuit is 240Ω. (ii) the value of the regulator resistance to be added or subtracted from the field circuit to allow the generator to self-excite to 237.5V. (iii) the value of the critical resistance at this speed.

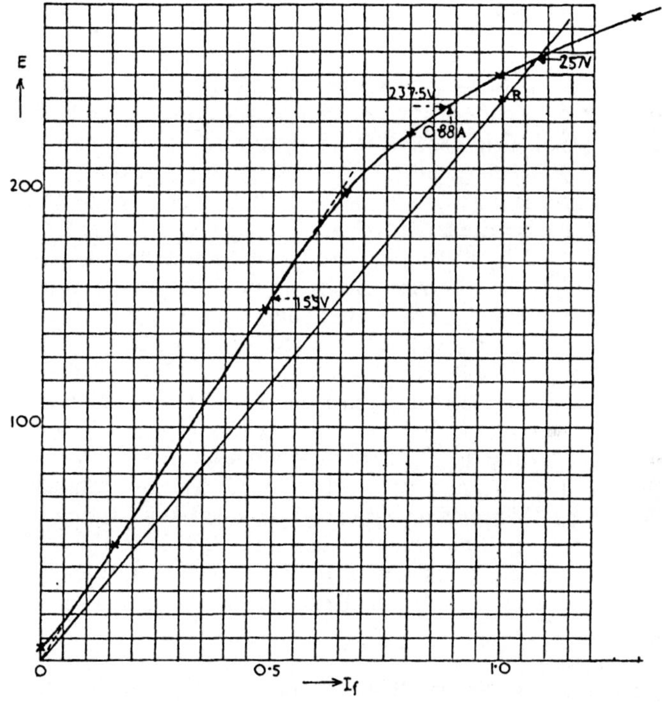

Fig 183

(i) Plot the O.C.C. as shown in the diagram (Fig 183) and using the graph, take any value of field current, *viz* 1 ampere. The voltage across the field circuit with 1A flowing would be 1 × 240 = 240V. Plot this point (R) and draw the field voltage-drop line through the origin as shown. Then the O.C. voltage to which the machine self-excites is 257V.

(ii) For the machine to excite to 237.5V, note this value on the O.C.C. and join it to zero to obtain the new field resistance voltage-drop line. Note the field current for 237.5V; this is 0.88A. Then from Ohm's law, the field-circuit resistance is

$$\frac{237.5}{0.88} = 269.9 = 270\Omega$$

$\therefore$ Resistance to be added $= 270 - 240 = 30\Omega$.

(iii) Neglecting the start of the graph (due to residual magnetism), draw the tangent through the origin. Read any voltage value on this tangent and the corresponding field current. Example: 155V and 0.5A. Then the critical resistance would be $\frac{155}{0.5}$ ohms $= 310\Omega$

CRITICAL RESISTANCE. The effect of altering the shunt-field resistance or regulator can be seen by referring to the example and the graph (Fig 183). Reduce R_f and the slope of the field resistance voltage-drop line becomes less and the point of intersection with the O.C.C. moves higher up, *ie* the O.C. voltage of the generator is raised. The converse occurs with R_f increased. If R_f is increased until the field voltage-drop line lies outside the magnetisation curve, there will be no point of intersection and the generator will not self-excite. Thus for any point on such a voltage-drop line, the voltage required is not available, as can be seen from the O.C.C. If R_f is reduced, the slope of the field voltage-drop line decreases until the line lies along or becomes tangential to the O.C.C. The resistance value deduced from the field voltage-drop line will thus fall until it attains the value given by the line tangential to the O.C.C. The resistance value obtained for this condition is called the Critical Resistance. The term explains itself and is dependent on speed. Thus for any one speed, if the field resistance is made less than the critical resistance, the machine will self-excite if the other conditions are satisfied.

THE LOAD CHARACTERISTIC. The test circuit and the resulting graph is shown by the diagram (Fig 184). The armature resistance R_a is measured by a separate test.

The External Load Characteristic is plotted from the test results obtained and the Internal Characteristic is drawn by construction already described for the separately-excited machine. Due allowance should be made for field current since armature current and not line current is used for the armature voltage drop.

Features of the Load characteristics are (i) the rapid fall off of terminal voltage (ii) the bend-back of the characteristic on itself.

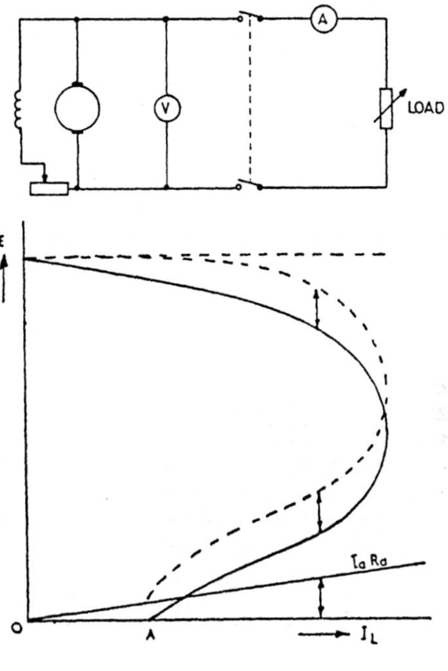

Fig 184

(i) This condition is explained as follows. When the external circuit is connected to a load, there is a voltage drop in the armature. The terminal voltage falls, resulting in a decrease of field exciting current. This in turn causes the external characteristic to droop more than it does for the separately-excited machine. The armature reaction effect is as for the separately-excited machine, *ie* it is responsible for a decrease of generated voltage — equivalent to an increased armature voltage drop. Thus this can be taken into account by crediting the armature with a R_a value greater than its ohmic resistance.

(ii) As the load resistance is decreased, load current will increase at first with a resulting fall in terminal voltage. This effect tends to slow up the increase of load current. At first the decrease of external load resistance with consequent rise of load current predominates and a rising current with falling terminal voltage is shown on the graph. At a certain value of current the demagnetising effect of armature reaction, the armature resistance voltage drop and the loss of field current due to reduced voltage, combine in effect, to produce a terminal voltage which results in less load current even though the load resistance is decreased and the curve bends back on itself. The armature may thus be short-circuited — a self-protecting effect being produced. OA is caused by residual magnetism, but a sudden short-circuit may cause an excess armature reaction effect which tends to cancel residual magnetism, thus demagnetising the machine which may then fail to self-excite when the short-circuit is removed. The machine will then require to be remagnetised before it can be put back into operation.

The shunt-connected machine can be used for most purposes where a simple generator is required. Examples would be for battery chargers and small lighting-sets, such as motor-car dynamo and electrical systems.

(C ii) THE SERIES-CONNECTED GENERATOR

The diagram (Fig 185) shows the connections for this machine, which is only used for specialised work. It is however of sufficient importance for the student to spend some little time in studying its behaviour and characteristics, which are used to advantage in the compound generator.

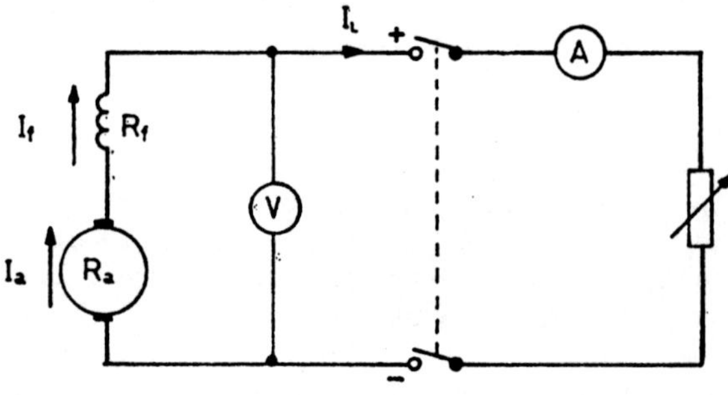

Fig 185

The series field of this generator is designed to be connected in the main armature circuit to the load. The field coils are wound with a few turns of thick cable, *ie* the field ampere-turns are produced by a large current and a small number of turns. Thus $I_a = I_f = I_L$. The terminal voltage on load is V and the generated e.m.f. E is greater than V by the internal voltage drops in the armature and series field. Thus:

$$E = V + I_f R_f + I_a R_a = V + I_a(R_a + R_f)$$

SELF-EXCITATION. The theory involved is as for the shunt-connected machine and the same conditions apply. It should be noted that the load resistance constitutes the field regulating resistance and thus for any particular speed there is a critical resistance value. If the load resistance, *ie* the field-circuit resistance, is less than the critical value, the machine will self-excite. If the circuit resistance is above the critical value for that particular machine speed, self-excitation and voltage build-up will not occur.

THE LOAD CHARACTERISTIC. Consider the circuit shown in the diagram (Fig 185). The circuit switch is closed with the load resistance set at maximum value. Load resistance is then gradually reduced until, at some point in the test, the machine is observed to self-excite. Load current and terminal voltage settle at some definite value but if the load is next altered, new voltage and current values will be obtained, which should be noted. This should be done for both decreasing and increasing load resistance values, until the full external load characteristic is obtained. The diagram (Fig 186) shows the characteristic.

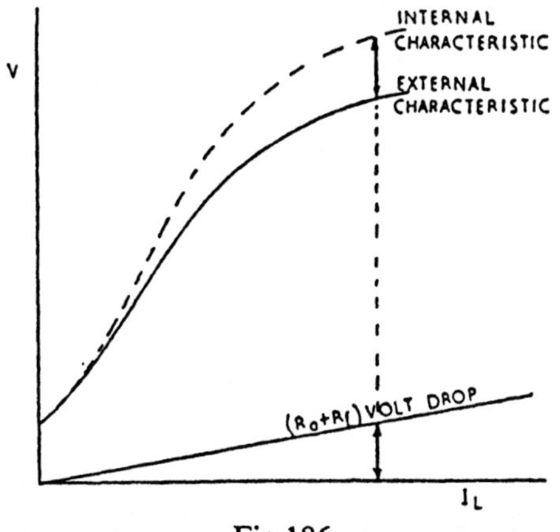

Fig 186

The machine is next shut down and R_a and R_f are measured separately. The armature and series-field resistance voltage-drop lines are next drawn and the internal load characteristic can be deduced. The effects of armature reaction could also be investigated if an O.C.C. (obtained by separate excitation at the correct speed), was superimposed on the characteristics.

The machine has the following disadvantages. (i) It cannot self-excite until the load circuit is completed and its resistance value is made less than the critical resistance. (ii) The voltage to which it self-excites is dependent on the load current and very little control of this voltage is possible. (iii) The load characteristic is a rising one and is unsuitable, in fact dangerous, since it could result in the 'burn-out' of the load.

The series generator is never used for normal generating purposes, but only for special applications such as a series booster. The machine in this form or variations, has been used for marine electrical systems but only for specialised applications. Examples would be for particular types of electric propulsion and winch control arrangements.

(c iii) THE COMPOUND-CONNECTED GENERATOR

As this generator utilises both series and shunt fields, its characteristic can be considered as being made up of shunt and series-machine characteristics. The ultimate shape of the characteristic will depend on the relative strengths of the individual fields, but it must be noted that the shunt field is the basic requirement and thus it is the performance of a shunt generator, which is being improved upon.

TYPES OF ELECTRICAL CONNECTION. The diagram (Fig 187) shows how a machine can be connected in either 'short' or 'long'

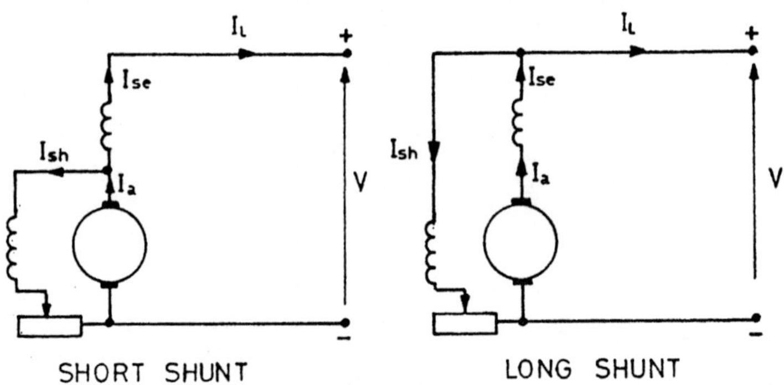

Fig 187

shunt. There is no appreciable difference in the resulting generated voltage as will be seen from the example.

Example 106. A 110V, compound generator has armature, shunt and series-field resistances of 0.06Ω, 25Ω and 0.04Ω respectively. The load consists of 200 lamps each rated at 55W, 110V. Find the generated e.m.f. and the armature current, if the generator is connected (a) long shunt, (b) short shunt.

(a) LONG SHUNT. Load current $I_L = \dfrac{200 \times 55}{110}$

$= 100A$

Shunt-field current $I_{sh} = \dfrac{110}{25} = 4.4A$

Series-field current I_{se} and armature current $I_a = 100 + 4.4$

$= 104.4A$

The generated voltage $E = V + I_a(0.06 + 0.04)$

$= 110 + 104.4(0.1)$

or $E = 120.44V$.

It will be noted that since there are two fields with different values of current, symbol I_f has not been used for field current, but instead symbols I_{sh} and I_{se} have been used in both Fig 187 and this example.

(b) SHORT SHUNT. Load current I_L (as before) $= 100A = I_{se}$

The voltage drop in the series field $= I_{se}R_{se}$

$= 100 \times 0.04 = 4V$

The voltage applied to the shunt field = terminal voltage + voltage drop in series field

$= 110 + 4 = 114V$

Shunt-field current $I_{sh} = \dfrac{114}{25} = 4.56A$

Armature current $I_a = 100 + 4.56 = 104.56A$

$E = V$ + voltage drop in series field + voltage drop in armature

$E = 110 + 4 + (104.56 \times 0.06) = 120.27V$.

TYPES OF FIELD ARRANGEMENT. The series field is usually so connected that the flux it produces adds to the shunt-field flux. For such an arrangement, which is the most common, the machine is said to be *cumulatively* connected. All generators, used for supplying lighting and power for electrically driven auxiliary machinery aboard ship, have this connection. If the series field is connected so as to weaken the shunt field, the generator is *differentially* connected. This arrangement is used for specialised work only and is not common. An example of its application would be for certain types of welding generator.

THE LOAD CHARACTERISTIC. The diagram (Fig 188a) shows the graphs which would be obtained by directly loading the machine in the manner already described. Curve (a) shows the characteristic which would be obtained if the shunt field only was used. Curve (b) would be obtained with the series field only and curve (c) would result from the use of both fields. Any point on this load characteristic could also be obtained by adding the voltages obtained from graphs (a) and (b), for any one value of load current.

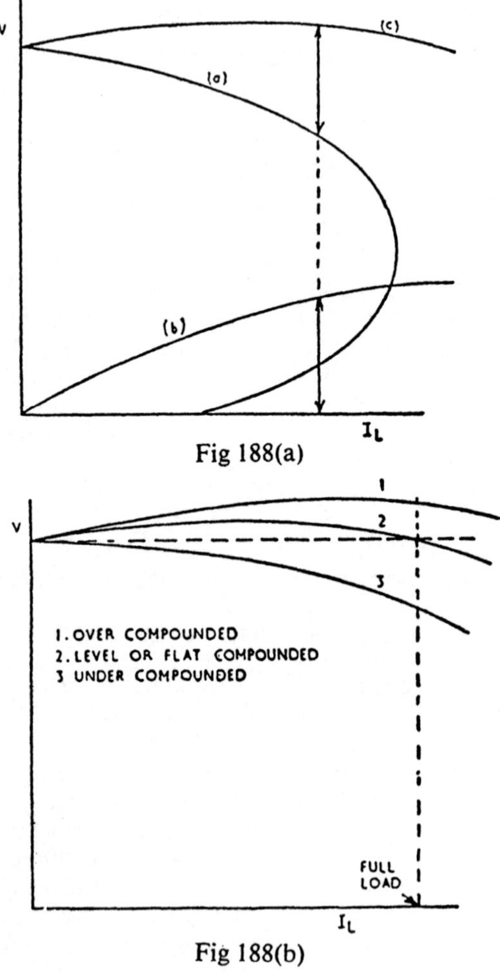

Fig 188(a)

Fig 188(b)

Fig (188b) shows how the load characteristics of a compound generator is varied by altering the relative strength of the series field. Flat-compounding is required by most Regulations. The curve is not quite flat and the rise in voltage between no load and full is called 'the hump'. It may be 6 to 7 per cent for small generators, but for normal cases it is about 2 to 3 per cent. Over-compounding is used to compensate for the voltage drop in a supply line, as was done in the past for land systems when direct current was used extensively. This is shown by the example.

Example 107. A factory is sited some distance from the generating-station and takes 100A at 200V. The resistance of the supply cable is 0.02Ω/core. Find the percentage compounding required for the generator.

The voltage drop in the line on full load = 100 × 2 × 0.02
= 4V

To supply 200V to the factory, the generated voltage should be 204V. An overcompounded characteristic as shown in the diagram (Fig 189) would be suitable and the compounding is

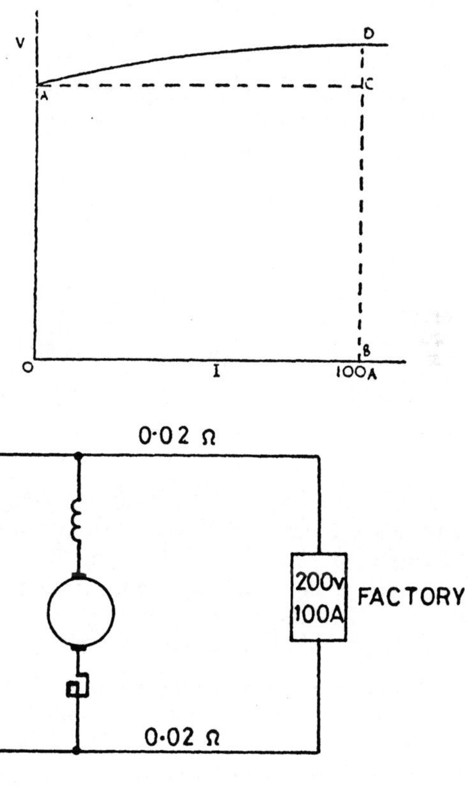

Fig. 189

usually expressed as the rise in voltage on full load to the O.C. voltage.

Thus percentage compounding

$$= \frac{DC}{CB} = \frac{DC}{AO} = \frac{204 - 200}{200}$$

$$= \frac{4}{200} = 0.02$$

$$\text{or} = 0.02 \times 100 = 2 \text{ per cent}$$

Thus the generator requires to be 2 per cent overcompounded.

Further work on the d.c. generator will be necessary for more advanced studies, but for the present sufficient knowledge has been acquired to allow some study of the motor to be made. The following examples form a useful conclusion to this chapter.

Example 108. A shunt generator is to be converted into a compound generator by the addition of a series-field winding. From a test on the machine with shunt excitation only, it is found that a field current of 3A gives 440V on no load and that 4A gives 440V at the full-load current of 200A. The shunt winding has 1600 turns/pole. Find the number of series turns required/pole.

Ampere-turns/pole required to give 440V on O.C.

$= 3 \times 1600 = 4800\text{At}$

Ampere-turns/pole required to give 200A at 440V on load

$= 4 \times 1600 = 6400\text{At}$

Full-load ampere-turns must be increased by 6400 — 4800

$= 1600\text{At}$

But these 1600At/pole are to be obtained from the series field which passes 200A

Thus the required number of series turns/pole $= \frac{1600}{200} = 8.$

Example 109. A four-pole, compound generator has a lap-wound armature and is connected in short shunt. The resistances of the armature and fields are 0.1Ω and 50Ω(shunt), 0.08Ω (series). The machine supplies a load consisting of sixty 100V, 40W lamps in parallel. Calculate the total armature current, the current/armature path and the generated e.m.f.

Since this is a lap-connected armature $A = P$

For one lamp, since $P = VI \quad \therefore \quad I = \frac{40}{100} = 0.4\text{A}$

The load current $I_L = 60 \times 0.4 = 24\text{A}$

Voltage drop in series field $= 24 \times 0.08 = 1.92\text{V}$

Voltage across shunt field $= 101.92\text{V}$

$$\text{Shunt-field current} = \frac{101.92}{50} = 2.04\text{A}$$

$$\text{Armature current} = 24 + 2.04 = 26.04\text{A}$$

$$\text{Current per armature path} = \frac{26.04}{4} = 6.51\text{A}$$

$$\begin{aligned}\text{Generated Voltage} &= \text{terminal voltage} + \text{voltage drop in series field} + \text{voltage drop in armature}\\ &= 100 + 1.92 + (26.04 \times 0.1)\\ &= 104.52\text{V}.\end{aligned}$$

CHAPTER 12

PRACTICE EXAMPLES

1. The armature of a four-pole, shunt generator is lap wound and generates 216V when running at 600 rev/min. The armature has 144 slots with 6 conductors/slot. If the armature is rewound to be wave connected, find the e.m.f. generated at the same speed and flux/pole.

2. A compound-wound, long shunt d.c. generator has an output of 250A at 220V. The equivalent resistances of the armature, series and shunt windings are 0.025, 0.015 and 176Ω respectively. If there is a 2V voltage drop across the brushes, find the induced voltage.

3. The curve of induced e.m.f. against excitation current for a separately-excited generator when run on no load at 1200 rev/min is given by:

E.m.f.(V)	15	88	146	196	226	244	254
Excitation Current (A)	0	0.4	0.8	1.2	1.6	2.0	2.4

   Deduce the voltage to which the machine would self-excite if the shunt-field resistance was set at 90Ω and the machine was run at 900 rev/min.

4. A 220V, four-pole, wave-wound, shunt generator has an armature resistance of 0.1Ω and a field resistance of 50Ω. Calculate the flux/pole, if the machine has 700 armature conductors, runs at 800 rev/min and is supplying a 38kW load.

5. In a 250kW, 440/480V, overcompounded generator, the flux/pole required to generate 440V on no load is 0.055Wb at 620 rev/min. The resistances of the armature, interpoles and series field are 0.01, 0.005 and 0.005Ω respectively. Find the flux/pole required at full load, the speed now being 600 rev/min. Neglect the current taken by the shunt field.

6. Estimate the series-turns/pole required for a 50kW, compound generator required to develop 500V on no load and 550V on full load. Assume a long-shunt connection and that the ampere-turns required per pole on no load are 7900 whereas the ampere-turns required per pole on full load are 11 200.

7. A four-pole machine has a lap-wound armature with 90 slots each containing 6 conductors. If the machine runs at 1500 rev/min and the flux/pole is 0.03Wb, calculate from first principles the e.m.f. generated.

If the machine is run as a shunt generator with the same field flux, the armature and field resistances being 1.0Ω and 200Ω respectively, calculate the output current when the armature current is 25A.

If due to a fall in speed the e.m.f. becomes 380V, calculate the load current in a 40Ω load.

8. A d.c. generator gave the following O.C.C. when driven at 1000 rev/min.

Field Current (A)	0.2	0.4	0.6	0.8	1.0	1.2	1.4	1.6
Armature Voltage (V)	32	58	78	93	104	113	120	125

If the machine is run as a shunt generator at 1000 rev/min, the shunt-field resistance being 100Ω, find (a) the O.C. voltage (b) the critical value of the shunt-field resistance, (c) the O.C. voltage if the speed was raised to 1100 rev/min, the field resistance being kept constant at 100Ω.

9. Calculate the input power required to drive a shunt generator when giving an output of 50kW at 230V, if under these conditions the bearing, friction, windage and core loss is 1.6kW and the total voltage drop at the brushes is 2V. The resistance of the armature is 0.034Ω and that of the field circuit 55Ω

10. A d.c. generator when separately excited and run at 200 rev/min gave the following test results:

Field Current (A)	0	1	2	3	4	5	6	7	8	9
O.C. Voltage (V)	10	38	61	78	93	106	115	123	130	135

The field is then shunt-connected and the machine is run at 400 rev/min. Determine, (a) the e.m.f. to which the machine will excite when the field-circuit resistance is 36Ω, (b) the critical value of the field-circuit resistance, (c) the additional resistance required in the shunt-field circuit to reduce the e.m.f. to 220V, (d) the critical speed when the field-circuit resistance is 36Ω.

CHAPTER 13

THE D.C. MOTOR

A d.c. machine will run as a motor if its field and armature are connected to a suitable supply. The 'motoring' action is based on the fundamental law described in Chapter 5, which stated that a force is set up on a conductor which lies in a magnetic field and carries current. The diagram (Fig 190) shows the basic arrangement for revision purposes.

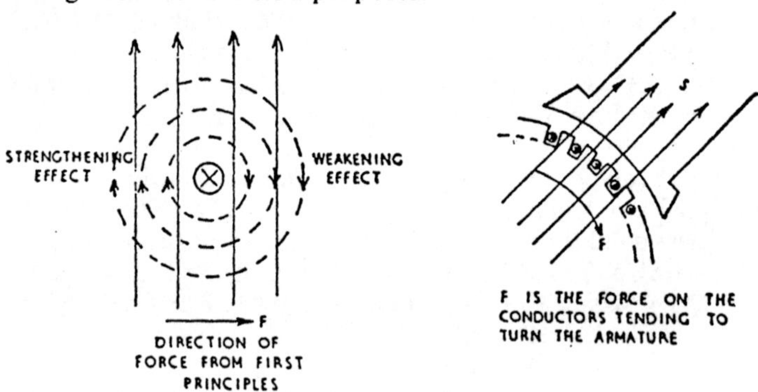

Fig 190

DIRECTION OF FORCE

The four small diagrams (Fig 191) show that, in order to reverse the direction of the force and thus the direction in which

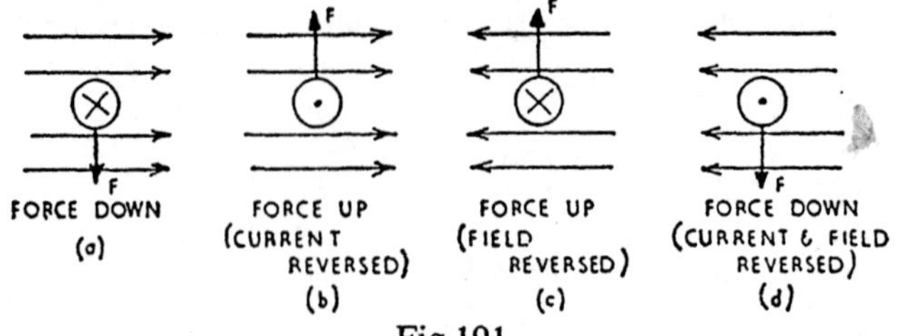

Fig 191

the armature will rotate, it is necessary to reverse the current in the conductor with respect to the magnetic flux.

The practical aspect of this rule should be remembered if a motor is found to run in the incorrect direction when first connected up. Reversal of rotation can be obtained by interchanging the supply leads to the armature circuit. A hand rule has been developed to help memorise motor action and is comparable with that enunciated in Chapter 12 for the generator.

LEFT-HAND RULE (Fleming's). The diagram (Fig 192), shows the practical interpretation. The first and second fingers are made to represent the flux and current respectively, as for the right-hand rule. The direction of force on the conductor will then be represented by the thumb. *Note.* As for the right-hand rule, the thumb, index finger and second finger must be placed at right angles to each other.

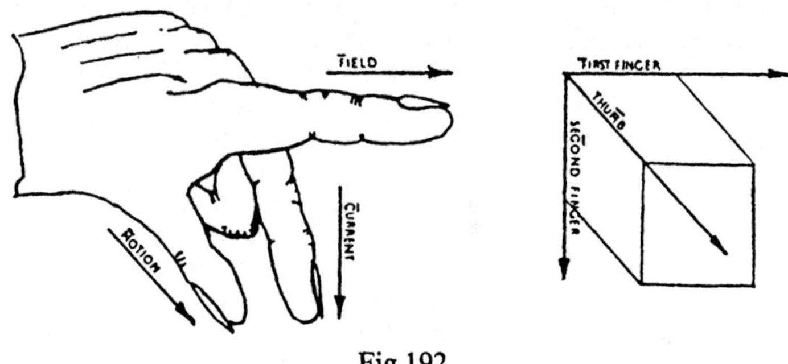

Fig 192

MAGNITUDE OF FORCE

From the first principles set out in Chapter 5, it was shown that the force acting on a conductor in a magnetic field, is proportional to the flux density, the current and the active length of the conductor in the field. The law was summarised by the fundamental formula $F = BIl$ newtons but, it is pointed out here that, the magnitude of the force also depends on the inclination of the conductor to the direction of the field. It is a maximum when they are at right angles.

Example 110. Calculate the force in newtons, as established on a conductor, 0.5m long, carrying a current of 500A in and at right angles to a magnetic field of uniform density 0.8T.

Since $F = BIl$

Then $F = 0.8 \times 500 \times 0.5 = 200$ newtons.

It should be noted that, if the conductor is situated on an armature at a radius of r metres, then the torque produced on the shaft can be expressed as $F \times r$ newton metres.

BACK E.M.F. OF A MOTOR

If the motor is allowed to rotate due to the torque produced by the armature conductors, then these same conductors will cut a magnetic field and from Faraday's law it is known that an e.m.f. will be induced, the magnitude of which will be given by the generator expression, as it has been developed in Chapter 12, namely:

$$E = \frac{Z\Phi N}{60} \times \frac{P}{A} \text{ volts}$$

From first principles, it can be reasoned that the direction of the induced e.m.f. will be such as to oppose the applied voltage and a condition of balance must result. This is also supported by the fact that since, the direction of rotation would be opposite to that for a generator which is to be operated under the same directions of flux and current in the armature conductors, then the induced e.m.f. opposes the current flow and can be termed a 'back e.m.f.'. This e.m.f. must always be less than the terminal voltage V, so as to allow the motoring condition. Thus the armature is seen to start as a passive load, but as it rotates, it accelerates until the condition of balance is attained when the supply voltage is equal to the voltage drop in the armature plus the back e.m.f. being generated. This balance condition is expressed by the voltage equation set out below and the motor armature operates as an active load.

VOLTAGE EQUATION

$V = E_b + I_aR_a$. This equation explains the voltage conditions as they occur for the armature circuit. Here V is the voltage applied to the armature, E_b is the back e.m.f. being generated and I_aR_a is the armature voltage drop caused by the armature current I_a passing through the armature resistance R_a. If a problem is encountered where the brush voltage drop is given, then due allowance must be made for this.

It should be noted that the equation is comparable with the generator terminal voltage equation $V = E - I_aR_a$ and a thought about the difference in the two equations will summarise the basics of generator and motor action.

CURRENT EQUATION

Since $V = E_b + I_aR_a$ then $I_aR_a = V - E_b$

and $I_a = \dfrac{V - E_b}{R_a}$

The equation in this form shows how the motor current is dependent on the value of the back e.m.f. being generated.

The starting conditions are also illustrated.

At start $E_b = 0 \quad \therefore I_{as} = \dfrac{V}{R_a}$

But R_a is usually very small so as to minimise the armature-resistance voltage drop for working conditions and thus I_a will be very large. For example, a 220V motor having an armature resistance of 0.4Ω may take a full-load current of 52A, but if started without making special arrangements, the starting current I_{as} would be given by $I_{as} = \dfrac{220}{0.4} = 550\text{A}$. Such a large starting current could give rise to undesirable starting conditions. It could 'blow' a fuse, or cause too rapid acceleration — resulting in mechanical or electrical damage through excessive sparking at the commutator and, it is for this reason that, the starting current I_{as} is limited by the use of a 'starter'. The basic feature of the starter is a variable resistance which is inserted into the armature circuit at starting and is gradually reduced or cut out as the motor accelerates up to speed.

At the 'instant of starting' $I_{as} = \dfrac{V}{R_{as} + R_s}$. Here R_s is the full value of the starting resistance.

SPEED EQUATION

The equation is also essential for understanding the action of a motor. It is most conveniently obtained by rearranging the terminal voltage equation and using the generator expression thus:

Since $V = E_b + I_aR_a$. Then $E_b = V - I_aR_a$.

But $E_b = \dfrac{Z\Phi N}{60} \times \dfrac{P}{A}$. E_b being a generated e.m.f., its magnitude can be determined from the generator formula.

Hence $\dfrac{Z\Phi N}{60} \times \dfrac{P}{A} = V - I_aR_a$

$$\text{or } N = \frac{(V - I_aR_a)}{Z\Phi} \times \frac{60A}{P} \text{ revolutions per minute}$$

Example 111. Calculate the full-load speed of a motor operating from a 440V supply, given: $R_a = 0.75\Omega$, full-load armature current is 55A, the flux/pole is 0.02Wb and that it is a four-pole machine with a simple wave-wound armature with 43 slots and 12 conductors per slot.

Number of armature conductors $Z = 43 \times 12 = 516$

For a wave-wound armature $A = 2$.

Also $P = 4$ and $\Phi = 0.02\,Wb$

$$\text{Then } N = \frac{440 - (55 \times 0.75)}{516 \times 0.02} \times \frac{60 \times 2}{4}$$
$$= 1160 \text{ rev/min}$$

SPEED CONTROLLING FACTORS

The deductions set out below, as derived from the speed equation, are of the utmost importance and should be considered in detail by the student. He should ensure that he fully understands the implication of each deduction.

Since $N = \dfrac{V - I_aR_a}{Z\Phi} \times \dfrac{60A}{P}$, it is obvious that for any particular machine only certain variables affect the expression. Thus 60, A, Z and P are all constants and can be written as k.

Then we have $N = \dfrac{k(V - I_aR_a)}{\Phi}$

or $N = k\dfrac{E_b}{\Phi}$ since $E_b = V - I_aR_a$.

If the expression $N = \dfrac{k(V - I_aR_a)}{\Phi}$ is considered, then for the purposes of approximation, the voltage drop I_aR_a, being small, can be neglected and we now have $N = \dfrac{kV}{\Phi}$ or $N \propto \dfrac{V}{\Phi}$ (approx).

Thus speed can be controlled by varying V *or* Φ and

Variation of V gives *direct* speed control, whereas
Variation of Φ gives *inverse* speed control.

In deducing the above it should be remembered that the true relation is $N \propto \dfrac{E_b}{\Phi}$ but under working conditions the value of E_b is not very different from that of V, the I_aR_a voltage drop being small. The practical application of the deduction leads to the basic systems of motor speed control in that:

Variation of the voltage across the armature terminals produces a direct variation of speed, *ie* raise the armature voltage and speed rises, lower the armature voltage and speed falls.

In contrast;

Variation of the field flux produces an inverse variation of speed, *ie* lower or weaken the flux and speed rises, strengthen flux and speed falls.

The relationship $N \propto \dfrac{V}{\Phi}$ (approx) will also be used to determine the shape of the motor speed characteristics, when these are being considered.

Example 112. The armature resistance of a 200V shunt motor is 0.4Ω. The no-load (this is the term used when the motor is running light, *ie* not loaded) armature current is 2A. When loaded and taking an armature current of 50A, the motor speed is 1200 rev/min. Find the approximate no-load speed.

On No-load. Back e.m.f. $E_{b0} = V - I_{a0}R_a$

$$= 200 - (2 \times 0.4)$$
$$= 199.2\text{V}$$

On Load. Back e.m.f. $= E_{b1} = V - I_{a1}R_a$

$$= 200 - (50 \times 0.4)$$
$$= 180\text{V}$$

Also since $N = k\dfrac{E_b}{\Phi}$. Then $N_0 = \dfrac{kE_{b0}}{\Phi_0}$ and $N_1 = \dfrac{kE_{b1}}{\Phi_1}$

Since this is a shunt motor, the field is unaffected by the loading of the armature and $\Phi_1 = \Phi_0$.

$$\therefore \frac{N_0}{N_1} = \frac{kE_{b0}}{\Phi_0} \Big/ \frac{kE_{b1}}{\Phi_1} = \frac{E_{b0}}{E_b} \text{ since } k,\ \Phi_1 \text{ and } \Phi_0 \text{ cancel}$$

$$N_0 = N_1 \times \frac{E_{b0}}{E_{b1}} = 1200 \times \frac{199.2}{180}$$

$$\text{or } N_0 = 1328 \text{ rev/min}$$

TYPES OF D.C. MOTOR

As for the generator, the motor-field windings can be connected in shunt, series or a combination of both to give a compound arrangement. The main point to remember is that the motor is a machine which at all times, is taking current from the supply and that the fields are a load, additional to the armature circuit.

(a) THE SHUNT MOTOR

The arrangement is shown in the diagram (Fig 193a) and it will be seen that $I_L = I_a + I_{sh}$. The supply voltage V is applied to both the armature and the field circuits but here is an instance where the 'equivalent resistance' treatment for a parallel circuit cannot be supplied to find I_L because, although R_{sh} is a passive load, the armature is an active load when the machine is running. The shunt motor is substantially a constant-speed machine, used for most duties.

Here $I_{sh} = \dfrac{V}{R_{sh}}$ and $I_a = I_L - I_{sh}$.

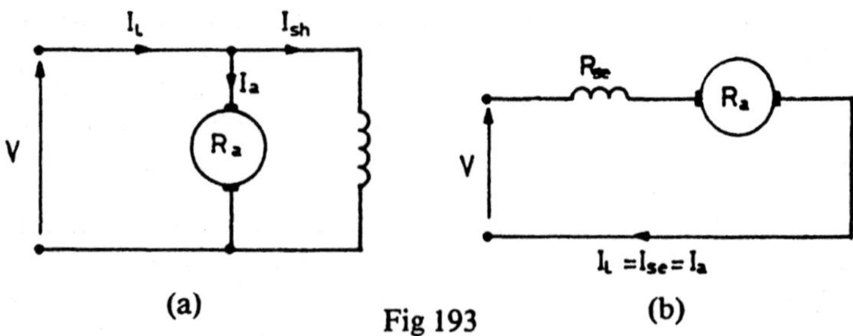

(a) Fig 193 (b)

(b) THE SERIES MOTOR

The arrangement is shown in the diagram (Fig 193b). Here $I_L = I_{se} = I_a$. The voltage equation is modified slightly in that, if V is taken as the supply voltage then allowance must be made for the voltage drop in the series field and the equation should be written as:

$$V = I_{se} R_{se} + I_a R_a + E_b = E_b + I_a R_a + I_a R_{se}$$

or $V = E_b + I_a(R_a + R_{se})$.

The voltage as applied to the armature is equal to V minus the voltage drop in the series field and any voltage drop at the brushes if this is mentioned. As the machine current rises with increase of load, the voltage across the armature falls and speed is affected accordingly. By design, the ohmic value of R_{se} is kept as small as possible for this type of machine. This is a variable speed motor, used mainly for traction, hoist, crane and winch work.

(c) THE COMPOUND MOTOR

As shown by the diagram (Fig 194), this motor, like the generator, can be connected as a long-shunt or short-shunt machine. Again the two fields can also be connected to assist or oppose each other magnetically. If the resultant flux is strengthened by the arrangement, the fields are said to be 'cumulatively' connected. If the fields are however connected to weaken each other, then the motor is 'differentially' connected — an arrangement which is rarely used.

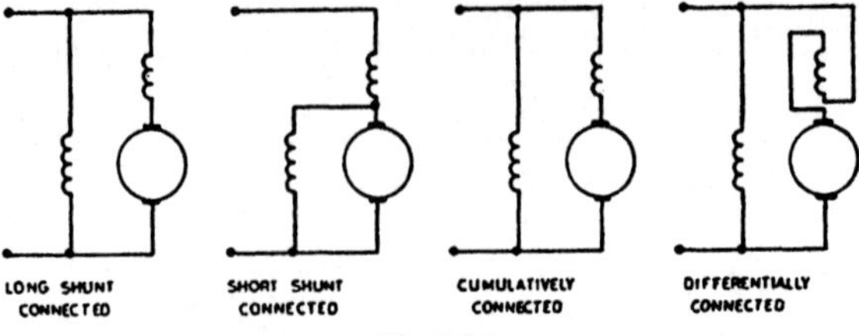

Fig 194

Most marine motors are cumulatively-compounded machines. The relative strengths of the shunt and series fields are decided by the type of performance required and this statement will be considered further when the characteristics are studied in detail.

THE POWER EQUATION

This equation is important because it shows the conversion from electrical to mechanical power and the cause of electrical losses. It is also used for deducing the Torque Equation. Procedure is simple and will give the student no difficulty provided he understands and has mastered the voltage equation.

Since $V = E_b + I_aR_a$ and the armature is the agent by which the electrical energy supplied is converted into mechanical energy, then the following deduction is possible;

Multiply the expression by I_a and study the result.

Thus $V = E_b + I_aR_a$ becomes

$$VI_a = E_bI_a + I_a^2R_a$$

Obviously VI_a is a measure of the power input to the armature circuit $I_a^2R_a$ indicates a resistance loss and is the power lost by being converted into heat in the armature itself. It is known as a Copper Loss and is due to the resistance of the armature. E_bI_a must be a measure of the power developed by the armature. This can be seen if the expression is arranged thus:

$$\frac{VI_a}{\text{Input Power}} - \frac{I_a^2R_a}{\text{Copper Loss}} = \frac{E_bI_a}{\text{Output Power}}$$

Note. The Output Power E_bI_a is in watts and is the mechanical power developed by the armature conductors and is not a true measure of the shaft output until the machine's mechanical losses, such as those due to friction and windage, have been subtracted. For a problem, when data concerning the mechanical losses is not given, then only an estimate of the shaft output power can be obtained in terms of the electrical output.

Example 113. A four-pole motor has a wave-wound armature with 594 conductors. The armature current is 30A and the flux per pole is 0.009Wb. Calculate the total power developed when running at 1400 rev/min. Estimate the shaft output power if the mechanical losses are assumed to absorb 10 per cent of the developed power.

For this machine $P = 4, A = 2, Z = 594$ and $\Phi = 0.009$ Wb

$$\text{Also } E_b = \frac{Z\Phi N}{60} \times \frac{P}{A} = \frac{594 \times 0.009 \times 1400}{60} \times \frac{4}{2}$$

$$= 249.48\text{V}$$

The power developed $= E_bI_a = 249.48 \times 30 = 7484.4\text{W}$
$= 7.5\text{kW}$ (approx)

Since mechanical power loss = 10 per cent of 7.5kW
= 0.75kW

then shaft output power = 7.5 — 0.75 = 6.75kW.

THE TORQUE EQUATION

This is an important expression, often required to be developed from first principles for examination purposes. The method used here involves the power and voltage equations and is considered to be the simplest.

Since the electrical power output of the armature $= E_bI_a$ watts and the mechanical power developed is given by:

$$\frac{2\pi \times \text{speed (rev/min)} \times \text{torque (newton metres)}}{60}$$

Then we can write: $E_bI_a = \dfrac{2\pi NT}{60}$

$$\text{or } T = \frac{60}{2\pi N} \times E_bI_a = \frac{60}{2 \times 3.14} \times \frac{E_bI_a}{N}$$

Substituting for E_b in terms of machine data, we have;

$$T = \frac{60}{2 \times 3.14} \times \frac{Z\Phi N}{60N} \times \frac{P}{A} \times I_a$$

$$= \frac{60}{2 \times 3.14 \times 60} \times Z\Phi I_a \frac{P}{A} = 0.159Z\Phi I_a \frac{P}{A}$$

or $T = 0.159Z\Phi I_a \dfrac{P}{A}$ newton metres.

TORQUE CONTROLLING FACTORS

As for the speed equation so for the torque equation, the factors which influence the torque can be determined. Thus for any one particular machine 0.159, Z, P and A are all constants and when considered together can be written as k.

Thus we have the expression $T = k\Phi I_a$ or $T \propto \Phi I_a$. This means that the torque developed varies directly with either the flux and/or the armature current and this fact will also be made use of for problems and when considering machine characteristics. As a preliminary point of importance, it can be stressed here that, for a shunt motor for different conditions of loading Φ is substantially constant and so $T \propto I_a$. For a series motor however, Φ is not constant and is frequently taken as being proportional to I_a. Therefore, if $\Phi \propto I_a$ and $T \propto \Phi I_a$, we can write for a series motor $T \propto I_a^2$. This deduction is used in the example.

Example 114. A series motor when running at a speed of 600 rev/min develops 3kW and takes a current of 40A. If the starting current is limited by means of the starter to 60A, find the starting torque. Neglect the effects of armature reaction and assume that the magnetic circuit is unsaturated.

Since the magnetic circuit is unsaturated, it can be assumed that $\Phi \propto I_{se} \propto I_a$. Thus we have $T \propto \Phi I_a$ or $T = kI_a^2$.

There are also, for this problem two torque conditions.

Thus: when running $T_1 = kI_{a1}^2$.

At starting $T_2 = kI_{as}^2$ or kI_{a2}^2.

When running at 600 rev/min, the output = 3kW and thus T_1 is given by:

$$3000 = \frac{2\pi N T_1}{60} \text{ or } T_1 = \frac{3000 \times 60}{2 \times \pi \times 600} = 47.8\text{Nm}.$$

$$\text{Also } \frac{T_2}{T_1} = \frac{kI_{a2}^2}{kI_{a1}^2} \text{ or } T_2 = T_1\left(\frac{I_{a2}}{I_{a1}}\right)^2 = 47.8\left(\frac{60}{40}\right)^2$$

And starting torque $T_2 = 107.6$Nm.

MOTOR CHARACTERISTICS

The behaviour of shunt, series and compound motors can be illustrated by means of characteristics, which can be considered under (a) Electrical Load Characteristics (b) Mechanical Characteristic. The Electrical Characteristics show speed and torque in terms of armature current whereas the Mechanical Characteristic shows speed related to torque, assuming a constant applied terminal voltage. The Electrical Characteristics are important, in that they show the performance of the machine when loaded. The Mechanical Characteristic shows the suitability of the motor for any particular application.

The characteristics may be checked by making a load test on a type of motor, but the theoretical performance may be reasoned from the two expressions already deduced, namely $N \propto \frac{V}{\Phi}$ (approx) and $T \propto \Phi I_a$.

THE SHUNT MOTOR

(a) ELECTRICAL CHARACTERISTICS. SPEED. If flux Φ is constant, assuming a constant applied voltage V, then N may be considered as constant over the load range, since $N \propto V$ and V is constant. Speed is unaffected by I_a and the theoretical graph is shown dotted in the diagram (Fig 195), as N_1 against I_a. This motor is deemed to be a constant speed machine although, in practice the speed does fall slightly with load, as shown by the graph N. This is explained by the fact that the back e.m.f. does reduce slightly (the fall from no load to full load being some 2 per cent for large machines and some 6 per cent for small machines), due to the armature voltage drop I_aR_a increasing. Although the field current I_{sh} is constant and flux Φ is constant, the armature reaction effect causes the overall resulting flux Φ to drop slightly. Since $N \propto \frac{E_b}{\Phi}$, it should be constant if the E_b and Φ variations are proportional. Weakening of flux however means a rise in armature current due to the corresponding drop in E_b. The I_aR_a drop increases as a result, and thus the speed lowering effect of a reduced E_b is greater than the speed raising effect of a falling Φ. The net result is that the speed falls slightly over the load range of I_a.

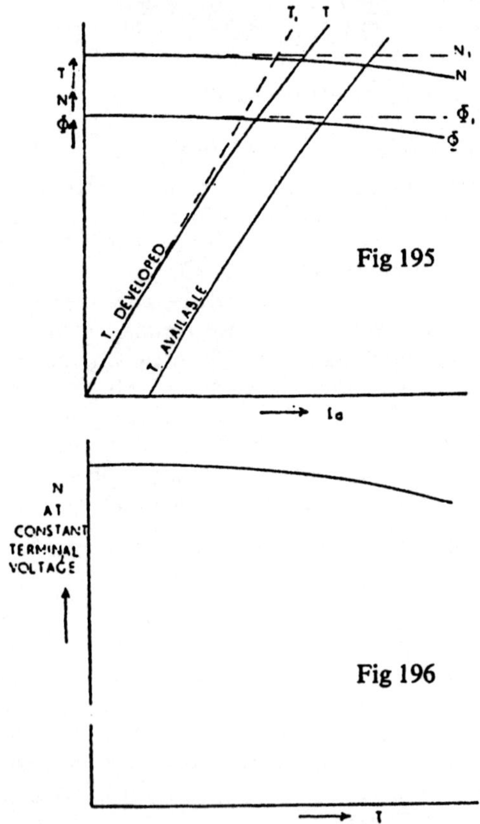

Fig 195

Fig 196

TORQUE. T varies as I_a giving a straight line through the origin, since Φ, is assumed constant. In practice Φ is weakened by armature reaction and T drops as a result, departing from the theoretical straight line T_1 as shown. The torque available at the shaft is everywhere lower because of the lost torque due to rotational losses. Thus two torque characteristics are shown in Fig 195.

(b) MECHANICAL CHARACTERISTIC. As illustrated by the diagram (Fig 196), this is obtained by plotting N against T and is seen to be slightly drooping.

Shunt motors are considered to be constant-speed machines and have only about a 4 per cent drop in speed from no load to full load. As stated earlier they are used for all constant speed drives such as for machine tools, centrifugal pumps, purifiers, etc.

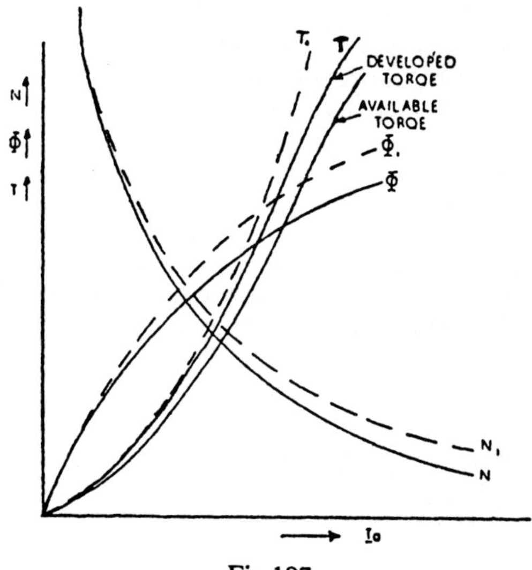

Fig 197

THE SERIES MOTOR

(a) ELECTRICAL CHARACTERISTICS. SPEED. For this machine, the load current value is also that of the field current and, allowing for the effect of armature reaction, it is seen from the diagram (Fig 197) that the useful flux Φ is only slightly less than that given by the magnetisation curve Φ_1. Since Φ increases with load and N varies as $\frac{1}{\Phi}$, it follows that the speed must drop and the curve will conform to that for inverse variation (a rectangular hyperbola), flattening out as saturation of Φ occurs. The no-load flux is small and speed can be excessive. It is for this reason that a series motor should never by run 'light'. It is liable to 'race' and be destroyed by centrifugal force. Like the shunt motor, N is lower than N_1 for reasons already described.

TORQUE. Saturation of the field is not normally achieved over the working load range and Φ is assumed proportional to I_a. Then since torque is proportional to $\Phi \times I_a$ we have, the deduction made earlier that, $T_1 \propto I_a^2$. The curve therefore follows a parabola. On heavy loads, as Φ commences to saturate, $T \propto I_a$ and the graph tends to follow a straight line passing through the origin. This is shown by Fig 197. As for the shunt motor, due to machine losses, the torque available at the shaft is less than the developed torque. At start $T \propto I_a^2$ and the starting torque is very high. This is one of the advantages of this type of motor.

(b) MECHANICAL CHARACTERISTIC. This is shown by the diagram (Fig 198) and is given by plotting the N and T values, for the same armature-current value, as obtained from the electrical characteristics. The result is a curve similar in shape to the speed-current curve as shown by the diagram (Fig 197).

Series motors are variable-speed machines, giving a low speed on heavy loads. They are ideal for traction, winch, hoist and fan work. Their excellent starting-torque characteristic can be used to advantage where heavy masses have to be accelerated quickly, as for lifting or traction.

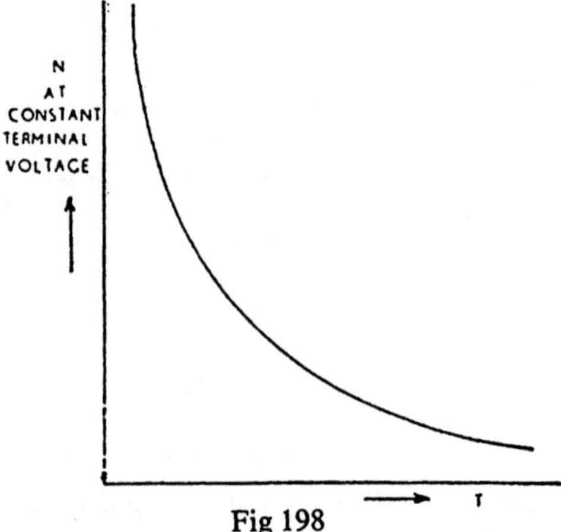

Fig 198

THE COMPOUND MOTOR

As has already been stated, the field connections can be such as to give a cumulative or differential flux result. The former is usual and the latter is used for only exceptional motor duties.

The shunt and series motor have in themselves such good characteristics, that compounding is only used as a means to minimise disadvantages which may occur in the basically connected machine. Thus for example, the series motor tends to race on no load. This effect can be limited by providing a stabilising shunt field, and the compounding for any machine can thus be arranged to give either a *strong shunt, weak series* effect or a *strong series, weak shunt* field combination. The characteristics will therefore be considered with these two arrangements in mind.

CUMULATIVE CONNECTION OF FIELDS. Here the two fields assist each other to give a resultant strengthening of flux. The machine characteristics will depend on the relative strengths of the fields.

(1) *Strong shunt—weak series.* The characteristics of the shunt motor are so good, that in practice it is quite suitable for most drive duties. The provision of a weak series field will not materially alter the load characteristics, but this field does give an improved starting torque, — as explained below.

(a) ELECTRICAL CHARACTERISTICS. SPEED. The diagram (Fig 199) shows the characteristics. Since the net flux rises due to the series

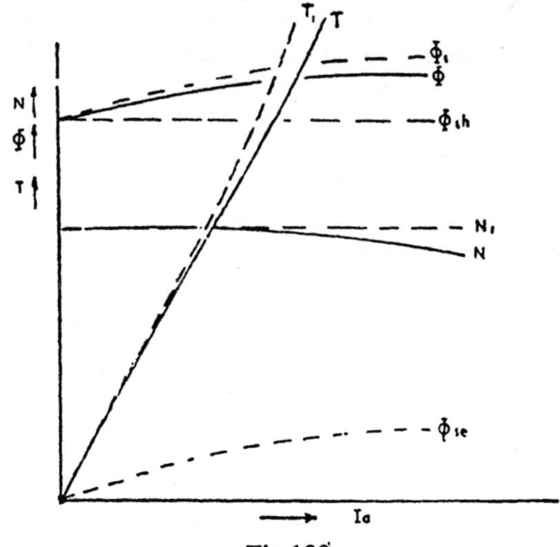

Fig 199

field, this will have a speed-lowering effect since $N \propto \frac{V}{\Phi}$ (approx). The speed will tend to sit down slightly more than it would for the same machine without a series field. If the series field is weak, its effect is not appreciable on the speed characteristic, which differs little from that of the shunt motor. However, when the machine is coupled to a flywheel, a stronger series field can be used, so that sudden application of load causes momentary slowing down with a rise of I_a. The motor speed tends to 'sit down' and the required driving power is obtained from the flywheel which, due to its momentum, gives up some of its energy and does work. This arrangement enables the motor and the electrical system to be protected from undue shock and is used in connection with motors driving specialised loads, such as the rolls in steel-works, presses and hammers, some types of compressors etc.

TORQUE. During starting, when voltage is applied to the shunt field, due to its self-inductance — it being a winding of thin wire and many turns, a back e.m.f. is induced which tends to oppose the shunt field current. Thus the shunt field current builds up very slowly and the torque ($T \propto \Phi I_a$) is small in spite of the large armature current. A series field arranged to pass the starting current I_{as}, will produce a flux to strengthen the shunt flux. Thus the net flux at starting will be very much larger and an improved starting torque would be obtained which may be used for starting against heavy loads, such as those encountered for compressors, centrifugal pumps, certain machine tools, etc. Once the machine accelerates, the characteristic will follow that of a shunt motor, and the effect of armature reaction will alter the theoretical characteristic from T_1 to T as shown (Fig 199).

(b) MECHANICAL CHARACTERISTIC. This characteristic is generally similar to that for a shunt motor.

(2) *Strong series—weak shunt.* Here again the characteristic of the series motor makes it so suitable for its appropriate applications that, it is obvious, its basic performance features will be retained. Its major disadvantage, *eg* the tendency to race on light load, requires to be removed and this is the main function of the shunt field.

(a) ELECTRICAL CHARACTERISTICS. SPEED. It will be seen from the diagram (Fig 200) that although the net flux varies, *ie* it follows the magnetisation curve, yet it never falls to zero as it does in the case of the series motor. In effect the shunt field predominates on light loads and the machine will run as a shunt motor at a predetermined speed. Once load is applied, the series field asserts itself and the speed characteristic passes from that of the shunt machine into that of the series machine. The tendency for racing on no load has thus been removed and this is the typical characteristic for a ship's d.c. winch. *Note.* The effect of armature reaction and voltage drop on E_b is seen at the higher current values and speed N is lower than the theoretical value N_1.

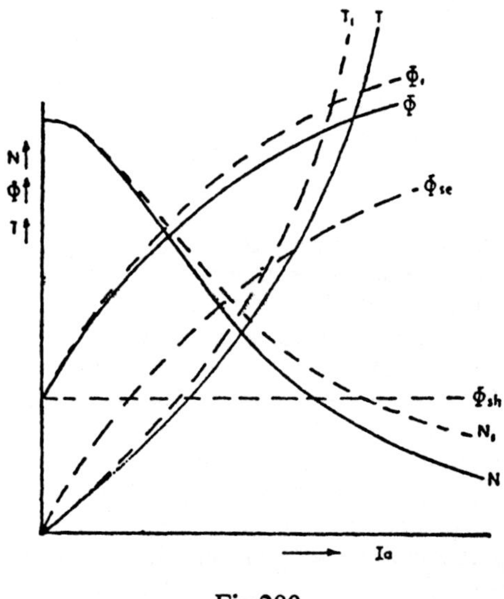

Fig 200

TORQUE. Since the motor behaves like a shunt machine on light loads, the torque characteristic commences as a straight line through the origin and then becomes parabolic as the series field increases in strength. The armature reaction effect gives a slight reduction of net flux with the consequent falling off of torque T from the theoretical graph T_1 (shown dotted).

(b) MECHANICAL CHARACTERISTIC. The characteristic for this motor is shown by the diagram (Fig 201). It is seen to be similar to the electrical speed characteristics and can be deduced, as was described for the series and shunt machines. It is apparent that the exact shape and position of the graph will depend on the relative strengths of the shunt and series fields.

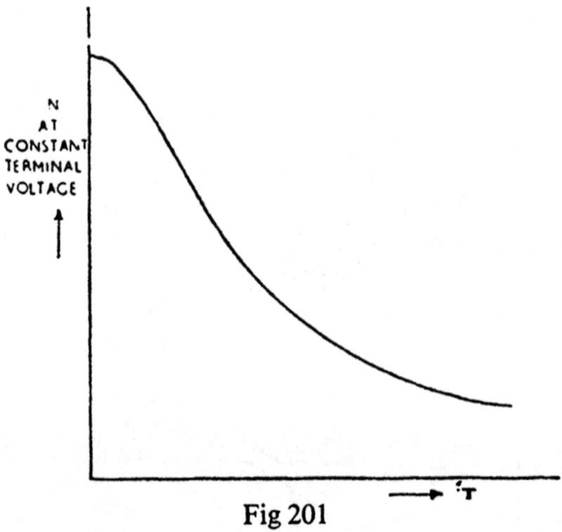

Fig 201

DIFFERENTIAL CONNECTION OF FIELDS. This can be used to maintain a constant speed, *eg* as for an alternator drive. Increase of load results in an increase in the series flux and, as the fields are in opposition, the resultant flux is decreased. The speed, being inversely proportional to flux, increases to compensate for the fall due to the application of load. Thus the machine speed and alternator frequency tends to remain constant but the armature current increases appreciably to provide the required torque with a reduced flux value. A great disadvantage of the arrangement is the field cancelling effect at starting and the fact that, due to the series field establishing itself quicker, the machine may start to run in reverse. Special arrangements have therefore to be made when starting a motor with a differentially-connected field system.

Example 115. A 220V shunt motor runs on light load at a speed of 1250 rev/min and takes a current of 2.8A. On full load the current taken from the mains is 40A and owing to armature reaction, the flux per pole is 4 per cent less than the no-load value. Calculate the speed on full load if the armature resistance is 0.29Ω and the field resistance is 165Ω.

No load. Voltage across shunt field = 220V

$$\text{Current through shunt field} = \frac{220}{165} = 1.332\text{A}$$

$$\text{Armature current} = 2.8 - 1.33 = 1.47\text{A}$$

$$\text{Voltage drop across armature} = I_{a0}R_a = 1.47 \times 0.29$$

$$= 0.426\text{V}$$

and $E_{b0} = 220 - 0.426 = 219.574V$

Full load. Current through shunt field as before = 1.332A

Armature current = 40 − 1.332 = 38.67A

Voltage drop across armature = $I_{a1} R_a = 38.67 \times 0.29 = 11.23V$

and $E_{b1} = 220 - 11.23 = 208.77V$

Now since $E_b \propto \Phi N$ $\quad \therefore E_b = k\Phi N$

and $\dfrac{E_{b0}}{E_{b1}} = \dfrac{\Phi_0 N_0}{\Phi_1 N_1}$ $\quad$ But $\Phi_1 = 0.96\Phi_0$

$$\therefore \frac{E_{b0}}{E_{b1}} = \frac{\Phi_0 N_0}{0.96\Phi_0 N_1} \text{ or } N_1 = \frac{N_0 \times E_{b1}}{0.96 \times E_{b0}} = \frac{1250 \times 208.77}{0.96 \times 219.57}$$

Thus $N_1 = 1238$ *ie* speed on full load = 1238 rev/min.

Example 116. A 220V series motor is working with an unsaturated field taking a current of 100A and running at 800 rev/min. Calculate at what speed the motor will run when developing half the torque? The total resistance of the motor is 0.1Ω.

Here $T = kI_a^2 \therefore \dfrac{T_1}{T_2} = \dfrac{I_{a1}^2}{I_{a2}^2}$ But $T_2 = 0.5T_1$

So $\dfrac{T_1}{0.5T_1} = \dfrac{100^2}{I_{a2}^2}$ or $I_{a2}^2 = 100^2 \times 0.5 = 5000$

$\therefore I_{a2} = \sqrt{5000} = 70.7A$

Also under the first condition $E_{b1} = V - I_{a1}(R_a + R_{se})$
$= 220 - (100 \times 0.1)$
$= 210V$

Under the second condition $E_{b2} = V - I_{a2}(R_a + R_{se})$
$= 220 - (70.7 \times 0.1)$
$= 212.93V$

But $E_b = k\Phi N$ or $\dfrac{E_{b1}}{E_{b2}} = \dfrac{\Phi_1 N_1}{\Phi_2 N_2}$ also $\Phi \propto I_a$

$$\text{so } \frac{E_{b1}}{E_{b2}} = \frac{I_{a1}N_1}{I_{a2}N_2} \text{ or } N_2 = \frac{I_{a1}N_1E_{b2}}{I_{a2}E_{b1}} = \frac{100 \times 800 \times 212.93}{70.7 \times 210}$$

and $N_2 = 1147$ rev/min

Thus speed at ½ torque = 1147 rev/min.

MOTOR STARTERS

The need for a starter to work in conjunction with a motor was mentioned earlier in the chapter, when it was seen that, at the instant of starting since the machine is not rotating, there is no back e.m.f. The current is consequently limited by the armature resistance alone, unless some arrangements are made to include additional resistance in the armature circuit. Thus for all but 'fractional output power' motors, which have quite an appreciable resistance, a resistor is inserted into the armature circuit and then removed in steps, as the motor accelerates up to its correct running speed. The arrangement used is incorporated in a unit, called a 'motor starter' or more simply a 'starter' and consists of a tapped resistor and a switching device which enables the resistance to be gradually reduced and finally cut out altogether. The starter may also incorporate other special attachments which may be considered necessary for the safe operation of the motor. Thus it may include protective arrangements to safeguard the motor against the adverse effects of a reduced working voltage or an overcurrent.

Although motor starters will be studied later in more detail, it is appropriate to mention here that the form of starter necessary for any particular machine is mainly decided by the duty for which the motor is being used. Thus it may be of the manually-operated or automatic type. It may be designed for merely starting and stopping the motor and this may require to be done only once a day. In contrast, the duty may be such as to require the motor to be started and stopped almost continually for long periods, as is necessary when working a winch or hoist. Such a starter is more frequently referred to as a 'controller'. A further point of importance to note is that the starting resistor is not cut out in equal sections from the armature circuit, but that the resistance values of these sections follow a Geometrical Progression.

The foregoing observations indicate that the starter is of sufficient importance to require detailed attention. It is an item of equipment which requires both careful and routine maintenance and a thorough knowledge of its function is necessary both from a theoretical and practical point of view.

SPEED CONTROL

As for the starter, so for the full treatment of speed control, much additional study has yet to be made. It is proposed here, only to deal with the basic methods whereby the speed of a d.c. motor can be controlled and, in this connection, the reader is reminded of the basic deduction $N \propto \frac{E_b}{\Phi}$ or $N \propto \frac{V}{\Phi}$ (approx).

Thus varying the voltage applied to the motor armature and keeping the flux constant will vary the speed in direct proportion. This is termed 'Voltage Control'. Varying the flux of the machine and keeping voltage constant will vary the speed in inverse proportions and is termed 'Field Control'.

FIELD CONTROL. This is introduced first, since it is the most usual type of control. When a motor is loaded, its speed will vary with load. It may be desired to adjust the speed for any load condition *ie* keep it constant throughout the working range or to raise it above the normal running speed. Field control is used because its adaptation into the field circuit is easily achieved, control is smooth and effective and little energy is wasted as heat.

It must be remembered that this type of control will give speed variation in an *upward* direction only. It is used for raising speed above normal and as flux is weakened, for the same driving torque, armature current will rise. *Note.* $T \propto \Phi I_a$. Thus the motor may be of larger dimensions, if speed variation is required and interpoles (compoles) must be fitted to ensure good commutation throughout the working range.

VOLTAGE CONTROL. This is achieved in various ways for the different kinds of d.c. motor but the fundamental requirement is to reduce the voltage applied to the machine armature. Thus a large variable rheostat may be connected in series with the armature or the latter may be supplied from a variable voltage supply. The method is always used to lower speed and control is in a *downward* direction only.

A wide range in the adjustment of motor speed can be obtained by combining field and voltage control and the methods of applying these are sufficiently important to require further detailed study. To meet the requirements of the duty for which the motor is required, the starter and speed controller may be incorporated into one unit which, though simplifying the electrical circuit requirements, appears to complicate the theory of the control. Since the correct application and use of a motor is of prime importance to the practical engineer, it is hoped that the additional treatment, given to the d.c. machine in the next book, will be regarded as a necessary continuation of theory and that too long a break is not introduced into the period of study before the necessary advancement is attempted. If such a break does occur in the student's studies of Electrotechnology then, he would be well advised to revise the work undertaken in this chapter before proceeding to the work in Volume VII.

Example 117. The armature of a motor has 660 conductors whose effective length is 410mm; of these, only 0.7 are simultaneously in the magnetic field. The flux density is 0.65T, the effective diameter of the armature is 300mm, and each conductor carries a current of 80A. If the armature speed is 800 rev/min calculate the output power developed.

Force on one conductor is given by $F = BIl$ newtons

$\therefore F = 0.65 \times 80 \times 410 \times 10^{-3}$

$F = 21.32\text{N}$

Number of conductors in the field at any given instant

$= 0.7 \times 660$

$\therefore$ Total force $= 21.32 \times 0.7 \times 660 = 9.85\text{kN}$

Torque = force × radius or $T = 9850 \times \dfrac{0.3}{2}$ newton metres

Thus $T = 9850 \times 0.15 = 1477.5\text{Nm}$

And power developed $= \dfrac{2 \times 3.14 \times 800 \times 9850 \times 0.15}{60}$

$= 124\text{kW}$

Example 118. A shunt motor takes 180A. The supply voltage is 400V, the resistance of the shunt field is 200Ω, and that of the armature 0.02Ω. If there is a voltage drop of 2V at the brushes, calculate (a) the back e.m.f. of the motor (b) the output power developed (c) the efficiency, neglecting all losses for which information is not given.

Shunt-field current $= \dfrac{400}{200} = 2\text{A}$

Armature current $= 180 - 2 = 178\text{A}$

Armature voltage drop $= 178 \times 0.02 = 3.56\text{V}$

(a) Back e.m.f. $= 400 - 3.56 - 2$ (voltage drop at brushes)

$= 394.44\text{V}$

(b) Output power developed $= \dfrac{394.4 \times 178}{1000} = 70.2\text{kW}$

Efficiency $= \dfrac{\text{output}}{\text{input}} = \dfrac{394.4 \times 178}{400 \times 180}$

or $\eta = 0.975$ or 97.5 per cent

Example 119. A four-pole d.c. motor with a lap winding is connected to 200V supply mains. The armature carries 600 conductors and has a resistance of 0.3Ω. The resistance of the shunt-field circuit is 100Ω, the flux per pole is 0.02Wb. On no load, the armature current to 3A. If the normal full-load current in the armature is 50A, determine the drop in the speed of the motor from no load to full load. Neglect the effect of armature reaction.

Back e.m.f. on no load $E_{b0} = 200 - I_{a0} R_a$

Shunt-field current $= \dfrac{200}{100} = 2\text{A} \quad I_{a0} = 3 - 2 = 1\text{A}$

$$\therefore E_{b0} = 200 - (1 \times 0.3) = 200 - 0.3 = 199.7\text{V}$$

No-load speed is given by N_0 where:

$$E_{b0} = \frac{Z\Phi_0 N_0}{60} \times \frac{P}{A}$$

$$\text{or } 199.7 = \frac{600 \times 0.02 \times N_0}{60} \times \frac{4}{4}$$

$$\text{Thus } 199.7 = 0.2 \times N_0$$

$$\text{or } N_0 = \frac{199.7}{0.2} = 998.5 \text{ rev/min.}$$

Back e.m.f. E_{b1} on full load is given by:

$$E_{b1} = 200 - I_{a1}R_a = 200 - (50 - 2)\,0.3$$

$$E_{b1} = 200 - (48 \times 0.3)$$

$$\text{or } E_{b1} = 185.6\text{V}$$

Since $E_{b1} = k\Phi_1 N_1$ and assuming a constant flux, then

$$\Phi_0 = \Phi_1$$

$$\text{or } \frac{E_{b1}}{E_{b0}} = \frac{k\Phi_1 N_1}{k\Phi_0 N_0} \text{ whence } N_1 = \frac{N_0 E_{b1}}{E_{b0}}$$

$$\text{Thus } N_1 = \frac{998.5 \times 185.6}{199.7}$$

Full-load speed = 928 rev/min.

Example 120. Calculate the first resistance step of a starter for a 240V shunt motor having an armature resistance of 0.5Ω, if the maximum current limit is 60A and the lower limit about 45A.

Let R_s = the total resistance of the series resistor put into the armature circuit. Then if I_{as} is the armature current at start I_{as} = 60A.

$$\text{and also } I_{as} = \frac{240}{R_a + R_s} = \frac{240}{0.5 + R_s}$$

$$\text{or } R_s + 0.5 = \frac{240}{60} \text{ giving } R_s = 4 - 0.5 = 3.5\Omega$$

As the motor starts and accelerates up to speed, the starter handle is kept in position until the current falls to 45A. Thus the starting resistance is still in circuit, but a back e.m.f. is building to a final value given by E_{b1}.

$$\text{Here } E_{b1} = 240 - 45(3.5 + 0.5)$$

$$= 240 - 180 = 60\text{V}$$

At this stage the handle is moved and a section of the starting resistor is cut out. Let R_1 be the new value of the total starter resistance. The current rises to 60A but the back e.m.f. does not change until the motor speed changes. Thus at the instant of moving the handle

$$240 = E_{b1} + I_a (R_a + R_1)$$

$$\text{or } 240 = 60 + 60\,(0.5 + R_1) \text{ whence}$$

$$0.5 + R_1 = \frac{240 - 60}{60} = 3$$

$$\text{or } R_1 = 3 - 0.5$$

$= 2.5\Omega$. Thus the resistance removed during the first movement of the handle after switching on, is $3.5 - 2.5 = 1\Omega$

The first resistance step is thus 1Ω.

ESTIMATION OF D.C. MACHINE EFFICIENCY

The efficiency of a d.c. machine can be assessed by measuring its output power and comparing this with the input power. Thus in general:

$$\text{Efficiency} = \frac{\text{Power Output}}{\text{Power Input}}$$

For small motors the output can be measured using a calibrated brake or a dynamometer and the input power measured by electrical instrumentation. However, on large motors it is more difficult to measure the output power with any reasonable degree of accuracy. Similarly the mechanical input power to a generator is difficult to measure, and, also with large generators, the electrical output involves the dissipation of large amounts of energy, usually in the form of heat.

A technique has therefore been devised, referred to as the Swinburne Test, whereby the losses are assessed and the efficiency at any load estimated from:

$$\eta = \frac{\text{Power input} - \text{losses}}{\text{Power input}} \text{ (for a motor)}$$

$$\eta = \frac{\text{Power input}}{\text{Power output} + \text{losses}} \text{ (for a generator)}$$

The losses which occur can be divided into two groups, those which vary with load and those which remain substantially constant at all loads. These can be summarised as follows:

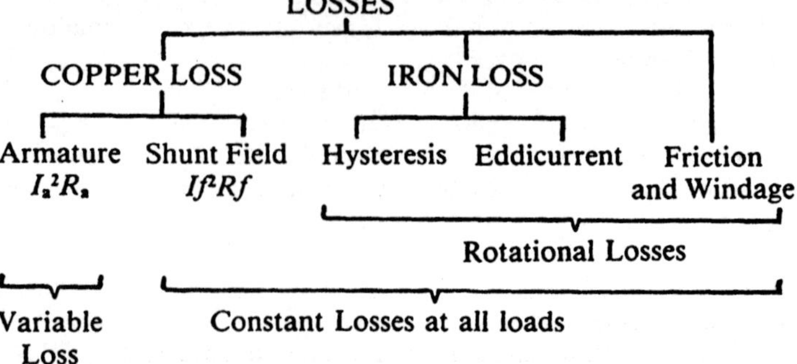

The Rotational Losses due to friction and windage are constant if the speed is constant. The *Iron Losses* are due to the magnetic properties of the core material (hysteresis) and the eddicurrents which are minimised by laminating the cores. Hysteresis and Eddicurrent losses were dealt with in detail in Chapter 6. The iron losses do in fact vary with load, but such variation has little significant effect on the overall constant losses. Losses due to the resistance of the windings, usually referred to as *Copper Losses*, occur in the armature and in the field windings. The field copper loss is also constant providing the supply voltage remains unaltered and the winding is shunt connected. The only significant losses which do vary with load are therefore those due to the armature circuit resistance — the armature copper losses Ia^2Ra.

Hence for a motor:

$$\eta = \frac{\text{Power input} - \text{variable loss} - \text{constant loss}}{\text{Power input}}$$

$$\eta = \frac{VI - Ia^2Ra - If^2Rf - P_{WIF}}{VI} \times 100 \text{ per cent}$$

For a generator:

$$\eta = \frac{\text{Power output}}{\text{Power output} + \text{variable loss} + \text{constant loss}}$$

$$\eta = \frac{VI}{VI + Ia^2Ra + If^2Rf + P_{WIF}} \times 100 \text{ per cent}$$

To assess the constant losses the machine is run as a motor on no-load. Since there is no output power developed, then all the input power is overcoming the machine constant losses. On a shunt connected machine, this no load input power supplies the shunt field loss (VIf or If^2Rf), as well as the armature loss. This no-load armature loss comprises the iron, windage and friction losses, together with a very small armature copper loss which may usually be ignored.

To make an even more accurate estimate of a machine's efficiency it is necessary to take into account the change in copper losses due to resistance change as temperature rises. This involves calculating the new value of resistance at the higher temperature using the Temperature Coefficient of Resistance equation $R_T = R_o(1 + \alpha T)$, the rotational losses remaining unaltered.

Example 121. A 250V d.c. shunt motor takes a current of 7 amp when running on no-load. The armature resistance and shunt field circuit resistance are found to be 0.15Ω and 125Ω respectively. Find the efficiency of the machine when it (a) runs as a shunt motor taking a total current of 52 amp, and (b) is driven as a shunt generator delivering 15kW at its output terminals. On no-load:

$$\text{Shunt field current } If = \frac{V}{Rf} = \frac{250}{125} = 2\text{A}$$

Armature current $Ia_o = I_L - If = 7 - 2 = 5\text{A}$

No load armature power input = W.I.F. losses (+ small Ia^2Ra negligible loss).

$$\therefore P_o = VIa_o = 250 \times 5 = 1250\text{W}$$

$$\text{Field Copper loss } If^2Rf = VIf = 250 \times 2 = 500\text{W}$$

(a) At 52A load $Ia = 52 - 2 = 50\text{A}$ (If constant).

$$\text{Armature Copper loss} = Ia_2Ra = 50^2 \times 0.15 = 375\text{W}$$

$$\eta = \frac{VI - Ia^2Ra - If^2Rf - \text{W.I.F.}}{VI} \times 100 \text{ per cent}$$

$$\eta = \frac{(250 \times 52) - 375 - 500 - 1250}{(250 \times 52)} \times 100 \text{ per cent}$$

$$\eta = 83.65 \text{ per cent}$$

(b)

$$\text{Load current } I_L = \frac{15 \times 10^3}{250} = 60\text{A}$$

$$\text{Shunt Field current } If = 2\text{A}$$

$$\text{Armature current } Ia = 60 + 2 = 62\text{A}$$

$$\text{Armature Cu Loss } Ia^2Ra = 62^2 \times 0.15 = 576.6\text{W}$$

$$\eta = \frac{VI}{VI + Ia^2Ra + If^2Rf + \text{W.I.F.}} \times 100 \text{ per cent}$$

$$\eta = \frac{(15 \times 10^3)}{(15 \times 10^3) + 576.6 + 500 + 1250} \times 100 \text{ per cent}$$

$$\eta = 86.57 \text{ per cent}$$

The efficiency at any load can therefore be estimated and a graph of Efficiency against Load Current plotted. This graph will indicate the efficiency trend and give a fairly accurate assessment of the load current at which maximum efficiency occurs.

CHAPTER 13

PRACTICE EXAMPLES

1. A 110V series motor has a resistance of 0.12Ω. Determine its back e.m.f. when developing a shaft output of 7.5kW when the efficiency is 85 per cent.

2. A 500V d.c. shunt motor has an input of 90kW when loaded. The armature and field resistances are 0.1Ω and 100Ω respectively. Calculate the value of the back e.m.f.

3. A 460V, d.c. motor takes an armature current of 10A at no load. At full load the armature current is 300A. If the resistance of the armature is 0.025Ω, what is the value of the back e.m.f. at no load and full load?

4. An armature winding of a d.c. motor consists of 240 conductors arranged in four parallel paths on an armature whose effective length and diameter are 400mm and 300mm respectively. Assuming that the average flux density in the air gap is 1.2T and that the input to the armature is 40A, calculate (a) the force in newtons and the torque in newton metres developed by one conductor (b) the total torque developed by the complete winding, assuming that all the conductors are effective (c) the power output of the armature in watts, if the speed is 800 rev/min.

5. A marine shunt motor is used for driving a 'fresh water' pump and is found to take an armature current of 25A at 220V, when running on full load. The speed is measured to be 725 rev/min and the armature resistance is 0.2Ω. If the field strength is reduced by 10 per cent by means of the speed regulator and the torque remains unchanged, determine the steady speed ultimately attained and the armature current.

6. A shunt generator delivers 50kW at 250V and 400 rev/min. The armature and field resistances are 0.02Ω and 50Ω respectively. Calculate the speed of the machine when running as a shunt motor taking 50kW input at 250V. Allow 2V for brush-contact drop.

7. A 105V, 3kW d.c. shunt motor has a full-load efficiency of 82 per cent. The armature and field resistances are 0.25Ω and 90Ω respectively. The full-load speed of the motor is 1000 rev/min. Neglecting armature reaction and brush drop, calculate the speed at which the motor will run at no load if the line current at no load is 3.5A. Calculate the resistance to be added to the armature circuit, in order to reduce the speed to 800 rev/min, the torque remaining constant at full-load value.

8. A shunt motor runs at 1000 rev/min when cold, taking 50A from a 230V supply. If the armature and field windings both increase in average temperature from 15°C to 60°C, as the motor warms up; determine the speed when the motor is warm, given that the armature resistance is 0.2Ω and the field resistance 200Ω at 15°C and that the total current drawn from the supply remains constant. Neglect brush drop and armature reaction and assume the magnetic circuit to be unsaturated. (Resistivity temperature coefficient 0.40 per cent from and at 15°C.)

9. A four-pole, shunt motor has a wave-wound armature having 294 conductors. The flux per pole is 0.025Wb and the resistance of the armature is 0.35Ω. Calculate (a) the speed of the armature (b) the torque developed, when the armature is taking a current of 200A from a 230V supply.

10. A shunt motor runs at 600 rev/min from a 230V supply when taking a line current of 50A. Its armature and field resistances are 0.4Ω and 104.5Ω respectively. Neglecting the effects of armature reaction and allowing a 2V brush drop, calculate (a) the no-load speed if the no-load line current is 5A. (b) the resistance to be placed in the armature circuit in order to reduce the speed to 500 rev/min when taking a line current of 50A. (c) the percentage reduction in the flux per pole in order that the speed may be 750 rev/min, when taking an armature current of 30A with no added resistance in the armature circuit.

CHAPTER 14

ELECTRONICS AND THERMIONIC DEVICES

A knowledge of electronic fundamentals is now considered an essential requirement for all engineers, and the student is reminded that, although electronic engineering is, on its own, a fast developing technology nevertheless, with the advent of semi-conductor devices, its impact on power engineering is a factor of major consequence. Large solid-state motor starters, controlled rectifiers and static inverters have been introduced into the ship's engine-room, as items of electrical equipment to complement the smaller but highly sophisticated apparatus relating to instrumentation and control such as; monitoring and alarm systems, supervisory units and data loggers. Because of the complexity of such equipment, specialist knowledge is required for major breakdowns and malfunctioning but the manufacturers have gone some way to assist in everyday fault-finding and repair by evolving procedures for checking by substitution and methods of routine maintenance. Even such tasks would only be performed effectively by an engineer with some basic knowledge of the art and, it is to this end that, the author would encourage an approach to the study of electronics.

An introduction to electronics is made in this chapter, with a consideration of the thermionic diode, the triode, the cathode ray oscilloscope and the application of thermionics to illumination.

ELECTRON EMISSION

The electrons asssociated with the atoms constituting a metal, are considered to be normally confined to the shells, as was described earlier in Chapters 1 and 2. At the surface of a metal, the 'free' electrons in the outermost shells may, by virtue of increased velocity or deflection by collision, tend to leave the surface. They are however immediately attracted back by the unbalance of the electric field which is caused and tends to set up a potential barrier. If the free electrons acquire sufficient velocity to overcome the attraction of the electric field and thus break through the potential barrier, the process is termed *electron emission.* Electron emission can result from various causes and is therefore described by classification under one of the following headings; (a) thermionic emission, (b) cold or field emission already described in Chapter 5, (c) secondary emission, (d) photo-electric emission.

Thermionic emission will be considered here in detail. The others will be mentioned only when such reference is required.

THERMIONIC EMISSION

When a metal is heated, energy is transferred to it and the electrons are considered to acquire increased and random velocities which may be sufficient for some to break through the surface potential barrier. Such electrons can be likened to the vapour globules given off from the surface of water when it is boiling and it is found that, unless some provision is made for drawing off the electrons as they are emitted, they will lose their velocities and tend to collect together to form a *space charge.* Such a cloud of electrons or space charge will, in effect, give rise to a −ve charge which will tend to repel further electrons and we can thus visualise electrons leaving the surface of the metal but, being repelled by the space charge which has built up, returning to the metal. To enable the electrons to leave the heated metal surface, an additional electrode is placed adjacent to, but insulated from, the metal. This electrode — called an *anode* — is made +ve with respect to the heated surface — called the *cathode.* The arrangement has already been introduced in earlier chapters and such an arrangement is essential for the operation of radio valves, cathode ray tubes, mercury-arc rectifiers and discharge lamps.

For most modern electronic equipment, the use of thermionic valves has been discontinued in favour of semiconductor devices but a basic understanding of the operation of the former will be useful in appreciating the functioning of other thermionic devices for which no substitute has been evolved. An example is the cathode-ray tube as used in an oscilloscope, a computer video display unit or a television set. Again the principles of rectification and amplification were developed during the evolution of the various types of valve and although semiconductor technology has produced solid-state components around which corresponding circuitry can be built up, nevertheless a knowledge of the characteristics and limitations of the former will be found to be of assistance when the latter are studied in Chapter 15.

THE VACUUM DIODE

The technique of placing a charged plate near the source of thermionic electron emission is made use of in the valve. The plate is made positive with respect to the electron emitter so that the electrons will be attracted to it and a space charge will not accumulate. The valve in this basic form is called a vacuum diode. It consists of two electrodes sealed into an evacuated glass envelope and constitutes a simple form of electronic device. One electrode is made in the form of a wire which is heated by a current flowing through it. This results in thermionic emission from the wire or *filament*. The other electrode is in the form of a cylinder which is made to surround the filament and is called the *anode*. The anode is made +ve with respect to the filament or *cathode* and attracts the emitted electrons. Thus we have current flow from anode to cathode inside the valve as long as the anode is made +ve with respect to the cathode, but it should be noted that if the anode is made −ve with respect to the cathode, the emitted electrons will be repelled and current will stop. Thus the diode is a unidirectional conducting device, *ie* it will allow current to flow in one direction only and functions like a one-way valve — hence the term. The diagrams (Fig 202(a) and (b)) show the simple construction and the equivalent electrical circuit.

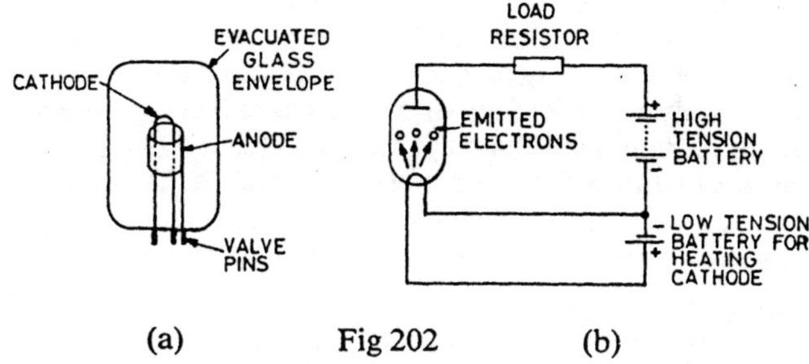

(a) Fig 202 (b)

In practice, the cathode can be heated either by passing a separate current through the filament electrode itself or through a heater placed adjacent to the cathode. The diagrams (Fig 203) show the circuit symbols of the diode valve and its position in the circuit.

For the indirectly heated arrangement, the cathode consists of a nickel tube on to which is sprayed barium or strontium oxide. The heater wire is looped inside it and suitably insulated from it. For a.c. radio sets, a low-voltage heater-transformer is used to energise this heater and those of other valves.

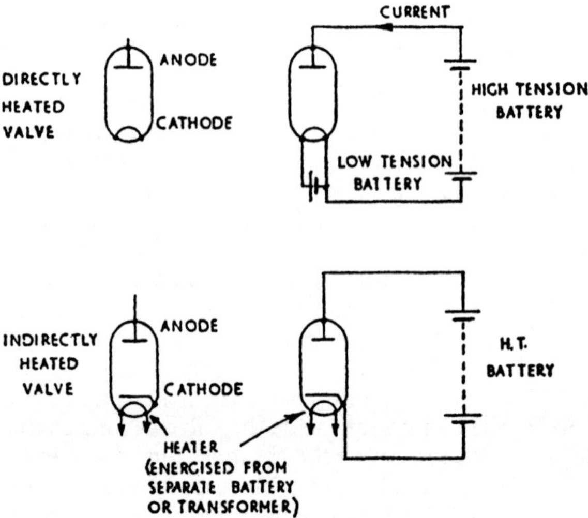

Fig 203

The diagrams (Fig 204) show the test circuit used to determine the characteristics of the diode valve. As already stated, conduction is only possible when the anode is +ve with respect to the cathode and that the more +ve the anode is made, the greater the number of electrons reaching it, *ie* the larger the anode current. The anode current cannot increase indefinitely, since there are only a finite number of electrons leaving the cathode per second. The anode current thus reaches a saturation value which is shown on the characteristic at the point where, when anode voltage is increased beyond a certain value, no further increase of anode current is obtained and the curve bends over to

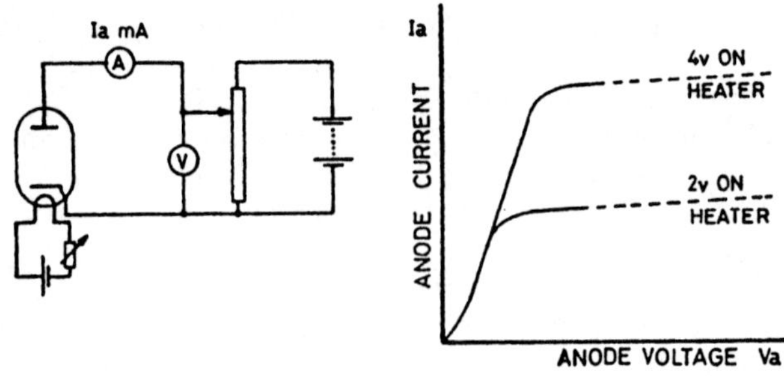

Fig 204

become horizontal. To obtain more anode current, the cathode temperature must be increased to give an increase of electron emission.

The valve characteristic, *ie* relationship between anode voltage V_a and anode current I_a, is obtained by varying the *high-tension* voltage applied to the anode, by means of the adjustable contact on the potentiometer resistor. The corresponding readings of the voltmeter and milliammeter are noted and plotted to give the graph shown. Certain factors for the diode can be derived from the characteristics in order to use the valve for a particular application. Thus the diode can be used as a rectifier, but its internal resistance in the conducting direction would require to be known. This value can be obtained from the characteristic but it is necessary, at this point, to distinguish between *static* and *dynamic* operating conditions. In the circuit of Fig 204, the valve anode is not connected in series with a load, such as a resistor. The voltage across the anode and cathode is set to a known value when doing the test — a condition different from that when the valve is loaded or operating dynamically. This second condition will also be examined and so the characteristic is considered as (a) static and (b) dynamic.

(a) STATIC CHARACTERISTIC. The departure from Ohm's Law, in the behaviour of the diode, shows that its internal resistance is not constant but varies with voltage. If an enlarged part of the static characteristic, as obtained from the test circuit, is examined it is seen from the diagram (Fig 205) that, for any

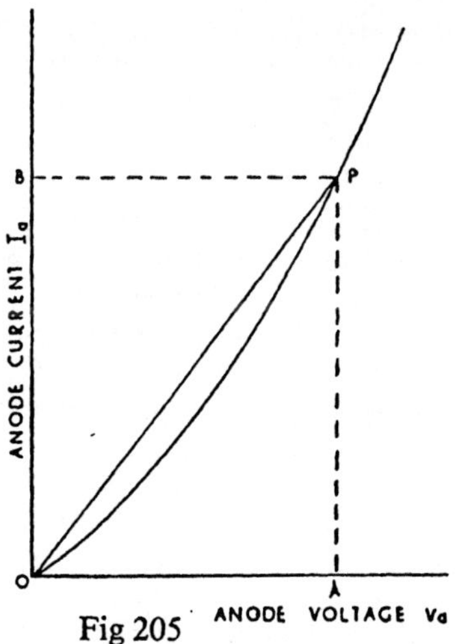

Fig 205

particular anode current value there is a definite ratio of voltage to current. This is the *anode d.c. resistance.* Thus for point P the d.c. resistance $= \frac{BP}{PA}$ or $= \frac{OA}{OB}$. This value would also be obtained from the reciprocal of the slope of the line from P through the origin, *ie* $1 \bigg/ \frac{OB}{OA}$. Point P is known as the 'operating point' or 'quiescent point' and the d.c. resistance value is relative to this point.

Although operation of the diode under a.c. conditions will not be considered to any extent, it is appropriate, at this stage to introduce the term *a.c. resistance* or *slope resistance.* This term — symbol r_a, will be reconsidered when operation of the triode valve is being described, and here we merely consider it in relation to the d.c. value.

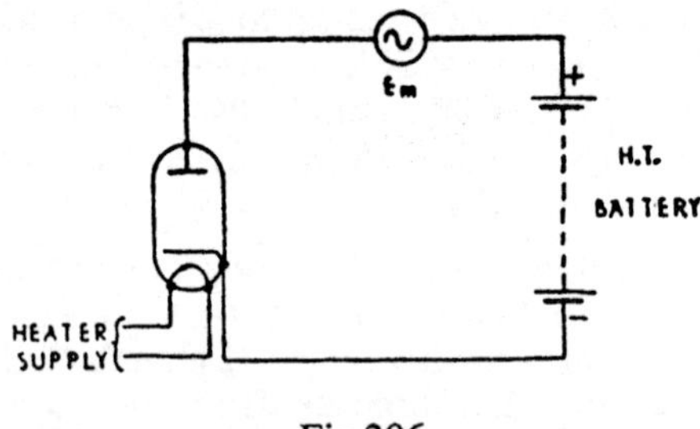

Fig 206

Assume that the alternator, shown in the diagram (Fig 206), generates a sinusoidal voltage of maximum value E_m, which is small compared with the battery high-tension voltage, as applied to the anode of the diode.

The battery is assumed to offer negligible resistance to the a.c. voltage and so an alternating current will flow which is superimposed on the direct current, and the total current fluctuates between the limits of a maximum and minimum value. As will be deduced from the graph (Fig 207), the valve offers a resistance value which is different to that for the d.c. working. Summarising, it can be said that, if the diode is used in a circuit with a steady d.c. current then the anode d.c. resistance is used, but if the anode current is varying between limits then the a.c. value is used. Using the characteristic, note the limits *O* and *Q* between which the operating point oscillates. The changes in the voltage and current values can be found by drawing the small triangle XYZ. Hypotenuse ZX is the tangent to the curve at point P.

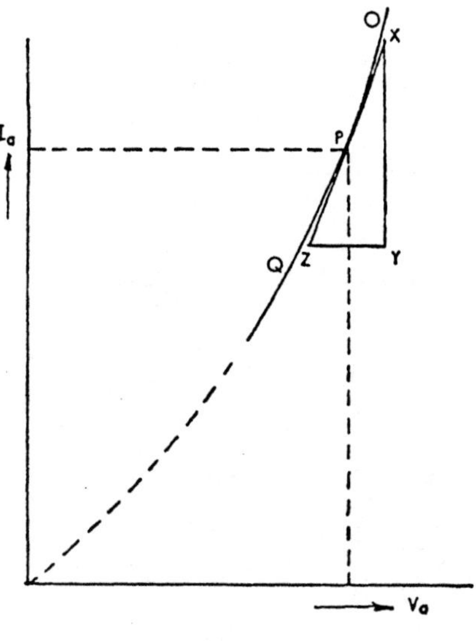

Fig 207

The slope of the graph would give a conductance $\left(\frac{\text{current}}{\text{voltage}}\right)$ but it is more practical to use resistance or the reciprocal of the slope. Thus the reciprocal of the slope or the a.c. resistance value (r_a) is given by $\frac{ZY}{XY}$ *ie* $\left(\frac{\text{voltage}}{\text{current}}\right)$.

Thus r_a

$$= \text{the ratio given by } \frac{\text{small change in anode voltage}}{\text{resulting small change in anode current}}$$

or $r_a = \frac{\delta V_a}{\delta I_a}$.

The sign δ is used here to signify a small change and in the limit it can be written as $\frac{dV_a}{dI_a}$.

The a.c. resistance is less than the d.c. resistance over the working range and the following example shows comparative values.

Example 122. The anode current-voltage static characteristic of a diode can be drawn from the following test values. Deduce the values of a.c. and d.c. resistance for anode current values of 5mA and 10mA.

Anode current (mA)	0	2	4	6	8	10
Anode voltage (V)	0	30.5	46	58	68	77.5

The I_a/V_a characteristic is plotted as shown by the graph (Fig 208).

Then for the 5mA value:

$$\text{D.C. resistance—Point A} = \frac{52.5}{5 \times 10^{-3}}$$
$$= 10.5 \times 10^{-3} \text{ ohms or } 10.5\text{k}\Omega$$

A.C. resistance. Assume an a.c. voltage of 2.5V superimposed on the 52.5V for a standing current of 5mA. Then the anode voltage varies between 50V and 55V and the corresponding currents would be 4.6 and 5.5mA.

$$\text{or } r_a = \frac{55 - 50}{(5.5 - 4.6) \times 10^{-3}} = 5.55 \text{ k}\Omega$$

For the 10mA value:

$$\text{D.C. resistance—Point B} = \frac{77.5\text{V}}{10\text{mA}} = 7.75\text{k}\Omega$$

A.C. resistance. As before, assume an a.c. voltage of 2.5V. For a standing current of 10mA the corresponding voltages would be 77.5 ± 2.5 = 80V and 75V. The currents would be 10.6 and 9.5mA

$$\text{or } r_a = \frac{80 - 75}{(10.5 - 9.5) \times 10^{-3}} = 4.55 \text{ k}\Omega$$

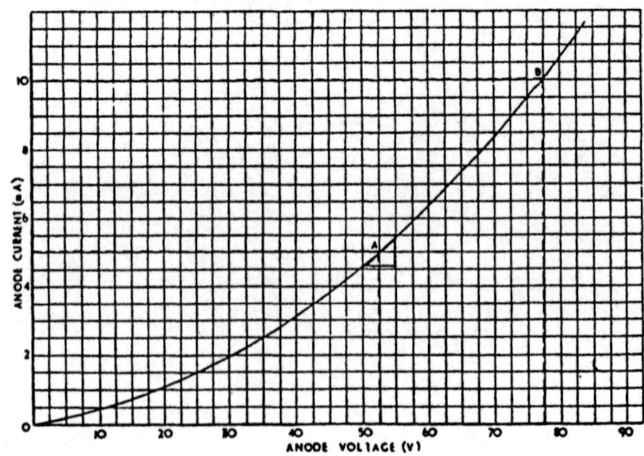

Fig 208

(b) DYNAMIC CHARACTERISTIC—LOAD LINE. A valve is usually used with a load which, in the simplest form, can consist of a resistor in the anode circuit. It is evident that, even though the high-tension voltage V is kept constant, when the anode current varies the voltage drop across the resistor R will vary and the resulting voltage, as applied to the anode will vary. The valve is now operating under dynamic conditions, *ie* with a load and, in order to investigate the correct circuit conditions, it is necessary to use the dynamic characteristic rather than the static characteristic. The former can be obtained from a test circuit similar to that shown by Fig 204 except that a load resistor would be included and the voltage drop across this would also be measured for each value of anode current.

A constructional method can also be used to enable the dynamic characteristic to be obtained from the static characteristic. Thus refer to the diagram (Fig 209). Let P be a point on the static characteristic. Then I_a = OB and the valve voltage drop equals OA. For the same current, the resistance voltage drop = I_aR. Call this V_R, then the high-tension voltage V would be $V_a + V_R$ and the point P[1] would be obtained on the dynamic characteristic. If the procedure is repeated, the complete characteristic can be obtained and then used to find the current and voltage values for any operating condition of high-tension voltage. It must be noted that, there is a separate dynamic characteristic for each value of R and that the characteristic

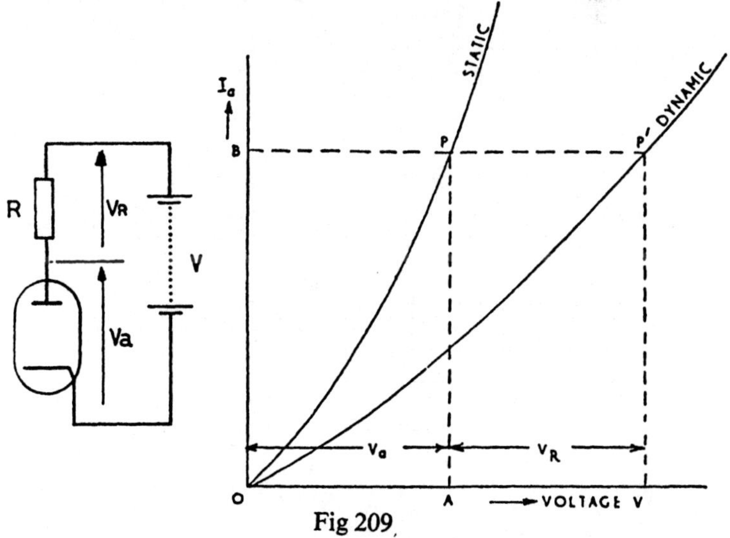

Fig 209

approximates to a straight line as the load-resistor value is increased.

LOAD LINE. As shown above, the voltage-drop conditions in the anode circuit of a diode with a resistor in circuit, can be expressed as $V = V_a + V_R$ or $V_a = V - V_R$. Since the voltage drop across the resistor will be proportional to the anode current, the relationship can be represented by the straight-line graph which would result from the expression:

$$V_a = V - I_a R \text{ or } I_a = \frac{V - V_a}{R} = \frac{V}{R} - \frac{V_a}{R}$$

If the line is superimposed on the static characteristic (Fig 210), it is seen to have a 'negative slope', *ie* it slopes in the opposite direction to the valve charcteristic because an increase in I_a means a decrease in anode voltage. Such a line will therefore

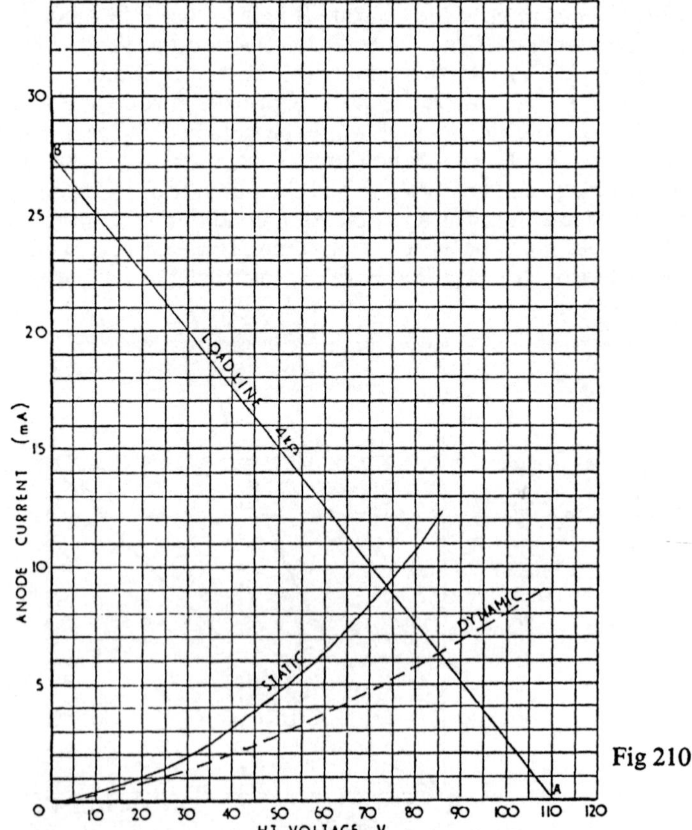

Fig 210

cut the static characteristic to give a point which shows the respective voltage-drop conditions across the valve and resistor for the indicated current value. The straight-line graph or 'load line' will be different for various values of R and can be used to advantage for a specific problem such as that shown below. Thus when the loading condition for one particular value of R at a specified high-tension voltage is required, it is not necessary to construct a dynamic characteristic but merely to draw the load line which is obtained as follows.

Referring to Fig 210, point A is obtained by assuming $I_a = 0$. Under this condition the voltage drop across R would be zero and V_a would be the full high-tension voltage V. Point B is obtained by finding the value of current which would flow if the anode voltage was reduced to zero, *ie* if the diode developed a short-circuit between anode and cathode. Under this condition anode current would be given by $I_a = \frac{V}{R}$. Example 123 shows the treatment in full and with Fig 210, the two methods of solution.

Method 1 has involved plotting the static characteristic and the load line to give the required answer.

Method 2 has involved deducing the dynamic characteristic — its values being determined from the table compiled. This method is obviously of more value if the operating conditions for various values of high-tension voltage are to be considered.

Example 123. Using the test values of Example 122, deduce the dynamic characteristic for a load-resistor of 4kΩ. Find the circuit current for a high-tension voltage of 110V and the voltage drop across the resistor.

Method 1. If the dynamic characteristic had not been required, this method would give the required answers. Consider Fig 210. Plot the static characteristic and obtain the load line thus:

Point A at 110V, since $I_a = 0$

Point B $= \frac{110}{4 \times 10^3} = 27.5 \times 10^{-3} = 27.5$mA. Here the valve resistance is assumed to be zero. The point of intersection gives the answers. Circuit current $=$ 9.1mA. Voltage drop across resistor $= 9.1 \times 10^{-3} \times 4 \times 10^3 = 36.4$V.

Also $V_R = V - V_a = 110 - 36.4 = 73.6$V.

As read from the graph: $V_R = 74$V and $V_a = 36$V.

Method 2. This involves deducing the dynamic characteristic

which is shown plotted. One point only is considered. Assume a current of 4mA. Then V_a = 46V and voltage drop across the resistor = $4 \times 10^{-3} \times 4 \times 10^3$ = 16V. High-tension voltage = 46 + 16 = 62V, *ie* a value for the dynamic characteristic. The following table can be deduced.

Anode current (mA)	0	2	4	6	8	10
Anode voltage (V)	0	30.5	46	58	68	77.5
Resistor voltage drop (V)	0	8	16	24	32	40
High-tension voltage (V)	0	38.5	62	82	100	117.5

For a high-tension voltage of 110V, the circuit current is seen to be 9.1mA. The voltage drop across the resistor = 110 − 74V = 36V.

THE VACUUM TRIODE

Here we consider the effect of introducing a third electrode into the diode assembly and varying its potential. It is found that the electron flow can be closely controlled by this electrode which is usually in the form of an open spiral of wire wound closely round the cathode. The arrangement is shown in Fig 211 and this additional electrode is called the grid. The circuit symbol for the vacuum triode is also shown. The grid is positioned in the region of the "space charge" and is usually made −ve with respect to the cathode. By altering the grid potential the space charge effect can be modified and the anode current can be controlled.

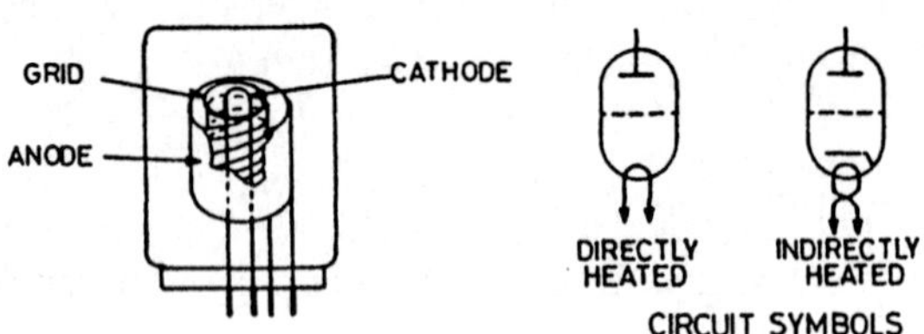

Fig 211

Consider the diagram of Fig 212 which shows the potential gradient between the cathode and anode for three different values of grid voltage. For Curve 1, the grid is made only slightly −ve with respect to the cathode. For Curve 2, the −ve grid potential is increased slightly and for Curve 3, the potential has been increased further to make the grid highly −ve with respect to the cathode. By using the term potential gradient, it will be remembered that we can be dealing with a measure of the electric field intensity or the magnitude of the electric force which acts on a charge which is placed in the field between the two electrodes.

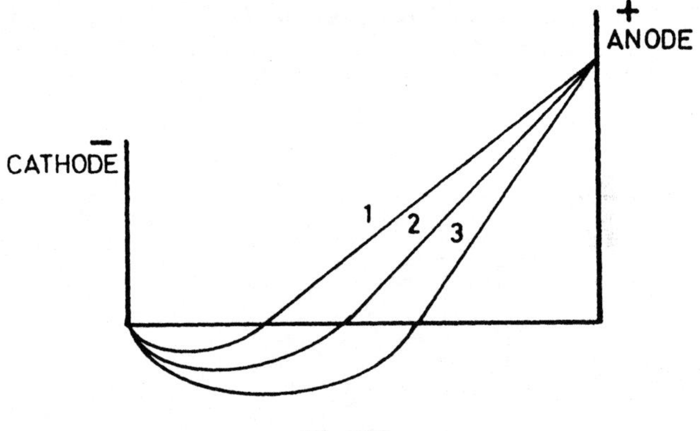

Fig 212

An electron has a −ve charge and, when emitted from the cathode, it will experience a force which moves it from a point of low potential to one of high potential. The magnitude of the force is directly proportional to the potential gradient and if the latter is +ve, the force is towards the anode and if −ve, the force is towards the cathode. For Curve 1 most of the emitted electrons from the cathode, pass the zero potential gradient value and are then accelerated towards the anode to give a large anode current. For Curve 3 conditions, only a few electrons will have sufficient velocity to pass the point of zero potential, the remainder being urged to return back to the cathode. The few electrons which do pass the point of zero potential gradient will proceed to the anode to give a relatively small anode current. It will now be seen how, by merely altering the potential of the grid, the anode current can be controlled. Furthermore, since the grid is invariably kept at a potential which is −ve to the cathode, electrons do not enter it and therefore no current is drawn from the grid voltage supply source. The power required by the grid to control the anode current is zero and thus, although comparatively large powers can be controlled by this method yet the power required to achieve this control is zero.

STATIC CHARACTERISTICS

The circuit set out in Fig 213, shows how the triode can be tested and the control of the anode current by (i) variation of grid voltage and (ii) variation of anode voltage, can be investigated. The curves when plotted from the test results are divided into two groups of static characteristics known as (a) the *Mutual Characteristics,* which show the change in anode current I_a for a change in grid voltage V_g, the anode voltage V_a remaining constant and (b) the *Anode Characteristics* which show the variation of anode current I_a with a change in anode voltage V_a, the grid voltage V_g remaining constant.

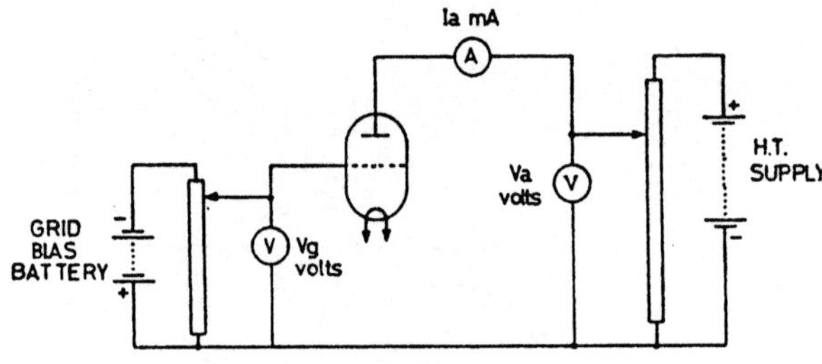

Fig 213

MUTUAL OR TRANSFER (I_a/V_g) CHARACTERISTICS

By keeping the anode voltage constant, the effect on the anode current of a variable grid voltage can be investigated by the test circuit of Fig 213. If the results of such tests are plotted a family of curves, such as those shown by Fig 214(a) are obtained which provide information similar to that available from the anode characteristics although in a more direct form for particular purposes.

Valve manufacturers provide information, on any particular type of valve, in the form of the above curves which also allow three *coefficients* or *parameters* to be determined. These specify the behaviour of a triode when introduced into a circuit and each parameter — sometimes called a "valve constant," represents the relation between the changes in any two of the variables, the third being kept constant.

ANODE OR OUTPUT (I_a/V_a) CHARACTERISTICS

By keeping the grid at cathode potential, *ie* $V_g = 0$, a characteristic curve relating anode current to increasing anode voltage can be obtained from the tests. This curve is shown by Fig 214(b). If the grid is next given a small −ve potential and the test is repeated, a curve of similar shape is obtained which is displaced towards the right. Such a curve is marked as $V_g = -2$ V and the displacement is due to the fact that, as has been explained, the anode must attain a certain minimum +ve potential to overcome the repelling force of the grid on the emitted electrons. The higher the −ve potential on the grid, the more the anode characteristic is moved to the right. A complete family of such curves at various −ve values of grid voltage can be obtained in the manner described.

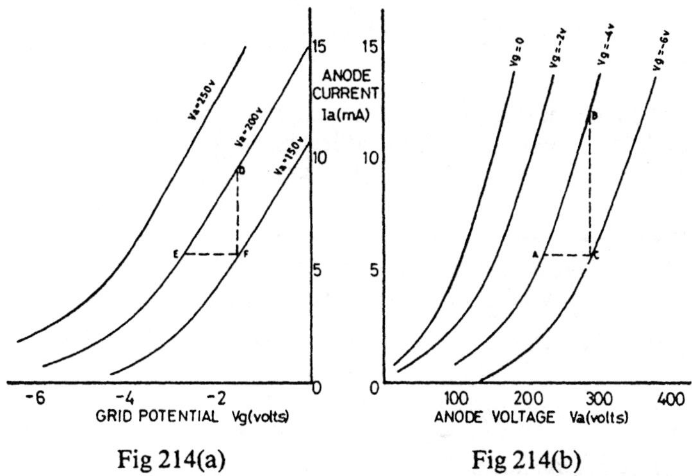

Fig 214(a) Fig 214(b)

VALVE PARAMETERS

These are derived from the slopes or change conditions of the graphs and, as an introduction, are considered as being obtained by constructing small triangles as shown. Since a straight position of a graph is required, the sides of any such triangle are kept as small as possible and are referred to as 'a small change of' or the sign δ is used. A mention of this convention has already been made when the diode was being described.

(1) INTERNAL RESISTANCE

As for the diode, the slope of an anode characteristic, such as $\frac{BC}{AC}$ (Fig 214b) is seen to give a conductance, being $\frac{\text{current}}{\text{voltage}}$. In practice it is more convenient to use the resistance of the valve

and the reciprocal of this slope $= \frac{AC}{BC}$ is known as the *slope* or *anode resistance.* Other alternative terms are the *internal resistance* or *a.c. resistance.* The symbol used is r_a and if a definition is considered we have: Slope or a.c. resistance (r_a) is the ratio of the small change of anode voltage to the small change of anode current produced, the grid voltage being kept constant.

$$\text{Thus } r_a = \frac{\text{small change of anode voltage}}{\text{resulting small change of anode current}}$$

$$\text{or } r_a = \frac{\delta V_a}{\delta I_a},\ V_g \text{ being constant.}$$

In the limit, mathematically the above can be written as

$$r_a = \frac{dV_a}{dI_a}.$$

The internal resistance of a triode for any particular operating condition can be found from either the anode or mutual characteristics and is illustrated by the examples which follow.

(2) MUTUAL CONDUCTANCE

The influence of the control-grid voltage on the anode current can be measured most readily by determining the slope of a mutual characteristic curve. Thus for example, $\frac{DF}{EF}$ (Fig 214a) gives the second paramater, namely $\frac{\delta I_a}{\delta V_g}$ — referred to as the *mutual conductance.*

The symbol g_m is used and the definition would be: Mutual conductance is the ratio of the small change of anode current to the small change of grid voltage producing it, the anode voltage being kept constant.

$$\text{Thus } g_m = \frac{\text{small change of anode current}}{\text{small change of grid voltage}}$$

$$\text{or } g_m = \frac{\delta I_a}{\delta V_g},\ V_a \text{ being constant.}$$

Here again either an anode or mutual characteristic can be used to determine g_m. The examples show the unit as $\frac{\text{Milliamperes}}{\text{Volts}}$ or millisiemens (mS) but the use of milliamperes per volt of mA/V is more general.

(3) AMPLIFICATION FACTOR

This is the third parameter and is used to represent the maximum theoretical *voltage gain* which can be obtained from any particular valve. Note that as it is a ratio of two voltages, a unit is unnecessary. The symbol used is μ and it is defined thus: Amplification factor is the ratio of the change in anode voltage to the change in grid voltage when both are concerned with the same change of anode current.

$$\text{Thus } \mu = \frac{\text{change in anode voltage}}{\text{change in grid voltage}}$$

$$\text{or } \mu = \frac{\delta V_a}{\delta V_g}, \ I_a \text{ being constant.}$$

As for the two previous parameters, amplification factor can be determined from either the anode or mutual characteristics. A relationship connecting r_a, g_m and μ will be deduced shortly and thus the third constant can always be determined if the other two are known. The examples extend practice in the use of the parameters and characteristics and are instructive.

PARAMETER RELATIONSHIPS

Mention has been made of the relation between the three valve parameters and this is now considered.

Since $\mu = \dfrac{\delta V_a}{\delta V_g}$ equality still exists if written as

$$\mu = \frac{\delta V_a}{\delta V_g} \times \frac{\delta I_a}{\delta I_a}$$

Then $\mu = \dfrac{\delta V_a}{\delta I_a} \times \dfrac{\delta I_a}{\delta V_g}$ giving $\mu = r_a \times g_m$ or

Amplification factor = anode a.c. resistance × mutual conductance.

Example 124. In a certain triode, the anode current is 5mA with an anode potential of 220V and a grid potential of −3V. When the anode potential is increased to 260V, the current rises to 7mA and a change of grid potential to −4V restores the current to its original value. Determine the valve constants.

$$\text{Here } r_a = \frac{\delta V_a}{\delta I_a} = \frac{(260 - 220)\text{V}}{(7 - 5) \times 10^{-3}\text{A}} = \frac{40 \times 10^3}{2}$$

$$= 20\,000 \text{ ohms or } 20\text{k}\Omega.$$

$$g_m = \frac{\delta I_a}{\delta V_g} = \frac{(7 - 5)\text{mA}}{-3 - (-4)\text{V}}$$

$$= 2\text{mA/V or } 2 \times 10^{-3} \text{ A/V}.$$

$$\mu = \frac{\delta V_a}{\delta V_g} = \frac{(260 - 220)\text{V}}{-3 - (-4)\text{V}} = \frac{40}{1} = 40.$$

or using the relationship

$$\mu = r_a \times g_m$$

$$= 20 \times 10^3 \times 2 \times 10^{-3} = 40.$$

Example 125. The static characteristic test values for a triode are as follows.

$V_g = 0V$.	Then for V_a	50	100	150	200	(Volts)
	I_a	3	6.3	9.6	13	(Milliamperes)
$V_g = -3V$.	Then for V_a	150	200	250	300	(Volts)
	I_a	1.4	5	8.5	12.1	(Milliamperes)

Plot the anode characteristic and determine r_a, g_m and μ.

The small lengths of anode characteristics are shown by Fig 215. Where data is given in this form, the most convenient method of solution is to erect a triangle which makes contact with an adjacent curve. Solution is then as follows.

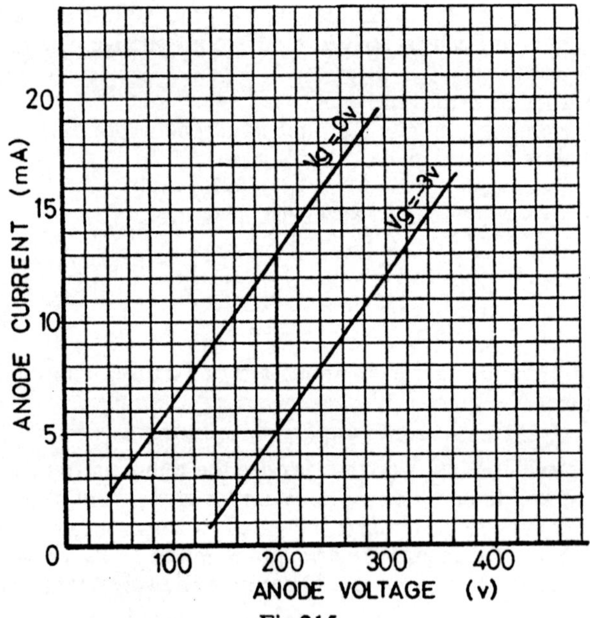

Fig 215

$$r_a = \frac{\delta V_a}{\delta I_a} = \frac{200 - 80}{(13 - 5)10^{-3}} = 15\text{k}\Omega$$

$$g_m = \frac{\delta I_a}{\delta V_g} = \frac{(13 - 5)10^{-3}}{0 - (-3)} = 2.667\text{mA/V}$$

$$\mu = \frac{\delta V_a}{\delta V_g} = \frac{200 - 80}{0 - (-3)} = 40.$$

Check $\mu = r_a g_m = 15 \times 10^3 \times 2.667 \times 10^{-3} = 40.$

IONISATION

Earlier in Chapter 1, ionisation was mentioned, when it was stated that if an electron is removed from an atom, the latter acquires a +ve charge and becomes a +ve ion. The ionisation of a gas atom is accompanied by an emission of light, which is monochromatic, *ie* the colour or wavelength of the light depends on the gas being ionised. It is unlike that given off by the tungsten-filament lamp which covers the wide spectrum of visible white light, consisting of seven colours. Ionisation of a gas is mainly the result of the collision of fast, free-moving electrons with gas atoms. We have seen how electrons can be given off from the surface of a metal by thermionic emission, and how these electrons can be made to travel in a stream with considerable velocity towards an electrode, if the latter is made +ve to the emitter. The higher the p.d. between anode and cathode, the greater the velocity of the electrons, and if these collide with gas atoms with a velocity sufficient to remove further electrons from the parent gas atoms, ionisation takes place with the accompanying of light. The additional electrons, being subjected to the electric field between anode and cathode, accelerate towards the anode, joining the electron stream and collide with further gas atoms to maintain the ionisation. We then have a stream of electrons making their way to the anode or a current flowing from anode to cathode. The electronic device, whether in the form of a gas-filled valve or discharge tube, is thus dependent on ionisation for its conducting properties. When the electrons reach the anode, they displace further electrons in the conducting metal of the circuit and this process explains the current taken from the mains, which is necessary to maintain the device in operation. The +ve ions of the ionised gas, being very much heavier than electrons, move slowly to the cathode to combine with the electrons moving round the circuit. The current stream, in an ionised gas, thus consists of electrons moving from cathode to anode and +ve ions moving from anode to cathode. It should be noted that the +ve ions, in moving to the cathode, neutralise any space charge effect and, if the voltage across the electrodes is too large, they may bombard the cathode with sufficient velocity to destroy it rapidly. It should be remembered that the mass of a proton, which constitutes the major part of the +ve ion, is alone some 1850 times that of an electron and that the heat generated at the cathode by ion bombardment, under conditions of normal operation, is sufficient to ensure that the temperature is high enough for electron emission to be maintained.

We now consider the practical applications in which electron emission and ionisation feature. The cold-cathode discharge tube has already been described in Chapter 3 but most electronic devices, other than those involving semi-conductors, rely on thermionic emission. The light given off, with all forms of discharge, is mainly in the ultra violet range of the spectrum and is largely invisible, thus the tubes or glass envelopes are coated internally with a *phosphorescent powder.* This phosphor glows or *fluoresces* when subjected to the ultra-violet radiations from the ionisation.

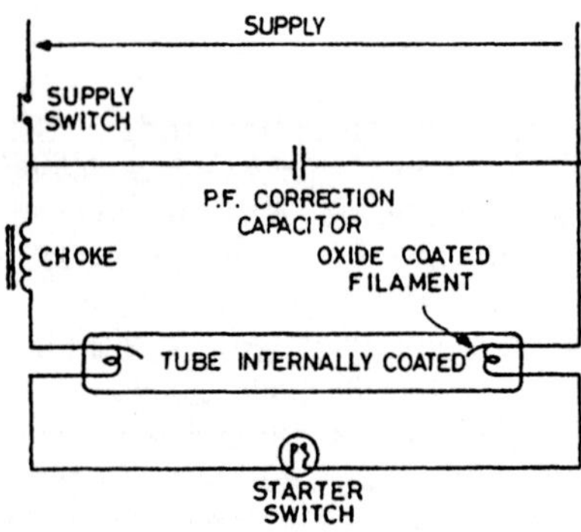

Fig 216

THE HOT-CATHODE DISCHARGE LAMP (Low-pressure)

The diagram (Fig 216) shows the basic construction. The lamp envelope or tube is long compared with its diameter. Lengths vary from 300mm to 2.4m, with respective diameters from 15mm to 38mm. The inside of the tube is coated with the phosphor, and mounted at each end are the electrodes in the form of small tungsten wire spirals. These spirals may be oxide-coated to assist electron emission, and additional guard electrodes may be fitted to form the roots for the main discharge. The tube contains a small quantity of mercury and argon gas which, when warm under working conditions, is at about 10^{-6} bar or 0.1Pa. The low working pressure allows the lamp to run at a temperature which does not affect the phosphor coating. The argon gas helps to initiate the discharge through the low pressure mercury vapour, and although the amount of white light given off is small, some 65 per cent of the input energy is converted into radiant energy in the ultra-violet range. The ultra-violet energy acts upon the fluorescent powder in such a way that the latter radiates visible light.

At start, current is passed through both electrodes, thereby heating them, making them suitable for electron emission and reducing the voltage value required for starting. The passage of current also sets up a small potential between the two ends of each filament, ionising the gas and vapour in the vicinity of each electrode, thereby assisting striking of the main discharge. When the electrodes are hot, the heating current is broken by the starter-switch and a momentary surge voltage of 700-1000V is set up by the choke-coil. This voltage is sufficient to start the main discharge, which is then maintained by the normal mains voltage. It will be seen that the choke-coil serves a dual purpose. Initially it gives the high-voltage impulse to cause the tube to strike, and then acts as a voltage dropper to maintain the correct p.d. across the tube.

Although 'instant-start' fluorescent lamps are widely used ashore, marine systems will mainly prefer circuits which employ a starter-switch. The starter-switch may be of the 'thermal' or 'glow' type but its function is basically to allow preheating of the lamp electrodes, and then to apply the striking voltage. Action of the lamp and starter-switch may therefore be summarised as follows. When the control-switch is closed, the supply is applied to the circuit and as the starter-switch is closed, the current flows through the filaments which give copious electron emission and start the ionisation. At a pre-determined time, the starter-switch opens automatically to interrupt the main circuit. The collapsing field around the choke induces a high-voltage across the tube, causing it to strike. Once the main discharge current flows, the voltage drop across the choke results in the correct p.d. being applied across the tube, and a stable running condition.

It is not proposed to describe the action of the starter-switch here, since this is more a matter of engineering knowledge than basic theory. A measure of the efficiency of a lamp may be made by comparing the lumens put out for every watt put in. Although it is not necessary here, to define 'the lumen', the efficiency of the fluorescent lamp can be compared with that of the tungsten filament lamp, when its efficiency is given as 40-50 lumens/watt. The figure for the filament lamp can be taken as 14-16 lumens/ watt.

THE HOT-CATHODE DISCHARGE LAMP (High-pressure)

This type of lamp is more compact and is mainly used for outdoor installations on docksides, wharfs and, to a limited extent for deck illumination on board ship.

The most common types of hot cathode discharge lamps are

the sodium and mercury-vapour types. The sodium discharge lamp and control circuit is illustrated in Fig 217. When the supply is switched on a voltage of about 450V is applied from the secondary winding of a high reactance transformer connected across the lamp electrodes. This causes a low discharge through the neon gas in the discharge tube, which glows red. The heat produced gradually vaporises the metallic sodium, also contained in the tube, and the sodium vapour produced ionises to give a bright golden-yellow discharge light. The vacuum jacket is necessary to thermally insulate the discharge tube to improve vaporisation of the sodium.

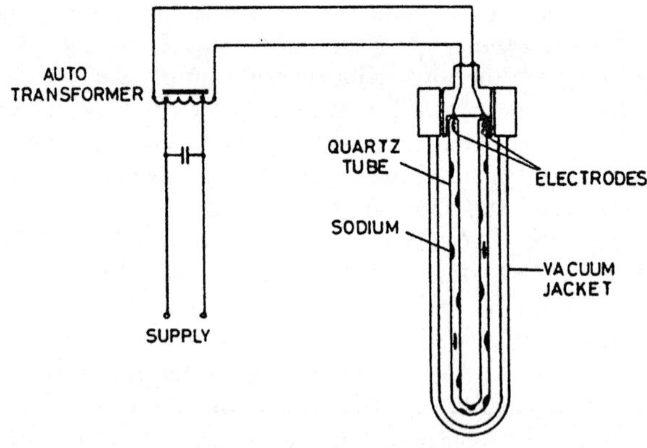

Fig 217

Since the light given out is monochromatic, a fluorescent coating would not achieve colour correction. The high-pressure mercury-vapour lamp is more frequently used than the sodium type to illuminate deck working-spaces and is briefly described below in conjunction with the diagram (Fig 218).

Since this lamp will operate at the normal mains voltage, no step-up transformer is required. The inner tube is made of quartz glass to withstand the high operating temperature and contains mercury with a small amount of neon or argon gas. The inner tube is contained in an evacuated, outer, glass envelope which may be coated on the inside with a fluorescent phosphor. There are two main electrodes and a starting electrode adjacent to one of them.

When the supply is switched on discharge occurs in the inert gas (neon or argon) between the main and auxiliary electrode. This preliminary discharge is sufficient to 'trigger' the discharge between the two main electrodes by ionising the mercury

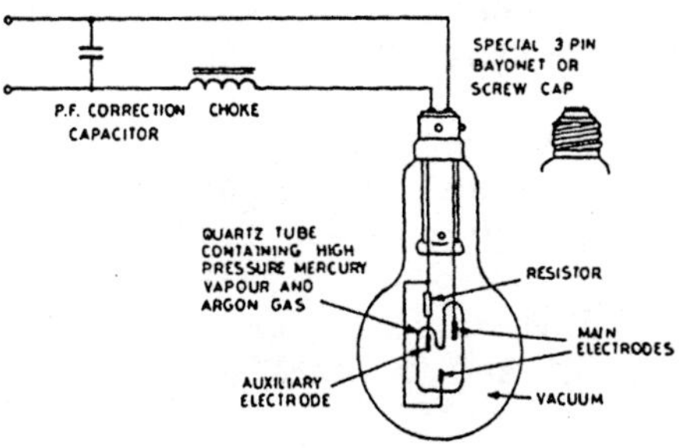

Fig 218

vapour. Once the main discharge strikes, the electrodes are maintained at their working or emitting temperature by the bombardment of the +ve mercury ions. Having started the lamp, the auxiliary electrode has no further effect, the current flowing through it being kept to a minimum by a series resistor of high ohmic value.

A stabilising choke in the main circuit prevents excessive build up of current after discharge commences. The light emitted is mostly in the ultra-violet range, hence the need for colour correction by the fluorescent coating. The main electrodes, though not heated by a preheating current, are usually made in the form of a wire spiral which contains a pellet of barium or strontium oxide. This assists in electron emission when the electrodes are heated.

The lamp takes some 10 minutes to warm-up, as the mercury vaporises and the pressure builds up to about standard atmospheric pressure (1 bar). A disadvantage of the high working pressure is that the lamp will not restrike immediately after being switched off, and must be allowed to cool down, to enable the pressure to fall, before the discharge will re-start.

THE CATHODE-RAY OSCILLOSCOPE

This instrument has now been developed to suit a wide variety of applications but it should be appreciated that, in its fundamental form, the cathode-ray oscilloscope (C.R.O.) is a device for measuring voltage both in magnitude and time. By displaying a luminescent, mobile spot on a fluorescent screen, a trace of the deflection and/or waveform can be observed and measured — a graticule suitably scaled, is usually provided for superposition on the screen. By the use of ancillary equipment such as amplifiers and transducers, any measurable quantity can be converted into a voltage and is so displayed. Suitable calibration will then allow the quantity, being checked, to be measured directly in its appropriate units.

THE CATHODE-RAY TUBE (C.R.T.)

The construction of a cathode-ray tube is described in relation to the diagram (Fig 219). It may be considered to consist of four distinct parts:

(1) A system of electrodes, to produce an electron beam of high velocity and a means of controlling the beam intensity.
(2) A means of concentrating the beam, by electrodes to give a sharply focused spot on the screen.
(3) A deflecting system, to move the focused beam to any part of the screen.
(4) A screen, coated with fluorescent material, which emits light when energised by a stream of high velocity electrons.

The whole arrangement is enclosed in an evacuated glass envelope. The system of electrodes and the focusing electrodes, *ie* parts 1 and 2 are sometimes referred to as the 'electrode gun'. A coating of colloidal graphite is connected to the final anode and terminates at the screen. Its purpose is to allow the path for electrons to be completed back to the supply source, once they have leaked through the glass and screen phosphors from the point of beam impact.

Note. The voltage values, shown for the resistor chain, are approximate only. They do however, indicate the order of such voltages.

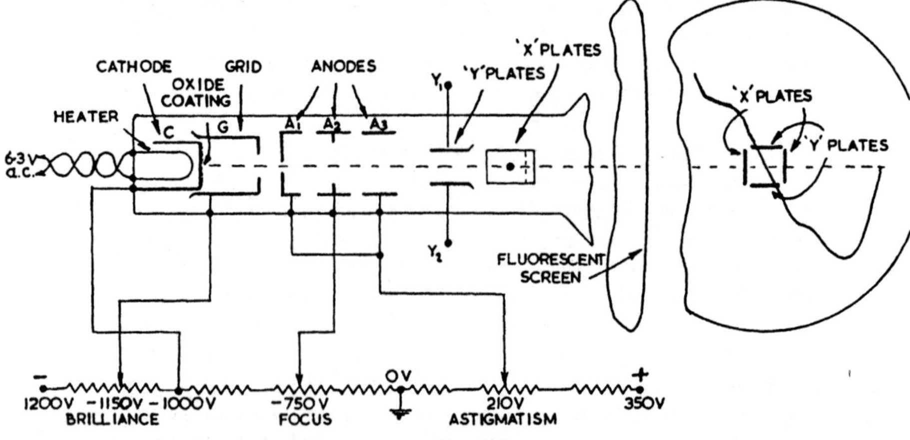

Fig 219

OPERATION. As for the thermionic valve, electrons are produced from a heated cathode C, and attracted to an anode A_1 to which a high, +ve voltage is applied. The cathode is oxide-coated and heated by a separately energised heating element. Interposed between the cathode and anode is a cylindrical electrode called the *grid* or *modulator* G. The name has been borrowed from the triode valve where the grid is in the form of an open-wire spiral or mesh and here, although a complete cylinder, it performs a similar function. The grid, which is used to control the number of electrons leaving the cathode, is given a highly negative potential with respect to the cathode. Electrons are repelled by the −ve charge on the grid, the degree of repulsion depending on the magnitude of the negative potential. Complete 'cut-off' of the electron beam can thus be achieved and varying the grid potential therefore controls the beam intensity, *ie* its 'Brightness or Brilliance'.

The anode A_1 can be in the form of a disc with a small central hole, but the more usual arrangement consists of the disc being built into the end of a small cylinder. This cylinder assists in forming the electrostatic focusing field which will be described shortly. The electrons, emitted by the cathode, are attracted to the anode by its high +ve potential and some pass through the hole at high velocity, thus constituting a beam. If uncontrolled, the electrons, being −ve charges, would repel each other and would diverge to reach the screen in a spreading beam and would give a relatively large, dull spot. The beam, after emerging from anode A_1 is therefore to be focused, and this is achieved by the second and third anodes (A_2 and A_3). These are at high but differing d.c. potentials, and produce an electric field, of such a configuration, that the electron beam is compressed and thus focused. The high d.c. voltages on electrodes A_2 and A_3 further accelerate the electrons.

An introduction was made to fluorescence when discharge lamps were considered. As for the lamps, the screen material consists of phosphors which emit visible light when excited by electron bombardment. The electron beam thus results in a luminous spot which draws the pattern to be observed. Due to the 'after-glow' — the time-lag in the extinction of the fluorescent glow, and the persistence of vision of the human eye, the light spot, when moving, appears to trace a line and thus produce an image which can be observed. Such an image can be a straight-line deflection, a waveform or a recurring figure.

FOCUSING. The full function of anodes A_2 and A_3 can now be examined. In Chapter 8, electrostatic fields were considered and it was seen that their forms could be depicted by lines of flux. The shape of such fields can be altered by varying the potential differences between adjacent electrodes. When an electron moves in a region of changing potential, a force acts on it, at any point in its path, which is perpendicular to the planes of equal potential. Such planes can also be depicted by lines, similar to the flux lines and, if drawn, the overall effect will be to reveal an electrical lens pattern which functions in a manner similar to an optical lens system focusing a beam of light. Thus the three anodes A_1, A_2 and A_3 form a converging lens in the gaps between the electrodes and, by adjusting this lens, focusing can be carried out in this region. It will be seen that the appropriate potentiometer in the diagram (Fig 219) is marked 'Focus'.

There is also a control marked 'Astigmatism'. This works with the focus and brightness controls and allows adjustment of the spot so that its size, intensity and regularity is uniform in any position on the screen. The quality of the trace and the resulting picture is thus capable of adjustment.

DEFLECTION. To display a picture it is necessary to deflect the beam in both a vertical and horizontal direction. The object of the X and Y plates will now be seen. These deflector plates are arranged in parallel in pairs, each pair being at right angles and located between the 'gun' and the screen. Viewed along the axis of the C.R.T., the area enclosed by the deflector plates is square in cross-section, the length of the sides, *ie* width of plates, being sufficient to allow a full-screen deflection of the electron beam on both the X and Y axes. The beam is deflected by the electric field between the plates when a potential is applied across them, — the +ve plate attracting, the −ve plate repelling. An alternating voltage applied to the Y plates will deflect the spot up and down to give a vertical line.

In order to examine waveforms and transient changes in circuit conditions, it is also necessary to move the spot sideways. A deflective effect occurs if the horizontal plates have a d.c. potential applied across them. This is achieved by providing a *time-base.*

TIME-BASE

This can be regarded as a repetitive pulse of voltage which can be adjusted. The time-base impulse is usually provided by a saw-tooth voltage sequence, built up by special circuitry. This voltage increases uniformly with time until it reaches a value when it collapses, to restart building up again. The spot accordingly moves uniformly across the screen, flies back and again commences moving at uniform rate. Additional circuitry allows 'synchronising', by feeding some of the Y amplifier's output to the X time-base. This arrangement allows a trace pattern to be held stationary.

The time-base for a modern C.R.O. is a specialised, involved item of circuitry, arranged in modular form to allow different time-range requirements to be met. The criteria would be stability, reliability and accuracy but a simple circuit to illustrate the principle can be considered. This would be made up from a small cold-cathode lamp or tube connected across a capacitor in series with a resistor. When a d.c. voltage is applied across the combination, the capacitor charges up, with the voltage across the capacitor plates rising uniformly with time, until the breakdown voltage value of the tube is attained. The tube then strikes and the capacitor discharges, with its p.d. falling to zero. The process of charge and discharge recommences and a graph, if plotted with time, will be of a sawtooth shape. Thus the voltage rises uniformly, falls to zero, rises again, falls and so on. If connection is made to the X plates, the spot will advance across the screen from left to right, flies back, moves to the right and returns, the sequence giving a horizontal straight line. A sinusoidal voltage applied to the Y plates will give a vertical deflection and by adjusting the 'time-constant' of the time-base circuit, to equal the periodic time of the alternating quantity for an exact multiple of the periodic time, the waveform will be displayed upon the fluorescent screen as a graph of voltage or current against time. The time-constant of the time-base circuit can be adjusted by varying the size of capacitor and series resistor.

CHAPTER 14

PRACTICE EXAMPLES

1. The anode power dissipation of a diode valve is 336mW. The anode to cathode voltage is 160V. Find the anode current.

2. Example 122 shows that when an anode voltage of 58V is applied to a two-electrode valve the anode current is 6mA. Calculate the power dissipated at the anode.

3. When the linear part of the I_a/V_a characteristic of a diode valve is examined, the slope is found to be 1.64mA/8.13V. Find the a.c. resistance of the valve.

4. The I_a/V_a characteristic of a diode valve shows that an increase of anode voltage from 75V to 129V increases the current from 12mA to 22mA. Find the a.c. resistance of the valve and the d.c. resistance for each voltage condition.

5. The I_a/V_a curve of a diode valve is given by:

Anode voltage (V)	5	10	15	20	25	27.5
Anode current (mA)	0.6	2.0	4.2	7.25	10.5	12

Plot the curve and find the a.c. resistance over the straight line region.

6. A vacuum diode has the following forward characteristic:

Anode-cathode voltage (V)	0	10	20	30	40	50
Anode current (mA)	0	25	60	100	145	200

This diode is placed in series with a 300Ω resistor load and the combination is fed from a 60V d.c. supply, connected so as to make the diode conduct. Determine the current that flows and the power dissipated in the load resistor.

7. The current in a certain thermionic vacuum diode is related to its anode-cathode voltage by the expression $I_a = V_a^{3/2}$. The current is in milliamperes when V_a is in volts. The diode is connected via a 1kΩ resistor to a d.c. supply. Plot the dynamic characteristic showing the variation of anode current against supply voltage for the range 0-10V. What is the value of anode current when the p.d. across the circuit is 8V?

8. A diode valve has the following characteristic:

Anode current (mA)	0	5.5	13	23	33	43	52	59	63
Anode voltage (V)	0	25	50	75	100	125	150	175	200

The valve is connected in series with a resistor of 10kΩ to a 240V d.c. supply. If a resistor of 50kΩ is connected between the anode and cathode of the diode, determine the current through the diode.

9. When the anode voltage of a triode valve is kept constant at 200V and the grid voltage is varied from −3V to −1.25V, the anode current rises from 10mA to 18mA. Find the mutual conductance of the valve.

10. The following readings were obtained from the linear portions of the static characteristics of a triode valve.

Va (volts)	120	120	80
Vg (volts)	−1.3	−3.8	−1.3
Ia (mA)	10	4	6.2

Find the a.c. resistance, mutual conductance and amplification factor of the valve.

CHAPTER 15

SOLID-STATE ELECTRONICS AND DEVICES

SEMICONDUCTORS

In this chapter, an introduction is made to the *semiconductor diode.* This device is comparable to the thermionic diode and as such is required to function in a manner prescribed by the associated electronic circuit. It should be appreciated however, that semiconductor materials have properties which allow their usage for devices other than those mentioned in this chapter and to date they have been developed to respond to the effects of temperature, pressure, stress, light-energy, magnetic and electric fields. As such they can be used as sensors and transducers.

In general the term 'semiconductors' can be taken to include all solids whose electrical conductivity lies between that of metals and that of insulators. Germanium and silicon are the most important semiconductor materials but before proceeding to investigate their electrical properties, it is of interest to recall the history of semiconductor development for electrical work.

The rectifying properties of certain substances were known during the last century and, early in this century, the advent of radio saw both the thermionic valve and rectifying crystal being used. The latter was a form of natural semiconductor diode, but was displaced in favour of the valve which, in time had been evolved into its triode form and, could be used for detection, rectification and amplification. The crystal again became popular as a detector in the early days of broadcasting because of its cheapness and the device using it, was most commonly known as a 'cats-whisker'. Use of the crystal lapsed again as valves became more readily available and were adapted for 'mains' use, but it was redeveloped for use in radar.

The crystal diode as used till then was known as the 'point-contact rectifier' and the first 'junction' diode was evolved in 1941. Impetus was given to the use of semiconductors when the crystal triode was developed in the Bell Telephone Laboratories in America in 1948. Since then intense research projects investigating semiconductor theory, development and applications have been launched by all electrical firms and scientific bodies, resulting in an ever-expanding field of usage. To-date the most remarkable achievements have been in the fields of electronic automation equipment, electrical power control devices and in the miniaturisation of electronic apparatus.

To understand the action of a semiconductor device it is necessary to repeat work already covered in Chapter 3, since reference will be made to the electron theory and the conception of an electric current as being due to the movement of electrons. The essentials of the fundamental concepts are therefore summarised as an introduction to the basic semiconductor theory.

BASIC THEORY

The atom of a material consists of a nucleus (+vely charged) and planetary electrons (−vely charged). The nucleus has nearly all the mass of the atom and consists of protons and neutrons. Neutrons have no electrical properties, but since forces of repulsion exist between +ve charged protons, then the neutrons have the function of holding the nucleus together. An atom contains equal numbers of protons and electrons with charges cancelling to make the atom neutral. The atoms of various elements contain different numbers of protons, neutrons and electrons and the greater the number of protons and neutrons in the nucleus, the greater the density of the substance. The electrons move in orbits and for the larger atoms, these electrons are arranged in shells. The chemical properties of an atom may be explained in terms of the number and grouping of the planetary electrons. The shells — called 'quantum shells', are considered to be concentric and seven in number, distinguished by the letters K, L, M, N, O, P, Q. The K shell is closest to the nucleus and the appropriate maximum number of electrons possible in each shell is 2, 8, 18, 32, 18, 13 and 2. Since germanium will be of particular interest in our studies, it should be noted that its 32 electrons are arranged in the K, L, M, and N shells thus: 2, 8, 18 and 4. Silicon, also of interest, has 14 electrons arranged in the K, L, and M shells thus: 2, 8, 4.

A solid in its smallest particle form, if investigated, is found to consist of a large number of crystals joined together and the crystals are built up from a regular structure of atoms which repeats itself continually to form a lattice. The electrons in the innermost shells of an atom do not appear to play an active part in the crystal lattice structure, but those in the outer shells are important, in that they decide the electrical and chemical properties of the substance and are known as 'valence electrons'. The chemical nature of an atom is revealed by the manner in which it combines with other atoms to form molecules. Valency is a chemical term and is, explained as, the property of the atom of an element which enables it to enter into chemical combination with another atom. Thus when the molecules of a substance are formed from two atoms, this is done by the electrons in the external quantum shells or by the valence electrons. The valency of an element is determined by the electron disposition in its two outermost shells. Note that two shells are concerned because a combination of atoms is achieved by the gaining or losing of electrons, so that it appears that all shells are complete. Thus we see that the planetary electrons of an atom can be divided into two classes: (1) valency electrons or those available for linking atom to atom and (2) the core or those which retain their configurations in all compounds of the element. These are not affected by the formation of linkages.

CO-VALENT BONDING. One way in which an atom may combine with other atoms and bring about a change in the number of atoms in its valency shells, is by co-valent bonding. This means the sharing of a pair of electrons by two atoms, each atom contributing one electron to form a shared pair. A powerful bond between atoms thus results.

Consider hydrogen which has one planetary or valence electron. The K shell electrons should be 2. Thus the hydrogen molecule contains two atoms, with the nucleii linked by the valency electrons to form a pair. The arrangement is illustrated by (a) the first diagram of Figure 220.

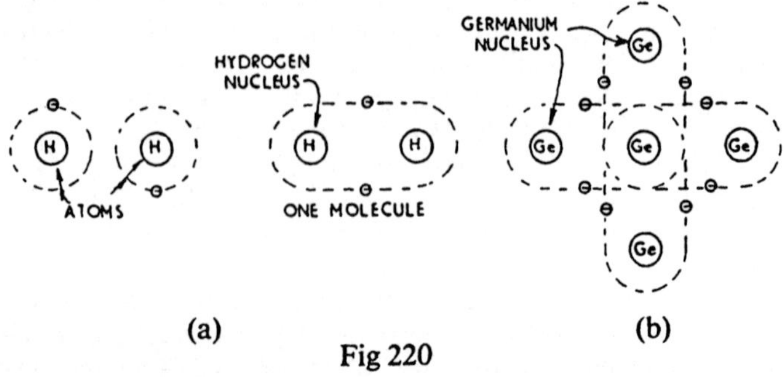

Fig 220

Consider germanium where the outermost shell of an atom contains four electrons. To obtain stability, the atoms of the element build up into a molecular structure giving combinations as shown by (b), the second diagram of Fig 220. It will be seen that one valence electron of the four goes towards making up a 'shared pair' between two adjacent atoms.

The shared pairs of electrons can be depicted by double lines as shown by (a), the first diagram of Fig 221 or alternatively the crystal structure which leads to a regular arrangement of atoms throughout the lattice (in three dimensions), can be represented by (b), the second diagram of Fig 221. Germanium is considered as an obvious example.

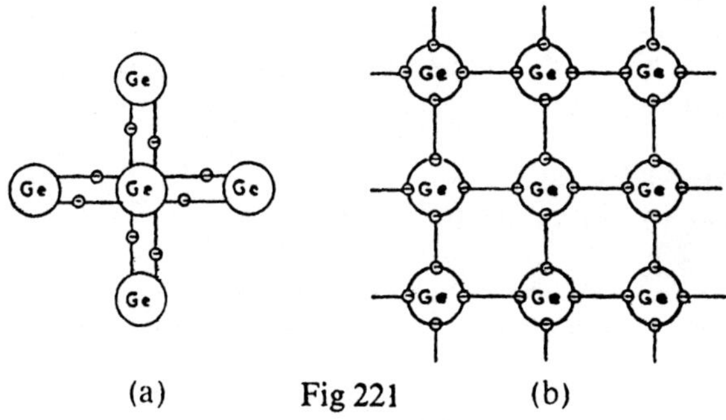

(a) Fig 221 (b)

CONDUCTION CONTROL

Germanium and silicon, in the pure form, have crystalline structures and the atoms, arranged in a crystal lattice — as depicted in Fig 221b, show these materials to be similar to carbon in its diamond form. The arrangement, as illustrated, is frequently called a 'diamond lattice' and it is seen that there are no free electrons except for any shaken free by the application of energy in the form of heat or light. This effect comes under the general heading of ionisation, which is considered later. The conductivity of pure germanium, for example, is thus poor — except when heated, but it can be improved by injecting an impurity. For semiconductors the fact that conduction is a function of (a) temperature and (b) impurity content, is a most important property. In general an increase of (a) or (b) will increase conductivity which is now considered with these effects in mind.

INTRINSIC CONDUCTIVITY. At −273°C pure germanium (still being taken as an example), has no free electrons since all the valence electrons are attached to their respective atoms. If the crystal is supplied with energy such as heat or light, the bound electrons can absorb some energy and escape from the atom. Through this 'dissociation', electrons leave the atoms and wander around the crystal and conductivity will increase as more energy is supplied. The places from which the electrons leave are known as *holes*. If the crystal is subjected to a potential gradient, a current will flow, the electrons and holes acting as charge carriers. The idea of the holes moving will be considered shortly, but using this assumption, the holes are regarded as +ve charge carriers moving towards the negative terminal while the electrons, being −ve charge carriers, move towards the positive terminal. Thus although electrons move in the opposite direction to that of conventional current flow, hole movement is in the same direction. Intrinsic conductivity is therefore associated with the movement of equal numbers of oppositely charged carriers.

IMPURITY (EXTRINSIC) CONDUCTIVITY. As already stated, the conductivity of a pure semiconductor material can be controlled by the addition of a known impurity such as antimony or indium. The process is known as 'doping' and gives a result which can be considered under one of the two following headings.

N-TYPE GERMANIUM. When a pentavalent (5 valence electrons) atom, such as that of antimony or arsenic is introduced into the pure germanium (or silicon) crystal lattice, one of the electrons of the impurity atom is free to become a −ve charge carrier. This is shown by diagram (a) of Fig 222. The prevalent atom, when bonding with germanium, only utilises 4 valence electrons and thus one electron of the impurity atom is not held by covalent bonding. This electron has sufficient energy to migrate

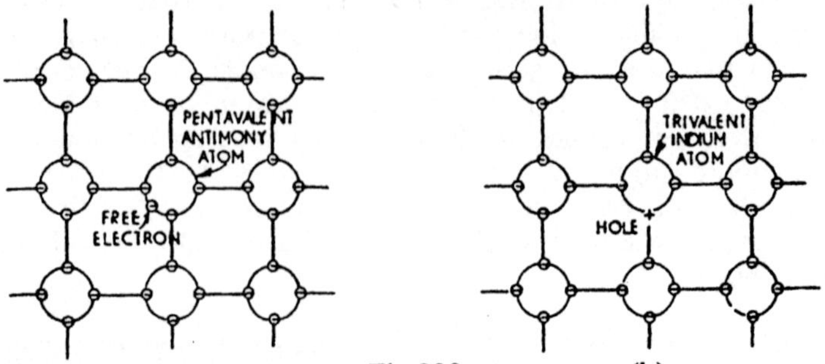

(a) Fig 222 (b)

through the lattice structure as a charge carrier. The impurity atom is a *donor* or an N-type impurity since it provided an electron. The adulterated germanium is known as N-type germanium.

P-TYPE GERMANIUM. If a trivalent impurity atom such as that of indium or aluminium is introduced into the crystal lattice, then bonding is imperfect in that a 'hole' is left in the lattice. The term +ve hole, is given to the deficiency which is created in the bonding arrangement, since the valence electrons of the impurity atom only pair up with three of the electrons in the neighbouring 4 germanium atoms. A hole exerts an attractive force on and will 'capture' any electron which, having been liberated from a germanium atom by some ionising effect, happens to pass near the hole. Thus, as an electron moves from the orbit of one germanium atom to fill a deficiency in another, it leaves a similar 'hole' in its original orbit. The hole will then appear to move in the opposite direction to the electron and as the process is continuous throughout the germanium it constitutes an effective flow of +ve charge carriers. The arrangement is shown by the diagram (b) of Fig 222. A 'hole' has the characteristics of a +ve charge and behaves as such and can be regarded as a free, mobile, +ve charge. Trivalent impurity atoms are called *acceptors,* since they accept electrons from covalent bonds and in doing so they create holes or +ve charge carriers. They are termed P-type impurities and the germanium, containing such an acceptor impurity, is known as P-type germanium.

It should be noted that what has been said about germanium also applies to silicon since the latter also has 4 valence electrons.

IONISATION. In addition to the charge carriers made available by 'doping', the reader is reminded of the fact already discussed, that electrons can be liberated by the action of heat, light or other radiations. This process comes under the general heading of ionisation. It must be remembered that the total charge in a given piece of material is always zero, since for every charge carrier produced by 'doping' or ionisation, there is an incomplete atom or ion carrying a charge of opposite sign. Some dissociation or breakdown of the covalent bonds can occur for both P- and N-type materials, even at room temperature and thus a few holes could occur in N-type material with a corresponding number of extra free electrons. Free electrons would thus predominate in N-type material and are called the *majority* carriers, whilst the holes are known as the *minority* carriers. Similarly for P-type material, holes would be the majority carriers and any free electrons the minority carriers.

THE P-N JUNCTION

By such a junction is not meant the joint between two pieces of germanium, one being P-type and the other N-type. To produce an effective junction, one piece of germanium is grown to ensure a suitable continuous crystal lattice. During the growing process, P- and N-type zones are created by 'doping', with a definite line of demarcation between them. Since one side has a preponderance of holes and the other of electrons and, because both sides are electrically neutral, there is no tendency for electrons to flow and neutralise the holes. Majority carriers will however tend to drift across the junction in an attempt to even out the concentration. As soon as electrons flow from the N-type to the P-type material or holes pass from the P-type to the N-type material, the latter becomes positively charged and the P-type material becomes negatively charged. This effect immediately prevents any further flow of charge carriers across the junction and can be looked upon as the build up of a potential barrier. This potential barrier can be likened to a battery connected in the sense shown by the diagram of Fig 223.

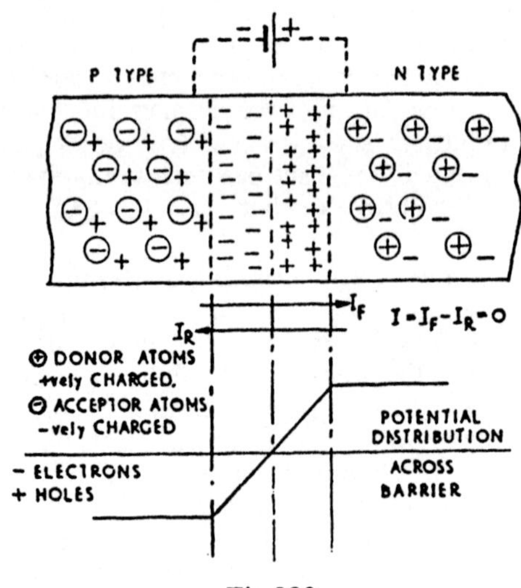

Fig 223

For the arrangement shown, few charges can cross the barrier unless they have an energy level high enough to overcome the potential. The +ve charge of the N-type will however attract the minority carrier electrons from the P-type side and the −ve charge will attract the minority carrier +ve holes from the N-type side. The resulting current $I = I_F - I_R$ will be zero and since there is no external voltage applied, the junction will be in equilibrium with charges and potentials as shown.

THE JUNCTION DIODE

There are two types of semiconductor diode, the junction diode and the point-contact diode. An example of the latter was the crystal — cat's whisker, as used in early radio receivers and, although a modern version of the arrangement is used, it will not be considered in any detail as its applications are specialised. Attention here, is therefore given to the junction diode only.

FORWARD BIAS. (Good conduction). If an actual battery is connected across the junction with opposite polarity to that of the barrier, the barrier potential is reduced, the effective resistance decreases and a current of several milliamperes can flow. A forward voltage of 0.2V producing 4 to 5mA. The barrier potential having been reduced, the majority carriers in the P-type material are reinforced by the +ve potential and the forward current increases, overcoming the reverse current. The forward current is thus considerably larger than the reverse current or $I = I_F - I_R$. The conditions being discussed are shown by the diagrams of Fig 224.

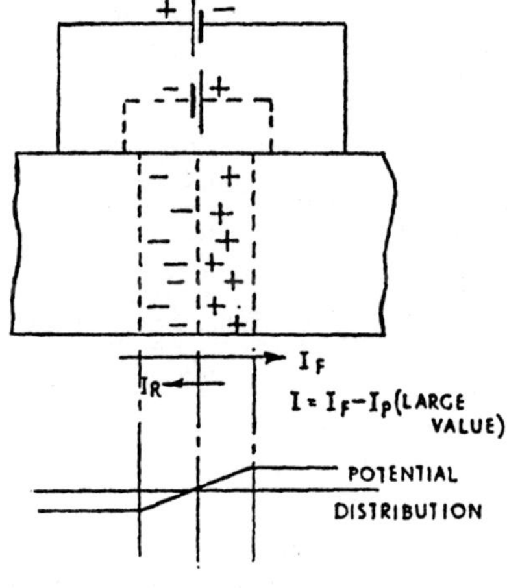

Fig 224

REVERSE BIAS. (Very poor conduction). If the external battery is reversed, *ie* connected across the P-N junction so as to assist the barrier potential, then the latter will be increased. As shown by the diagram (Fig 225), the action is equivalent to increasing the effective resistance of the junction and it is 'reverse' or back biased by the battery. The reverse current is only of the order of a few microamperes due to electrons released by ionisation, *ie* 'intrinsic conductivity'. For this reverse-bias connection, the minority carriers are reinforced and the reverse current, although small, predominates and remains at a substantially constant value for a wide range of impressed reverse potentials.

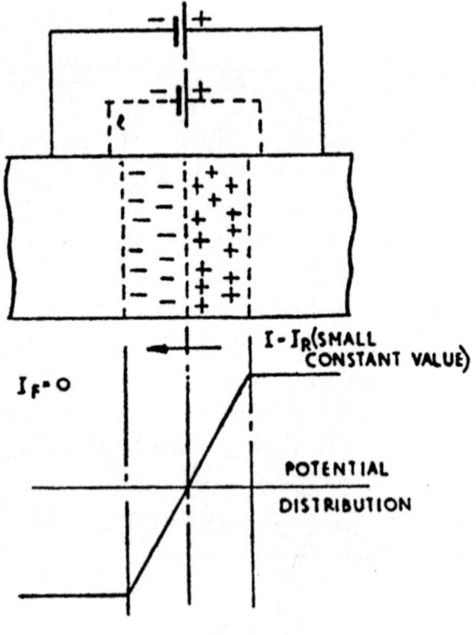

Fig 225

At a large reverse voltage a breakdown region occurs where the reverse biased P-N junction would conduct, this value being known as the PEAK INVERSE VOLTAGE (P.I.V.) the value of which could be very large. There are two causes of this sudden conduction — the avalanche effect and zener effect. The avalanche effect occurs when minority carriers, travelling at high velocity through the barrier layer, dislodge electrons from the parent atoms which, in turn, dislodge other electrons from their atoms. The zener effect is caused by the potential difference (or electric field) across the narrow barrier layer becoming large enough to force electrons away from their parent atoms. Breakdown is not necessarily permanent and if the voltage is reduced to below the P.I.V. then the reverse biased diode will operate as before. However, if the applied potential greatly exceeds the P.I.V. then irreparable breakdown could occur. When a diode is used as a one way conduction device then it is important to pay particular attention to the maximum circuit voltage and the P.I.V. rating of the diode.

Avoiding involved details of the junction diode action, a clear explanation is possible if we consider one side of the junction to have only holes as charge carriers and the other side to have only electrons.

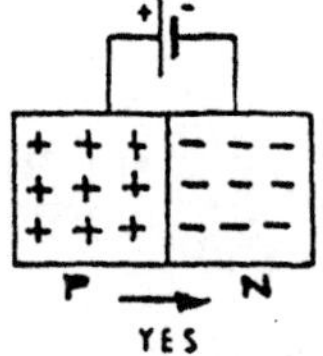

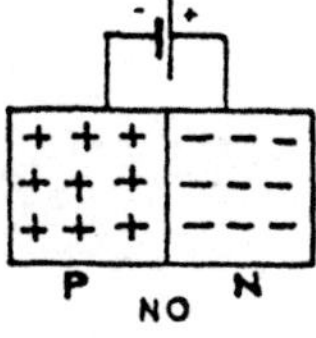

Fig 226

The arrangement is shown by the diagrams (Fig 226). If a potential is applied so that the P-end is +ve with respect to the N-end then a current due to this potential consists of either electrons moving from right to left or holes moving from left to right. Since the N-side contains free electrons and the P-side contains holes, current readily flows across the junction.

If the applied voltage is reversed, the appropriate direction of current would occur either by electrons moving from left to right and there are no electrons available, or by holes moving from right to left. There are no such holes available and therefore current cannot flow.

DIODE CHARACTERISTIC. If a test circuit is arranged, as in the diagram of Fig 227a, and the results of applied voltage are plotted against diode current, the semiconductor junction is seen to act like a thermionic diode when the latter is conducting. The general shape of the current/voltage characteristic is also shown by Fig 227b. It resembles that of the vacuum diode, but it should be noted that when the applied voltage is reversed, a minute reverse current will flow. As mentioned earlier this reverse current is due to the release of electrons by ionisation, *ie* the thermal agitation gives the P-type side a few electrons and the N-type side a few +ve holes. If the reverse voltage becomes too great, there is a breakdown of the covalent bond structure and the current increases very rapidly.

Note. For a semiconductor diode, the electrode to be connected to the +ve terminal of the supply for forward conduction, is always clearly marked. As for the diode valve the terms anode and cathode are used accordingly.

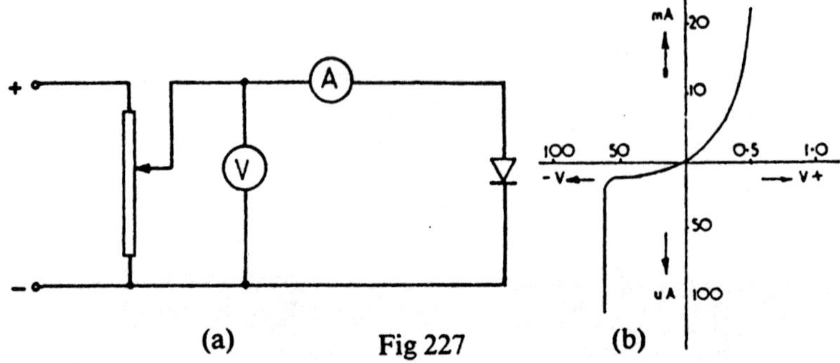

(a) Fig 227 (b)

The characteristic shown is typical for a small semiconductor diode. Note the change of current scale for the reverse graph, which shows the limit where the saturation value of the reverse current is broken down — the 'zener effect', and the semiconductor becomes a conductor.

Unlike the vacuum diode this reverse operation of the semiconductor diode is exploited for various circuits and the Zener Diode, as it is called, is used for voltage regulation and stabilisation, for meter protection, cut-off, limiting and clipping circuits which is considered later in this chapter.

RECTIFIER OPERATION

Since the junction diode requires no heater supply, it can be introduced directly into circuit arrangements to allow either half-wave or full-wave rectification. The characteristic shows the device to be suitable for a rectifier and if an alternating voltage of about 1 volt (peak to peak) is applied to a P-N junction, the potential barrier is alternatively strengthened and weakened to allow the rectifying action. The reverse current can be considered as negligible. Silicon is used in preference to germanium for power rectifiers since it can carry large currents and can operate at higher temperatures. Its reverse current is also lower than that for germanium with similar forward current values.

The advantages of semiconductor rectifiers over the earlier type of 'metal' rectifiers such as the copper-oxide and selenium, lie chiefly in their smaller size, longer life and greatly improved regulation and efficiency. The latter features are the result of the low forward resistance and voltage drop and the fact that fewer elements are required in series to handle a given voltage.

Example 126. The following values refer to a germanium diode.

Forward current (mA)	0	0.1	0.18	0.22	0.4	0.6	0.8	1.1	1.6	2.3	3.45
Forward voltage (mV)	100	140	160	180	200	220	240	260	280	300	320

Reverse current (μA)	−0.2	−0.3	−0.4	−0.4
Reverse voltage (mV)	−200	−400	−600	−800

Plot the anode characteristics for the above diode and determine from it (a) the 'forward' d.c. resistance when the current is 3mA. (b) the forward anode voltage when the d.c. resistance is 200Ω.

(a) From the graph of Fig 228 the forward d.c. resistance

$$= \frac{OB}{OA} = \frac{315 \times 10^{-3}}{3 \times 10^{-3}} = 105\Omega$$

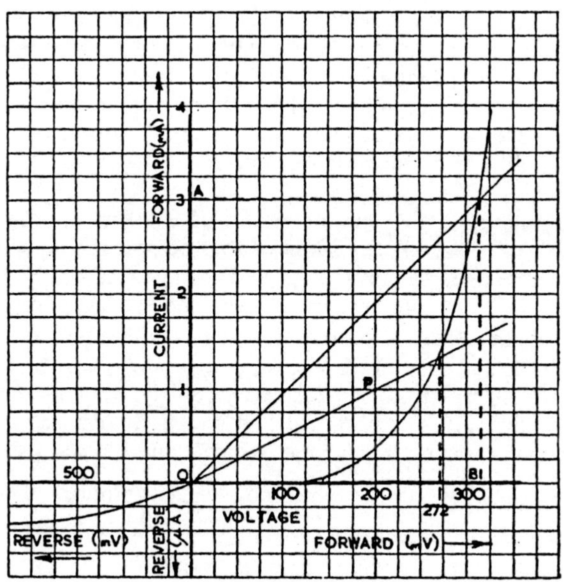

Fig 228

(b) Construct the 200Ω d.c. resistance lines thus. Assume a current of 1mA. Then the applied voltage would be 1 × 200 = 200mV. Plot this point P. Draw the voltage drop line to cut the curve. The forward anode voltage would be 272mV.

STATIC AND DYNAMIC OPERATION. It has been seen from the I_a/V_a curve that, as for the thermionic diode there is a definite ratio of voltage to current for any particular anode current value. As shown by the preceding example, this is the anode d.c. resistance, determined for any one point on the characteristic by dividing the anode-voltage value by the corresponding anode-current value. Operation of the semiconductor diode for a.c. conditions, as when used as a R.F. demodulator or detector, will not be considered here but, mention is made in passing that, the slope or a.c. resistance value for any working range can be determined, as for the vacuum diode, by the ratio:

$$\frac{\text{small change in anode voltage}}{\text{resulting small change in anode current}}$$

When used as a rectifier, the diode is loaded or is operating dynamically. It is therefore apparent that a dynamic characteristic should be obtained either by direct testing or by deduction in a manner similar to that used for the diode valve. Since a separate dynamic characteristic is required for each value of R — the load resistance, then use of a 'load line' would be often more appropriate for solving some problems. This is illustrated by the following example.

Example 127. The characteristic of a germanium diode is shown by Fig 229. If the value of load reisistance is to be 100Ω and the average value of the applied voltage is to be 2V, find the average value of the terminal voltage and the load current.

The characteristic is plotted as shown and the load line drawn for 100Ω thus:
With no current flow the full 2V would be applied across the diode and point A is obtained. With a diode d.c. resistance of zero, the maximum current which could flow would be $\frac{2}{100}$ = 0.02A or 20mA. Thus point B is obtained. The point of intersection between the load line and characteristic shows that the terminal voltage would be about 2 − 0.48V = 1.52V and the load current would be 15.25mA.

RECTIFIER CIRCUITS

Most electronic circuits operate from a d.c. supply which could be obtained from batteries. However, as batteries

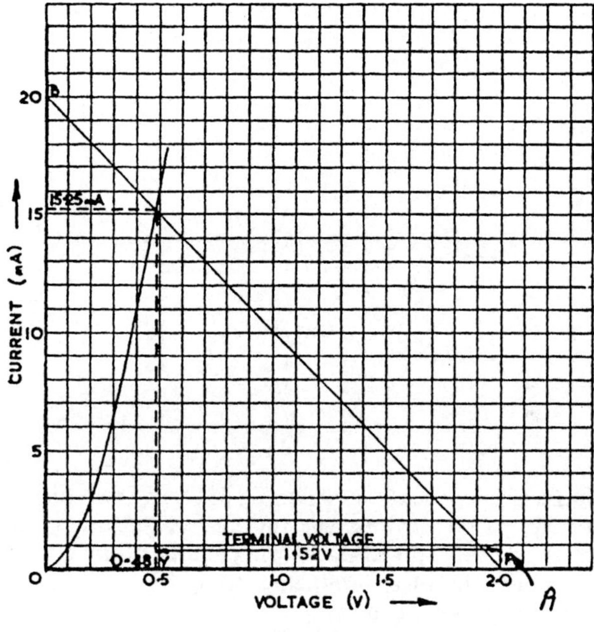

Fig 229

discharge, continuity of supply cannot be assured and reliability of the equipment falls. Rectifier circuits, using the unidirectional conduction properties of junction diodes are often incorporated in electronic units which are to be connected to a.c. supplies. The rectified output can be either half or full wave arrangements which are now considered.

HALF WAVE

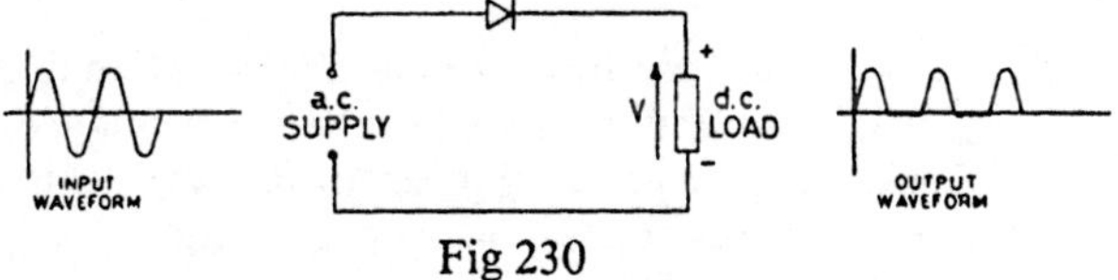

Fig 230

When a diode is connected in series with an a.c. supply and a resistive load, current will only flow in one direction. Every alternate half cycle of a.c. is 'blocked' and a series of positive pulses occur as shown in the output waveform (Fig 230). A small reverse current does flow but for most practical purposes this can be ignored. If a low voltage d.c. output is required from a mains a.c. supply, a transformer with a suitable step-down ratio must be incorporated; the rectifier circuit being connected to the secondary output.

FULL WAVE

1. Bi-Phase Circuit.

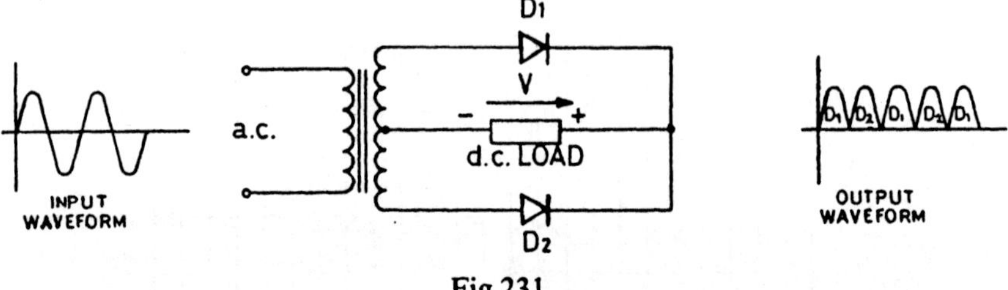

Fig 231

With the arrangement shown in Fig 231 using two diodes and a centre tapped transformer the full a.c. waveform is made to conduct in a positive direction. During alternate half cycles each diode conducts in turn producing an output as a series of unidirectional pulses as shown.

2. Bridge Circuit.

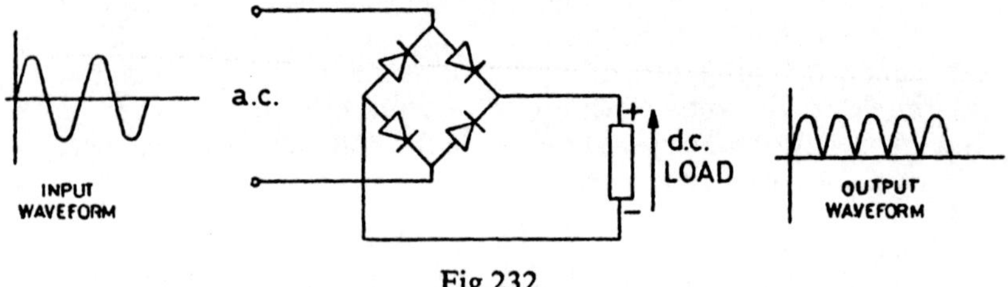

Fig 232

The full wave bridge rectifier circuit illustrated in Fig 232 eliminates the need for a centre tapped transformer but uses four diodes. In this arrangement, during each half cycle alternate pairs of diodes conduct once again producing a series of unidirectional pulses.

The alternating nature of the unidirectional output to the load is undesirable for many electronic circuits. To reduce this problem, smoothing circuits must be incorporated. The simplest method is to connect a capacitor in parallel with the load (Fig 233). This capacitor must have a high capacitance value and thus is usually of the electrolytic type.

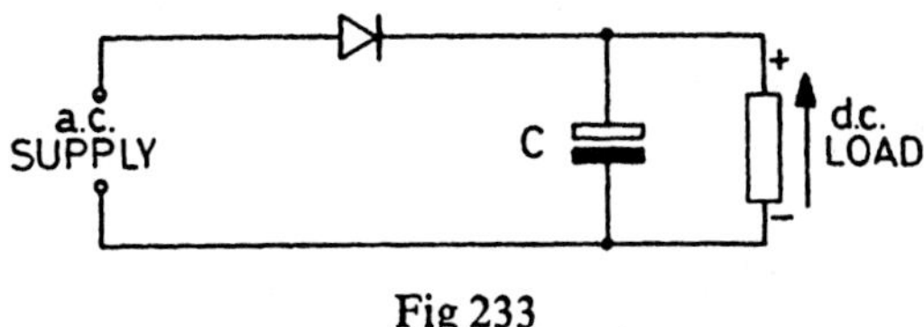

Fig 233

During the half cycle in which the diode conducts the capacitor C charges up to the peak value of the d.c. load voltage. During the next half cycle, when the diode is not conducting, the capacitor discharges through the load. Mention was made in Chapter 8 of the charge stored by a capacitor being dependent upon the voltage and capacitor value. Hence, for a given d.c. voltage, if a larger capacitor is used, more charge can be stored to be available to maintain the load voltage. Fig 234 shows the output waveform for such an arrangement.

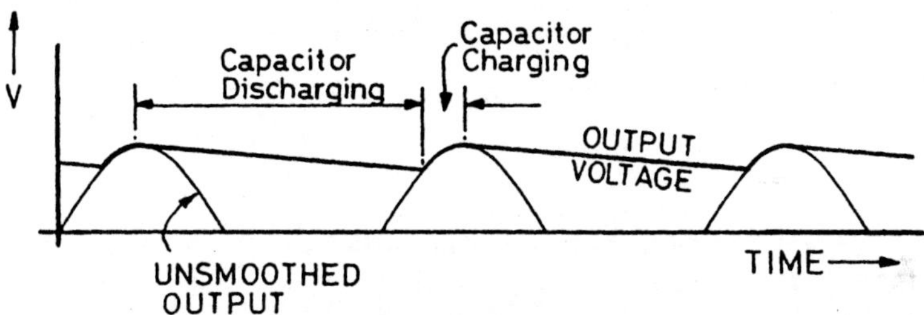

Fig 234

It should be noted that the output voltage still has a definite ripple, referred to as the *ripple voltage* which reduces with increasing values of capacitance. The time taken to fully charge the capacitor is very small and also decreases as the capacitor value increases, thus for practical purposes, it can be assumed to charge up instantaneously. In a half wave rectifier circuit the capacitor may therefore be assumed to discharge in the time taken for a complete cycle $\left(t = \frac{1}{f}\right)$. Similarly with a full wave rectifier, circuit discharge occurs in half this time $\left(t = \frac{1}{2f}\right)$ as shown in Fig 235.

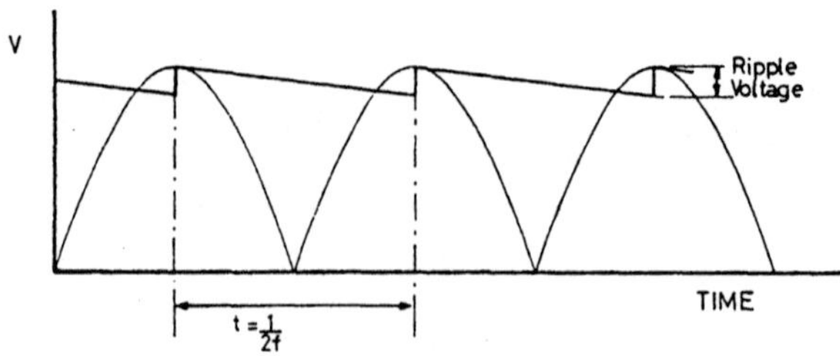

Fig 235

Example 128. A full wave rectifier with capacitor smoothing, operating from a 50Hz a.c. supply, produces a d.c. output of maximum value 60 volts. The capacitor has a value of 2000μF.

Find the peak to peak value of the ripple voltage when delivering 250mA to a resistive load.

Quantity of Electricity stored during charge Q = CVr

Quantity of Electricity given out during discharge Q = It

(refer Chapter 8)

ie, CVr = It

$$I = \text{Discharge current} = 250\text{mA}$$

$$t = \text{Discharge time} = \frac{1}{2f} = \frac{1}{2 \times 50} = 0.01 \text{ s}$$

$$\text{Ripple Voltage Vr} = \frac{250 \times 10^{-3} \times 0.01}{2000 \times 10^{-6}}$$

$$\therefore \text{Ripple Voltage} = 1.25 \text{ volts}$$

In many cases this ripple does not affect the operation of the load but in some electronic circuits its presence is highly undesirable and must be eliminated completely.

Connecting a capacitor across the load affects the *peak inverse voltage* developed across the diode. The reason for this is that when the diode is in the non-conducting mode, the voltage developed across it is the peak value of the supply PLUS the peak value of the capacitor voltage.

Example 129. A 240/20V transformer feeds a half wave rectifier circuit which has capacitor smoothing. Calculate the maximum output voltage and the minimum P.I.V. rating of the diode.

Note. A.C. voltages are usually expressed in R.M.S. values

$$ie,\ V_{RMS} = 0.707\ Vm = \frac{Vm}{\sqrt{2}}$$

$$Vm = \sqrt{2}\ V_{RMS}$$

$$= \sqrt{2} \times 20$$

Output Voltage Vm = 28.28V
P.I.V. = Peak value of supply + Peak value on capacitor
= 28.28 + 28.28 volts
= 56.56V
Diode P.I.V. rating = 56.56V mimimum

FILTER CIRCUIT

The storage or reservoir capacitor may not completely eliminate ripple and, if total ripple suppression is required, another capacitor and an iron cored inductor must be connected as shown in Fig 236.

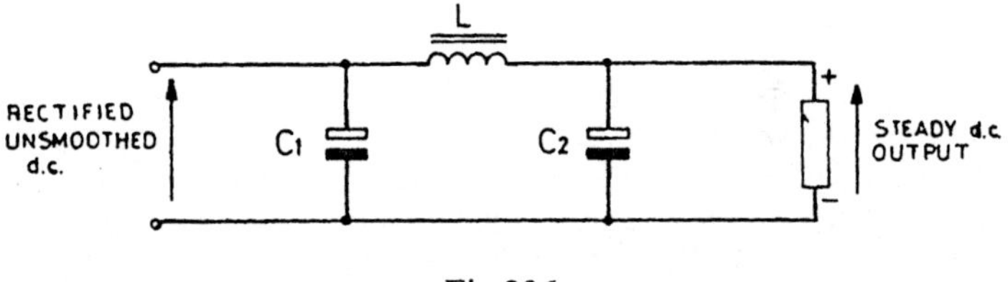

Fig 236

The capacitor C_1 acts as a smoothing capacitor, as explained earlier, which produces a unidirectional potential difference across it with a definite ripple. The inductor L and capacitor C_2 act as a filter to the circuit, separating the steady d.c. voltage from the ripple a.c. component of the voltage across C_1. This occurs because the inductor offers a much greater impedance to the a.c. component but little resistance to the d.c. component, thus most of the unwanted ripple voltage is developed across it. Since the resistance of the capacitor is very high the d.c. component is established between its plates. A small d.c. voltage drop does however occur across the inductor due to its resistance.

VOLTAGE DOUBLER CIRCUIT

Another circuit using the storage capacity of a capacitor together with semi conductor diodes is the voltage doubler circuit. This arrangement can be used in circuits requiring a high voltage — low current d.c. from a lower voltage a.c. supply.

The circuit required two diodes and two capacitors connected as shown in Fig 237 (a) and (b).

During the positive half cycle of the a.c. voltage, C_1 charges up to the peak value of the a.c. voltage Vp. During the negative half cycle C_2 charges and C_1 discharges. However the voltage now developed across the load comprises the discharge voltage

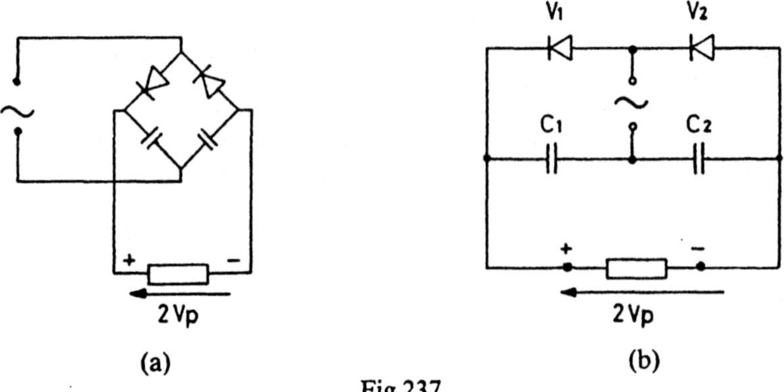

(a) (b)

Fig 237

of C_1 and the voltage V_2 (*ie,* $V_L = V_2 + V_{C1} = 2Vp$). During the next positive half cycle C_2 discharges and C_1 charges, hence the d.c. load voltage once again comprises discharge voltage of C_2 and voltage V_1 (*ie,* $V_L = V_1 + V_{C2} = 2Vp$).

STABILISED POWER SUPPLIES

The d.c. output from a rectifier circuit, even though it may be smoothed using a capacitor, may vary due to variations of the peak value of the input voltage Vm. This variation is often encountered as the a.c. mains voltage changes due to supply variations. Similarly as the d.c. load is increased the increased current causes the output voltage to fall due to voltage drops in the circuit. These effects are undesirable in many electronic control and monitoring devices and can be reduced by a stabilising circuit. This circuit incorporates a *Zener diode* as a voltage reference to provide a steady value of output voltage.

THE ZENER DIODE

Earlier in this chapter reference was made to the fact that a pn junction diode will breakdown at a definite voltage when reverse biased. This breakdown, usually referred to as the zener effect, occurs at a definite voltage which is predetermined by the manufacturer. Thus a junction diode which is continuously operated with reverse bias is referred to as a Zener Diode.

Fig 238

Below the breakdown or Zener Voltage the diode resistance is very high hence the reverse current is extremely small, and is usually ignored in calculations. However beyond the breakdown voltage the resistance falls to a low value and a rapid increase of current occurs. Hence the reverse biased diode will conduct above the predetermined Zener Voltage as illustrated in Fig 239. Beyond the breakdown point the voltage developed across the diode is the Zener Voltage Vz plus the voltage due to resistance volt drop in the diode.

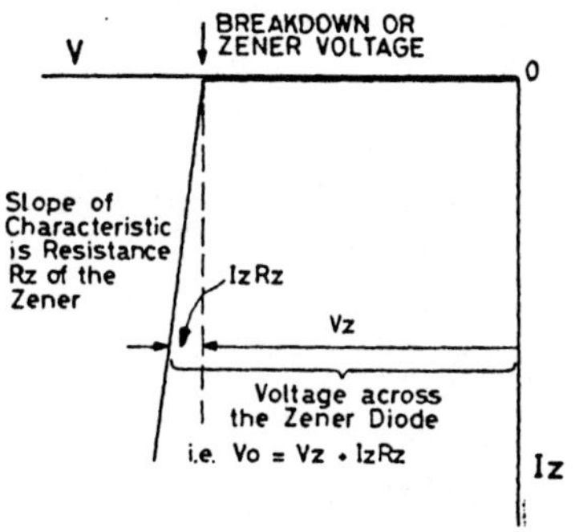

Fig 239

SERIES STABILISATION

To produce a stabilised voltage the Zener Diode is connected across the load as shown in Fig 240. The voltage rating of the Zener Diode used is determined by the d.c. voltage required across the load.

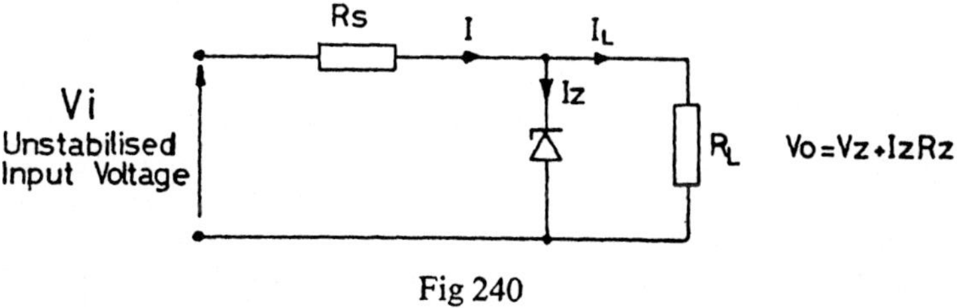

Fig 240

Under all conditions when the zener diode is conducting Vi = Vo + IRs. The zener will conduct when the zener voltage Vz is reached. At this position the zener will draw extra current from the supply, hence increasing the volt drop across Rs, thus bringing the output voltage back to near Vz value depending upon the zener resistance and its current. As Vi rises Iz increases and the series resistance volt drop increases maintaining Vo constant. As the load current changes the current Iz changes inversely but the voltage across the series resistor stays near constant and hence the load voltage stays sensibly constant. However this will only occur providing a maximum load current is not exceeded since beyond this value the volt drop across the series resistor Rs would be excessive and the output voltage will fall.

Example 130. A zener diode has a breakdown value of 9.1V and, beyond this voltage, has a resistance of 15Ω. It is connected in a stabilising circuit, with a series resistor of 500Ω to a supply voltage of 30V d.c. unstabilised. When the load current is 30mA calculate the diode (zener) current, the load voltage and the power dissipated in the diode. Refer to Fig 240.

$$Vo = Vz + IzRz$$

$$Vo = 9.1 + 15Iz$$

$$Vi = IRs + Vo$$

$$30 = 500(0.03 + Iz) + 9.1 + 15Iz$$

$$30 = 15 + 500Iz + 9.1 + 15Iz$$

$$Iz = \frac{30 - 24.1}{515} \text{ amperes}$$

Diode Current $Iz = 11.46\text{mA}$

$$Vo = 9.1 + (15 \times 11.46 \times 10^{-3})$$

Load Voltage $Vo = 9.272$ volts

$$Pz = VoIz$$

$$= 9.272 \times 11.46 \times 10^{-3}$$

Power dissipated in Diode = 106mW

CHAPTER 15

PRACTICE EXAMPLES

1. The following data represents the forward characteristics of a p-n junction diode.

I(mA)	0.25	0.55	1.1	2.5
V(volts	1	2	3	4

Estimate (a) the forward d.c. resistance when the current is 1.5mA and (b) the voltage and current for a forward d.c. resistance of 1.75kΩ.

2. On test, a full-wave, silicon semiconductor rectifier is found to give a constant forward voltage drop of 0.7V, independent of current. Calculate the power dissipated in the rectifier diode for a d.c. current of 5A. If the rectifier dissipates heat, at the rate of 0.1J/s/°C rise in temperature, find the maximum d.c. output current rating if the temperature rise is to be limited to 70°C.

3. The characteristics of a typical semiconductor junction diode is as follows:

Forward voltage (V)	0	0.25	0.5	0.75	1.0
Forward current (mA)—at 25°C	0	0.5	1.5	4.8	8.75
Forward current (mA)—at 60°C	0	1.0	3.0	6.5	11.5
Reverse voltage (V)	0	0.2	0.4	0.6	
Reverse current (μA)—at 25°C	0	1.5	1.5	1.6	
Reverse current (μA)—at 60°C	0	10	10.25	10.5	

(a) If a voltage of 1V is applied across the diode and a series resistor of 100Ω, find the load current variation between the temperature limits shown by the graph.

(b) Find the variation of its d.c. resistance over the range of 25°C to 60°C for the above load condition.

4. A half wave rectifier with capacitor smoothing is fed from a 230V; 50Hz supply via a 12:1 step down transformer. The volt drop across the diode when conducting is 0.75V. Calculate the peak value of the d.c. output voltage and the Peak Inverse Voltage rating of the diode.

5. An electronic circuit requires 350mA d.c. from a half wave rectifier circuit which incorporates a 1800μF capacitor for smoothing. If the supply voltage is 50V; 60Hz, calculate (a) the peak value of the d.c. output voltage, and (b) the peak to peak value of the output ripple voltage.

6. A d.c. load is to be supplied from a 250/40V; 50Hz transformer and bridge rectifier circuit with a 2000μF capacitor for smoothing. Determine (a) the P.I.V. Rating of each diode, and (b) the peak to peak value of the ripple voltage when supplying 250mA.

7. An a.c. supply of 50 volts is connected to a voltage doubler circuit as illustrated in Fig. 237. Calculate the peak voltage across the load resistor assuming the diodes to have negligible forward voltage drop.

8. The output from a rectifier circuit with capacitor smoothing is 25V ± 20%. A zener diode series stabilising circuit is connected to this supply to provide a stabilised output from which currents up to 50mA will be drawn. The zener diode has a breakdown value of 9.7V and for stability it must carry a minimum current of 1mA. The diode resistance above 9.7V is 12Ω. Determine (a) a suitable value of series resistance, and (b) the stabilised voltage.

9. Using the previous example calculate (a) the zener current (b) the stabilised voltage, and (c) the power dissipated in the zener diode when the supply voltage is maximum and the load current is maintained at 50mA.

10. The following diagram illustrates a rectifier circuit incorporating capacitor smoothing and zener stabilisation. The zener diode has a breakdown value of 24 volts and requires a minimum current of 1mA for stable operation. For values of voltage above 24 volts the slope resistance is 14Ω. For a stabilised d.c. load current of 60mA calculate (a) the peak to peak ripple voltage measured across the capacitor, (b) the required P.I.V. Rating of the diodes, (c) the minimum value of Rs for stable operation, and (d) the power rating of the zener diode.

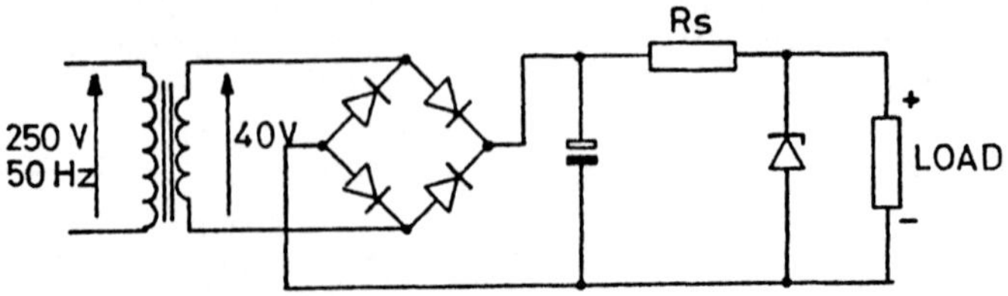

CHAPTER 16

MISCELLANEOUS CIRCUIT CONDITIONS AND METHODS OF SOLUTION SPECIAL APPLICATIONS

Certain networks problems may be much too difficult for solving by the direct application of the fundamental rules for series and parallel circuits. Solutions can then only be effected by the application of one of several well-known Theorems or Methods of Solutions. These are based on first principles, but the simplest only are considered and are confined to d.c. applications. Although most of the methods are suitable for application to instantaneous and phasor quantities, since impedances may be involved instead of resistances, nevertheless due allowance must be made for phase, and considerable experience is required, before confidence is attained. For this reason the methods set out below show d.c. examples only.

To enable the more involved a.c. circuit problems to be solved, the reader is introduced to Conductance, Susceptance and Admittance. It will be seen that the terms and methods of resolving are based on well-developed principles. The reader should regard the new work as an aid to solution only and as the first of several methods, devised to facilitate progression into more advanced theory and problems.

D.C. NETWORKS

1. APPLICATION OF KIRCHHOFF'S LAWS

The laws are revised in association with the diagram (Figs 241a and b).

Law 1 states that in any closed circuit or mesh, the algebraic sum of all e.m.fs. is equal to the algebraic sum of the voltage drops.

Thus $E_1 + E_2 = IR_1 + IR_2 + IR_{i1} + IR_{i2}$ as shown in the diagram.

Law 2 states that the algebraic sum of all the currents at a

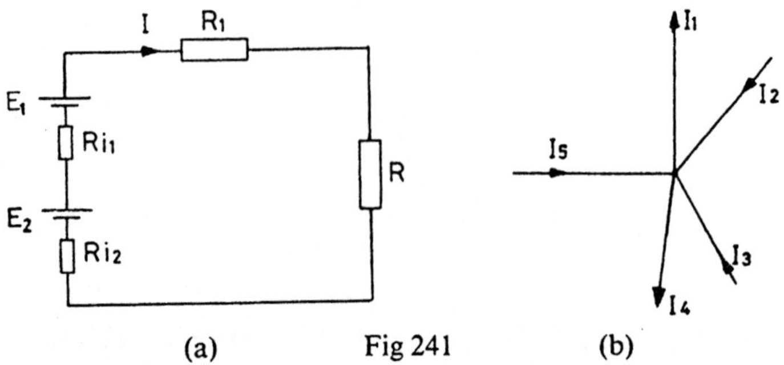

(a) Fig 241 (b)

point is zero. Current flowing to the point being reckoned as +ve and current in the reverse direction as −ve.

In the diagram $I_2 + I_3 + I_5 - I_1 - I_4 = 0$.

Problems are solved by the application of the above two laws and by building up equations which are then solved algebraically. The examples illustrate the procedure.

Example 131. Three cells each having an e.m.f. of 2V and negligible resistance are connected in series. A resistor of 100Ω and one of 25Ω are connected in series across the battery, the 100Ω resistor being connected to the +ve terminal. Calculate the current flowing in a 10Ω resistor connected between a tapping on the battery, 4V from the +ve end and the junction of the two resistors. The arrangement is illustrated by the diagram (Fig 242).

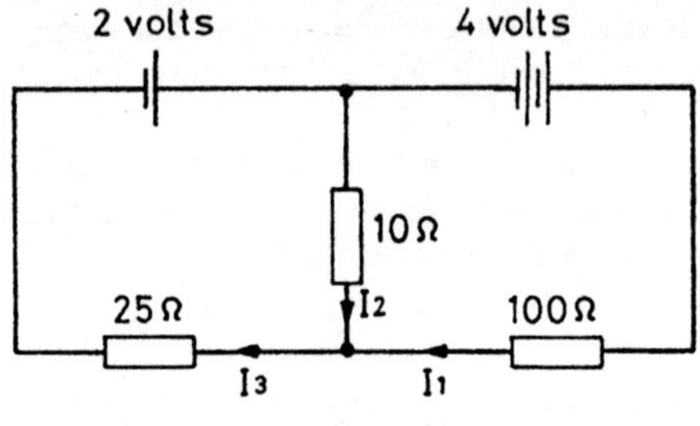

Fig 242

Let currents I_1, I_2 and I_3 be assumed to flow in the directions shown. Then $I_3 = I_1 + I_2$. This involves applying Kirchhoff's current law.

Consider any closed circuit or mesh, such as the right-hand rectangle. Then applying the voltage law we have:

$E = IR_1 + IR_2$ etc or $4 = (I_1 \times 100) - (I_2 \times 10) \ldots$ *(a)*

Note the $- I_2 \times 10$. Since I_2 has been assumed in the direction shown, the voltage drop in $I_2 \times 10$ would be in the opposite direction to the fall of p.d. as assumed for the closed circuit or mesh, hence the −ve sign.

Equation (*a*) contains two unknowns and thus a second equation is required to allow solution. This can be obtained by considering a second mesh, such as the left-hand rectangle.

Then $2 = (I_2 \times 10) + (I_3 \times 25)$ or $2 = (I_2 \times 10) + (I_1 + I_2) \times 25$ . . . *(b)*

Solving *(a)* and *(b)* together we have:

$$100I_1 - 10I_2 = 4 \quad \text{or} \quad 100I_1 - 10I_2 = 4$$

$$\text{and } 10I_2 + 25I_2 + 25I_1 = 2 \qquad 25I_1 + 35I_2 = 2$$

$$\text{whence } 100I_1 - 10I_2 = 4$$

$$\text{and } 100I_1 + 140I_2 = 8$$

$$\text{By subtraction } -150I_2 = -4 \qquad \therefore I_2 = \frac{4}{150}$$

Thus $I_2 = 0.0266$A and the assumed direction of flow is correct, *ie* it is actually downwards towards the junction of the resistors. Current in 10Ω resistor = 26.6mA. I_1 can be obtained by substituting back in equation *(a)*.

2. MAXWELL'S CIRCULATING-CURRENT THEOREM

For certain problems, the direct application of Kirchhoff's laws may lead to a solution which may be laborious, and in some instances this 'cyclic current' method may be used to advantage. It will be seen that to each closed mesh is assigned a clockwise circulating current and that the voltage law is applied to build up a set of simultaneous equations. *Note.* The current flow must be indicated and assumed to be the +ve direction. Voltage drops in this direction will be +ve, in the opposite direction, −ve.

Consider Example 131, solved by this method. Then the diagram (Fig 243) would be as shown.

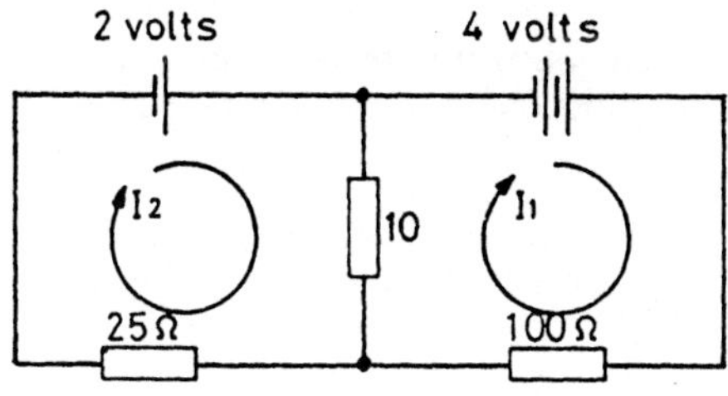

Fig 243

Consider the right-hand mesh, then $100I_1 + 10(I_1 - I_2) = 4(a)$. It will be seen that the current in the 10Ω resistor has been taken as $I_1 - I_2$. Next consider the left-hand mesh, then $10(I_2 - I_1) + 25I_2 = 2$ (b). Here cyclic current I_2 being clockwise, is taken as +ve.

Setting down equations *(a)* and *(b)* gives:

$$100I_1 + 10(I_1 - I_2) = 4 \text{ or } 110I_1 - 10I_2 = 4$$

Also

$$10(I_2 - I_1) + 25I_2 = 2 \text{ or } -10I_1 + 35I_2 = 2$$

Thus we have:

$$110I_1 - 10I_2 = 4$$

$$\text{and } -110I_1 + 385I_2 = 22$$

$$\text{Adding gives, } 375I_2 = 26$$

$$\text{or } I_2 = \frac{26 \times 10^{-2}}{3.75} = 0.0694\text{A}$$

Substituting back; then $35I_2 - 10I_1 = 2$

$$\text{or } (35 \times 0.0694) - 10I_1 = 2 \text{ and } I_1 = \frac{(35 \times 0.0694) - 2}{10}$$

$$\text{giving } I_1 = 0.0427\text{A}$$

The current in the 10Ω resistor is then given by:
0.0694 − 0.0427 = 0.0267A (in the downward direction).

3. THE SUPER-POSITION OF CURRENT THEOREM

This is based on the fact that, the current in any branch of a network is the sum of the currents due to each battery or generator taken separately, all other generators or batteries being replaced meanwhile by their internal resistances. This theorem avoids the use of simultaneous equations. The magnitude and direction of current for each section is calculated, assuming one supply source, and all other sources having their e.m.f.s suppressed. Their internal resistances however, are taken into account. This is repeated for each source of supply and the current, in magnitude and direction through each resistance, is finally obtained by adding algebraically all the appropriate values.

The solution, shown below, refers to Example 131 and should be studied in conjunction with the diagram (Fig 244 a, b, c and d).

Condition 1. 4V battery e.m.f. suppressed. Fig 244a.

To illustrate procedure, the diagram has been set out as shown in Fig 244b.

Let R = resistance of parallel section

$$\text{then } \frac{1}{R} = \frac{1}{100} + \frac{1}{10} = \frac{11}{100} \text{ or } R = 9.09\Omega$$

The total circuit resistance would be 25 + 9.09 = 34.09Ω and the current flowing $= \frac{2}{34.09} = 0.0587$A

Voltage drop across the parallel section = 0.0587 × 9.09 volts and the current in the 10Ω resistor $= \frac{0.0587 \times 9.09}{10}$

$= 0.0533$A.

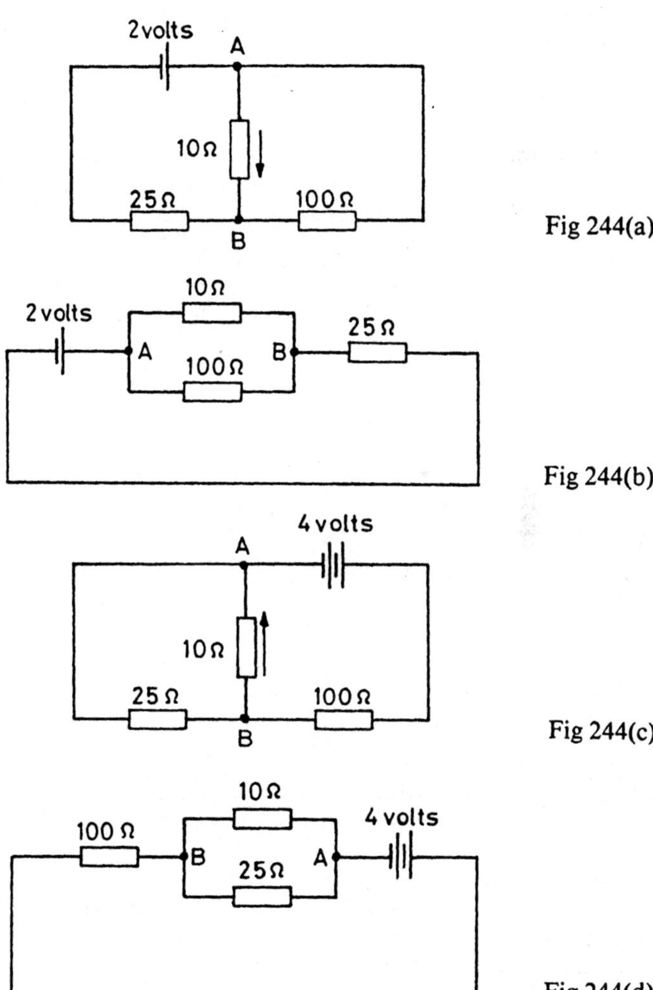

Fig 244(a)

Fig 244(b)

Fig 244(c)

Fig 244(d)

Condition 2. 2V battery e.m.f. suppressed.

Let R = resistance of parallel section

then $\frac{1}{R} = \frac{1}{25} + \frac{1}{10} = \frac{4 + 10}{100} = \frac{14}{100}$ or $R = \frac{100}{14} = 7.14\Omega$

The total circuit resistance would be $100 + 7.14 = 107.14\Omega$

and the current flowing $= \frac{4}{107.14} = 0.037\ 35$A

To illustrate procedure, the diagram has been re-arranged as shown in Fig 244d.

The voltage drop across the parallel section $= 0.037\ 35 \times 7.14$ volts.

$$\text{Current in } 10\Omega \text{ resistor} = \frac{0.037\ 35 \times 7.14}{10} = 0.0267\text{A}$$

The current in the 10Ω resistor is given by the joint effects of conditions 1 and 2. Thus $I = 0.0533 - 0.0267 = 0.0266$A.

Current flow is towards the junction of the resistors — as shown for the alternative solutions.

Example 132. The diagram (Fig 245) shows a Wheatstone-bridge network. Find the current, and its direction, in the galvanometer circuit.

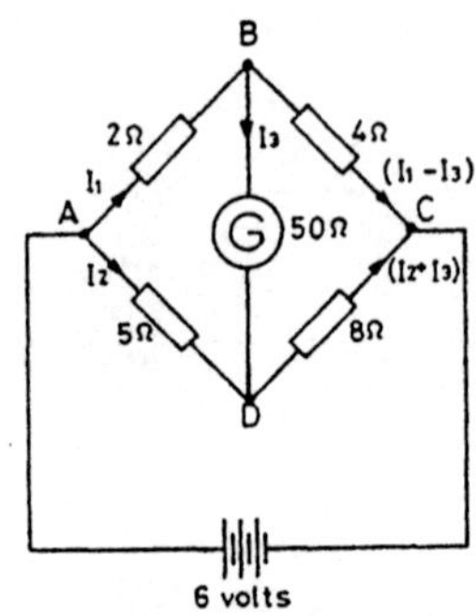

Fig 245

Method 1. Application of Kirchhoff's laws.

After allocating currents as shown, we apply the voltage law to:

Mesh ABD $2I_1 + 50I_3 - 5I_2 = 0$ . . . *(a)*

Mesh BCD $4(I_1 - I_3) - 8(I_2 + I_3) - 50I_3 = 0$ . . . *(b)*

Circuit ABC and battery $2I_1 + 4(I_1 - I_3) = 6$ . . . *(c)*

Simplifying *(a)*, *(b)* and *(c)*, we have:

$2I_1 - 5I_2 + 50I_3 = 0$ . . . *(a)*

and $4I_1 - 4I_3 - 8I_2 - 8I_3 - 50I_3 = 0$

or $4I_1 - 8I_2 - 62I_3 = 0$ ... *(d)*

also from *(c)* $2I_1 + 4I_1 - 4I_3 = 6$

or $6I_1 - 4I_3 = 6$ giving $I_1 = 1 + \frac{2}{3}I_3$ or $I_1 = 1 + 0.66I_3$

Substituting in *(a)*

$2(1 + 0.66I_3) - 5I_2 + 50I_3 = 0$

or $2 + 1.32I_3 - 5I_2 + 50I_3 = 0$

giving $-5I_2 + 51.32I_3 = -2$ ... *(e)*

Substituting in *(d)*

$4(1 + 0.66I_3) - 8I_2 - 62I_3 = 0$

or $4 + 2.64I_3 - 8I_2 - 62I_3 = 0$

giving $-8I_2 - 59.36I_3 = -4$ ... *(f)*

Solving *(e)* and *(f)* Multiply *(e)* by 8 to get

$-40I_2 + 410.56I_3 = -16$ ... *(g)*

Multiply *(f)* by 5 to get

$-40I_2 - 296.80I_3 = -20$ ... *(h)*

Subtract *(h)* from *(g)* then $707.36I_3 = 4$

or $I_3 = \frac{4}{707.36}$ amperes

Thus $I_3 = \frac{4 \times 10^{-3}}{0.707} = 5.65 \times 10^{-3} = 0.0056\text{A}$ (downwards).

Method 2. Application of Maxwell's Circulating-Current Theorem

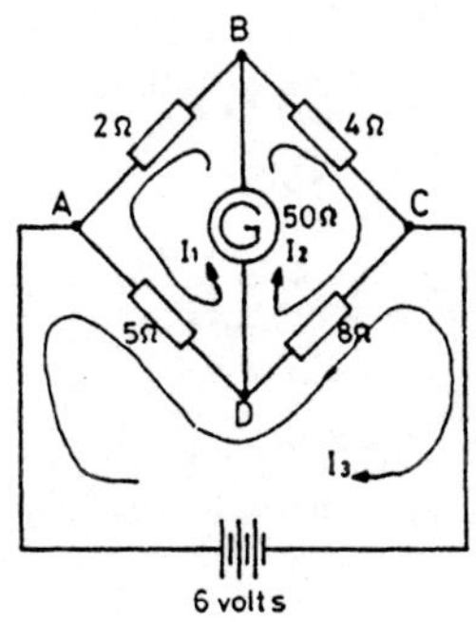

Fig 246

Mesh ABD

$2I_1 + 50(I_1 - I_2) + 5(I_1 - I_3) = 0$

or $2I_1 + 50I_1 - 50I_2 + 5I_1 - 5I_3 = 0$

and $57I_1 - 50I_2 - 5I_3 = 0$ ... *(a)*

Mesh BCD

$$4I_2 + 8(I_2 - I_3) + 50(I_2 - I_1) = 0$$
$$\text{or } 4I_2 + 8I_2 - 8I_3 + 50I_2 - 50I_1 = 0$$
$$\text{and } -50I_1 + 62I_2 - 8I_3 = 0 \qquad \ldots (b)$$

Multiply *(a)* by 8 and *(b)* by 5

$$\text{Then } 456I_1 - 400I_2 - 40I_3 = 0$$
$$\text{and } -250I_1 + 310I_2 - 40I_3 = 0$$
$$\text{Subtracting } 706I_1 - 710I_2 = 0 \qquad \ldots (c)$$

Mesh ADC and battery

$$5(I_3 - I_1) + 8(I_3 - I_2) = 6$$
$$\text{or } 5I_3 - 5I_1 + 8I_3 - 8I_2 = 6$$
$$\text{and } -5I_1 - 8I_2 + 13I_3 = 6 \qquad \ldots (d)$$

Multiply *(a)* by 13 and *(d)* by 5

$$\text{Then } 741I_1 - 650I_2 - 65I_3 = 0$$
$$\text{and } -25I_1 - 40I_2 + 65I_3 = 30$$
$$\text{Adding } 716I_1 - 690I_2 = 30 \qquad \ldots (e)$$

Solving *(c)* and *(e)* Multiply *(c)* by 6.9 and *(e)* by 7.1

$$\text{Then } 6.9 \times (7.06I_1 - 7.1I_2 = 0)$$
$$\text{or } 48.714I_1 - 48.99I_2 = 0$$
$$\text{and } 7.1 \times (7.16I_1 - 6.9I_2 = 0.3)$$
$$\text{or } 50.836I_1 - 48.99I_2 = 2.13$$

Subtracting $-2.122I_1 = -2.13$ or $I_1 = 1.003\ 77\text{A}$ (downwards)

$$I_2 \doteq \frac{706}{710} I_1 = 0.994 \times 1.003\ 77 = 0.998\ 16\text{A (upwards)}$$

$\therefore$ Resulting current $= 1.003\ 77 - 0.998\ 16 = 0.0056\text{A}$ (downwards)

It will be seen that the Super-position of Current Theorem cannot be applied to this network since there is only one source of e.m.f.

CONDUCTANCE, SUSCEPTANCE AND ADMITTANCE
(FOR A.C. CIRCUITS ONLY)

SERIES CIRCUIT

REVISION $\overline{V} = \overline{V}_1 + \overline{V}_2 + \overline{V}_3$

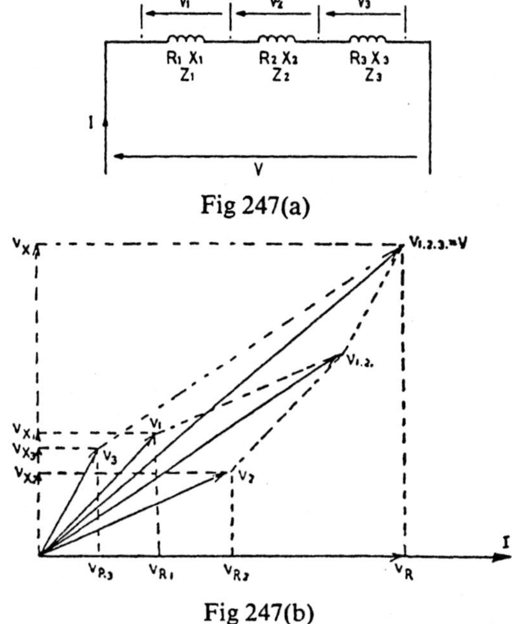

Fig 247(a)

Fig 247(b)

The circuit diagram (Fig 247a) and phasor diagram (Fig 247b) summarise the conditions, from which it can be deduced that:

Since V

$$= \sqrt{V_R^2 + V_X^2}$$

$$= \sqrt{V_{R1} + V_{R2} + V_{R3}\ldots)^2 + (V_{X1} + V_{X2} + V_{X3}\ldots)^2}$$

$$= \sqrt{(IR_1 + IR_2 + IR_3\ldots)^2 + (IX_1 + IX_2 + IX_3\ldots)^2}$$

$$= \sqrt{(R_1 + R_2 + R_3\ldots)^2 + (X_1 + X_2 + X_3\ldots)^2}$$

But $\dfrac{V}{I} = Z$—where Z is the equivalent impedance

Then $Z = \sqrt{(R_1 + R_2 + R_3\ldots)^2 + (X_1 + X_2 + X_3\ldots)^2}$

Also since

$$R_1 = Z_1 \cos\phi_1 \quad X_1 = Z_1 \sin\phi_1 \text{ etc}$$

$$\text{and } R_2 = Z_2 \cos\phi_2 \quad X_2 = Z_2 \sin\phi_2 \text{ etc}$$

So $Z = \sqrt{Z_1\cos\phi_1 + Z_2\cos\phi_2\ldots)^2 + (Z_1\sin\phi_1 + Z_2\sin\phi_2\ldots)^2}$

PARALLEL CIRCUIT

Revision $\bar{I} = \bar{I}_1\bar{I}_2 + \bar{I}_3$.

The circuit diagram (Fig 248a) and phasor diagram (Fig 248b) summarise the conditions, from which it can be deduced that:

$$I = \sqrt{I_a^2 + I_r^2} = \sqrt{(I_{a1} + I_{a2} + I_{a3} \ldots)^2 + (I_{r1} + I_{r2} + I_{r3} \ldots)^2}$$

$$= \sqrt{\left(\frac{V}{Z_1}\cos\phi_1 + \frac{V}{Z_2}\cos\phi_2 + \frac{V}{Z_3}\cos\phi_3 \ldots\right)^2 + \left(\frac{V}{Z_1}\sin\phi_1 + \frac{V}{Z_2}\sin\phi_2 + \frac{V}{Z_3}\sin\phi_3 \ldots\right)^2}$$

$$\text{or } I = V\sqrt{\left(\frac{1}{Z_1}\cos\phi_1 + \frac{1}{Z_2}\cos\phi_2 + \frac{1}{Z_3}\cos\phi_3 \ldots\right)^2 + \left(\frac{1}{Z_1}\sin\phi_1 + \frac{1}{Z_2}\sin\phi_2 + \frac{1}{Z_3}\sin\phi_3 \ldots\right)^2}$$

But if Z is the equivalent impedance, then $Z = \frac{V}{I}$ or $\frac{I}{V} = \frac{1}{Z}$

whence

$$\frac{I}{V} = \frac{1}{Z} = \sqrt{\left(\frac{1}{Z_1}\cos\phi_1 + \frac{1}{Z_2}\cos\phi_2 + \frac{1}{Z_3}\cos\phi_3 \ldots\right)^2 + \left(\frac{1}{Z_1}\sin\phi_1 + \frac{1}{Z_2}\sin\phi_2 + \frac{1}{Z_3}\sin\phi_3 \ldots\right)^2}$$

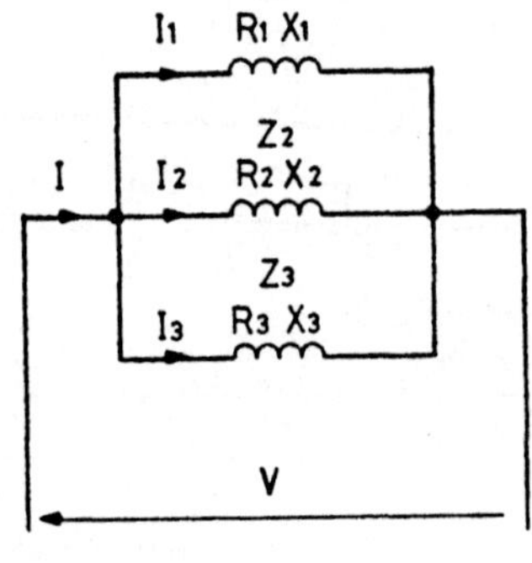

Fig 248(a)

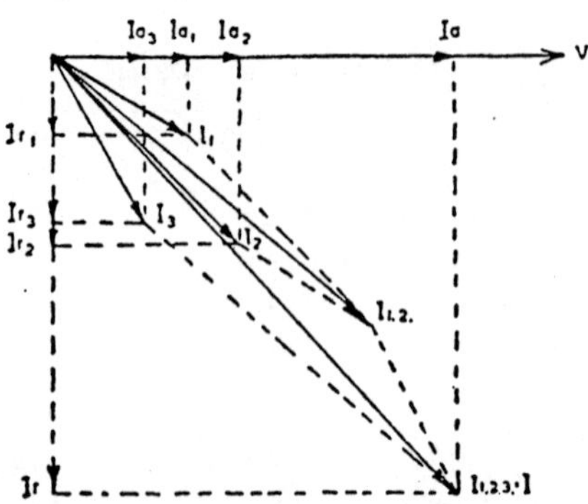

Fig 248(b)

From the expressions for the equivalent impedances of series and parallel circuits, as deduced above, we see a certain similarity and note that the resultant is obtained from the summation of in-phase and quadrature components.

For the series circuit, the components involve the known values of resistance and reactance and thus no difficulty is presented for obtaining the resultant.

For the parallel circuit, the reciprocals of impedance are required and it would appear that, if a new name was given to these reciprocals of impedances and to their in-phase and quadrature components then, some simplification of the relationships is possible. Furthermore, since the new terms are concerned with rectangular co-ordinates and their resultant, they must follow known relationships and the pattern of the right-angle triangle. This will be shown, once the new terms have been defined.

$\frac{1}{Z}$ is called the *Admittance* of the circuit, expressed in siemens (symbol S) and is denoted by Y. Just as impedance is resolved into the two components, resistance and reactance, so admittance is resolved into *Conductance,* denoted by G and *Susceptance* denoted by B. Again 'the siemens' is the unit for both components.

The conductance of a circuit is the extent to which a circuit admits the power component of current.

$$\text{Thus } G = \frac{\text{Power component of current}}{\text{Voltage}} = \frac{I \cos \phi}{V}$$

$$\text{or,} \quad G = \frac{1}{Z} \sin \phi = \frac{R}{Z^2}$$

The Susceptance of a circuit is the extent to which a circuit admits the wattless component of current.

$$\text{Thus } B = \frac{\text{Wattless component of current}}{\text{Voltage}} = \frac{I \sin \phi}{V}$$

$$\text{or,} \quad B = \frac{1}{Z} \sin \phi = \frac{X}{Z^2}$$

The Admittance is the extent to which the circuit admits current being $Y = \frac{\text{current}}{\text{voltage}} = \frac{I}{V} = \frac{1}{Z}$

$$\text{Also since } \frac{1}{Z^2} = \frac{1}{Z^2} (\sin^2 \phi + \cos^2 \phi)$$

$$\text{then } \quad \frac{1}{Z^2} = \left(\frac{1}{Z} \cos \phi\right)^2 + \left(\frac{1}{Z} \sin \phi\right)^2$$

$$\text{or} \quad \frac{1}{Z} = \sqrt{\left(\frac{1}{Z}\cos\phi\right)^2 + \left(\frac{1}{Z}\sin\phi\right)^2}$$

$$\text{and} \quad Y = \sqrt{G^2 + B^2}$$

Also, from the expression deduced for the parallel circuit:

$$\frac{1}{Z} = \sqrt{\left(\frac{1}{Z_1}\cos\phi_1 + \frac{1}{Z_2}\cos\phi_2 + \frac{1}{Z_3}\cos\phi_3\right)^2 + \left(\frac{1}{Z_1}\sin\phi_1 + \frac{1}{Z_2}\sin\phi_2 + \frac{1}{Z_3}\sin\phi_3\right)^2}$$

$$\text{or } Y = \sqrt{(G_1 + G_2 + G_3 \ldots)^2 + (B_1 + B_2 + B_3 \ldots)^2} \text{ etc}$$

Since there is a right-angled relationship between G, B and Y, it follows that this can be represented by an admittance triangle, as shown in the diagram (Fig 249a and 249b).

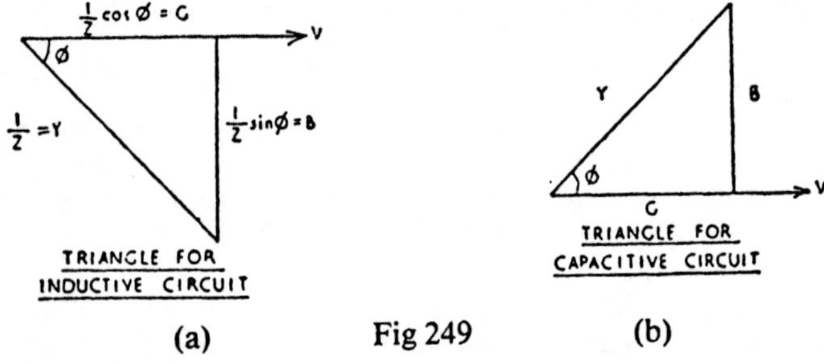

(a) Fig 249 (b)

It will be noted that as reciprocals are involved, and if the phasor diagrams of Fig 250a, 250b and 250c, are studied, then the deduced current triangle when modified by dividing all sides

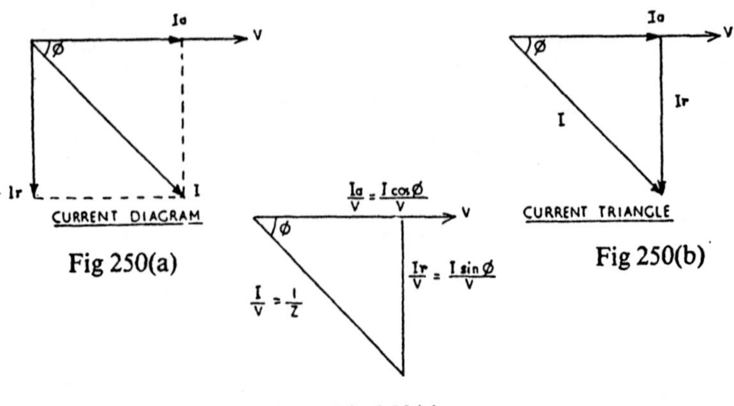

Fig 250(a) Fig 250(b)

Fig 250(c)

by V, will give the admittance triangle. This, for a parallel circuit containing resistance and inductive reactance, is inverted when compared with the impedance triangle. This is shown by the diagram (Fig 249a and 249b). For parallel circuits made up from resistance and capacitive reactance branches, the opposite is the case and due allowance must be made for the sign of B when solving problems. This will be illustrated in the examples which follow, but it must be stressed here that:

the conductance, susceptance and admittance method of solution is only applicable to parallel circuits or to the parallel section of a series-parallel circuit.

Example 133. Consider the parallel circuit shown by the diagram (Fig 251). Find the total current and its power factor.

Here $Z_1^2 = R_1^2 + X_1^2 = 2^2 + 3^2 = 4 + 9 = 13$

$Z_2^2 = R_2^2 + X_2^2 = 3^2 + 4^2 = 9 + 16 = 25$

$Z_3^2 = R_3^2 + X_3^2 = 3^2 + 2^2 = 9 + 4 = 13$

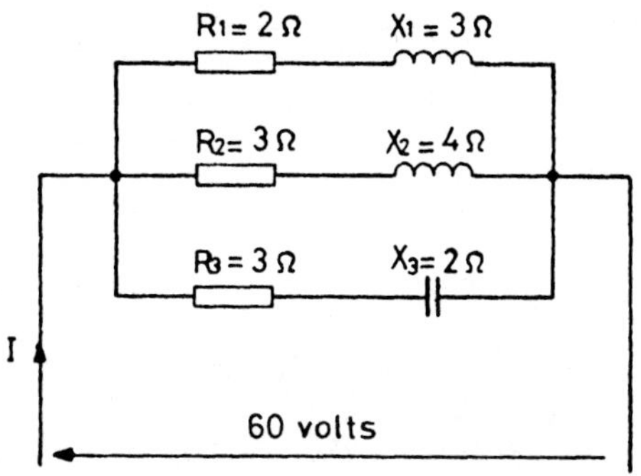

Also

Fig 251

$$G_1 = \frac{R_1}{Z_1^2} = \frac{2}{13} = 0.154\text{S} \qquad B_1 = \frac{-X_1}{Z_1^2} = \frac{-3}{13} = -0.231\text{S}$$

$$G_2 = \frac{R_2}{Z_2^2} = \frac{3}{25} = 0.12\text{S} \qquad B_2 = \frac{-X_2}{Z_2^2} = \frac{-4}{25} = -0.16\text{S}$$

$$G_3 = \frac{R_3}{Z_3^2} = \frac{3}{13} = 0.231\text{S} \qquad B_3 = \frac{+X_3}{Z_3^2} = \frac{+2}{13} = +0.154\text{S}$$

$$\text{Total } G = 0.505\text{S} \qquad \text{Total } B = -0.237\text{S}$$

$$Y = \sqrt{0.505^2 + 0.237^2}$$
$$= 0.56\text{S}$$

Also since, $I = \dfrac{V}{Z}$ and $\dfrac{1}{Z} = Y \quad \therefore I = VY = 60 \times 0.56$
$= 33.6\text{A}$

$$\cos\phi = \frac{0.505}{0.56} = 0.9 \text{ (lagging)}$$

The lagging condition is determined from the fact that the inductive effect of the circuit predominates, as indicated by the total value of $B = -0.237$S.

Example 134. A series-parallel circuit ABC is made up as follows: a 5Ω resistor connected in series with a 0.02H inductor with a 100μF capacitor connected in parallel across these components. This constitutes the parallel section of the circuit. The series section AB consists of a 10Ω resistor in series with a 0.02H inductor, which in turn is connected in series with the parallel section. The complete circuit is connected across a 100V, 50Hz supply. Find the voltage across the parallel section BC. The problem is illustrated by the diagram (Fig 252).

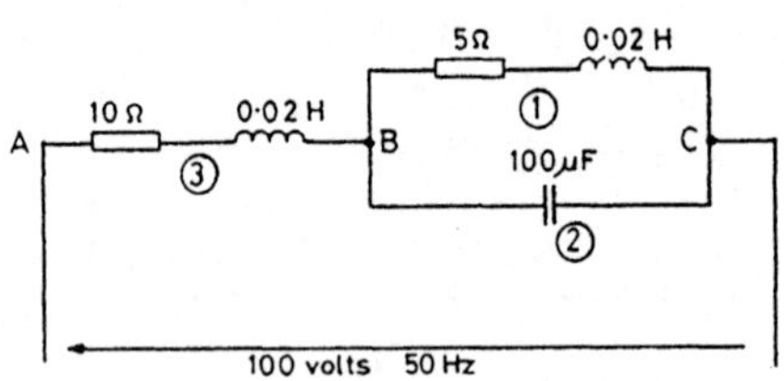

Fig 252

Branch 1 $X_1 = 2 \times \pi \times 50 \times 0.02 = 6.28\Omega$

$R_1 = 5\Omega$

$Z_1^2 = 5^2 + 6.28^2 = 64.5$

$$G_1 = \frac{R_1}{Z_1^2} = \frac{5}{64.5} = 0.0775\text{S} \quad B_1 = \frac{-X_1}{Z_1^2} = \frac{-6.28}{64.5} = -0.0975\text{S}$$

Branch 2

$$X_2 = \frac{10^6}{2 \times \pi \times 50 \times 10^2} = 31.85\Omega$$

$$Z_2^2 = 0^2 + X_2^2 \quad \text{so } Z_2^2 = X_2^2$$

$$G_2 = \frac{R_2}{Z_2^2} = \frac{0}{Z_2^2} = 0\text{S} \qquad B_2 = \frac{X_2}{Z_2^2} = \frac{1}{31.85} = +0.0314\text{S}$$

Total $G = 0.0775$S Total $B = -0.0661$S

Thus:

$$Y = \sqrt{0.0775^2 + 0.0661^2}$$
$$= 0.1017\text{S}$$

$$Z \text{ of parallel section} = \frac{1}{0.1017} = 9.84\Omega$$

$$\text{Equivalent } R \text{ of parallel section} = 9.84^2 \times 0.0775$$
$$= 7.52\Omega$$
$$\text{Equivalent } X \text{ of parallel section} = 9.84^2 \times 0.0661$$
$$= 6.42\Omega \text{ (inductive)}$$
$$\text{Total } R \text{ of whole circuit} = 10 + 7.52 = 17.52\Omega$$
$$\text{Total } X \text{ of whole circuit} = 6.28 + 6.42 = 12.70\Omega$$
$$\text{Total } Z \text{ of whole circuit} = \sqrt{17.52^2 + 12.7^2}$$
$$= 21.65\Omega$$
$$\text{Circuit current} = \frac{100}{21.65} = 4.62\text{A}$$
$$\text{Voltage drop across section BC} = 4.62 \times 9.84 = 45.4\text{V}.$$

THE WHEATSTONE BRIDGE

Although this network has already been introduced as an example for 'methods of problem solution', it should be noted that the condition of no current in the centre arm is of particular importance. The bridge in this basic form has been adapted for methods of measurement and control but the original application, used by H. Wheatstone — a railway engineer, indicated the condition of unbalance and was used for signalling. The now best known adaptation enables resistance measurements to be made and is considered here in connection with the diagram (Fig 253).

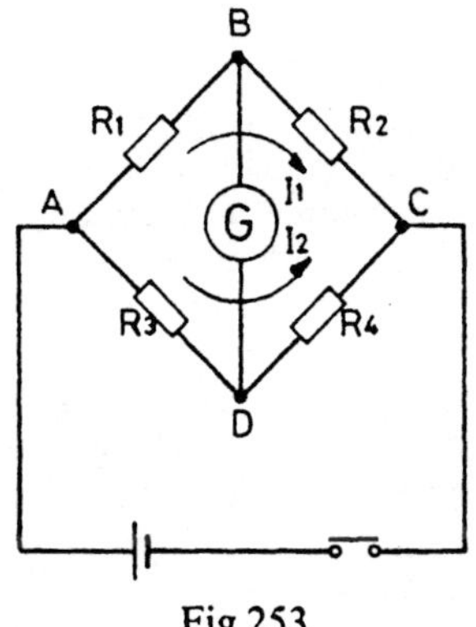

Fig 253

MEASUREMENT OF RESISTANCE

The bridge consists of a network of four resistors which, by the choice of values for three, can be used to measure the value of the fourth resistor. R_1, R_2 and R_3 are the known resistance values. R_4 is the unknown resistor and G is the centre-zero ammeter or milliammeter, called a galvanometer. A cell, battery or 'mains' power unit can be used as the supply source. The bridge, as stated, is used in the 'balanced' mode and, for this condition, gives a measurement in terms of the known resistors and is accurate irrespective of variations of the supply voltage.

To understand the theory, letters have been given to the network and currents I_1 and I_2 are assumed to flow. The bridge is balanced by varying resistors R_1, R_2 and R_3. Under the balanced or 'null-deflection' condition, connection BD carries no current. Thus the assumption of only currents I_1 and I_2 is valid, even though these current values may be dissimilar. Further, it is seen that since there is no current between B and D, then these points are at the same potential. Also point A being common to branches AB and AD then:

the voltage drop across AB = the voltage drop across AD

$$\text{or } I_1R_1 = I_2R_3 \qquad \ldots (a)$$

Again point C being common to branches BC and DC then;

the voltage drop across BC = the voltage drop across DC

$$\text{or } I_1R_2 = I_2R_4 \qquad \ldots (b)$$

Dividing *(b)* by *(a)*, we have:

$$\frac{I_1R_2}{I_1R_1} = \frac{I_2R_4}{I_2R_3}$$

$$\text{or } R_4 = R_3 \times \frac{R_2}{R_1}$$

It will be seen that, if R_4 is an unknown resistor — now termed R_X, resistor R_3 can be made a variable standard whilst R_1 and R_2 can also be standard fixed resistors of decade value. R_1 and R_2 are frequently termed the 'ratio arms' and, by suitable choice, the value of R_4 can be accurately determined for a wide range.

The Wheatstone-bridge method of measuring resistance is utilised for commercial instruments which may appear in various forms. Generally the resistance value of the variable resistor R_3 — now termed R_S (say 1000Ω) is shown against a calibrated scale, whilst R_1 and R_2 values are switched together to give ratios of say 1 : 100, 10 : 100, 100 : 100, 100 : 10 and 100 : 1, *ie* times 0.01, 0.1, 1, 10 and 100. Thus, if the switch is on 'times 0.01' and the balance occurs at 340Ω, then the unknown resistance = 0.01 × 340 = 3.4Ω.

The range of measurement is thus extended by using ratio arms of three or four decades and its accuracy is dependent only on the accuracy of the components used in the bridge. The range resistors R_1 and R_2 and the variable resistor R_S would be wound from constantan wire and would be calibrated with precision. Great care would be taken to minimise the contact resistance of the sliding contact on R_S and at the terminals of the selector or range switch.

Whilst considering the Wheatstone-bridge principle, it should be noted that other forms of 'direct-reading' ohmmeters are also available. These show an ohmic value direct on the meter scale and a variation can employ the bridge network. Thus if the terminals for connecting R_X are open circuited, the value of R_X is described as 'infinite'. The meter would be arranged to give full-scale deflection — a side-zero instrument is used for this application. If standard resistors of known decreasing value are substituted for R_X, the meter pointer deflection would decrease and the scale could be marked accordingly until a minimum is shown when R_X is zero. The scale would not be linear but the meter would be suitable for everyday work and factory production 'checks'. More accurate direct-reading ohmmeters, employing electronic principles, have now been devised, but the bridge 'out-of-balance' mode is still favoured for other measurements such as temperature, gas content, strain, etc. Note that for this arrangement, accuracy is dependent on the voltage of the supply.

TEMPERATURE MEASUREMENT

Although it is not proposed to describe here, the many methods available for measuring temperature, it is necessary to point out to the engineer that present-day methods utilise either a resistance-thermometer probe or a thermocouple probe. The former only is described here, as the 'thermo-electric' effect has not been introduced as yet.

The fact that the resistance value of a resistor varies with temperature is already known, the relationship being a straight line — ref Fig 20. If therefore, a thermometer 'head', *ie* probe is connected into a Wheatstone-bridge arrangement, then the resistance value for any temperature can be measured and either reference can be made to the appropriate graph or the scale of R_S can be directly calibrated in temperature units. The resistance thermometer head can consist of a spool of platinum wire wound onto a ceramic former, the whole being enclosed in an envelope of special steel. Various forms of 'head' or probe have been evolved for extreme conditions such as furnace or refrigerator work and it should be noted that, thermistors have also been introduced as alternative probe elements for temperature measurement.

Here again, the 'out-of-balance' bridge method can be used in conjunction with a direct-reading instrument. The scale is calibrated in temperature units and by means of a selector switch various probes can be connected-in and readings taken. As before the readings are dependent on a constant voltage being available to the bridge and, to check accuracy, a standardising facility is usually provided. Thus, by means of a two-position switch, a known standard resistor can be connected in place of a probe. The pointer should then come to rest at a calibration line on the scale. If not, a variable resistor in series with the battery or power-pack can be adjusted to achieve the required condition. The switch can next be moved from the 'test' to the 'run' position when readings, as obtained, can be judged to be accurate.

THE POTENTIOMETER

The word is derived from 'potential meter' and thus the instrument performs the same functions as a voltmeter but with greater accuracy because, when a 'reading' is being taken, no current is being drawn from the potential or e.m.f. source which is being measured. The potentiometer is not an indicating meter in the true sense, but consists of a resistance arrangement, galvanometer and cells which can be used for measurement, once the apparatus has been set up. Like the Wheatstone bridge, commercial versions are available but here also measurements must be made rather than 'readings' taken.

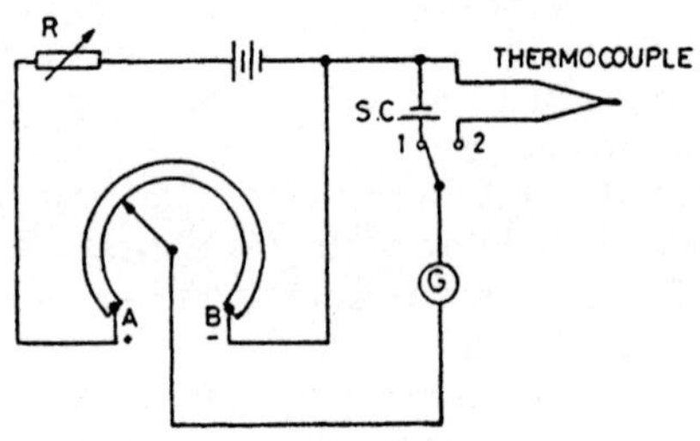

Fig 254

Consider the diagram (Fig 254). AB is a resistor of high accuracy provided with a sliding contact C. AB is wound with a length of uniform resistance wire so that, when a current flows through it, the potential dropped across a portion of it is proportional to the length of the wire contained therein. Thus if 2V is applied across AB then, if the sliding contact C is placed a quarter of the way from B, the potential across CB would be $2 \times \frac{1}{4} = 0.5$V. Similarly with C midway between AB, the potential CB would be 1V and so on. The polarity of the supply across AB is most important. Thus if A is positive, with respect to B, then C for the quarter way position would be 0.5V above

B. For the half-way position, C would be 1V above B.

Consider next, an e.m.f. source of standard voltage value to be connected across CB, strict attention being paid to polarity. Such an e.m.f. source is available — an example being the Weston Standard Cell. This cell is manufactured with extreme care, only the purest of materials being used for the electrodes and electrolyte. The construction and chemicals used are not of importance here, but the cell can be recognised, in any type of potentiometric indicator or recorder, by noting that it is enclosed in a sealed glass envelope and mounted so that it is subject to the minimum of disturbance. A constant e.m.f. of 1.018V can be guaranteed under no-load conditions. Referring to the diagram, with the change-over switch in position 1, the standard cell is connected across CB, a galvanometer being included in the circuit. The position of sliding contact C will then determine the deflection on the meter. If C is midway and the p.d. across CB is assumed to be 1V, due to the current from the main energising battery then, as the standard cell is applying an opposing e.m.f. of 1.018V, a difference of 0.018V will exist and a current will result, flowing from the standard cell through the meter to point C, down the resistor — usually called the 'slide-wire,' to point B and then back to the cell. The resistance section CB will thus carry a current additional to that already due to the applied voltage across AB. If C is moved up the slide-wire resistor, the meter deflection will decrease until a condition of no-deflection or balance is attained. This means the potential across CB is now equal to 1.018V and, if an evenly graduated scale is provided, this can be marked in voltage, both at the standardising point and also over the whole scale range in values proportional to length. Note that the diagram (Fig 254) shows a series resistor *R* in the slide-wire circuit. This can be varied to adjust the main current and so alter the potential drop along the wire and thus bring the scale markings to their correct values.

We can now consider the normal operational procedure when working with a potentiometer. The first step would be to standardise the slide-wire. The change-over switch is put in position 1 and the slide-wire contact C is set against 1.018V on the scale. The meter deflection is noted and R is then adjusted until a balanced condition is obtained — the scale markings, in voltage, are then correct. The next operation would be to move the switch to position 2. Another external e.m.f. can now be measured, provided its polarity is correct with respect to the slide-wire. A typical application would be the measurement of a thermocouple e.m.f. and hence its temperature.

A description of the thermocouple and its associated theory follow later in the chapter but, it can be taken that, when the arrangement is used as a thermometer, an e.m.f. is produced which is proportional to temperature rise. If this e.m.f. is measured accurately, the temperature can be determined. Thus in the diagram (Fig 254), the thermocouple shown, can be taken to relate to a temperature indicator or recorder and is connected in when the switch on the instrument is moved into the position 2 or 'run' setting. Position 1 would be the 'test or standardise' position. Since, for a thermocouple, the relationship between e.m.f. and temperature is straight line, the scale of the potentiometer can be marked in °C and is accurate, if standardising is checked periodically. It should be noted that, a potentiometric type of temperature indicator is calibrated directly for use with thermocouples made up from specified metal pairs, such as copper and constantan, and such thermocouples must always be used. Most modern multipoint indicators and recorders operate on the potentiometer principle, the slider contact being motor driven, the drive power coming from an amplifier which is fed from the error voltage. Thus the greater the unbalance the larger the power drive and the slider C is driven fast to the balance point. Once this is reached the drive ceases. Standardising is carried out automatically at regular intervals to maintain accuracy.

THE THERMOCOUPLE

A method of making temperature measurements by electrical means, has already been considered when the resistance thermometer was described. It will be noted that for such an arrangement, a battery or power-pack is needed to energise the Wheatstone bridge network. The advantage of using a thermocouple is that, for fundamental operation, no such additional e.m.f. source is required and basic instrumentation consists of a thermocouple and voltmeter. For accuracy however, the thermocouple is usually used with a potentiometric indicator.

TEMPERATURE MEASUREMENT

The thermo-electric effect was discovered by Seebeck in 1826. He joined together one end of a piece of iron wire and a piece of copper wire by soldering and connected the free ends to a sensitive galvanometer (Fig 255). On heating the copper-iron junction a current was shown to flow, *ie* an e.m.f. was generated. Seebeck experimented with various metals and arranged them in an order which showed the current across the hot junction from the earlier to the later metal. Amongst these metals are: bismuth, platinum, copper, lead, tin, silver, zinc, iron, antimony, etc. For practical purposes thermocouples are classed into two main groups (a) rare or precious metals and (b) base metals. Because of the obvious reason of cost, the latter are used most extensively in industry and, with experience, alloys have been developed which satisfy the requirements of a working temperature range. Thus a common usage for measuring the temperature of electrical machines is the copper/constantan or copper/eureka couple. The latter is an alloy of nickel (40 per cent) and copper (60 per cent), the temperature range is 0-400°C. An iron/constantan couple is suitable for a 0-800°C range, whilst nickel-chromium/nickel-aluminium alloys, when used for a thermocouple, can cover temperatures up to 1200°C.

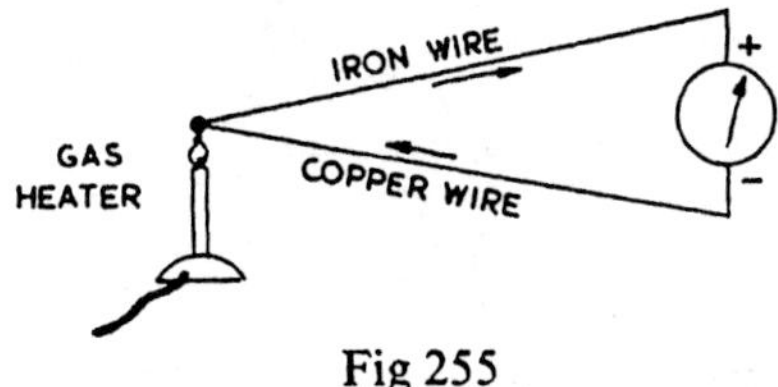

Fig 255

In practice the hot (common) junction is kept in good electrical and thermal contact by welding together the thermocouple wires and the free ends are kept at a common temperature, as is the measuring instrument. The thermocouple may be used bare but for most applications it is inserted into a metal pocket which must be strong enough to withstand corrosion effects, heat and pressure. The arrangement is shown by the diagram (Fig 256).

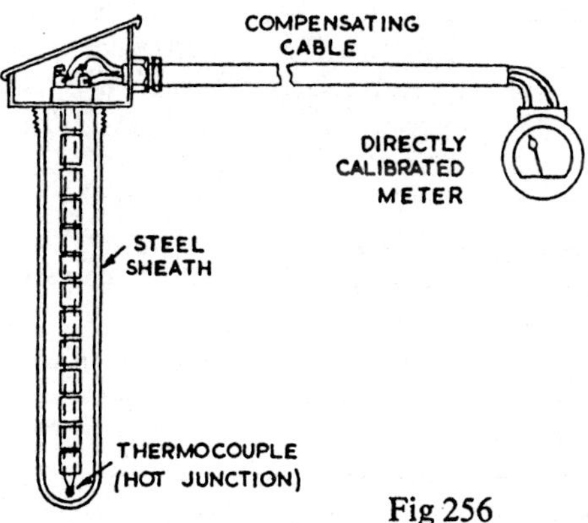

Fig 256

The e.m.f. generated is a function of the temperature difference between the hot junction (H.J.) and the instrument terminals — called the cold junction (C.J.). The graphs (Fig 257) show the relationships to be linear over a part of the temperature range. Since the linearity is based on the assumption of a constant cold junction temperature, allowance is made for this by circuits which compensate for C.J. temperature variation or by the provision of bimetal components in the instrument movement.

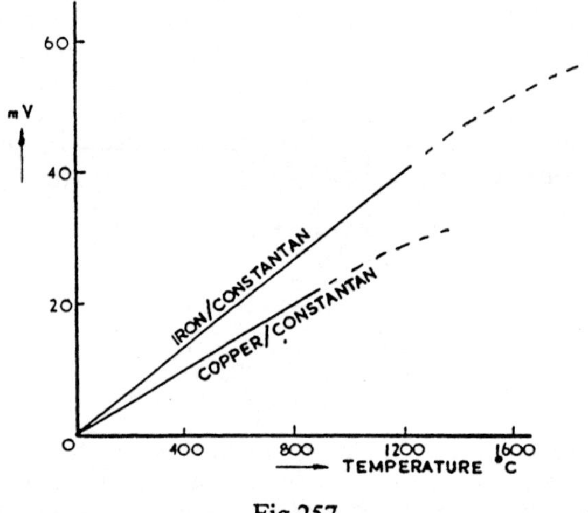

Fig 257

COMPENSATING LEADS OR CABLES. Consider the first of the diagrams (Fig 258). Since all parts of the instrument and terminals B and C are at the same temperature, this section constitutes an electrical connection only and the cold junction can be considered to be at point *B*. If the instrument is located near the hot junction then the cold junction temperature will rise and the temperature difference between A and B will be reduced thus giving an error in measurement. The obvious method of reducing this, is to mount the instrument at a point remote from the hot junction. This is also an advantage for instrumentation since the indicator can be placed at a convenient point or grouped with other instruments.

If the leads shown in the second diagram of Fig 258 are made of copper, the cold junction would still be at B and any improvement is only achieved if the thermocouple metals are extended by making the leads of appropriate metals. The leads between BC and DE are thus known as 'compensating' leads' and vary to

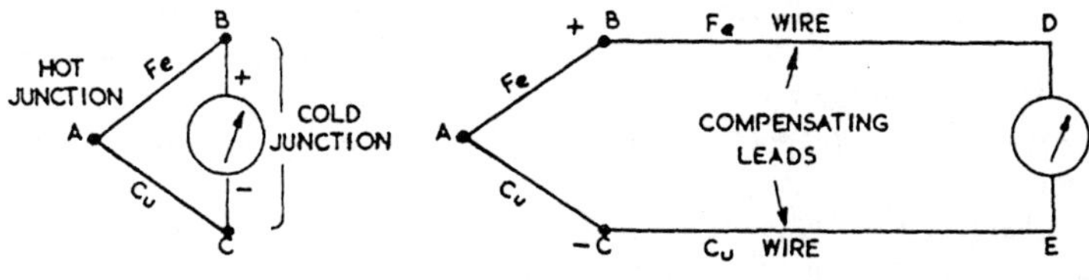

Fig 258

suit the thermocouple. Thus a copper/constantan couple will use copper and constantan wire in the compensating leads. Similarly iron and constantan are the wires for an iron/constantan couple.

INSTRUMENTATION. As was stated earlier, measurement can be made by using either a meter or a potentiometer. For the former, a sensitive milli-voltmeter of the moving-coil type is used since the e.m.f. is unidirectional and a thermocouple has polarity. At the cold junction, the wire from which the current flows, is called the positive element and that into which the current flows is the negative element. Polarity is also given to the compensating leads.

For a copper/constantan couple; the +ve element is copper and the −ve element is constantan. For the compensating leads the +ve wire is copper and the −ve wire is constantan. To achieve accuracy, the thermocouple responses need to be registered by a highly sensitive instrument movement, *ie* one which requires only a small current for full-scale deflection. Resistance voltage drops are thus minimised and measurement errors are reduced. Such indicators are delicate and not extensively used for marine work. With the advent of reliable semi-conductor amplifiers however, the sensor signals can be magnified and applied to a more robust instrument with a more substantial movement. The indicator would be calibrated directly in temperature units and if the e.m.f. is to be a true measure of the temperature difference ($T_H - T_C$), between the hot and cold junctions, it is essential to keep T_C constant or to compensate for variations of T_C. Such automatic cold junction compensation is a feature of direct-reading indicators and is achieved by introducing a bimetal corrector into the movement suspension system.

The potentiometric or null-deflection method has already been described (Fig 254) and has the advantage that 'at balance,' it does not draw current and hence registers the true e.m.f. — all resistance voltage drops being eliminated. The method, when operated manually, is much used in industrial laboratories but is not favoured for marine applications. However, when adapted to function automatically, it forms the basis of many familiar ship-board installations incorporating well-known makes of recorder and controller.

Basic theory states that, provided the slide-wire is of uniform section and material then, the voltage drop along it is proportional to its length. Once the current has been adjusted to give a known voltage drop over a length BC (Fig 254), the wire can be scaled over its length in voltage and even in temperature if the thermocouple characteristic is known. The calibration and operational procedure of a potentiometer have also been described but, it should be noted that, for temperature measurement, certain additional refinements are introduced into the instrument. Thus automatic temperature compensation for the cold junction can be achieved by special circuitry and in line with modern electronic developments a zener diode arrangement can be used to replace the standard cell.

CHAPTER 16

PRACTICE EXAMPLES

1. The sides of a triangle ABC are made up as follows: AB, resistance of 3Ω, AC, resistance of 8Ω and BC is a battery with a constant e.m.f. of 10V and an internal resistance of 1Ω. The points A, B and C are joined to a common point D by resistors BD = 6Ω, DC = xΩ, AD = yΩ. Find the value of x and the battery current when the current in AD is zero.

2. Two batteries X and Y, having e.m.f.s of 8V and 4V and internal resistances of 1.5Ω and 3Ω respectively, are connected in parallel across a resistor of 6Ω. Calculate the current flowing through each battery and the terminal voltage.

3. A 10V battery of resistance 2Ω is connected in parallel with a 12V battery or resistance 3Ω. Both batteries are connected to a 10Ω resistor. Calculate the current flowing in the resistor and the current supplied by each battery. Solve by (1) Maxwell's Cyclic-Current Theorem (2) Superposition of Current Method.

4. A network is arranged, as shown by the diagram. Calculate the value of the current in the 8Ω resistor (by **Kirchhoff's laws).**

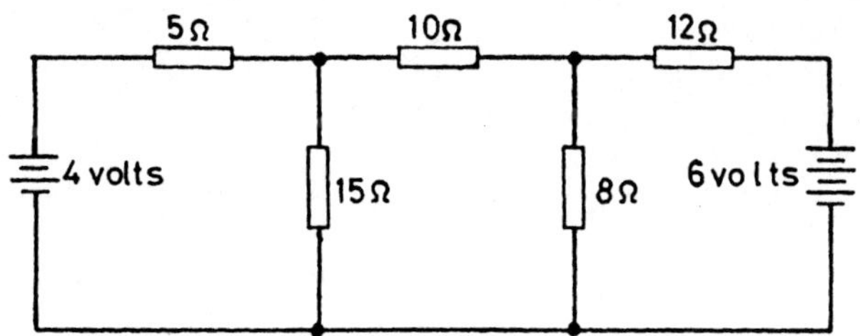

5. Solve the problem of Q4 by the Superposition of Current Theorem.

6. For the circuit shown below, calculate the value and direction of the current in resistor A.

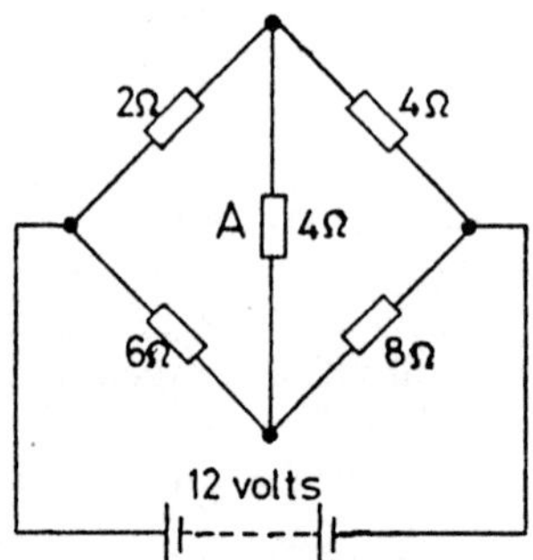

7. A parallel a.c. circuit has two branches A and B. A consists of an inductive coil, of which the resistance is 30Ω and the inductance is 0.1H. B consists of a 30μF capacitor. Calculate (a) the joint impedance at 50Hz. (b) the power taken from 230V mains.

8. A circuit consists of two sections AB and BC in series. AB consists of two parallel branches, the first being formed by a 60Ω resistor in series with a 50μF capacitor, the second by a 60Ω resistor in series with an inductance of 0.25H. BC consists of a 100Ω resistor in series with a 0.3H inductor. If the frequency is 50Hz and the voltage across AC is 500V, find the voltage across BC.

9. A parallel circuit consists of three branches A, B and C. Branch A is made up of 3Ω resistance in series with 0.03H inductance. Branch B is made up of 100Ω resistance in series with 400μF capacitance. Branch C is made up of 7Ω resistance, 0.02H inductance and 300μF capacitance of all in series. Find the impedance and equivalent resistance and reactance of the complete circuit.

10. A parallel circuit consists of the following three branches all in parallel:
    (a) A non-inductive resistor of 20Ω
    (b) A coil of inductance 0.05H, resistance 5Ω
    (c) A capacitor of 50μF capacitance

The whole circuit is connected aross 100V, 50Hz mains. Find the total current taken from the mains and the operating power factor.

SOLUTIONS TO PRACTICE EXAMPLES

CHAPTER 1

1.

Let R be the equivalent resistance of the parallel arrangement.

Then $\frac{1}{R} = \frac{1}{2} + \frac{1}{4} + \frac{1}{5} + \frac{1}{10} = \frac{10 + 5 + 4 + 2}{20}$

$= \frac{21}{20} = \frac{2.1}{2}$ and $R = \frac{2}{2.1} = 0.952\Omega$

Voltage drop across the arrangement $= 8.6 \times 0.952$
$= 8.19\text{V}$ Ans.

Current I_1 in 2Ω resistor $= \frac{8.19}{2} = 4.095\text{A}$ Ans.

Current I_2 in 4Ω resistor $= \frac{8.19}{4} = 2.0475\text{A}$ Ans.

Current I_3 in 5Ω resistor $= \frac{8.19}{5} = 1.638\text{A}$ Ans.

Current I_4 in 10Ω resistor $= \frac{8.19}{10} = 0.819\text{A}$ Ans.

Check. Total current $= 8.6\text{A}$ Ans.

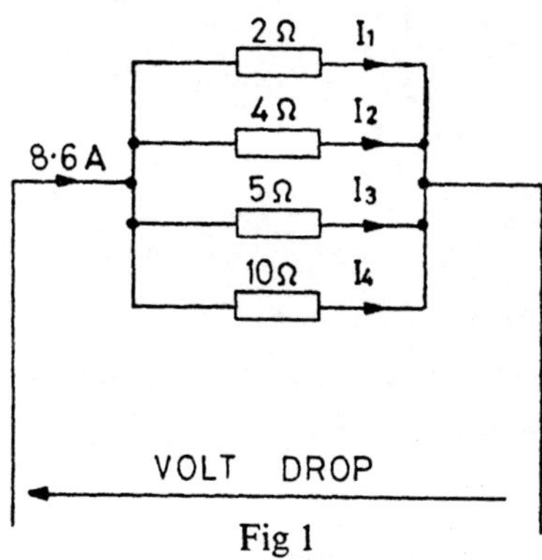

Fig 1

2.

For Group A. Let R_A = the equivalent resistance, then:

$\frac{1}{R_A} = \frac{1}{4} + \frac{1}{6} + \frac{1}{8} = \frac{6 + 4 + 3}{24}$ or $R_A = \frac{24}{13} = 1.85\Omega$

For Group B. Let R_B = the equivalent resistance, then:

$$\frac{1}{R_B} = \frac{1}{3} + \frac{1}{6} = \frac{2+1}{6} = \frac{3}{6} \text{ or } R_B = 2\Omega$$

Total circuit resistance $R = R_A + R_B = 1.85 + 2 = 3.85\Omega$

$$\text{Circuit current} = \frac{24}{3.85} = 6.23\text{A} \quad \text{Ans.}$$

Voltage drop across group A = 1.85 × 6.23 = 11.53V Ans.
Voltage drop across group B = 2 × 6.23 = 12.46V Ans.

Check. Total voltage drop = 23.99 = 24V Ans.

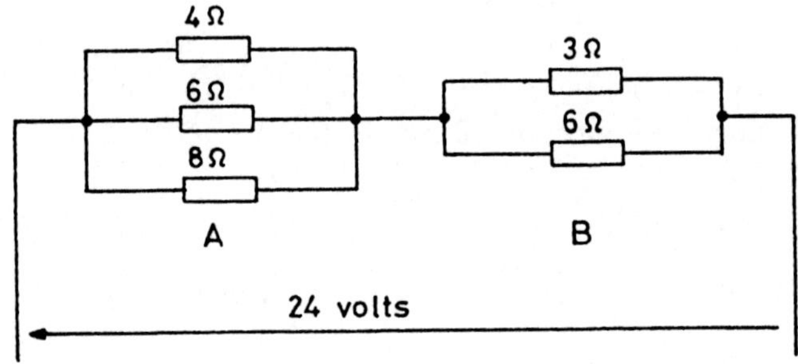

Fig 2

Current in resistors. Group A

$$\frac{11.53}{4} = 2.88\text{A} \quad \text{Ans.}$$

$$\frac{11.53}{6} = 1.91\text{A} \quad \text{Ans.}$$

$$\frac{11.53}{8} = 1.44\text{A} \quad \text{Ans.}$$

Check 6.23A Ans.

Current in resistors. Group B

$$\frac{12.46}{3} = 4.153\text{A} \quad \text{Ans.}$$

$$\frac{12.46}{6} = 2.076\text{A} \quad \text{Ans.}$$

Check 6.23A Ans.

3.
From Q1 the equivalent resistance R of the load = 0.95Ω
The total resistance of the circuit = 0.95 + 0.65 = 1.6Ω

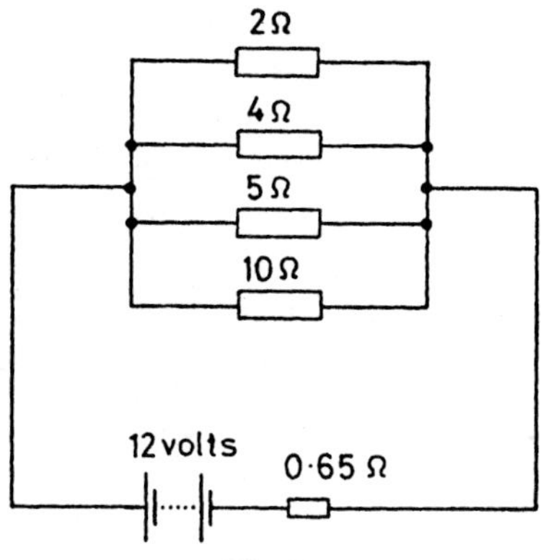

Fig 3

$$\text{The circuit current} = \frac{12}{1.6} = 7.5\text{A} \quad \text{Ans.}$$

$$\text{The terminal voltage} = 7.5 \times 0.95 = 7.125\text{V} \quad \text{Ans.}$$

$$\text{Current in } 5\Omega \text{ resistor} = \frac{7.125}{5} = 1.425\text{A} \quad \text{Ans.}$$

4. (a) Ammeter with shunt.

Voltage drop across parallel arrangement for full-scale deflection $= 10 \times 15 \times 10^{-3} = 0.15\text{V}$

Current to be carried by shunt

$$= 25 - (15 \times 10^{-3})$$
$$= 24.985\text{A}$$

$$\text{Resistance of shunt} = \frac{0.15}{24.985}$$
$$= 0.006\Omega \quad \text{Ans.}$$

(b) Voltmeter with series resistance.

Resistance of instrument circuit to drop 500V

$$= \frac{500}{15 \times 10^{-3}}$$
$$= 33\ 333\Omega$$

$\therefore$ Series resistance to be added $= 33\ 333 - 10$
$= 33\ 323\Omega$ Ans.

5. Parallel section BC has a resistance, given by:

$$\frac{1}{R} = \frac{1}{40} + \frac{1}{40} = \frac{2}{40} \text{ or } R = 20\text{k}\Omega$$

Total resistance of network $= 60 + 20 = 80\text{k}\Omega$

$$\text{Current taken by network} = \frac{240}{80\ 000}$$
$$= 3 \times 10^{-3}\text{A or 3mA}$$

Voltage drop across section BC

$= 3 \times 10^{-3} \times 20\,000$ volts
$= 60V$
$=$ reading on voltmeter Ans.

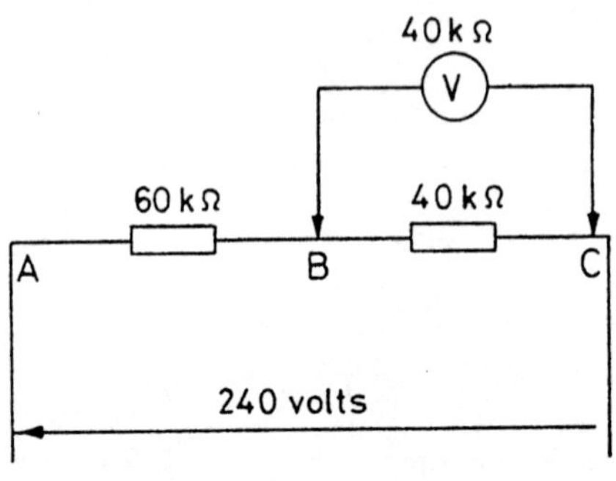

Fig 4

Note. When the voltmeter resistance is comparable with that of the resistor across which it is connected, then the above reading will give a false indication of the actual voltage drop in the working circuit. Thus, if the 60kΩ and 40kΩ resistors were part of a resistor chain in a radio set, the actual voltage drop across BC would be $\frac{40}{100} \times 240 = 96V$ and a special voltmeter, taking in effect no current such as a valve voltmeter, would be used.

The problem however, is applicable to this chapter's work, being a network problem as such.

6.

Let E = e.m.f. of the battery and R_i its internal resistance.
then $E = 0.18\,(10 + R_i)$ . . . a.
and $E = 0.08\,(25 + R_i)$ . . . b.

Equating (a) and (b) $0.18\,(10 + R_i) = 0.08\,(25 + R_i)$

$$\text{or } 1.8 + 0.18\,R_i = 2 + 0.08\,R_i$$
$$\therefore (0.18 - 0.08)\,R_i = 2 - 1.8$$
$$\text{or } 0.1\,R_i = 0.2$$
$$\text{or } R_i = 2\Omega \quad \text{Ans.}$$
$$\text{Substituting in } E = 0.18\,(10 + 2)$$
$$= 0.18 \times 12$$
$$= 2.16V \quad \text{Ans.}$$

7. PD across 4Ω resistor = 4 × 1.5 = 6V

This is also the voltage drop across the other resistors in group A

$$\text{Current in } 2\Omega \text{ resistor} = \frac{6}{2} = 3\text{A} \quad \text{Ans.}$$

$$\text{Current in } 6\Omega \text{ resistor} = \frac{6}{6} = 1\text{A} \quad \text{Ans.}$$

$$\text{Current in } 8\Omega \text{ resistor} = \frac{6}{8} = 0.75\text{A} \quad \text{Ans.}$$

$$\text{Current in } 4\Omega \text{ resistor} = \frac{6}{4} = 1.5\text{A} \quad \text{Ans.}$$

Total Current = 6.25A Ans.

The equivalent resistance R_B of parallel group B is obtained from

$$\frac{1}{R_B} = \frac{1}{10} + \frac{1}{15} = \frac{3+2}{30} = \frac{5}{30} \text{ or } R_B = \frac{30}{5} = 6\Omega$$

So voltage drop across group B = 6 × 6.25 = 37.5V Ans.

$$\text{Current in } 10\Omega \text{ resistor} = \frac{37.5}{10} = 3.75\text{A} \quad \text{Ans.}$$

$$\text{Current in } 15\Omega \text{ resistor} = \frac{37.5}{15} = 2.5\text{A} \quad \text{Ans.}$$

Check Total current = 6.25A Ans.

Voltage drop across group A = 6V Ans.

Supply Voltage = 6 + 37.5 = 43.5V Ans.

8. Open-circuit voltage of generator = 110V

Voltage drop in generator for 75A = 110 − 108.8 = 1.2V

$$\text{Internal resistance of generator} = \frac{1.2}{75} = 0.016\Omega \quad \text{Ans.}$$

Voltage drop in cables = 108.8 − 105 = 3.8V

$$\text{Resistance of cables} = \frac{3.8}{75} = 0.0507\Omega \quad \text{Ans.}$$

On 'short-circuit' the only limitation to the current, is the resistance of the generator and the cables.

= 0.016 + 0.057 = 0.0667Ω

$$\text{So S.C. current} = \frac{110}{0.0667}$$

= 1650A Ans.

9. Meter voltage drop for full-scale deflection = 1 × 0.12 volts

= 120mV.

Since shunt voltage drop for 300A is 150mV, then the meter resistance would have to be increased by an external resistor of value 0.03Ω. Obtained from 1 amp × (0.03 + 0.12) ohm = 1 × 0.15 = 0.15V or 150mV.

Under this condition the actual current being metered would be 301A. 300A would pass through the shunt and 1A through the ammeter.

Note. The resistor would have to be rated for this current *ie* 1 amp. Thus $1^2 \times 0.03 = 0.03$ watts—dealt with in Chapter 2.

10.

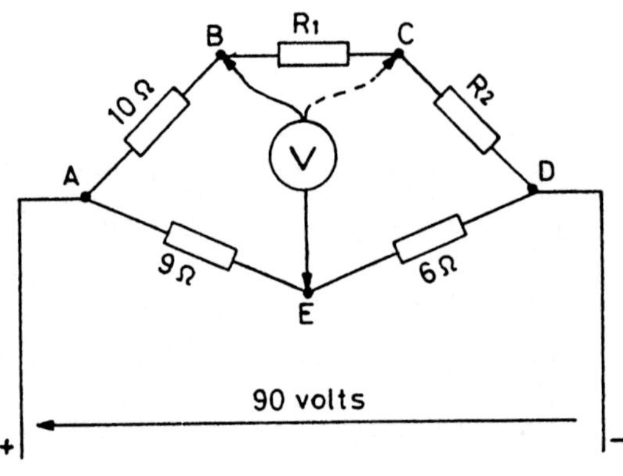

Fig 5

PD across BC = 34 + 6 = 40V, since E is 6V above C and B in turn is 34V above E.

Now the p.d. across AE and ED is proportional to their resistance values.

Thus p.d. across AE $= \frac{9}{15} \times 90 = 54$V with A +ve to E.

and p.d. across ED $= \frac{6}{15} \times 90 = 36$V with E +ve to D.

Since A is positive to E by 54V and B is positive to E by 34V (voltmeter reading), then A is +ve to B by (54 − 34) = 20V.

Similarly since B is above E by 34V and E is +ve to C by 6V then the p.d. across BC = 34 + 6 = 40V. Again E is above D by 36V and above C by 6V so C must be +ve to D by (36 − 6) = 30V.

Thus the p.d. across the resistors are $\left.\begin{array}{l} AB = 20V \\ BC = 40V \\ CD = 30V \end{array}\right\}$ Total 90V.

Also as the branches are series circuits, the ohmic values are proportional to the p.ds.

$$\therefore \frac{\text{p.d. across AB}}{\text{p.d. across BC}} = \frac{10}{R_1} \text{ or } R_1 = \frac{10 \times 40}{20} = 20\Omega \quad \text{Ans.}$$

$$\text{and } \frac{\text{p.d. across CD}}{\text{p.d. across AB}} = \frac{R_2}{10} \text{ or } R_2 = \frac{10 \times 30}{20} = 15\Omega \quad \text{Ans.}$$

$$\text{Current in branch ABCD} = \frac{90}{10 + 20 + 15} = \frac{90}{45} = 2\text{A Ans.}$$

$$\text{Current in branch AED} = \frac{90}{9 + 6} = \frac{90}{15} = 6\text{A} \quad \text{Ans.}$$

$$\text{Supply current} = 2 + 6 = 8\text{A} \quad \text{Ans.}$$

CHAPTER 2

1. Mass to be lifted = $(2 + 0.25) \times 10^3$ kilogrammes
Force to be exerted = $2.25 \times 10^3 \times 9.81$ newtons
Work to be done = $2.25 \times 10^3 \times 9.81 \times 30$ newton metres
$= 66.2175 \times 10^4$Nm or 662.175kJ

$$\text{Power output of hoist} = \frac{\text{work done (joules)}}{\text{time (seconds)}}$$
$$= \frac{662\ 175}{90} = 7.36\text{kW}$$

Power input = $220 \times 50 = 11$kW

$$\text{Efficiency} = \frac{7.36}{11} = 0.6689 \text{ or } 66.9\% \quad \text{Ans.}$$

2. Since battery voltage in the region of 20V is required, the cells must be connected in a series-parallel arrangement. It is apparent that 10 cells in series would give 22V and that this would be the e.m.f. of the battery irrespective of the number of identical parallel banks.

Thus an arrangement of 10 cells in series with three such banks in parallel would be a possible combination. This can be tried.

E.m.f. of battery = e.m.f. of 1 bank = 2.2×10
= 22V

Internal resistance of 1 bank = $0.3 \times 10 = 3\Omega$

$$\text{Internal resistance of battery} = \frac{3}{3} = 1\Omega$$

$$\text{Resistance of 1 lamp} = \frac{V^2}{P} = \frac{20^2}{10} = 40\Omega$$

$$\text{or lamp current} = \frac{P}{V} = \frac{10}{20} = 0.5\text{A}$$

$$\text{and lamp resistance} = \frac{20}{0.5} = 40\Omega$$

$$\text{Resistance of 3 lamps in parallel} = \frac{40}{3} = 13.33\Omega$$

Total resistance of complete circuit
$= 13.33 + 1 = 14.33\Omega$

$$\text{Circuit current} = \frac{22}{14.33} = 1.54\text{A}$$

(b) $\text{Current taken by 1 lamp} = \frac{1.54}{3} = 0.513\text{A}$ Ans.

(a) Voltage drop in battery = $1.54 \times 1 = 1.54$V

Battery terminal voltage = $22 - 1.54 = 20.46$V Ans.

(c) Power loss per cell = (current in 1 bank)^2 $\times$ resistance of a cell

$$= \left(\frac{1.54}{3}\right)^2 \times 0.3$$

$$= 0.513^2 \times 0.3 = 0.079\text{W} \quad \text{Ans.}$$

3. The equivalent head of water can be obtained thus:

A pressure of 15 bars is 15×10^5 newtons per square metre

Specific weight of water is $10^3 \times 9.81$ newtons per cubic metre

$$\text{Then head of water} = \frac{15 \times 10^5}{10^3 \times 9.81} = 152.85\text{m}$$

Force required to lift 12 700 litres or 12.7×10^3 kilogrammes is $12.7 \times 10^3 \times 9.81$ newtons = 124 587N

Work to be done per hour = $124\,587 \times 152.85$ Nm

= 19.064MJ

$$\text{Power output of pump} = \frac{19.064 \times 10^6}{3600} = 5.296\text{kW}$$

Input to pump or output of motor

$$= \frac{5.296}{0.82} = 6.465\text{kW}$$

$$\text{Input to motor} = \frac{6.465}{0.89} = 7.275\text{kW}$$

$$\text{Motor current} = \frac{7275}{220} = 33.1\text{A} \quad \text{Ans.}$$

4. E.m.f. of battery = $4 \times 2.2 = 8.8$V

Terminal voltage of battery = voltage drop across resistor

= $5 \times 1.4\text{V} = 7$V

Voltage drop in battery = $8.8 - 7 = 1.8$V

$$\text{Internal resistance of battery} = \frac{1.8}{1.4} = 1.29\Omega$$

$$\text{Internal resistance of 1 cell} = \frac{1.29}{4} = 0.32\Omega$$

For parallel working:

$$\text{Internal resistance of battery} = \frac{0.32}{4} = 0.08\Omega$$

E.m.f. of battery = e.m.f. of 1 cell = 2.2V

Total circuit resistance = $5 + 0.08 = 5.08\Omega$

Circuit current = $\frac{2.2}{5.08} = 0.43A$ Ans.

5. Output of winch

$= 5 \times 10^3 \times 9.81 \times 36.5$ newton metres per minute

$= 4.905 \times 36.5 \times 10^4$ joules per minute

$= \frac{4.905 \times 36.5 \times 10^4}{60}$ joules per second

$= \frac{179.033}{6} \times 10^3$ watts

$= 29.84kW$

Since the winch is 75 per cent efficient,

the input must be $29.84 \times \frac{100}{75}$ kilowatts

$= 29.84 \times \frac{4}{3} = 39.78kW$

Input to winch = output of motor

∴ power rating of motor = 39.78kW Ans.

If the motor efficiency is taken as 85 per cent,

The electrical input would be $\frac{39.78}{0.85} = 46.8kW$

Current taken from the mains = $\frac{46\,800}{220} = \frac{2340}{11}$

$= 212.7A$ Ans.

6. Lighting load = $100 \times 100 = 10\,000W$ and

$200 \times 60 = 12\,000W$

$= 10 + 12 = 22kW$

Heating load = 25kW

Miscellaneous loads = $30 \times 220 = 6.6kW$

Total load = $22 + 25 + 6.6 = 53.6kW$

Generator output = 53.6kW

Generator input = $\frac{53.6}{0.85} = 63.1kW$

Now generator input = engine output.

So engine must develop 63.1kW Ans.

7. O.C. e.m.f. of battery = 4.3V

O.C. e.m.f./cell = $\frac{4.3}{3} = 1.43V$

Value of load resistor = $\frac{4.23}{0.4} = 10.575\Omega$

Voltage drop in battery = $4.3 - 4.23 = 0.07V$

$$\text{Internal resistance of battery} = \frac{0.07}{0.4} = 0.175\Omega$$

$$\text{Internal resistance of 1 cell} = \frac{0.175}{3} = 0.058\Omega$$

With a cell reversed, the e.m.f. of two cells cancel each other and effective e.m.f. = that of 1 cell = 1.43V.

Let I be the current under this condition.

$$\text{Then } 1.43 = I(10.57 + 0.175) \therefore I = \frac{1.43}{10.75} = 0.134\text{A} \quad \text{Ans.}$$

Note. For the solution, the internal resistance of a cell has been assumed to be the same in both the forward and reverse direction, which is in order in the absence of any further detailed information.

8.

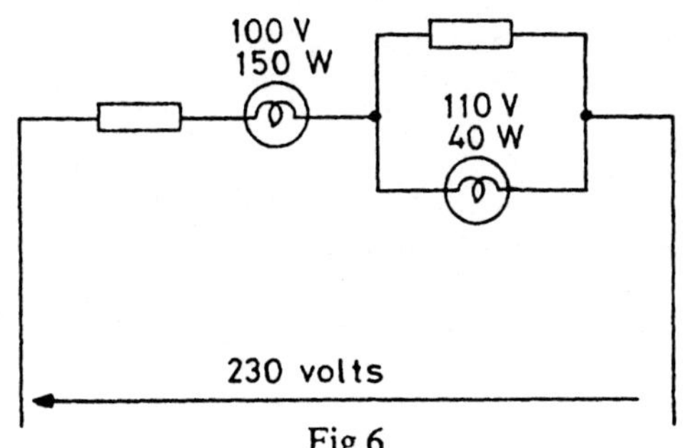

Fig 6

$$\text{Current for 40W lamp} = \frac{40}{110} = 0.363\text{A}$$

$$\text{Resistance of 40W lamp} = \frac{110}{0.363} = 303\Omega$$

$$\text{Current for 150W lamp} = \frac{150}{100} = 1.5\text{A}$$

$$\text{Resistance of 150W lamp} = \frac{100}{1.5} = 66.66\Omega$$

Parallel circuit has to carry 1.5A

$\therefore$ Current in shunt resistor = 1.5 − 0.363 = 1.137A

Voltage drop across shunt = 110V

$$\text{Resistance of shunt} = \frac{110}{1.137} = 96.8\Omega \quad \text{Ans.}$$

110V is dropped across the parallel circuit

100V is dropped across the series lamp

$\therefore$ 230 − 210 = 20V must be dropped across the series resistor which carries 1.5A

$$\therefore \text{Resistance value of series resistor} = \frac{20}{1.5} = 13.3\Omega \quad \text{Ans.}$$

9.

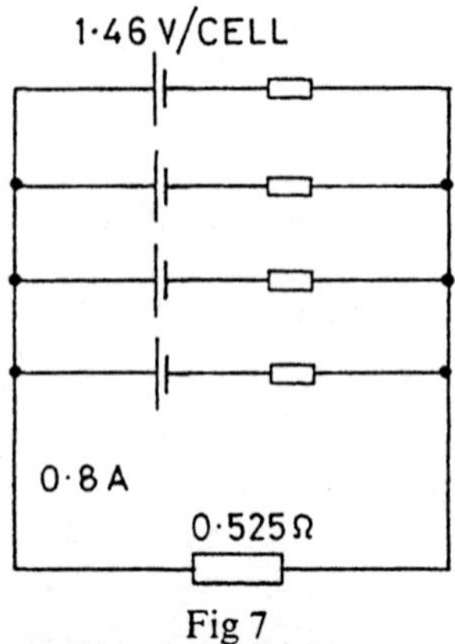

Fig 7

E.m.f. of battery = e.m.f. of 1 cell for parallel working
= 1.46V

Let R_i = the internal resistance of the battery
and R_c = the internal resistance of 1 cell.

The total resistance of the circuit = $0.525 + R_i$

Circuit voltage drop = $0.8\ (0.525 + R_i)$
= applied e.m.f. = 1.46V

$\therefore\ 1.46 = (0.8 \times 0.525) + 0.8R_i$
$= 0.42 + 0.8R_i$ or $0.8R_i = 1.46 - 0.42 = 1.04$

So $R_i = \dfrac{1.04}{0.8} = 1.3\Omega$

Now since the cells are in parallel, then

$$\frac{1}{R_i} = \frac{1}{R_c} + \frac{1}{R_c} + \frac{1}{R_c} + \frac{1}{R_c} = \frac{4}{R_c}$$

or $R_c = 4 \times R_i = 4 \times 1.3$

and internal resistance of 1 cell = 5.2Ω Ans.

10.

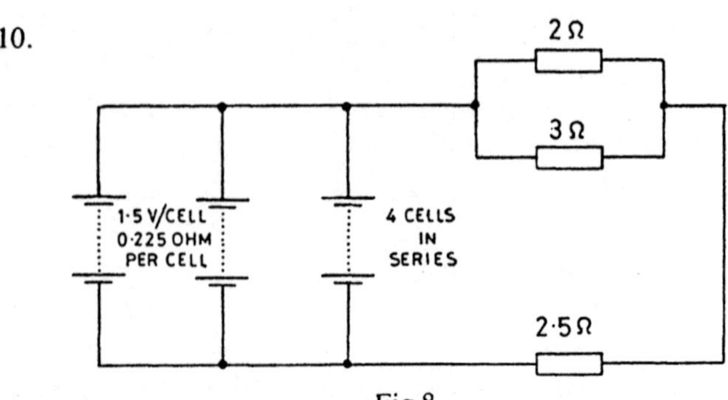

Fig 8

Battery e.m.f. = e.m.f. of 1 bank = $4 \times 1.5 = 6$V

Battery resistance = $\dfrac{\text{resistance of 1 bank}}{3} = \dfrac{4 \times 0.225}{3}$

= 0.3Ω

Load resistance = $2.5 + R$ (resistance of parallel section)

here $\dfrac{1}{R} = \dfrac{1}{3} + \dfrac{1}{2} = \dfrac{5}{6}$ or $R = \dfrac{6}{5} = 1.2\Omega$

Load resistance = $2.5 + 1.2 = 3.7\Omega$

Resistance of complete circuit = $3.7 + 0.3 = 4\Omega$

Circuit current = $\dfrac{6}{4} = 1.5$A

Voltage drop in battery = $0.3 \times 1.5 = 0.45$V

Battery terminal voltage = 6 − internal voltage drop

= $6 - 0.45 = 5.55$V Ans.

Power rating of 2.5Ω resistor = $I^2R = 1.5^2 \times 2.5$

= 5.625W Ans.

Current in 2Ω resistor = $\dfrac{\text{voltage drop}}{\text{resistance}} = \dfrac{\text{total current} \times R}{2}$

= $\dfrac{1.5 \times 1.2}{2} = 0.9$A

Power rating of 2Ω resistor = $0.9^2 \times 2 = 1.62$W Ans.

Current in 3Ω resistor = $1.5 - 0.9 = 0.6$A

Power rating of 3Ω resistor = $0.6^2 \times 3 = 1.08$W Ans.

Energy conversion = energy in external resistors + energy in battery

= time (total wattage of external resistors + battery resistance power wastage)

= $t\,(5.625 + 1.62 + 1.08 + 1.5^2 \times 0.3)$

= $3600\,(8.325 + 0.675) = 32\,400$ joules Ans.

This could also be obtained thus:

Energy = e.m.f. × current × time

= $6 \times 1.5 \times 3600$

= 32 400 joules or 32.4kJ Ans.

CHAPTER 3

1.

(a) Volume $=$ Area $\times$ Length or $A = \dfrac{V}{l} = \dfrac{10 \times 10^3}{100 + 10^3}$
$= 0.1\text{mm}^2$

Then $R = \dfrac{\varrho l}{A} = \dfrac{17 \times 10^{-6} \times 100 \times 10^3}{10^{-1}}$
$= 17\Omega$ Ans.

(b) Area of plate $= 100 \times 100 = 10^4\text{mm}^2$

Thickness of plate $= \dfrac{10 \times 10^3}{10^4} = 1\text{mm}$

This is the length in the expression $R = \dfrac{\varrho l}{A}$

$\therefore R = \dfrac{17 \times 10^{-6} \times 1}{10^4}$ ohms $= 1.7 \times 10^{-3}\,\mu\Omega$ Ans.

Alternatively using $\varrho = 1.7 \times 10^{-8}$ ohm-metres for (a)—as an example.

$$R = \frac{1.7 \times 10^{-8} \times 100}{0.1 \times 10^{-6}}$$
$= 17\Omega$ Ans.

2. Since $R_{20} = R_0(1 + \alpha 20)$ and $R_{60} = R_0(1 + \alpha 60)$

Then $\dfrac{R_{60}}{R_{20}} = \dfrac{R_0\,(1 + \alpha 60)}{R_0\,(1 + \alpha 20)}$

and $R_{60} = \dfrac{R_{20}\,[1 + (60 \times 0.004\,28)]}{[1 + (20 \times 0.004\,28)]}$

or $R_{60} = \dfrac{90(1 + 0.2568)}{1 + 0.0856} = \dfrac{90 \times 1.2568}{1.0856}$ ohms
$= 104.4\Omega$

Current taken by coil at 20°C $= \dfrac{230}{90} = 2.56\text{A}$

At 60°C to keep the current constant, the voltage must be 2.56 $\times$ 104.4 = 267.26V. So the voltage must be raised by 267.26 $-$ 230 = 37.26V Ans.

3. Assuming 1 litre of water to have a mass of 1 kilogramme
Mass of 0.75 litre of water
$= 0.75 \times 1 = 0.75\text{kg}$
Heat required $= 0.75 \times 4.2 \times (100 - 6)$
$= 296\text{kJ}$

The current taken by the heater is $\frac{220}{120} = 1.83A$

and the power rating of the heater $= 220 \times 1.83$
$= 403.3W$

Since the heater is only 84 per cent efficient, only 403.3×0.84 watts are available to heat the water.

$$\therefore \text{ time of heating } = \frac{296 \times 10^3}{403.3 \times 0.84} \text{ seconds}$$
$$= 873s$$
$$= \frac{873}{60} \text{ minutes} = 14\text{min } 33\text{s} \quad \text{Ans.}$$

4. Since $R = \frac{\varrho l}{A}$ then $l = \frac{RA}{\varrho}$

$$l = \frac{15.7 \times \pi \times 0.315^2}{407 \times 10^{-6} \times 4} \text{ millimetres}$$
$$= 2.99 \times 10^3 \text{ millimetres}$$
$$= 2.99 \text{ or } 3m \text{ (approx)} \quad \text{Ans.}$$

5. Since $\frac{R_2}{R_1} = \frac{R_0 (1 + \alpha T_2)}{R_0 (1 + \alpha T_1)}$

$$\therefore R_2 = \frac{R_1 (1 + \alpha T_2)}{1 + \alpha T_1}$$

Then $\frac{R_2}{R_1} \times (1 + \alpha T_1) = 1 + \alpha T_2$

$$\text{So } 1 + \alpha T_2 = \frac{240}{200} [1 + (0.0042 \times 15)]$$
$$= 1.2 (1 + 0.063)$$
$$\text{and } \alpha T_2 = 1.2 + 0.0756 - 1 = 0.2756$$
$$\text{thus } T_2 = \frac{0.2756}{0.0042} = 65.6°C$$
$$\text{Temperature rise} = 65.6 - 15 = 50.6°C \quad \text{Ans.}$$

6.

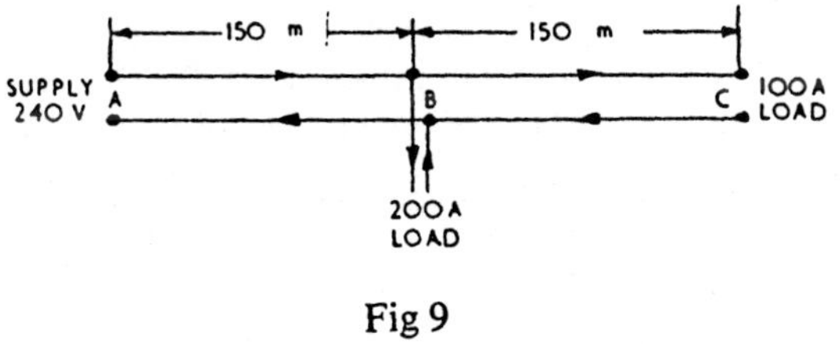

Fig 9

Since the resistance of a cable core 880m long and 50mm² in area is 0.219Ω

then the resistance of a cable core 880m long and 150mm²

in area $= \dfrac{0.219}{3} = 0.073\Omega$

and the resistance of a cable core 150m long and 150mm²

in area $= \dfrac{0.073 \times 150}{880} = 0.0124\Omega$

Current in length AB = 300A

Resistance of length AB = $2 \times 0.0124 = 0.0248\Omega$

Voltage drop in length AB = 300×0.0248

= 7.44V

Voltage at 200A load = $240 - 7.44 = 232.56$V Ans.

Voltage drop in section BC = $\frac{1}{3}$ of that in AB since the current is $\frac{1}{3}$, all else being the same

$\therefore$ Voltage drop in BC $= \dfrac{7.44}{3} = 2.48$V

Voltage at 100A load *ie* at C = $232.56 - 2.48$

= 230.08V Ans.

7. Resistance of 1mm diameter cable $R_1 = \dfrac{\varrho l_1}{A_1}$

$$\text{or } 2.47 = \frac{\varrho \times 100 \times 10^3 \times 4}{\pi \times 1^2}$$

Let R_2 = resistance of $800 \times \dfrac{105}{100} = 840$m of 1.5mm diameter cable

$$\text{Then } R_2 = \frac{\varrho \times 840 \times 10^3 \times 4}{\pi \times 1.5^2}$$

$$\text{So } \frac{R_2}{R_1} = \frac{\varrho \times 840 \times 10^3 \times 4}{\pi \times 1.5^2} \Big/ \frac{\varrho \times 100 \times 10^3 \times 4}{\pi \times 1^2}$$

$$= \frac{840 \times 1^2}{100 \times 1.5^2}$$

$$\text{or } R_2 = 2.47 \times 8.4 \times \left(\frac{1}{1.5}\right)^2$$

$$= 9.22\Omega$$

Since there are 19 strands in parallel, the resistance of the complete cable, being inversely proportional to area, will be reduced by 19

$\therefore$ Resistance of cable $= \dfrac{9.22}{19} = 0.485\Omega$ Ans.

8. Since

$R = R_0 (1 + \alpha T)$ or $15 = 10[1 + (\alpha \times 100)]$

$= 10 + (\alpha \times 1000)$

or $5 = 1000\alpha$ and $\alpha = 0.005$ Ans.

or using the definition

$$\alpha = \frac{\text{increase of resistance per } 0°\text{C rise in temperature}}{\text{resistance at } 0°\text{C}}$$

$$= \frac{15 - 10}{100} \Big/ 10 = \frac{5}{1000} = 0.005 \quad \text{Ans.}$$

Also since

$R = R_0(1 + \alpha T)$ then $30 = 10(1 + 0.005T)$

and $30 = 10 + 0.05T$

or $20 = 0.05T$ and $T = \frac{20}{0.05}$ °C

$= 400$°C Ans.

9. Heat required by brass $= 500 \times 0.39 \times (910 - 15)$ kilojoules

$= 5 \times 39 \times 895$

or energy required by brass $= 174.525$kJ

Energy taken from supply $= \frac{174.525 \times 10^3 \times 100}{80}$

$= 21.816 \times 10^4$ kilojoules

Time taken to expend this energy at the rate of 200kW

$= \frac{21.816 \times 10^4}{200}$ seconds

$= 1091$s

$= 18.18$ min or 18 min 11s Ans.

10. Electrical energy used $= \frac{744}{2} = 372$ units $= 372$kW h

$= 372 \times 3600$ kilojoules

Energy passed to heat water $= 372 \times 360 \times 0.8$ kilojoules

$= 1\,071\,360$kJ

Heat energy received by water $= 1.07 \times 10^6$ kilojoules

Temperature rise of water $= 82 - 16 = 66$

So quantity of water $= \frac{1.07 \times 10^6}{66 \times 4.2}$ kilogrammes

$= 3860$kg

Assuming 1 litre to have a mass of 1kg then quantity of water used $= 3860$ litres Ans.

CHAPTER 4

1. Input to accumulator $= 6 \times 18$ ampere hours

Output from accumulator $= 3.5 \times 28$ ampere hours

Ampere-hour efficiency $= \dfrac{3.5 \times 28}{6 \times 18} = 0.907$

$= 90.7$ per cent Ans.

2. Mass of deposit, $m = zIt$

or $(19.34 - 14.52) \times 10^{-3} = 330 \times 10^{-9} \times I \times 50 \times 60$

or $I = \dfrac{4.82 \times 10^{-3}}{50 \times 60 \times 330 \times 10^{-9}}$ amperes

$= 4.869\text{A}$

Error in reading $= 5.1 - 4.869$

$= 0.231\text{A}$ (high) Ans.

This may be better expressed as a percentage thus:

$= \dfrac{\text{Difference between false and true reading}}{\text{true reading}} \times 100$

$= \dfrac{5.1 - 4.869}{4.869} \times 100$

$= 4.75$ per cent (high) Ans.

3. E.m.f. of battery $= 40 \times 1.9 = 76\text{V}$

or $E_b = 76\text{V}$

Internal resistance of the battery $= 40 \times 0.0025$

$= 0.1\Omega$

Total resistance of circuit $= 1 + 0.1 = 1.1\Omega = R$

For charging $V = E_b + IR$

Thus $90 = 76 + (I \times 1.1)$

or $I = \dfrac{90 - 76}{1.1} =$ amperes

$= 12.72\text{A}$ Ans.

4. Area of nickel deposited $= \pi \times 100 \times 10^{-3} \times 150 \times 10^{-3}$

$= 0.0471\text{m}^2$

Volume of nickel deposited $= 47.1 \times 10^{-3} \times 0.5 \times 10^{-3}$

$= 23.55 \times 10^{-6}$ cubic metres

Mass of nickel deposited, $m = 23.55 \times 10^{-6} \times 8.6 \times 10^3$

$= 0.202\ 53\text{kg}$

$$\text{Now } m = zIt \text{ so } 0.202\,53 = 302 \times 10^{-9} \times I \times 8 \times 3600$$
$$\text{or } 202.53 \times 10^{-3} = 30.2 \times 8 \times 36 \times 10^{-6}I$$
$$\text{giving } I = \frac{202.53 \times 10^{3}}{30.2 \times 8 \times 36}$$
$$= 23.3\text{A} \quad \text{Ans.}$$

5. Discharge ampere hours = $6 \times 12 = 72$

Charge ampere hours = $4 \times 22 = 88$

$$\text{Ampere-hour efficiency} = \frac{72}{88}$$
$$= 0.82 \text{ or } 82 \text{ per cent} \quad \text{Ans.}$$

Discharge watt hours per cell = $6 \times 12 \times 1.2$ = 86.4

Charge watt hours per cell = $4 \times 22 \times 1.5$ = 132

$$\text{Watt-hour efficiency} = \frac{86.4}{132} = 0.65$$

or 65 per cent Ans.

6. Battery voltage at start of charge = $80 \times 1.8 = 144\text{V} = E_{b1}$

Battery voltage at end of charge = $80 \times 2.4 = 192\text{V} = E_{b2}$

No battery resistance is given and is therefore neglected.

Let R_1 = control resistance at start of charge

Then $V = E_{b1} + IR_1$ or $230 = 144 + 5R_1$

Thus $5R_1 = 230 - 144 = 86$

$$R_1 = \frac{86}{5} = 17.2\Omega \text{ (maximum value)} \quad \text{Ans.}$$

Let R_2 = control resistance at end of charge.

Then $230 = 192 + 5R_2$ or $5R_2 = 230 - 192 = 38$

$$R_2 = \frac{38}{5} = 7.6\Omega \text{ (minimum value)} \quad \text{Ans.}$$

$$\text{Charging time} = \frac{60}{5} = 12 \text{ hours (approx)}$$

Say 13h to allow for losses Ans.

7. Area of deposit = $2 \times 50 \times 150 = 15\,000\text{mm}^2$

Volume of deposit = $15 \times 10^{3} \times 10^{-6} \times 0.05 \times 10^{-3}$ = $0.75 \times 10^{-6}\text{m}^3$

Mass of deposit, $m = 0.75 \times 10^{-6} \times 8800$ = 0.0066kg

$$\text{Now } m = zIt \quad \text{so } I = \frac{m}{zt}$$

$$\text{or } I = \frac{6.6 \times 10^{-3}}{330 \times 10^{-9} \times 30 \times 60}$$

Thus I = 11.1A Ans.

8. Discharge ampere hours = 4 × 40 = 160
Charge ampere hours = 8 × 24 = 192

Ampere-hour efficiency of battery $= \frac{160}{192} = 0.833$
or 83.3 per cent Ans.

Discharge watt hours = 4 × 40 × 1.93 × 40
Charge watt hours = 8 × 24 × 2.2 × 40

Watt-hour efficiency of battery $= \frac{4 \times 40 \times 1.93 \times 40}{8 \times 24 \times 2.2 \times 40}$
= 0.731 or 73.1 per cent Ans.

9. Voltage drop in leads = 10 × 1 = 10V
Voltage drop due to battery internal resistance
= 10 × 0.01 × 30 = 3V

At start of charge, if R_1 is the external resistance
Then $200 = (30 \times 1.85) + 10 + 3 + 10R_1$
or $200 = 55.5 + 13 + 10R_1$
and $200 - 68.5 = 10R_1$ thus $R_1 = \frac{131.5}{10} = 13.15\Omega$

At end of charge, if R_2 is the external resistance
Then $200 = (30 \times 2.2) + 10 + 3 + 10R_2$
or $200 = 66 + 13 + 10R_2$
and $200 - 79 = 10R_2$ thus $R_2 = \frac{121}{10} = 12.1\Omega$

At start 13.15Ω are needed Ans.
At end 12.1Ω are needed Ans.

10. Here $m = zIt$ or $t = \frac{m}{zI} = \frac{4.2 \times 10^{-3}}{330 \times 10^{-9} \times 3.5}$ seconds

Thus t = 3636s = 60.6 min *ie* 60 min 36s Ans.

From the second law of electrolysis or by proportion:

$$\frac{\text{Mass of hydrogen liberated}}{\text{Mass of copper liberated}} = \frac{\text{Chemical Equivalent of hydrogen}}{\text{Chemical Equivalent of copper}}$$

Thus mass of hydrogen $= \frac{1 \times 4.2}{31.8} = 0.1321\text{g}$ Ans.

CHAPTER 5

1. $F = BIl$ newtons

$= 0.25 \times 100 \times 1 = 25$ newtons per metre length

Ans.

2. Magnetomotive force $F = 4 \times 250 = 1000\text{At}$

(a) Magnetising Force $H = \dfrac{F}{l}$ = m.m.f. per metre length

$$= \frac{1000}{500 \times 10^{-3}}$$

$= 2000\text{At/m}$ Ans.

(b) Flux Density $B = \mu_0 \times H = 4 \times \pi \times 10^{-7} \times 2000$

$= 8 \times \pi \times 10^{-4}$ teslas

Cross-sectional area of ring $= 400 \times 10^{-6}$ square metres

$\therefore$ Flux $\Phi = B \times A = 8 \times \pi \times 10^{-4} \times 400 \times 10^{-6}$

$= 1.0048 \times 10^{-6}$ webers

or $1.0048\mu\text{Wb}$ Ans.

3. Magnetomotive force F produced $= 3200 \times 1$

$= 3200\text{At}$

The magnetising force H or m.m.f./m $= \dfrac{F}{l} = \dfrac{3200}{800 \times 10^{-3}}$

$= 4000\text{At/m}$

Also since $B = \mu_0 \times H$

$B = 4 \times \pi \times 10^{-7} \times 4000$

$= 16 \times \pi \times 10^{-4}$ teslas

Area of solenoid $= \dfrac{\pi \times d^2}{4} = \dfrac{\pi \times 20^2 \times 10^{-6}}{4} = \pi \times 10^{-4}$

So $\Phi = B \times A = 16 \times \pi \times 10^{-4} \times \pi \times 10^{-4}$ webers

$= 1.58 \times 10^{-6}$ webers or $1.58\mu\text{Wb}$ Ans.

4. Magnetising force H of a long, straight conductor

$= \dfrac{I}{2\pi r} = \dfrac{2000}{2 \times \pi \times 0.8}$ ampere-turns/metre

or H at conductor X, due to current in Y,

$= \dfrac{1000}{0.8 \times \pi}$ ampere-turns/metre

and B at conductor X due to current in Y $= \mu_0 \times H$

$$= \frac{4 \times \pi \times 10^{-7} \times 1000}{0.8 \times \pi}$$

$$= \frac{10^{-3}}{2} \text{ teslas}$$

So $F = BIl$ newtons

$$= \frac{10^{-3}}{2} \times 2000 \times 1 = 1 \text{ newton/metre length}$$

Ans.

5. Current to give full-scale deflection

$$= \frac{50 \times 10^{-3}}{10} = 5 \times 10^{-3}\text{A}$$

Force exerted on 1 conductor

$$= BIl = 0.1 \times 5 \times 10^{-3} \times 25 \times 10^{-3}$$
$$= 12.5 \times 10^{-6} \text{ newtons}$$

Force exerted on all conductors on both sides of the coil

$$= 100 \times 2 \times 12.5 \times 10^{-6}$$
$$= 2500 \times 10^{-6} \text{ newtons}$$

Torque exerted by coil

$$= \text{Force} \times \text{radius}$$
$$= 2500 \times 10^{-6} \times \frac{30}{2} \times 10^{-3}$$
$$= 37.5 \times 10^{-6} \text{ newton metres}$$

Therefore the Controlling Torque of the spring

$= 37.5 \times 10^{-6}$Nm Ans.

or $= 37.5\mu$Nm Ans.

6. Flux density B in air gap $= \dfrac{0.05}{650 \times 10^{-6}}$

$$= \frac{5 \times 10^2}{6.5} \text{ teslas}$$

Also $B = \mu_0 \times H$ $\therefore$ $H = \dfrac{B}{\mu_0} = \dfrac{5 \times 10^2}{6.5 \times 4 \times \pi \times 10^{-7}}$

$= 6.12 \times 10^7$ ampere-turns/metre

Air gap $= 3\text{mm} = 3 \times 10^{-3}$

$\therefore$ Required ampere-turns $= 6.12 \times 10^7 \times 3 \times 10^{-3}$

$= 183\,600$At Ans.

7. $F = BIl$ newtons $= 0.6 \times 150 \times 1 = 90$N/m Ans.

Assuming current flowing away from the observer, then the force acts from right to left to move the conductor horizontally.

8. $F = BIl$ newtons
$= 0.5 \times 25 \times 400 \times 10^{-3}$
$= 5N$ Ans.

9. Force on 1 conductor $= 0.6 \times 0.8 \times 250 \times 10^{-3}$
Force on 800 conductor $= 0.6 \times 8 \times 250 \times 10^{-3} \times 8 \times 10^{2}$
$= 9.6 \times 10^{2}$ newtons
Torque on armature $= 9.6 \times 10^{2} \times 100 \times 10^{-3}$
$= 96$N m Ans.

Power developed is given by $\frac{2\pi NT}{60}$ watts

$$= \frac{2 \times \pi \times 1000 \times 96}{60}$$

$= 10.05$kW Ans.

10. Magnetising Force H of a long straight conductor

$$= \frac{I}{2\pi r} \text{ ampere-turns/metre}$$

$$= \frac{250}{2 \times \pi \times 25 \times 10^{-3}}$$

$$= \frac{10^{4}}{2 \times \pi}$$

Also $B = \mu_0 \times H$

or $B = 4 \times \pi \times 10^{-7} \times \frac{10^{4}}{2 \times \pi}$

$= 2 \times 10^{-3}$ teslas

Again $F = BIl = 2 \times 10^{-3} \times 250 \times 1$
$= 0.5$ newtons

Mutual Force per metre run $= 0.5$N Ans.

CHAPTER 6

1. (a) Total m.m.f., $F = 5 \times 500 = 2500\text{At}$

$$\text{Mean circumference} = \pi d = \pi \times 300 \times 10^{-3} = 0.942\text{m}$$

$$\text{So magnetising force, } H = \frac{F}{l} = \frac{2500}{0.942} = 2654\text{At/m} \quad \text{Ans.}$$

(b) Since $\mu_o = \dfrac{B}{H}$ then $B = \mu_o H$

$$\text{and } B = 4 \times \pi \times 10^{-7} \times 2654 = 0.0033\text{T} = 3.3\text{mT} \quad \text{Ans.}$$

(c) Total flux, $\Phi = BA$

$$= 0.0033 \times 1000 \times 10^{-6} \text{ weber}$$
$$= 3.3 \times 10^{-6}\text{Wb or } = 3.3\mu\text{Wb} \quad \text{Ans.}$$

2. $$B = \frac{\Phi}{A} = \frac{500 \times 10^{-6}}{400 \times 10^{-6}} = 1.25\text{T}$$

Also since $B = \mu H = \mu_r \mu_o H$ then

$$H = \frac{1.25}{2500 \times 4 \times \pi \times 10^{-7}} = \frac{1.25}{\pi \times 10^{-3}} \text{ ampere-turns/metre}$$

Length of iron $= 250 \times 10^{-3}$ metres

$$\text{So total m.m.f., } F = \frac{250 \times 10^{-3} \times 1.25}{\pi \times 10^{-3}}$$

or $F = 99.7\text{At}$

Required ampere-turns = 99.7, say 100. Ans.

3. (a) $B = \dfrac{\Phi}{A}$ then $B = \dfrac{400 \times 10^{-6}}{500 \times 10^{-6}} = \dfrac{400}{500} = 0.8T$

Also, as H is given by $\dfrac{\text{total magnetomotive force}}{\text{length}}$

$$\text{Then } H = \frac{F}{l} = \frac{500}{1} = 500\text{At/m}$$

Also, since $B = \mu H = \mu_o \mu_r H$ then

$$\mu_r = \frac{B}{\mu_o H} = \frac{0.8}{4 \times \pi \times 10^{-7} \times 500}$$

or relative permeability = 1275 Ans.

(b) Reluctance $= \dfrac{\text{Length}}{\mu \times \text{Area}} = \dfrac{l}{\mu_o\mu_r \times A}$ ampere-turns/weber

$$= \frac{1}{4 \times \pi \times 10^{-7} \times 1275 \times 500 \times 10^{-6}}$$

$= 1.25\text{MA/Wb}$ Ans.

4. (a) H = ampere-turns per metre $= \dfrac{F}{l} = \dfrac{400 \times 2.5}{1.25}$

Thus $H = \dfrac{1000}{1.25} = 800\text{At/m}$

Also $B = \dfrac{\Phi}{A} = \dfrac{0.000\,75}{1500 \times 10^{-6}} = 0.5\text{T}$

Again $B = \mu H$ or $\mu = \dfrac{B}{H} = \dfrac{0.5 \times 1.25}{1000}$

Also $\mu = \mu_r\,\mu_o$

$\therefore \quad \mu_r = \dfrac{\mu}{\mu_o} = \dfrac{0.625}{1000 \times 4\pi \times 10^{-7}}$

Thus relative permeability = 497.5 Ans.

(b) Reluctance, $S = \dfrac{l}{\mu A} = \dfrac{1.25 \times 10^3}{0.625 \times 1500 \times 10^{-6}}$ ampere-turns/weber

$= 1.33\text{MA/Wb}$ Ans.

(c) Since $F = Hl$

$= 800 \times 1.25 = 1000\text{At}$ Ans.

5. Area of air gap $= 1200 \times 10^{-6}$ square metres

$\therefore B_A = \dfrac{1.13 \times 10^{-3}}{12 \times 10^{-4}} = \dfrac{11.3}{12}$ teslas

also since $B = \mu_o H$

then H for air $= \dfrac{B}{\mu_o} = \dfrac{11.3}{12 \times 4\pi \times 10^{-7}}$

$= 75 \times 10^4$ ampere-turns/metre

The magnetomotive force for the air gaps is given by:

$75 \times 10^4 \times 2 \times 2 \times 10^{-3}$ ampere-turns

$= 3000\text{At}$

Area of iron is the same as for the air gaps

B value of the iron is the same

or $B_1 = \dfrac{1.13 \times 10^{-3} \times 10^4}{12} = 0.942\text{T}$

Using the graph of Fig 10, we see that, for a flux density of 0.942T, the ampere-turns per metre length of the iron = 850.

Since length of iron path = 0.6m

$\therefore$ M.M.F. for iron $= 0.6 \times 850 = 6 \times 85 = 510\text{At}$

Total magnetomotive force required = 3000 + 510 or 3510At Ans.

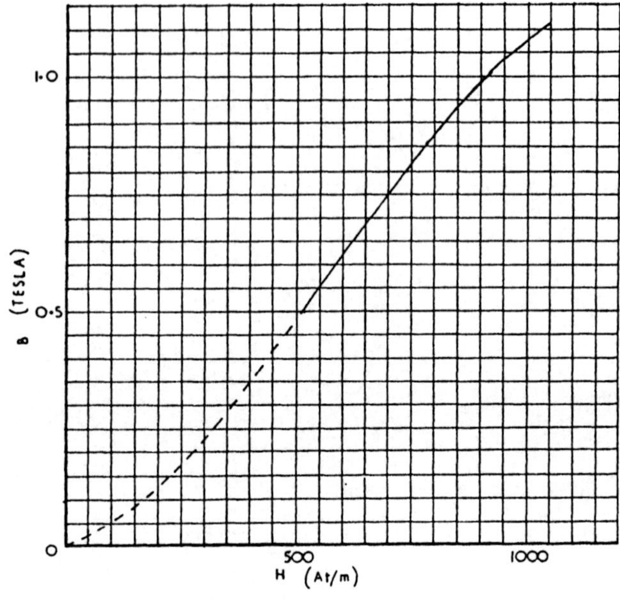

Fig 10

6. Circumference of flux path $= \pi \times 0.2 = 0.628$m

Length of air gap $= 2 \times 10^{-3} = 0.002$m

,, ,, ,, iron $= 0.626$m

This problem is best solved by trial and error thus:

Assume a flux density of 0.5T in the air gap and iron since these are of the same cross-sectional area. Then using the graph of Fig 10;

$$\text{M.M.F. for iron} = 520 \times 0.626 = 326\text{At}$$

$$\text{Since } B = \mu_0 H \quad \therefore H = \frac{B}{\mu_0} = \frac{0.5}{4\pi \times 10^{-7}}$$

$$\frac{0.5 \times 2 \times 10^{-3}}{4\pi \times 10^{-7}}$$

$$= 795\text{At}$$

Total m.m.f. would be $(326 + 795) = 1121$At. Thus too low a flux density has been assumed.

Again, assume a flux density 0.6T, then:

$$\text{M.M.F. for iron} = 585 \times 0.626 = 365.21\text{At}$$

$$\text{M.M.F. for air} = \frac{0.6 \times 2 \times 10^{-3}}{4\pi \times 10^{-7}} \text{ or } \frac{6}{5} \text{ of that required for 0.5T}$$

$$= \frac{6}{5} \times 795 = 954\text{At}$$

Total m.m.f. would be 365.2 + 954 = 1319At—still too low.
Assume a flux density of 0.7T. Then:

$$\text{M.M.F. for iron} = 660 \times 0.626 = 413.16\text{At}$$

$$\text{M.M.F. for air} = \frac{7}{5} \times 795 = 1113\text{At}$$

Total m.m.f. would be 413 + 1113 = 1526At

Thus for an exciting ampere-turn value of 3 × 500 = 1500, the estimated flux density in the air gap would be a little less than 0.7T Ans.

7. Since the B value in the cores is to be 1.2T then the At/m required will be 650. This is seen from the graph of Fig 11. The total magnetomotive force for the cores will be:

$$2 \times 160 \times 10^{-3} \times 650 = 208\text{At}.$$

In the yokes, the flux is the same as that for the cores and the flux density will therefore be different, as the areas are different.

$$\text{Thus flux, } \Phi = 1.2 \times \frac{\pi}{4} \times 50^2 \times 10^{-6} \text{ weber}$$

$$B \text{ in yokes} = 1.2 \times \frac{\pi}{4} \times \frac{25 \times 10^{-4}}{47 \times 47 \times 10^{-6}} \text{ teslas}$$

$$\therefore B = 1.066\text{T}$$

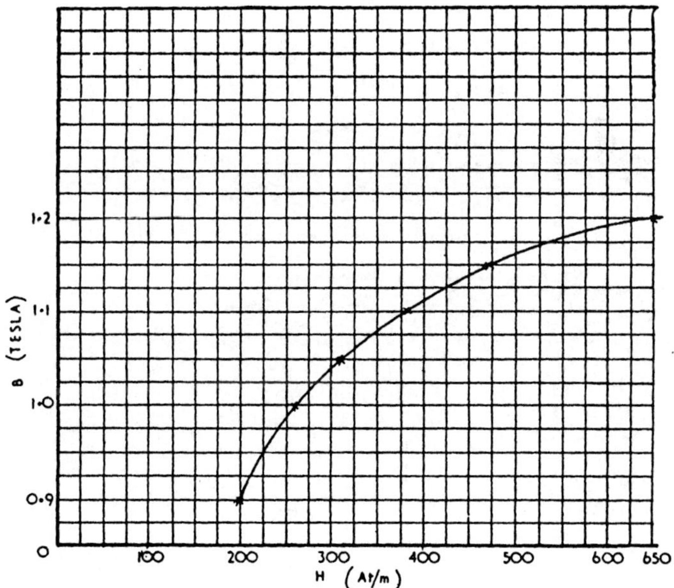

Fig 11

From the graph, the At/m for a density of 1.066T is 330

Mean length of flux path in yokes = (2 × 130) + (2 × 47)
= 354mm = 0.354m

Total m.m.f. for the yokes = $330 \times 354 \times 10^{-3}$
= 116.8At

Total m.m.f. for complete magnetic circuit
= 208 + 116.8
= 324.8 say 325At Ans.

8. (a) Pull of magnet
= 196.2N or 98.1N per contact face

$$\text{Also Pull} = \frac{B^2A}{2\mu_o} = \frac{B^2A}{2 \times 4\pi \times 10^{-7}} \text{ newtons}$$

$$\therefore 98.1 = \frac{B^2A}{8\pi \times 10^{-7}}$$

$$\text{Whence } B^2 = \frac{98.1 \times 8\pi \times 10^{-7} \times 4}{\pi \times 15 \times 15 \times 10^{-6}}$$

$$= 1.396$$

$$\text{and } B = \sqrt{1.396} = 1.185\text{T}$$

From the graph of Fig 11, for a B value of 1.185T, the H value = 560At/m.

$$\text{Area of one contact face} = \frac{\pi}{4} \times 15^2 \times 10^{-6} \text{ square metre}$$

$$\text{Flux, } \Phi = BA = 1.185 \times \frac{\pi}{4} \times 225 \times 10^{-6} \text{ weber}$$

$$= 2.09 \times 10^{-4}\text{Wb}$$

Since B for horse-shoe magnet = 1.185T
then, from Fig 11, the H value = 560 At/m

$$\text{Length of magnet path} = \pi \times \frac{115}{2} = 180.5 \text{ mm} = 0.1805\text{m}.$$

Note. Mean circumference of ring
= 2π × (radius of ring)
= 2π × (50 + radius of rod)
= 2π × (50 + 7.5)
= π × 115 millilitres

Then m.m.f. for magnet = 560 × 0.1805
= 101.1At

$$B \text{ value for armature} = \frac{\text{Flux}}{\text{Area}} = \frac{2.09 \times 10^{-4}}{15^2 \times 10^{-6}}$$

$$= 0.932\text{T}$$

and H value = 215At/m (from Fig 11)

Length of armature path = (115 + 15) = 130mm
= 130×10^{-3} metre

So m.m.f. required $= 215 \times 130 \times 10^{-3}$
$= 27.95$At

Total magnetomotive force required $= 101.1 + 27.95 =$ 129At

$$\text{Current} = \frac{129}{480} = 0.268\text{A} \quad \text{Ans.}$$

(b) In the air gap $B = 1.15$ teslas

$$\therefore \text{ Flux, } \Phi = 1.15 \times \frac{\pi}{4} \times 15^2 \times 10^{-6}$$
$$= 2.03 \times 10^{-4} \text{ weber}$$

$$\text{Flux density in core} = \frac{2.03 \times 10^{-4}}{\frac{\pi \times 15^2}{4} \times 10^{-6}} = 1.15\text{T}$$

and from curve, H value $= 470$At/m

Length of core path $= \pi \times 57.5 = 180.5$mm
$= 0.1805$m

M.M.F. for core $= 0.1805 \times 470 = 84.8$At

$$\text{Flux density in armature} = \frac{2.03 \times 10^{-4}}{15^2 \times 10^{-6}}$$
$$= 0.905\text{T}$$

From curve H value $= 205$At/m

Length of armature path $= 130$mm $= 0.13$m

M.M.F. for armature $= 0.13 \times 205 = 26.65$At

Flux density in 1 air gap $= 1.15$T, But $B = \mu_o H$

$$\therefore H = \frac{B}{\mu_o} = \frac{1.15}{4\pi \times 10^7} = \frac{1.15 \times 10^7}{12.56} \text{ ampere-turns/metre}$$

M.M.F. for 2 air gaps

$$= \frac{2 \times 0.5 \times 10^{-3} \times 1.15 \times 10^7}{12.56}$$

$= 9.125 \times 10^2$ ampere-turns $= 912.5$At

Total m.m.f. for circuit $= 84.8 + 26.65 + 912.5$
$= 1023.95$, say 1024At Ans.

9. (a) $$\text{Area of air gap} = \frac{\pi 100^2}{4} \times 10^{-6}$$

$$= \frac{\pi \times 10^{-2}}{4} \text{ square metre}$$

$$\text{Volume of air gap} = \frac{\pi \times 10^{-2}}{4} \times 2.5 \times 10^{-3}$$

$$= \frac{\pi}{16} \times 10^{-4} \text{ cubic metre}$$

$$\text{Flux density in gap} = \frac{0.004 \times 4}{\pi \times 100^2 \times 10^{-6}}$$
$$= 0.508\text{T}$$

$$\text{Energy stored in joules} = \frac{B^2}{2\mu_o} \times \text{Volume}$$
$$= \frac{0.508^2}{2 \times 4\pi \times 10^{-7}} \times \frac{\pi}{16} \times 10^{-4}$$
$$= 2\text{J} \quad \text{Ans}$$

(b) $$\text{Pull (newtons)} = \frac{B^2A}{2\mu_o} = \frac{0.508^2 \times \pi \times 10^{-2}}{2 \times 4\pi \times 10^{-7} \times 4}$$
$$= 806\text{N} \quad \text{Ans.}$$

10. Air gap. Useful flux = 0.05Wb/pole

$$\text{Flux density in air gap} = \frac{0.05}{60\,000 \times 10^{-6}}$$
$$= 0.833\text{T}$$
$$\text{Also } B = \mu_o H$$
$$\therefore H \text{ value for air} = \frac{B}{\mu_o} = \frac{0.833}{4\pi \times 10^{-7}}$$
$$= 66.2 \times 10^4 \text{ ampere-turns/metre}$$
$$\text{M.M.F. for air gap} = 66.2 \times 10^4 \times 5 \times 10^{-3}$$
$$= 3310\text{At}$$

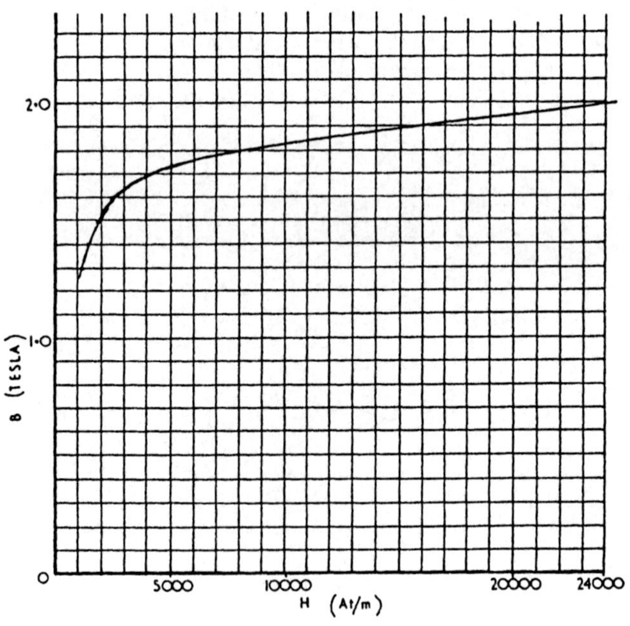

Fig 12

Pole. Total flux = $0.05 \times 1.2 = 0.06$Wb

$$\text{Flux density in pole} = \frac{0.06}{40\,000 \times 10^{6}}$$

$$\text{or } B = \mathit{1.5T}$$

From the magnetic characteristic, plotted for Fig 12, a flux density (*B*) of 1.5T gives an *H* value of 2000At/m.

Thus m.m.f. required for pole = $2000 \times 250 \times 10^{-3} = 500$At

Teeth. Total flux = 0.05Wb (same as the gap)

$$\text{Flux density in teeth} = \frac{0.05}{25\,000 \times 10^{-6}} = 2\text{T}$$

From the characteristic, a flux density of 2T, gives an *H* value of 24 000At/m.

Thus m.m.f. required for teeth = $24\,000 \times 45 \times 10^{-3}$
= 1080At

Total field-coil m.m.f. = 3310 + 500 + 1080
= 4890 At Ans.

CHAPTER 7

1. Dynamic Induction $E = Blv$

$$= 40 \times 10^{-6} \times 1.4 \times \frac{100 \times 10^3}{3600} \text{ volts}$$

$$= 1.55 \times 10^{-3} \text{ volts}$$

$$= 0.001\,55\text{V or } 1.55\text{mV} \quad \text{Ans.}$$

2. In 1 revolution, the flux cut $= 6 \times 64 \times 10^{-3}$ webers

The number of revolutions per second $= \frac{1000}{60} = \frac{100}{6}$

$\therefore$ Flux cut per second $= 6 \times 64 \times 10^{-3} \times \frac{10^2}{6}$

$= \text{generated voltage} = 6.4\text{V}$

The conductors in series per parallel path

$$= \frac{468}{6} = 78$$

So voltage per parallel path $= 6.4 \times 78 = 499.2\text{V}$

$= \text{e.m.f. of machine}$

Current per conductor $=$ current per parallel path

$= 50\text{A}$

Current in six parallel paths $= 50 \times 6 = 300\text{A}$

So power developed $= \frac{499.2 \times 300}{1000}$

$= 149.76\text{kW}$ Ans.

Check $E = \frac{Z\Phi N}{60} \times \frac{P}{A}$

Note. This formula is introduced in Chapter 12. Its use is shown here.

Thus $E = \frac{468 \times 64 \times 10^{-3} \times 1000}{60} \times \frac{6}{6}$

$= 499.2\text{V}$ Ans.

3. Change of flux $= (30 - 2)10^{-3}$ weber $= 28\text{mWb}$

Induced e.m.f. $=$ rate of change of flux-linkages

$$= \frac{2000 \times 28 \times 10^{-3}}{0.12}$$

$$= 466.6\text{V} \quad \text{Ans.}$$

4. Diameter of armature = 0.2m

Circumference = $\pi d = \pi \times 2 \times 10^{-1} = 0.628$m

In 1 second the armature turns $\frac{500}{60}$ revolutions

∴ In 1 second a coil side travels $\frac{500}{60} \times 0.628$ metres

or $v = 5.233$m/s

So $E = Blv = 1.2 \times 2 \times 0.4 \times 5.233$ volts

$= 5.024$V Ans.

5. Let E_1 Φ_1 and N_1 be the values under the original conditions and E_2 Φ_2 and N_2 be the values under the final condition.

Here N_1 and N_2 are the speed conditions.

Also for a generator $E \propto \Phi N$ or $E = k\Phi N$

We know that $\Phi \propto B$ and $N \propto v$ thus the more general form $E \propto \Phi N$ (introduced later) can be used. k is a constant.

$$\therefore \frac{E_2}{E_1} = \frac{k\Phi_2 N_2}{k\Phi_1 N_1} \text{ or } E_2 = \frac{E_1 \Phi_2 N_2}{\Phi_1 N_1}$$

$$\text{and } E^2 = \frac{200 \times 19 \times 10^{-3} \times 1100}{20 \times 10^{-3} \times 1000}$$

$E_2 = 209$V Ans.

6. Diameter of coil = 0.2m

Circumference of coil = $\pi \times 0.2 = 0.628$m

$$\text{Speed} = \frac{1200}{60} = 20 \text{ rev/s}$$

∴ velocity of coil side $= 20 \times 0.628$

$= 12.56$m/s

Now $E = Blv$ volts

$= 0.02 \times 0.3 \times 12.56$ volts per conductor

Total e.m.f. $= 2 \times 3 \times 10^{-3} \times 12.56 \times 400$

$= 30.144$V

The above is based on the conductors cutting the field at right angles and therefore the maximum value of e.m.f. generated is 30.144V Ans.

Time for 1 revolution $= \frac{1}{20}$ seconds. In one revolution a cycle is generated. So frequency of generated e.m.f. = 20 cycles per second or, 20 hertz Ans.

7. Average e.m.f. in volts = rate of change of flux-linkages

Change of flux = $(4 - 1.5)10^{-3}$
$= 2.5 \times 10^{-3}$ webers

Time for change = 0.04s

So rate of change of flux-linkages $= 1200 \times \dfrac{2.5 \times 10^{-3}}{4 \times 10^{-2}}$
= 75V Ans.

8. Flux per pole $= 0.09 \times 0.92 = 8.28 \times 10^{-2}$ webers
In one revolution a conductor cuts $4 \times 8.28 \times 10^{-2}$
$= 33.12 \times 10^{-2}$ webers

Also in one second the armature revolves $\dfrac{600}{60} = 10$ times

So by 1 conductor, the flux cut per second
$= 33.12 \times 10^{-2} \times 10 = 3.312$Wb

and the induced voltage per conductor = 3.312V Ans.
The number of armature conductors is $2 \times 210 = 420$
These are arranged in 4 parallel paths.

There are thus $\dfrac{420}{4} = 105$ conductors in series per parallel path

The e.m.f. of 1 parallel path $= 105 \times 3.312 = 347.76$V
The generated e.m.f. of the machine = e.m.f. of 1 parallel path
= 347.76V Ans.

9. (a) Ampere-turns of solenoid $= 400 \times 6 = 2400$At
The magnetising force H at the centre = ampere-turns/metre
$= \dfrac{2400}{1.5} = 1600$At/m

The flux density B at the centre of the solenoid and small coil
$= \mu_0 H = 4 \times \pi \times 10^{-7} \times 1600 = 64 \times \pi \times 10^{-5}$ tesla

Area of small coil $= \dfrac{\pi d^2}{4} = \dfrac{\pi \times (10 \times 10^{-3})^2}{4}$
$= \dfrac{\pi \times 10^{-4}}{4}$ square metre

So flux linked $= 64 \times \pi \times 10^{-5} \times \dfrac{\pi \times 10^{-4}}{4}$ weber
$= 0.158\mu$Wb Ans.

(b) Average induced e.m.f. = rate of change of flux-linkages
$= \dfrac{50 \times 16 \times \pi^2 \times 10^{-9}}{50 \times 10^{-3}}$
= 0.158mV Ans.

10. Coil A. Associated flux $\Phi = 18 \times 10^{-3}$ weber

Associated flux-linkages during reversal = turns × flux decrease to zero and then its build up to full value Φ in reversed direction.

$= 1000\,[0.018 - (-0.018)] = 1000(0.018 + 0.018)$
$= 1000 \times 2 \times 0.018 = 36$ weber-turns

Time of reversal = 0.1s

and induced e.m.f. = rate of change of flux-linkages

$= \dfrac{36}{0.1} = 360\text{V}$ (Average Value) Ans.

Coil B. Only 80 per cent flux is associated and turns are 500

∴ Associated flux-linkages during reversal

$= 500 \times 0.8 \times 2 \times 0.018$
$= 14.4$ weber-turns

Induced e.m.f. $= \dfrac{14.4}{0.1} = 144\text{V}$ (Average Value) Ans.

Alternatively:

Proportion of e.m.f. in Coil B to e.m.f. in Coil A

$= 360 \times \dfrac{500}{1000} = 180\text{V}$, if full flux is associated

For only 80 per cent flux, e.m.f. is reduced in proportion

$= 180 \times 0.8 = 144\text{V}$ Ans.

CHAPTER 8

1. For a series combination, the equivalent capacitance is given by C, where $\frac{1}{C} = \frac{1}{0.02} + \frac{1}{0.04}$

$$\text{or } C = \frac{0.04}{3} = 0.0133\mu\text{F} \quad \text{Also } Q = CV$$

$$\therefore \quad Q = 0.0133 \times 10^{-6} \times 10^{2} \text{ coulombs}$$

$$\text{Then } V_1 = \frac{1.33 \times 10^{-6}}{0.02 \times 10^{-6}} = 66.7\text{V and } V_2 = \frac{1.33 \times 10^{-6}}{0.04 \times 10^{-6}}$$

$$= 33.3\text{V}.$$

The voltage drops are respectively 66.7V and 33.3V Ans.

2. The final two parallel 5μF capacitors are equivalent to one unit of 10μF.

The capacitance C of the branch, consisting of 20μF, 10μF and 20μF in series, is given by:

$$\frac{1}{C} = \frac{1}{20} + \frac{1}{10} + \frac{1}{20} \qquad \text{or } C = 5\mu\text{F}$$

This series circuit is in parallel with a 5μF capacitor, making the equivalent capacitance = 10μF

The final arrangement between A and B is now equivalent to a 20μF, 10μF and 20μF capacitor in series. The equivalent capacitance is given by:

$$\frac{1}{C} = \frac{1}{20} + \frac{1}{10} + \frac{1}{20} \qquad \text{or } C = \frac{20}{4} = 5\mu\text{F} \quad \text{Ans.}$$

3. Since $Q = CV$. $\therefore$ quantity of electricity received initially is given by $Q = 1000 \times 10^{-6} \times 100 = 10^{5} \times 10^{-6}$

$$= 10^{-1} \text{ coulombs.}$$

Since the plates are separated by an insulated rod there is no loss of charge and hence Q remains the same.

Under the new condition since, as before, $Q = CV$ then

$$\text{Then } V = \frac{Q}{C} = \frac{10^{-1}}{300 \times 10^{-6}} = 333.3\text{V}$$

Thus the potential difference will have increased by 333.3 − 100 = 233.3V Ans.

4. The capacitor would be made up from 10 plates in parallel, making one assembly, interleaved with 9 plates in parallel forming the other plate assembly. There will be 18 mica separators or 18 electric fields and the total capacitance will be 18 times the capacitance between one pair of plates.

Thus C of one pair of plates $= \dfrac{\epsilon A}{d} = \dfrac{\epsilon_0 \epsilon_r A}{d}$

$$\text{or } C = \frac{8.85 \times 10^{-12} \times 7 \times 2580 \times 10^{-6}}{0.1 \times 10^{-3}}$$

$$= 1.6 \times 10^{-9} \text{ farads}$$

or with eighteen units in parallel

$$C = 18 \times 1.6 \times 10^{-9} \text{ farads}$$

$$= 0.0288\mu\text{F} \quad \text{Ans.}$$

5. Since $P = CV$, then $Q = 3 \times 10^{-4} \times 10^{-6} \times 10 \times 10^{3}$

$$= 3 \times 10^{-6} \text{ coulombs}$$

$$\therefore \text{ Flux density, } D = \frac{Q}{A} = \frac{3 \times 10^{-6}}{10\,000 \times 10^{-6}}$$

$$= 3 \times 10^{-4} \text{ coulomb per square metre}$$

Ans.

$$\text{Also, permittivity, } \epsilon = \frac{\text{electricity flux density}}{\text{electric force}} = \frac{D}{E}$$

$$\text{And electric force, } = \frac{V}{d} = \frac{10 \times 10^{3}}{1 \times 10^{-3}} = 10 \times 10^{6} \text{ volts per metre}$$

$$\text{Hence} \quad \epsilon = \frac{3 \times 10^{-4}}{10 \times 10} \quad \text{also } \epsilon = \epsilon_0 \times \epsilon_r$$

$$\text{and} \quad \epsilon_r = \frac{\epsilon}{\epsilon_0} = \frac{3 \times 10^{-4}}{10 \times 10^{6} \times 8.85 \times 10^{-12}}$$

$$\text{or} \quad \epsilon_r = 3.39 \quad \text{Ans.}$$

6. $$C = \frac{\epsilon A}{d} \text{ where } A = 6 \times 10^{4} \times 10^{-6} \text{ square metres}$$

$$= 6 \times 10^{-2} \text{ square metres}$$

$$d = 3.5 \times 10^{-3} \text{ metres}$$

$$\text{and} \quad \epsilon = \epsilon_0 \times \epsilon_r$$

$$= 8.85 \times 10^{-12} \times 3$$

$$\text{Hence} \quad C = \frac{8.85 \times 10^{-12} \times 3 \times 6 \times 10^{-2}}{3.5 \times 10^{-3}}$$

$$= 4.55 \times 10^{-10} \text{ F}$$

$$\text{Energy, } W = \tfrac{1}{2}CV^2 \text{ joules}$$

$$= \frac{1}{2} \times 4.55 \times 10^{-10} \times 300^2$$

$$= 20.475 \times 10^{-6} \text{ joules}$$

$$= 20.48\mu\text{J} \quad \text{Ans.}$$

7. A 10-plate capacitor is made up from two 5-plate assemblies interleaved with each other and separated by the dielectric. There are thus nine electric fields or the final capacitance is nine times that of one plate arrangement.

Hence $C = \frac{\epsilon A}{d}$ where $A = 1500 \times 10^{-6}$ square metres

$d = 0.3 \times 10^{-3}$ metres

$\epsilon = \epsilon_o \times \epsilon_r$

$$C = \frac{8.85 \times 10^{-12} \times 4 \times 15 \times 10^{-4}}{3 \times 10^{-4}}$$

$= 1.77 \times 10^{-10}$ farads $= 1.77 \times 10^{-4}$ microfarads

Total capacitance $= 9 \times 1.77 \times 10^{-4}$
$= 0.0016\mu F$ Ans.

8. Let C = capacitance of the series arrangement,

then $\frac{1}{C} = \frac{1}{20} + \frac{1}{30} = \frac{5}{60}$ or $C = 12\mu F$

The charge stored is given by $Q = CV = 12 \times 10^{-6} \times 600$
$= 72 \times 10^{-4}$ coulombs.

P.D. across $20\mu F$ capacitor A $= \frac{7.2 \times 10^{-3}}{20 \times 10^{-6}}$
$= 360V$ Ans.

P.D. across $30\mu F$ capacitor B $= 600 - 360 = 240V$ Ans.

If P.D. across B is 400V then P.D. across parallel arrangement would be 200V. Also the equivalent capacitance must be $60\mu F$ (double) since the voltage is half that across B.

C must be $30\mu F$, being in parallel with A

Also the energy stored, $W = \frac{1}{2}CV^2$
$= \frac{1}{2} \times 40 \times 10^{-6} \times 200^2$

Thus $W = 0.8J$ Ans.

9. Since

$Q = CV$ and $Q = It$ then $It = CV$

or $I = C\frac{V}{t}$ where V = the voltage change

then i. $I = 40 \times 10^{-6} \times \frac{100}{1 \times 10^{-3}}$ amperes
$= 4A$ Ans.

ii. $I = 40 \times 10^{-6} \times \frac{50}{1 \times 10^{-3}}$ amperes
$= 2A$ Ans.

iii. $I = 40 \times 10^{-6} \times \frac{0}{1 \times 10^{-3}}$ amperes $= 0A$ Ans.

$$\text{iv. } I = 40 \times 10^{-6} \times \frac{100}{1 \times 10^{-3}} \text{ amperes} = 4\text{A} \quad \text{Ans.}$$

$$\text{v. } I = 40 \times 10^{6} \times \frac{50}{1 \times 10^{-3}} \text{ amperes} = 2\text{A} \quad \text{Ans.}$$

The accompanying graph (Fig 13) shows the current and voltage conditions.

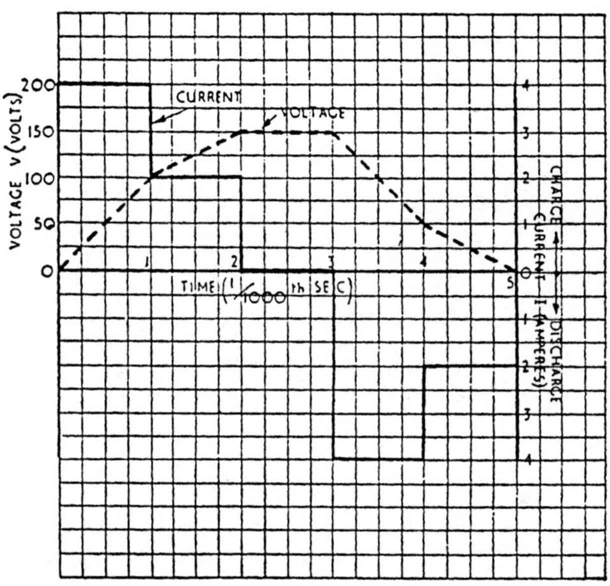

Fig 13

10. Reactance of cable per kilometre $= \dfrac{11\ 000}{1} = 11\ 000\Omega$

$$\text{Also } X_C = \frac{10^6}{2\pi f C} = \frac{10^6}{2 \times \pi \times 50 \times C} \text{ ohms}$$

$$\text{whence } 11 \times 10^3 = \frac{10^4}{\pi \times C} \text{ and } C = \frac{10}{11 \times \pi} \text{ microfarads}$$

$$\text{or } C = 0.289\mu\text{F}$$

$$\text{Again since permittivity} = \frac{\text{electric flux density}}{\text{electric force}}$$

$$\text{then } \epsilon = \frac{D}{E} \text{ and } D = \frac{Q}{A}$$

But $Q = CV$

$$\therefore D = \frac{CV}{A} = \frac{0.289 \times 10^{-6} \times 11 \times 10^3}{A} \text{ coulomb per square metre}$$

and $= \dfrac{V}{d} = \dfrac{11 \times 10^3}{10 \times 10^{-3}}$

$= 11 \times 10^5$ volts per metre

The mean diameter of the insulation = 10 + 12 = 22mm

The area of the dielectric

= mean circumference × length

= $\pi d \times 1000$ square metres

= $\pi \times 22 \times 10^{-3} \times 10^3$

= 69.1m^2

Also, from the above,

$$D = \frac{CV}{A} = \frac{0.289 \times 10^{-6} \times 11 \times 10^3}{69.1} \text{ coulomb per square metre}$$

$= 46 \times 10^{-6}$ C/m²

Again $\epsilon = \dfrac{D}{E}$

$= \dfrac{46 \times 10^{-6}}{11 \times 10^5}$

Also $\epsilon = \epsilon_0 \times \epsilon_r$

$\therefore \epsilon_r = \dfrac{46 \times 10^{-6}}{11 \times 10^5 \times 8.85 \times 10^{-12}}$

$\therefore \epsilon_r = 4.73$ Ans.

CHAPTER 9

1. A scale of 10mm = 1A is used and I_1 is the reference phasor drawn horizontally. The diagram, drawn geometrically to scale, shows the solution which gives I, the resultant current of value 9.23A lagging I_1 by 6°

If the above is checked mathematically

$$\begin{aligned}
I_H &= 4\cos 0 + 6\cos 30 + 2\cos 90 \\
&= (4 \times 1) + (6 \times 0.866) + (2 \times 0) \\
&= 4 + 5.196 = 9.196\text{A} \\
I_V &= 4\sin 0 - 6\sin 30 + 2\sin 90 \\
&= (4 \times 0) - (6 \times 0.5) + (2 \times 1) = 0 - 3 + 2 \\
&= -1\text{A} \\
I &= \sqrt{9.196^2 + 1^2} = 9.24\text{A} \quad \text{Ans.}
\end{aligned}$$

$$\text{Cos } \theta = \frac{9.2}{9.24} = 0.995 \text{ and } \theta = 6° \text{ (approx)} \quad \text{Ans.}$$

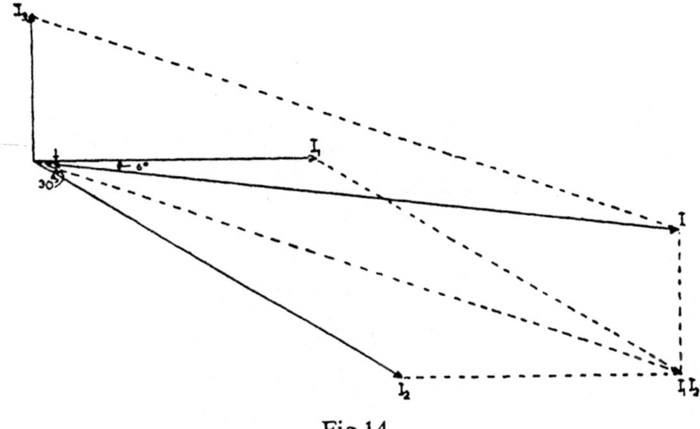

Fig 14

2. Since $v = V_m \sin(2\pi ft)$

$$\therefore \sin(2\pi ft) = \frac{200}{282.8} = 0.707$$

Now the angle whose sine is 0.707 = 45°

$$\therefore 2\pi ft = 45°$$

$$\text{or } t = \frac{45}{2 \times 180 \times 25}$$

$$= 0.005 \text{ seconds} = 5\text{ms}$$

(a) The first time is 5ms after zero value Ans.

(b) Time for 1 cycle $= \frac{1}{25} = 0.04$s

Time for $\frac{1}{2}$ cycle = 0.02 second

The second time is 0.02 − 0.005 = 0.015 seconds or 15ms after zero value Ans.

3. A simple phasor diagram, as shown, illustrates the problem and the mathematical solution.

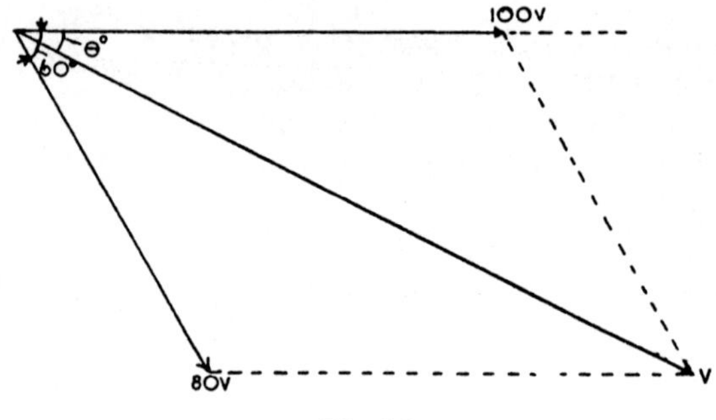

Fig 15

$$\text{Horizontal component } V_H = 100 + 80 \cos 60$$
$$= 100 + 80 \times 0.5 = 140\text{V}$$
$$\text{Vertical component } V_V = 0 - 80 \sin 60$$
$$= 0 - 80 \times 0.866$$
$$= -69.28\text{V}$$
$$\text{Resultant } V = \sqrt{140^2 + 69.28^2}$$
$$= 156.3\text{V} \quad \text{Ans.}$$
$$\cos \theta = \frac{140}{156.3} = 0.8955. \therefore \theta = 26° 26'$$

Since maximum values have been used for the phasor the resultant is a maximum value, which lags the 100V values by 26° 26′ Ans.

4. The waveform is plotted as shown in the diagram. Erecting mid-ordinates, measuring and squaring these gives the following columns.

$i_1 = 0.22 \qquad i_1^2 = 0.05$

$i_2 = 0.60 \qquad i_2^2 = 0.36$

$i_3 = 0.92 \qquad i_{32} = 0.85$

$i_4 = 1.25 \qquad i_4^2 = 1.56$

$i_5 = 1.55 \qquad i_5^2 = 2.40$

$i_6 = 1.8 \qquad i_6^2 = 3.24$

$i_7 = 1.97 \qquad i_7^2 = 3.87$

$i_8 = 1.92 \qquad i_8^2 = 3.68$

$i_9 = 1.56 \qquad i_9^2 = 2.44$

$i_{10} = 0.85 \qquad i_{10}^2 = 0.72$

Total of $i_2 = 19.17$

Average of $i^2 = \dfrac{19.17}{10}$

$= 1.917$

$\therefore$ r.m.s. value $= \sqrt{1.917}$

$= 1.385$A Ans.

Power dissipated obtained from

$I^2R = 1.917 \times 8 = 15.34$W Ans.

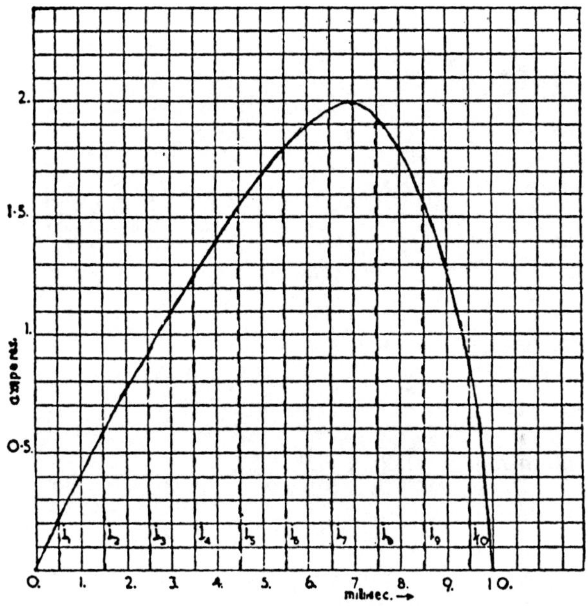

Fig 16

5. The phasor diagram is drawn as shown, but is not to scale since the resultant is obtained mathematically. For convenience, the second current has been used as the reference along the horizontal.

Sum of horizontal components

$$I_H = 17.32 \cos 0 + 20 \cos 60 + 10 \cos 90$$
$$= (17.32 \times 1) + (20 \times 0.5) + (10 \times 0)$$
$$= 17.32 + 10 + 0 = 27.32\text{A}$$

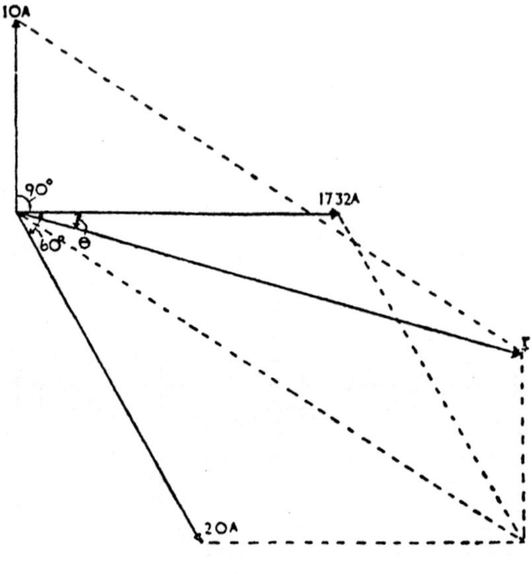

Fig 17

Sum of vertical components

$$I_V = 17.32 \sin 0 - 20 \sin 60 + 10 \sin 90$$
$$= (17 + 0) - (20 \times 0.866) + (10 \times 1)$$
$$= 0 - 17.32 + 10 = -7.32\text{A}$$

$$\text{Resultant } I = \sqrt{27.32^2 + 7.32^2}$$
$$= \sqrt{799.4} = 28.32\text{A}$$

$$\text{Cos } \theta = \frac{27.32}{28.32} = 0.965 \quad \text{So } \theta = 15°12'$$

Since peak or maximum values were used for the individual phasors, the maximum value of the resultant current is 28.32A and this lags 105° 12′ behind the 10A current. Ans.

6. For a sine wave voltage applied to a resistor, a sinusoidal current results, and the maximum value of this current is at the instant of maximum voltage.

$$\text{Maximum current} = \frac{\text{maximum voltage}}{\text{resistance}} = \frac{340}{24}$$
$$= 14.14\text{A}$$
$$\text{R.M.S. value of current} = 0.707 \times 14.14 = 9.996\text{A}$$
$$= 10\text{A} \quad \text{Ans.}$$

7. The phasor diagram is drawn as shown. For the construction a scale of 10mm = 25V is used. The phase relation between the various phasors are determined more clearly thus:

$$e_1 = 100 \sin \omega t$$

$$e_2 = 50 \cos \omega t = 50 \sin \left(\omega t + \frac{\pi}{2}\right)$$

$$= 50 \sin (\omega t + 90°)$$

$$e_3 = 75 \sin \left(\omega t + \frac{\pi}{3}\right) = 75 \sin (\omega t + 60°)$$

$$e_4 = 125 \cos \left(\omega t - \frac{2\pi}{3}\right)$$

$$= 125 \sin \left(\omega t - \frac{2\pi}{3} + \frac{\pi}{2}\right)$$

$$= 125 \sin \left(\omega t - \frac{\pi}{6}\right)$$

$$= 125 \sin (\omega t - 30°).$$

The diagram is drawn using the maximum value E_{1m} of the first voltage as reference. For the diagram E_{1m} = 100V

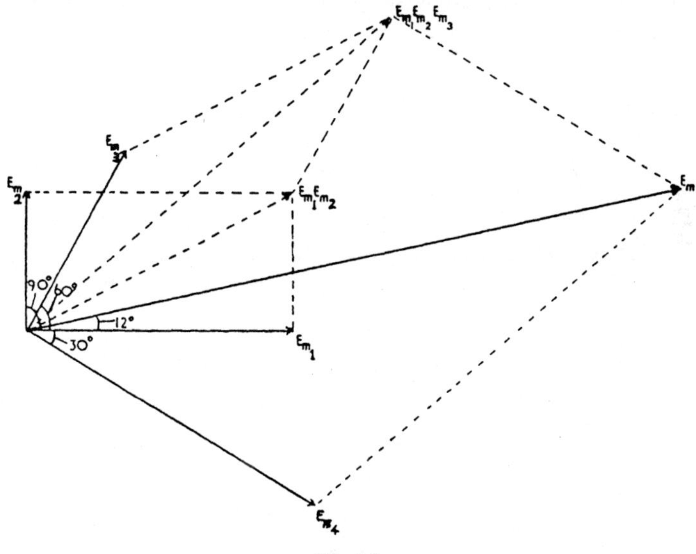

Fig 18

E_{2m} = 50v $\quad E_{3m}$ = 75V $\quad E_{4m}$ = 125V

For the resultant E = 100.5mm = 252V

The angle θ = 12° (approx)

Now $12° = \frac{180}{15}$ or $\frac{\pi}{15}$ radians (leading)

The required expression can be written

$$e_1 + e_2 + e_3 + e_4 \text{ or } e = 252 \sin\left(\omega t + \frac{\pi}{15}\right) \quad \text{Ans.}$$

8. (a) Alternators in step $V = 100 + 200 = 300\text{V}$ Ans.
(b) When phase displacement is 60°

$$V = \sqrt{100^2 + 200^2 + 2 \times 100 \times 200 \cos 60}$$
$$= \sqrt{10\,000 + 40\,000 + (40\,000 \times 0.5)}$$
$$= 264.8\text{V} \quad \text{Ans.}$$

(c) When phase displacement is 90°

$$V = \sqrt{100^2 + 200^2}$$
$$= 223.7\text{V} \quad \text{Ans.}$$

(d) When phase displacement is 120°

$$V = \sqrt{50\,000 + 40\,000 \cos 120}$$
$$= \sqrt{50\,000 + 40\,000\,(-\cos 60)}$$
$$= 173.2\text{V} \quad \text{Ans.}$$

(e) When phase displacement is 180°

$\text{Cos}\,\theta = -1$

$$V = \sqrt{50\,000 - 40\,000} = 100\text{V}$$

The above is obvious *ie* since the voltages oppose, the resultant is the arithmetical difference *ie* 100V Ans.

9. The waveform, when plotted, is seen to be of stepped shape but each half wave is regular and similar to its other half, except that it is reversed. The r.m.s. value can be obtained by considering a half wave only, since the reversal mentioned, will not affect this value.

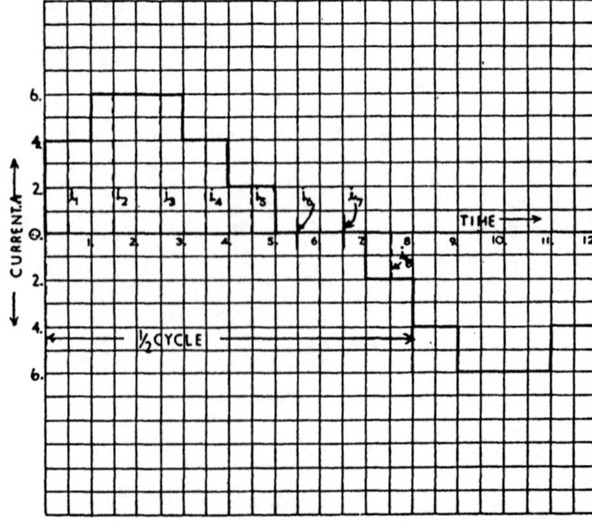

Fig 19

If the time interval 0 — 8 is considered as the base of the half wave, eight mid-ordinates can be used, giving:

$i_2 = 4$ and $i_{12} = 16$
$i_2 = 6$ and $i_2^2 = 36$
$i_3 = 6$ and $i_3^2 = 36$
$i_4 = 4$ and $i_4^2 = 16$
$i_5 = 2$ and $i_5^2 = 4$
$i_6 = 0$ and $i_6^2 = 0$
$i_7 = 0$ and $i_7^2 = 0$
$i_8 = -2$ and $i_8^2 = 4$

The sum of $i^2 = 112$

The mean of $i^2 = \frac{112}{8} = \sqrt{14}$

The r.m.s. value $= \sqrt{14}$
$= 3.75\text{A}$

The required value of d.c. would be 3.75A Ans.

10. The problem calls for a solution by phasor construction and is therefore worked by drawing; the scale chosen being 10mm = 2V and the first e.m.f. 0A being used as the reference.

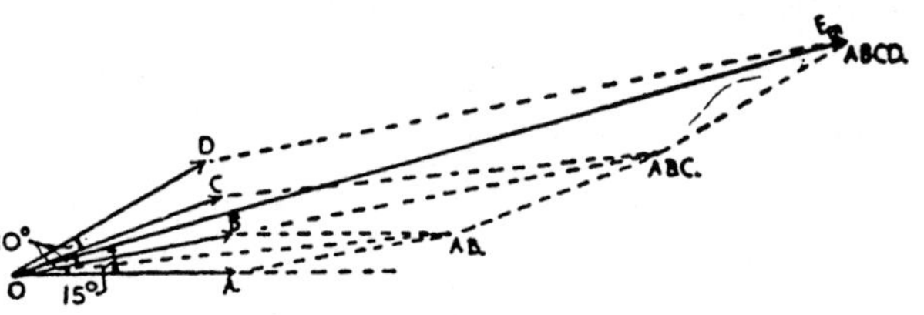

Fig 20a

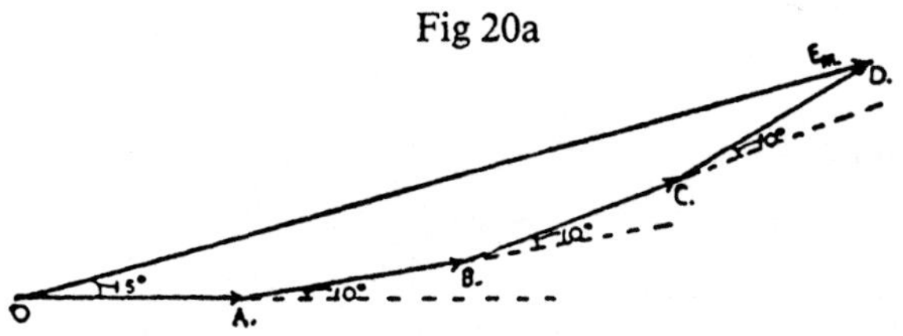

Fig 20b

The resultant OD or E_m is measured to be 15.6V and

$$\theta = 15° \text{ or } \frac{\pi}{12} \text{ radians}$$

$$\text{Thus } e = 15.6 \sin\left(\omega t + \frac{\pi}{12}\right) \quad \text{Ans.}$$

The above diagrams show alternative solutions. thus (a) uses the parallelogram method, (b) uses the polygon method.

CHAPTER 10

1. Inductive reactance of circuit $X_L = 2\pi fL$

$$= 2 \times \pi \times 50 \times 0.01$$
$$= 3.14\Omega$$

(a) Circuit impedance $= \sqrt{3^2 + 3.14^2}$

$$= 4.34\Omega \quad \text{Ans.}$$

(b) Power factor $= \dfrac{R}{Z} = \dfrac{3}{4.34} = 0.69$ (lagging) Ans.

(c) Power absorbed $= I^2 R = \left(\dfrac{V}{Z}\right)^2 R = \dfrac{V^2}{Z} \times \dfrac{R}{Z}$

$$= \frac{V^2}{Z} \cos \phi$$
$$= \frac{60^2}{4.34} \times 0.69$$
$$= 572.3\text{W} \quad \text{Ans.}$$

2. Lamp current $= \dfrac{P}{V} = \dfrac{100}{100} = 1\text{A}$

Lamp resistance $= \dfrac{100}{1} = 100\Omega$

(a) Total resistance to give 1A with 220V applied

$$= \frac{220}{1} = 220\Omega$$

$\therefore$ Series resistance $= 220 - 100 = 120\Omega$ Ans.

Power absorbed by circuit $= I^2R = 1^2 \times 220 = 220\text{W}$ Ans.

(b) When a coil (reactor) is used for voltage dropping

Impedance of circuit $Z = \dfrac{220}{1} = 220\Omega$

Reactance of circuit $= \sqrt{220^2 - 100^2}$

$$= 196\Omega$$

Also $X_L = 2\pi fL \;\therefore\; L = \dfrac{196}{2 \times \pi \times 50}$

$$= 0.624\text{H} \quad \text{Ans.}$$

Power absorbed by circuit, $P = I^2 R = 1^2 \times 100$

$$= 100\text{W} \quad \text{Ans.}$$

3. Power absorbed, $P = VI \cos \phi$

$$\therefore \cos \phi = \frac{P}{VI} = \frac{2500}{240 \times 15} = 0.694 \text{ (lagging)}$$ Ans.

Current in the circuit $I = 15\text{A}$ (this data is given).

Then since $P = I^2 R$

$$\therefore R = \frac{P}{I^2} = \frac{2500}{15^2} = 11.1\Omega$$ Ans.

$$Z = \frac{V}{I} = \frac{240}{15} = 16\Omega$$ Ans.

$$X = \sqrt{16^2 - 11.1^2}$$
$$= 11.6\Omega$$ Ans.

In Fig 21 $V_R = IR = 15 \times 11.1 = 166.5\text{V}$

$V_X = IX = 15 \times 11.6 = 174\text{V}$

$I = 15\text{A} \qquad V = 240\text{V}$ Ans.

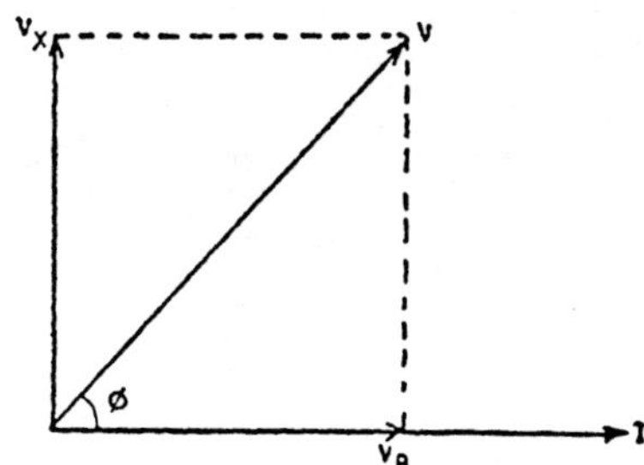

Fig 21

4. $R_A = 120\Omega$ $\qquad R_B = 100\Omega$

$X_A = 2\pi fL$ $\qquad X_B = 2\pi fL$

$= 2 \times \pi \times 50 \times 250 \times 10^{-3}$ $\qquad = 2 \times \pi \times 50 \times 400 \times 10^{-3}$

$= 78.5\Omega$ $\qquad = 125.6\Omega$

$Z_A = \sqrt{120^2 + 78.5^2}$ $\qquad Z_B = \sqrt{100^2 + 125.6^2}$

$= 143\Omega$ $\qquad = 160.5\Omega$

Total circuit $R = 120 + 100 = 220\Omega$

$X = 78.5 + 125.6 = 204.1\Omega$

$$Z = \sqrt{220^2 + 204.1^2}$$
$$= 300\Omega$$

(a) $I = \frac{230}{300} = 0.766\text{A}$ Ans.

(b) $\cos \phi = \frac{220}{300} = 0.733$ (lagging) and $\phi = 42° \, 46'$ Ans.

(c) Voltage across A $= 0.766 \times 143 = 109.6\text{V}$ Ans.
Voltage across B $= 0.766 \times 160.5 = 122.9\text{V}$ Ans.

(d) $\cos \phi_A = \frac{120}{143} = 0.838$ (lagging) or $\phi_A = 33° 7'$

$\cos \phi_B = \frac{100}{160.5} = 0.623$ (lagging) or $\phi_B = 51° 27'$

Thus phase difference $\phi = 51° 27' - 33° 7'$
$= 18°20'$ Ans.

5. D.C. condition $R_A = \frac{20}{2} = 10\Omega$ $\quad R_B = \frac{30}{2} = 15\Omega$

A.C. condition $Z_A = \frac{140}{2} = 70\Omega$ $\quad Z_B = \frac{100}{2} = 50\Omega$

$X_A = \sqrt{70^2 - 10^2}$
$= 69.3\Omega$
$X_B = \sqrt{50^2 - 15^2}$
$= 47.7\Omega$

Since X is proportional to frequency
Therefore at 50Hz $X_A = 69.3 \times \frac{5}{4} = 86.6\Omega$
$X_B = 47.7 \times \frac{5}{4} = 59.7\Omega$
For the total series circuit $R = 10 + 15 = 25\Omega$
$X = 86.6 + 59.7 = 146.3\Omega$
So $Z = \sqrt{25^2 + 146.3^2}$
$= 148.1\Omega$

Current $I = \frac{230}{148.1} = 1.55$A Ans.

6. Current in the line $I = \frac{P}{V_F \cos \phi_F} = \frac{750 \times 1000}{3300 \times 0.8} = 284$A
Resistance voltage drop in the line, $V_R = IR = 284 \times 1$
$= 0.284$kV
Reactance ,, ,, ,, ,, ,, , $V_X = IX = 284 \times 2.5$
$= 0.710$kV

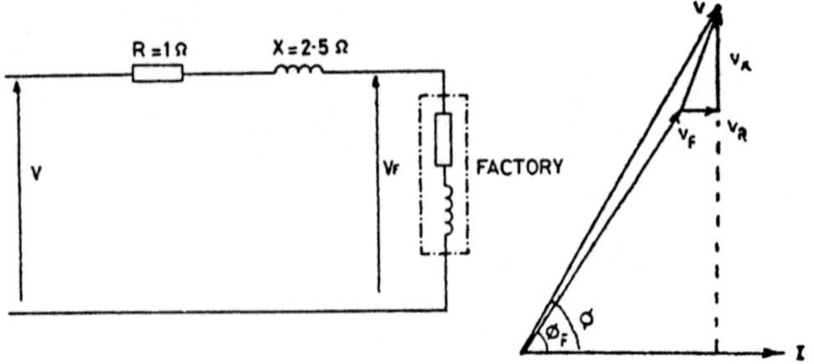

Fig 22

From the phasor diagram (Fig 22)

$$V = \sqrt{(3.3 \times 0.8 + 0.284)^2 + (3.3 \times 0.6 + 0.71)^2}$$
$$= \sqrt{2.64 + 0.284)^2 + (1.98 + 0.71)^2}$$
$$= \sqrt{2.924^2 + 2.69^2}$$
$$= 3.98\text{kV}$$

So voltage at the generator = 3.98kV Ans.

Generator power factor $\cos \phi = \frac{2.924}{3.98} = 0.73$ (lagging) Ans.

Generator output $= \frac{3980 \times 284 \times 0.73}{1000}$ kilowatts
$= 825\text{kW}$ Ans.

7. Current in 8Ω resistor $= \frac{64}{8} = 8\text{A}$
= current in circuit

(b) Power absorbed in resistor $= I^2R = 8^2 \times 8$
$= 64 \times 8 = 512\text{W}$ Ans.

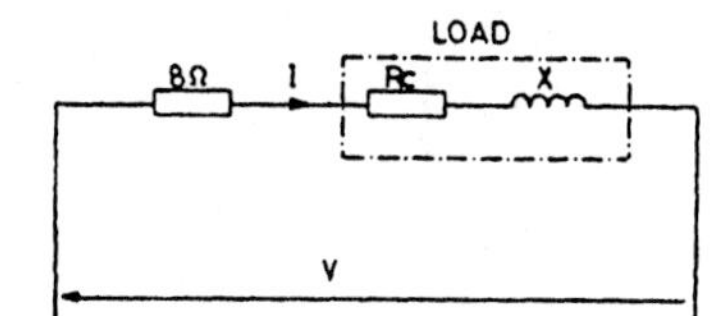

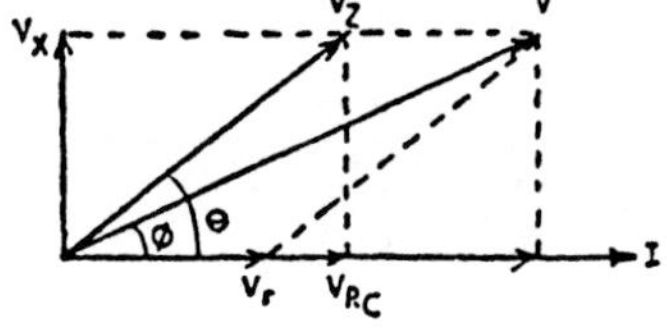

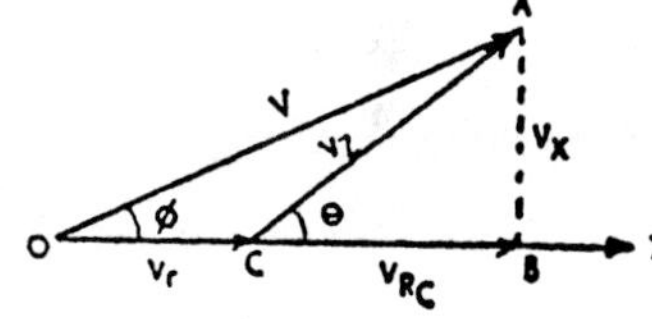

Fig 23

From the deduced diagrams (Fig 23)

$$OA^2 = OC^2 + CA^2 - 2 \times OC \times CA \times \cos(180-\theta)$$
$$\text{or } 100^2 = 64^2 + 48^2 + 2 \times 64 \times 48 \times \cos \theta$$
$$\cos \theta = \frac{10\,000 - 4096 - 2304}{128 \times 48}$$
$$= 0.586$$

(d) Power factor of load = 0.586 (lagging) Ans.

Voltage drop in resistance of inductive load $= V_{RC}$
$= V_Z \cos \theta = 48 \times 0.586 = 28.128\text{V}$
$= IR_C$

$$\therefore R_C = \frac{28.128}{8} = 3.52\Omega$$

(a) Power absorbed by load $= I^2 R_C = 8^2 \times 3.52$
$= 225.3\text{W}$ Ans.

(c) Total Power $= 512 + 225.3 = 737.3\text{W}$ Ans.

(d) Power factor of circuit or $\cos\phi = \frac{P}{VI}$

$$= \frac{737.3}{100 \times 8}$$

$= 0.92$ (lagging) Ans.

8. Let X_1 ohms = the reactance at 40Hz

$$\text{then } Z_1 = \sqrt{R_1^2 + X_1^2}$$

$$\text{and } Z_1 = \frac{200}{6.66} = 30.3\Omega$$

Let X_2 ohms = the reactance at 50Hz

$$\text{then } Z_2 = \sqrt{R_2 + X_2^2}$$

$$\text{and } Z_2 = \frac{200}{8} = 25\Omega$$

$$\text{Hence } 30.3^2 = R_1^2 + X_1^2$$

$$\text{and } 25^2 = R_2^2 + X_2^2$$

but, since $R_1 = R_2$ then:

Subtracting $30.3^2 - 25^2 = X_1^2 - X_2^2$

$$\text{or } (30.3 - 25)(30.3 + 25) = (X_1 - X_2)(X_1 + X_2)$$

$$\text{hence } 5.3 \times 55.3 = (X_1 - X_2)(X_1 + X_2)$$

$$\text{or } 293.09 = (X_1 - X_2)(X_1 + X_2)$$

$$\text{Also since } X = \frac{1}{2\pi fC} \quad \therefore X = \frac{k}{f} \text{ or } X \propto \frac{1}{f}$$

$$\text{Thus } X_1 = \frac{k}{f_1} \text{ and } X_2 = \frac{k}{F_2} \text{ or } \frac{X_1}{X_2} = \frac{k}{f_1} \Big/ \frac{k}{f_2} = \frac{f_2}{f_1}$$

$$\text{Hence } \frac{X_1}{X_2} = \frac{50}{40} \text{ and } X_1 = \frac{5}{4} X_2 \text{ or } X_1 = 1.25X_2$$

Substituting

$$293.09 = (1.25X_2 - X_2)(1.25X_2 + X_2)$$

$$= 0.25X_2 \times 2.25X_2$$

$$= 0.5625X_2^2$$

$$\text{Whence } X_2^2 = \frac{293.09}{0.5625} \text{ and } X_2 = \sqrt{520} = 22.8\Omega$$

$$X_1 = \frac{5}{4} \times 22.8 = 28.5\ \Omega$$

$$\text{Thus } R^2 = 25^2 - 22.8^2 = 105$$

$$R = \sqrt{105} = 10.25\Omega \quad \text{Ans.}$$

$$\text{Also } X_C = \frac{10^6}{2\pi fC} \text{ or } 22.8 = \frac{10^6}{2 \times \pi \times 50 \times C}$$

$$\text{Thus } C = \frac{10^6}{22.8 \times \pi \times 10^2}$$
$$= 139\mu F \quad \text{Ans.}$$

For the diagram Fig 24, as an example:
Resistance voltage drop $= IR = 6.66 \times 10.25 = 68.27V$
Reactance voltage drop $= IX_1 = 6.66 \times 28.5 = 190V$ etc.

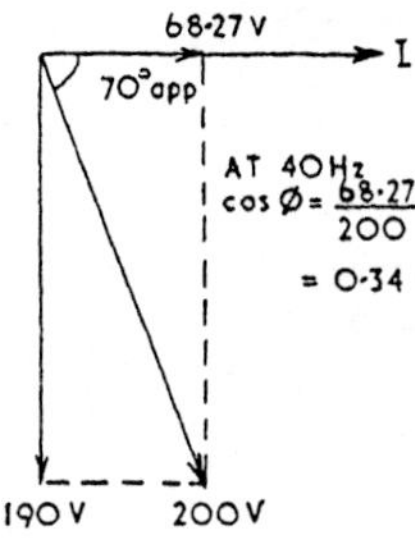

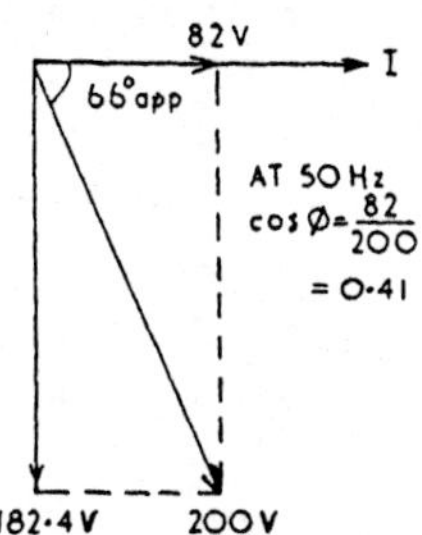

Fig 24

9.

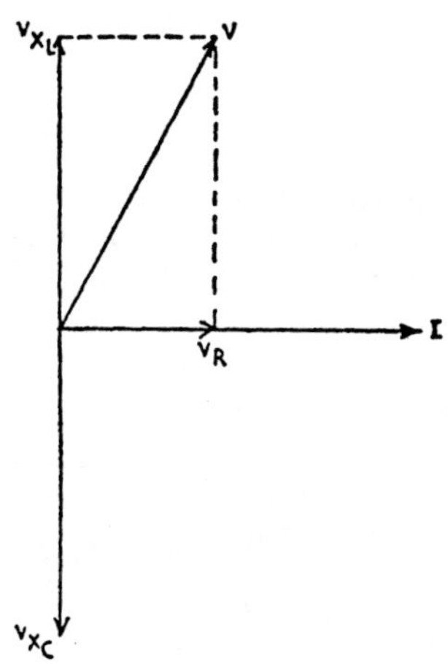

Fig 25

$$\text{At resonance } 2\pi fL = \frac{1}{2\pi fC}$$
$$\text{and } f^2 = \frac{1}{(2\pi)^2 LC} \text{ or } f = \frac{1}{2\pi\sqrt{LC}}$$

Here I can be obtained from $I = \dfrac{V}{X_C}$

$$\text{or } I = \frac{V}{\dfrac{1}{2\pi fC}} = V2\pi fC$$

$$\begin{aligned}
\text{Thus } I &= 100 \times 2 \times \pi \times f \times 10 \times 10^{-6} \\
&= 2\pi \times f \times 10^{-3} \\
&= 2\pi \times 10^{-3} \times \frac{1}{2 \times \pi\sqrt{LC}} \\
&= \frac{10^{-3}}{\sqrt{0.5 \times 10 \times 10^{-6}}} \\
&= 0.446\text{A} \quad \text{Ans.}
\end{aligned}$$

For Fig 25, if values were required then as an example:

$$V_R = IR = 0.446 \times 60 = 26.76\text{V}$$
$$X_L = X_C = 100\text{V}.$$

10. At 60Hz $R = 400\Omega \quad Z = 438\Omega$

$$\begin{aligned}
X_L &= \sqrt{438^2 - 400^2} \\
&= 178\Omega
\end{aligned}$$

$$\begin{aligned}
\text{At 50Hz } X_L &= 178 \times \frac{50}{60} \\
&= 148.3\Omega
\end{aligned}$$

Impedance of circuit

$$\begin{aligned}
Z &= \sqrt{400^2 + (X_L - X_C)^2} \\
&= \sqrt{400^2 + \left(148.3 - \frac{10^6}{2 \times \pi \times 50 \times 40}\right)^2} \\
&= \sqrt{400^2 + 68.7^2} \\
&= 405\Omega
\end{aligned}$$

$$I = \frac{200}{405} = 0.495\text{A}$$

$$\begin{aligned}
V_C &= 0.495 \times \frac{10^6}{2 \times \pi \times 50 \times 40} \\
&= 39.5\text{V} \quad \text{Ans.}
\end{aligned}$$

$$\begin{aligned}
\text{Impedance of coil } Z_L &= \sqrt{400^2 + 148.3^2} \\
&= 426.5\Omega \\
V_{ZL} &= IZ_L \\
&= 0.495 \times 426.5\Omega \\
&= 211\text{V} \quad \text{Ans.}
\end{aligned}$$

CHAPTER 11

1. For the d.c. circuit, $P = VI$ or $300 = 60 \times I$

$$\text{so d.c. current} = \frac{300}{60} = 5\text{A}$$

$$\text{and d.c. resistance} = \frac{V}{I} = \frac{60}{5} = 12\Omega$$

For the a.c. circuit, $P = I^2R$

$$\therefore 1200 = I^2 \times 12$$

$$\text{or } I^2 = 100 \text{ and } I = 10\text{A}$$

Since circuit taken by the a.c. circuit = 10A

$$\text{then impedance, } Z = \frac{V}{I} = \frac{130}{10} = 13\Omega$$

Also $X = \sqrt{Z^2 - R^2}$ or $X = \sqrt{13^2 - 12^2}$

Thus $X = 5\Omega$

Reactance of coil = 5 ohms Ans.

2. Impedance of Branch A. $Z_A = \sqrt{12^2 + 3^2}$

$$= 12.4\Omega$$

Current $I_A = \frac{V}{Z_A} = \frac{100}{12.4} = 8.08\text{A}$. Also $\cos\phi_A = \frac{12}{12.4}$

$$= 0.968 \text{ (lagging)}$$

$$\text{and } \sin\phi_A = \frac{3}{12.4} = 0.242$$

Impedance of Branch B. $Z_B = \sqrt{8^2 + 20^2}$

$$= 21.6\Omega$$

Current $I_B = \frac{100}{21.6} = 4.64\text{A}$ Also $\cos\phi_B = \frac{8}{21.6} = 0.372$ (lagging)

$$\text{and } \sin\phi_B = \frac{20}{21.6} = 0.928$$

Active components of current:

$$\begin{aligned} I_a &= I_A \cos\phi_A + I_B \cos\phi_B \\ &= (8.08 \times 0.968) + (4.64 \times 0.372) \\ &= 9.53\text{A} \end{aligned}$$

Reactive component of current:

$$\begin{aligned} I_r &= -\ I_A \sin\phi_A - I_B \sin\phi_B \\ &= -\ (8.08 \times 0.242) - (4.64 \times 0.928) \\ &= -\ 6.26\text{A} \end{aligned}$$

$$\begin{aligned} \text{Hence } I &= \sqrt{9.53^2 + 6.26^2} \\ &= 11.42\text{A} \quad \text{Ans.} \end{aligned}$$

3. Let the inductive circuit be circuit A, then:

$$X_A = 2\pi fL = 2 \times \pi \times 50 \times 0.02 = 6.28\Omega$$
$$R_A = 50\Omega$$
$$Z_A = \sqrt{50^2 + 6.28^2}$$
$$= 50.5\Omega$$
$$I_A = \frac{V}{Z_A} = \frac{200}{50.5} = 3.96\text{A}$$
$$\cos\phi_A = \frac{R_A}{Z_A} = \frac{50}{50.5} = 0.99 \text{ (lagging)}$$
$$\sin\phi_A = \frac{X_A}{Z_A} = \frac{6.28}{50.5} = 0.124$$

Let the capacitive circuit be circuit B, then:

$$X_B = \frac{10^6}{2 \times \pi \times 50 \times 25}$$
$$= 127\Omega$$
$$I_B = \frac{200}{127} = 1.575 \quad \cos\phi_B = 0 \quad \sin\phi_B = 1$$
$$\text{Then } I_a = (3.96 \times 0.99) + (1.575 \times 0) = 3.92\text{A}$$
$$I_r = -(3.96 \times 0.124) + (1.575 \times 1) = 1.084\text{A}$$
$$\text{or } I = \sqrt{3.92^2 + 1.084^2}$$
$$= 4.075\text{A}$$

Total current $= 4.075$A $\cos\phi = \dfrac{3.92}{4.075} = 0.962$ (leading)

Phase angle $\phi = 15° \; 50'$ Ans.

4. Let the branches be A, B and C respectively. Then:

$$X_A = 2\pi fL = 2 \times \pi \times 50 \times 0.02 = 6.28\Omega$$
$$Z_A = \sqrt{8^2 + 6.28^2}$$
$$= 10.2\Omega$$
$$\text{and } \cos\phi_A = \frac{8}{10.2} = 0.785 \text{ (lagging)} \quad \sin\phi_A = \frac{6.28}{10.2} = 0.616$$
$$I_A = \frac{100}{10.2} = 9.8\text{A}$$
$$X_B = 2 \times \pi \times 50 \times 0.05 = 15.7\Omega$$
$$Z_B = \sqrt{10^2 + 15.7^2}$$
$$= 18.6\Omega$$
$$\cos\phi_B = \frac{10}{18.6} = 0.537 \text{ (lagging)} \quad \sin\phi_B = \frac{15.7}{18.6} = 0.845$$
$$I_B = \frac{100}{18.6} = 5.37\text{A}$$
$$X_C = \frac{1}{2\pi fC} = \frac{10^6}{2 \times \pi \times 50 \times 80}$$
$$= 39.8\Omega$$

$$\text{Then } Z_C = \sqrt{20^2 + 39.8^2} = 44.54\Omega$$

$$\cos\phi_C = \frac{20}{44.54} = 0.449 \text{ (leading) } \sin\phi_C = \frac{39.8}{44.54} = 0.894$$

$$I_C = \frac{100}{44.54} = 2.24\text{A}$$

Adding the active and reactive current components.

$$I_a = I_A\cos\phi_A + I_B\cos\phi_B + I_C\cos\phi_C$$
$$= (9.8 \times 0.785) + (5.37 \times 0.537) + (2.24 \times 0.449)$$
$$= 11.59\text{A}$$

$$I_r = -I_A\sin\phi_A - I_B\sin\phi_B + I_C\sin\phi_C$$
$$= (9.8 \times 0.616) - (5.37 \times 0.845) + (2.24 \times 0.894)$$
$$= -8.54\text{A}$$

$$\text{Then } I = \sqrt{I_a^2 + I_r^2} = \sqrt{11.59^2 + 8.54^2}$$
$$= 14.38\text{A} \quad \text{Ans.}$$

$$\cos\phi = \frac{11.59}{14.38} = 0.805 \text{ (lagging) } \phi = 36° \text{ (approx) Ans.}$$

5. Apparent power $= VI = 240 \times 50.6 \times 10^{-3}$ kilovolt amperes

$= 12.144$kVA Ans.

$$\text{Power factor} = \frac{\text{true power}}{\text{apparent power}} = \frac{10}{12.144}$$
$$= 0.823 \text{ (lagging) Ans.}$$

$$\text{Efficiency} = \frac{\text{output (power)}}{\text{input (power)}} = \frac{9\text{kW}}{10\text{kW}} = 0.9$$

or η = 90 per cent Ans.

6. Output from motor = 1.5kW Efficiency = 80 per cent

$$\text{Input to motor} = \frac{1500}{80} \times 100 = 1875\text{W}$$

Also power input to motor, $VI\cos\phi$ = 1875W

$$\therefore 1875 = 230 \times 11.6\cos\phi$$

$$\therefore \cos\phi = \frac{1875}{230 \times 11.6} = 0.7 \text{ (lagging) } \sin\phi = 0.714$$

Power component of input current, $I\cos\phi = 11.6 \times 0.7$
$= 8.12$A

Reactive ,, ,, ,, ,, , $I \sin\phi = 11.6 \times 0.714$
$= 8.28$A

At the new power factor, the power component of current $= 8.12\text{A} = I_1\cos\phi_1$. Also since $I_1\cos\phi_1 = I\cos\phi$

$$\therefore I_1 \times 0.95 = 8.12 \text{ and } I_1 = \frac{8.12}{0.95} = 8.55\text{A}$$

Note. $\cos\phi_1 = 0.95 \quad \sin\phi_1 = 0.327$

The reactive component of input current, at the new power factor = $I_1 \sin \phi_1$ = 8.55 × 0.327 = 2.8A

So reduction of reactive current = 8.28 − 2.8 = 5.48A

and capacitor current = 5.48A

Capacitor reactance = $\frac{230}{5.48}$ ohms = $\frac{10^6}{2\pi fC}$

or $C = \frac{10^6 \times 5.48}{2 \times \pi \times 50 \times 230}$

= 76μF Ans.

Rating of capacitor = 230 × 5.48 × 10^{-3}

= 1.26VAr Ans.

7. Load (a) Apparent power, $S_a = \frac{\text{active power}}{\text{power factor}} = \frac{10}{1}$

= 10kVA

Reactive power, $Q_a = S_a \times \sin \phi$ = 10 × 0

= 0kVAr

Load (b) Apparent power, S_b = 80kVA at a power factor of 0.8 (lagging)

Active power, P_b = 80 × 0.8 = 64kW

Reactive power, Q_b = 80 × 0.6 = 48kVAr (lagging)

Load (c) Apparent power, S_c = 40kVA at a power factor of 0.7 (leading)

Active power, P_c = 28kW

Reactive power, Q_c = 40 × 0.7143 = 28.57kVAr

Total power taken from the supply, P = 10 + 64 + 28

= 102kW Ans.

Total reactive power, Q = 0 − 48 + 28.57 = − 19.43kVAr

Total apparent power from supply, $S = \sqrt{102^2 + 19.43^2}$

= 104kVA Ans.

Power factor of combined load, $\frac{P}{S} = \frac{102}{104}$ = 0.98 (lagging) Ans.

Mains current = $\frac{104\,000}{250}$ = 416A Ans.

8(a) Phase voltage = 100V Impedance per phase of load = 10Ω

∴ Load current per phase = $\frac{100}{10}$ = 10A

Line current = Phase current = 10A Ans.

Total power, $P = \sqrt{3}\, VI \cos \phi$

But $V = \sqrt{3} V_{ph}$ = 1.732 × 100 = 173.2V and I = 10A

$\therefore P = \frac{\sqrt{3} \times 173.2 \times 10 \times \cos 30}{1000}$

P = 2.598kW Ans.

(b) Line voltage $= \sqrt{3} \times 100$

$\therefore$ Voltage per phase of load $= \sqrt{3} \times 100$ volts

$$\text{Current per phase of load} = \frac{V_{ph}}{Z_{ph}} = \frac{\sqrt{3} \times 100}{10}$$
$$= \sqrt{3} \times 10 \text{ amperes}$$

Line current = phase current $= \sqrt{3} \times \sqrt{3} \times 10$

$= 30$ A Ans,

$$\text{Total Power, } P = \sqrt{3}\ VI \cos \phi$$
$$= \frac{\sqrt{3} \times \sqrt{3} \times 100 \times 30 \times 0.866}{1000}$$
$$= 7.794\text{W} \quad \text{Ans.}$$

(c) Line voltage $= 100$V

Voltage per phase of load $= 100$V

$$\text{Current ,, ,, ,, ,,} = \frac{100}{10} = 10\text{A}$$

Line current $= \sqrt{3}\, I_{ph} = 1.732 \times 10 = 17.32$A Ans.

$$\text{Total Power, } P = \sqrt{3}\ VI \cos \phi$$
$$= \frac{\sqrt{3}\ 1000\ \sqrt{3}\ 10 \times 0.866}{1000}$$
$$= 2.598\text{W} \quad \text{Ans.}$$

(d) Line voltage $= 100$V

$$\text{Voltage per phase of load} = \frac{100}{\sqrt{3}}$$

$$\therefore \text{Current ,, ,, ,,} = \frac{100}{\sqrt{3} \times 10} \text{ amperes}$$

$$\text{Line current} = \text{Phase current} = \frac{10}{\sqrt{3}}$$
$$= 5.77\text{A} \quad \text{Ans.}$$

$$\text{Total power, } P = \sqrt{3}\ VI \cos \phi$$
$$= \frac{\sqrt{3} \times 100 \times 10 \times 0.866}{1000 \times \sqrt{3}} \text{ kilowatts}$$
$$= 0.866\text{kW} \quad \text{Ans}$$

9. Output from motor $= 45$kW $= 45\ 000$W

Efficiency of motor $= 88$ per cent

$$\text{Input to motor} = 45\ 000 \times \frac{100}{88}$$
$$= 51\ 140\text{W}$$

Since $P = \sqrt{3}\ VI \cos \phi$

$$\text{(a) Line current, } I = \frac{51\ 140}{\sqrt{3} \times 500 \times 0.9}$$
$$I = 65.6\text{A} \quad \text{Ans.}$$

(b) Output from the alternator = input to motor
= 51.14kW Ans.

(c) Input to alternator $= \dfrac{51\,140 \times 100}{80}$ watts

or motor power of prime-mover = 64kW Ans.

10.

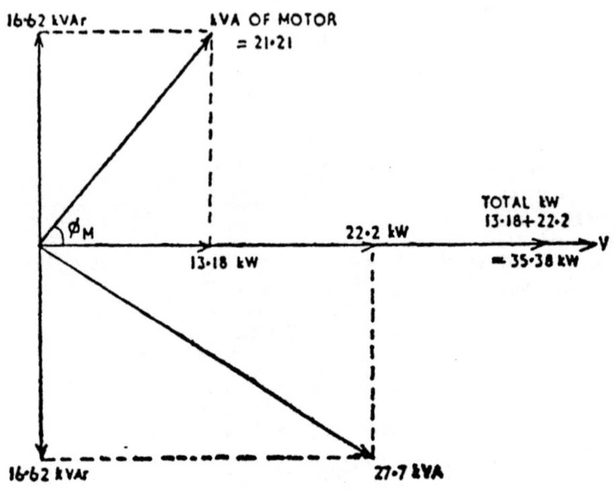

Fig 26

Input power to the system, $P_s = \sqrt{3}\ VI \cos \phi$

$= \dfrac{\sqrt{3} \times 400 \times 40 \times 0.8}{1000}$

$= 22.2$kW

Power output from the motor at 91 per cent efficiency
= 12kW

$\therefore$ Power input to the motor, $P_m = \dfrac{12 \times 100}{91} = 13.18$kW

Apparent power of system, $S_s = \dfrac{\sqrt{3} \times 400 \times 40}{1000}$
$= 27.7$kVA

Reactive power of system, $Q_s = S_s \sin \phi$
$= 27.7 \times 0.6 = 16.62$kVAr

To improve the power factor to unity, the reactive power of the motor must be equal to the reactive power of the system, so reactive power of motor, $Q_m = 16.62$kVAr.

Apparent power to motor, $S_m = \sqrt{P_m^2 + Q_m^2}$
$= \sqrt{13.18^2 + 16.62^2}$
$= 21.21$kVA Ans.

Power factor of motor, $\cos \phi_m = \dfrac{P_m}{S_m} = \dfrac{13.18}{21.21}$

$= 0.62$ (leading) Ans.

Total power taken from the mains = power supplied to the system + power supplied to the motor $= P_s + P_m$

$= 22.2 + 13.18 = 35.38$kW Ans.

The phasor diagram shows the method of solution. Even though the problem is a 3-phase one, the diagram, as drawn, can be applied, since balanced conditions can be assumed.

CHAPTER 12

1. E.m.f. generated $= \frac{Z\Phi N}{60} \times \frac{P}{A}$ volts

Here Z is $144 \times 6 = 864$ $\quad N = 600$ rev/min

$P = 4$ and $A = 4$ since this is a lap winding

$$\therefore 216 = \frac{864 \times \Phi \times 600}{60} \times \frac{4}{4}$$

$$\text{or } \Phi = \frac{216}{864 \times 10} \text{ webers}$$

If the armature is wave wound $A = 2$

Substituting the value of Φ

$$\text{then } E = 864 \times \frac{216}{864 \times 10} \times \frac{600}{60} \times \frac{4}{2}$$

$$= 432\text{V} \quad \text{Ans.}$$

2. Voltage applied to shunt field $= 220$V

Current through shunt field $= \frac{220}{176} = 1.25$A

Armature current $= 250 + 1.25$

$= 251.25$A

Voltage drop in armature and series field $= 251.25(0.05 + 0.015)$

$= 251.25 \times 0.04$

$= 10.05$V

Total voltage drop $= 10.05 + 2$V (brush voltage drop)

Induced e.m.f. $E = 220 + 10.05 + 2$

$= 232.05$V Ans.

3. It should be noted that a change of speed is involved in this problem and solution cannot be affected before the O.C.C. at 900 rev/min is obtained. Since $E \propto N$, the new values can be obtained by multiplying the original by $\frac{900}{1200} = \frac{3}{4}$

The table shows the adjustment for the 900 rev/min condition

Excitation Current I_f (amperes)	0	0.4	0.8	1.2	1.6	2.0	2.4
E.m.f. at 1200 rev/min E (volts)	15	88	146	196	226	244	254
E.m.f. at 900 rev/min E (volts)	$15 \times \frac{3}{4}$ 11.25	$88 \times \frac{3}{4}$ 66	etc 109.5	147	169.5	183	190.5

The field voltage drop line is drawn by taking any current value and multiplying it by 90Ω. Example: 2A × 90Ω = 180V. Join this point R to the origin. The required answer 185V is obtained from the point of intersection as shown.

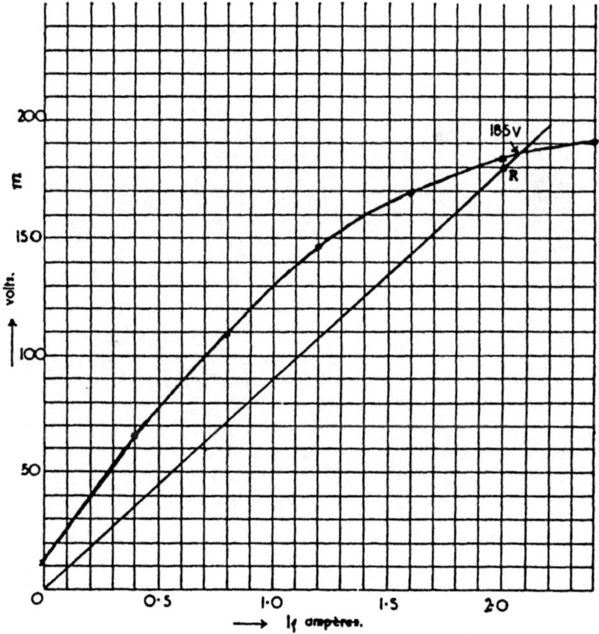

Fig 27

4. $I_f = \dfrac{220}{50} = 4.4\text{A}$ $\qquad I_L = \dfrac{38\,000}{220} = 172.72\text{A}$

$$I_a = 172.72 + 4.4 = 177.12\text{A}$$

$$\text{Also } E = V + I_aR_a = 220 + (177.12 \times 0.1) = 237.72\text{V}$$

$$\text{Again } E = \frac{Z\Phi N}{60} \times \frac{P}{A}$$

$$\therefore 237.72 = \frac{700 \times \Phi \times 800 \times 4}{60 \times 2}$$

$$\text{or } \Phi = \frac{237.71 \times 3}{56 \times 10^3}$$

$$= 0.0128\text{Wb or } 12.8\text{mWb} \quad \text{Ans.}$$

5. Open-circuit e.m.f $= 440\text{V} = k\Phi N$

$$\text{So } k = \frac{440}{0.055 \times 620}$$

$$\text{Full-load current} = \frac{250\,000}{480} = 520\text{A}$$

$$\text{Total voltage drop on full load} = 520\,(0.01 + 0.005 + 0.005) = 520 \times 0.02 = 10.4\text{V}$$

$$\text{So e.m.f. generated on full load} = 480 + 10.4 = 490.4\text{V}$$

$$\therefore 490.4 = k\Phi_2N_2 = \frac{440}{0.055 \times 620} \times \Phi_2 \times 600$$

$$\text{or } \Phi_2 = \frac{490.4 \times 0.055 \times 620}{440 \times 600}$$

$$= 0.0633\text{Wb or } 63.3\text{mWb} \quad \text{Ans.}$$

6. On no load, the e.m.f. generated is caused by the shunt-field ampere-turns = 7900. These give a no-load voltage of 500V. But since the load voltage rises to 550V, the shunt-field current will rise and the shunt-field ampere-turns, on no load, increase to

$$7900 \times \frac{550}{500} = 8690\text{At}$$

Actually on full load it is found, from a separate test, that 11 200At are required and the extra (above 8690At) must therefore be supplied by the series field.

Series field must supply 11 200 − 8690 = 2510At

$$\text{Now the full-load current} = \frac{500\,000}{550} = 910\text{A}$$

$$\therefore \text{The series turns required} = \frac{2510}{910} = 2.76$$

ie 3 turns Ans.

7. Here $Z = 90 \times 6 = 540$

$$\therefore E = \frac{540 \times 0.03 \times 1500}{60} \times \frac{4}{4}$$

$= 405$ volts Ans.

If $I_a = 25$A. The armature voltage drop $= 25 \times 1.0 = 25$ volts. Since the same field flux and speed are to be assumed, then the same e.m.f. is being generated or $V = E - I_aR_a$
$\therefore V = 405 - 25 = 380$ volts.

$$\text{So shunt-field current} = \frac{380}{200} = 1.9\text{A}$$

Machine output current $= 25 - 1.9 = 23.1$A Ans.

$$\begin{aligned}
\text{Let } I_L &= \text{the load current} \\
\text{Then } I_L \times 40 &= V \text{ (the terminal voltage)} \\
\text{Also } V &= E - I_aR_a \\
&= 380 - I_a \times 1.0 \quad \text{also } I_a = I_f + I_L \\
\therefore V &= 380 - 1.0\,(I_f + I_L) \\
\text{or } 40 \times I_L &= 380 - I_f - I_L \\
\text{and } 41 \times I_L &= 380 - I_f \\
\text{also } I_f &= \frac{V}{200} = \frac{40 \times I_L}{200} = \frac{I_L}{5}
\end{aligned}$$

So the above becomes

$$\begin{aligned}
41 \times I_L &= 380 - \frac{I_L}{5} \\
205 \times I_L &= 1900 - I_L \\
206 \times I_L &= 1900 \text{ and } I_L = \frac{1900}{206} = 9.22\text{A} \quad \text{Ans.}
\end{aligned}$$

8. (a) The 1000 rev/min O.C.C. is plotted and cut by the 100Ω field voltage-drop line which is plotted by drawing a straight line through any deduced point and the origin. Thus consider a I_f value of 1A, then a field voltage-drop line point would be $1 \times 100\Omega = 100$V. Join this point to the origin.

The point of intersection at 108V is the answer required.

(b) The tangent is drawn to the 1000 rev/min O.C.C. The critical resistance R_C is determined by taking any voltage value on this tangent and dividing by the current.

Thus $R_C = \dfrac{145\text{V}}{0.9\text{A}} = 161.1\Omega$. The critical resistance for this speed is 161Ω (approx.)

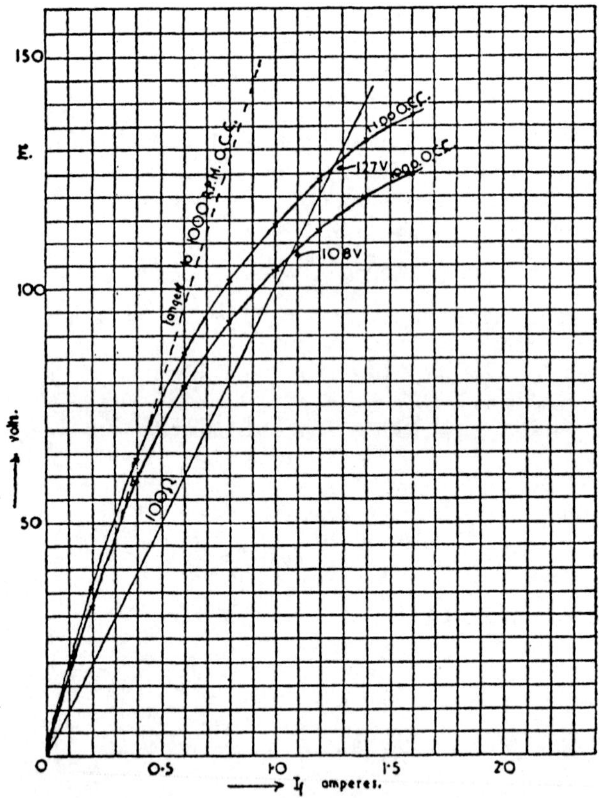

Fig 28

(c) The 1100 rev/min O.C.C. is obtained by multiplying the original 1000 rev/min values by $\frac{11}{10} = 1.1$ to give the new table:

Excitation Current I_f (amperes)	0.2	0.4	0.6	0.8	1.0	1.2	1.4	1.6
E.m.f. at 1000 rev/min E (volts)	32	58	78	93	104	113	120	125
E.m.f. at 1100 rev/min E (volts)	35.2	63.8	85.8	102.3	114.4	124.3	132	137.4

This 1100 rev/min characteristic when plotted is cut by the 100Ω field voltage drop line at 127 volts. This is the required answer.

9. Output current $= \dfrac{50 \times 1000}{230} = 217.39\text{A}$

Shunt-field current $= \dfrac{230}{55} = 4.18\text{A}$

Armature current $= 217.39 + 4.18 = 221.57\text{A}$

Armature voltage drop $= 221.57 \times 0.034$

$= 7.53\text{A}$

Induced e.m.f. $= 230 + 7.53 + 2 = 239.53\text{V}$

Electrical power required to be generated

$= 239.53 \times 221.57$ watts

$= 53.18$kW

Total input power = electrical power input + mechanical loss

$= 53.18 + 1.6 = 54.78$kW

Thus input power $= 54.78$kW Ans.

10. Since the answers required are at a different speed condition, the new O.C.C. at 400 rev/min is obtained by multiplying the original values by $\dfrac{400}{200} = 2$. Thus:

Excitation current I_f (amperes)	0	1	2	3	4	5	6	7	8	9
E.m.f. at 200 rev/min E (volts)	10	38	61	78	93	106	115	123	130	135
E.m.f. at 400 rev/min E (volts)	20	76	122	156	186	212	230	246	260	270

(a) E.m.f. to which machine self-excites = 243V Ans.
This is the point of intersection between the 400 rev/min O.C.C. and the 36Ω field voltage-drop line. The latter is drawn by taking any current, say 5A and finding the voltage drop $5 \times 36 = 180$V, and joining this point to the origin.

(b) Draw the tangent to 400 rev/min O.C.C. This may be difficult. Disregard the bottom part of the characteristic which is due to residual magnetism and assume that the characteristic would pass from the second point given, (76V at 1A) through the origin. A straight line drawn as shown through this point and the origin, would be sufficiently tangential to give a suitable answer.

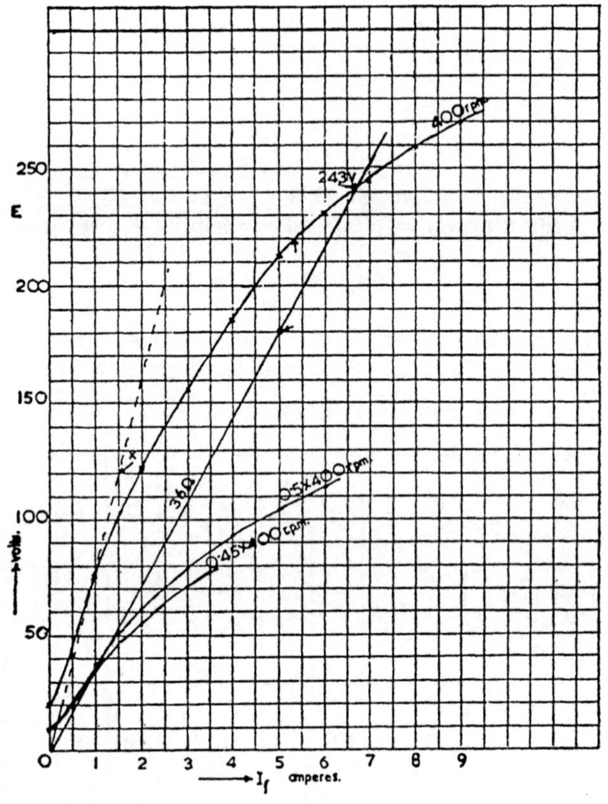

Fig 29

Consider a point such as X. Then critical resistance

$R_C = \frac{120}{1.5} = 80\Omega$ Ans.

(c) For the e.m.f. to reduce to 220V, the field voltage-drop line should cut the O.C.C. at this point. The field current would then be 5.3A. So field-current resistance would be $\frac{220}{5.3} = 41.5\Omega$. Thus additional resistance required $= 41.5 - 36 = 5.5\Omega$ Ans.

(d) This answer is obtained by assuming the machine slows down and is best obtained by trial and error. Multiply the O.C.C. values by various fractions of the speed to obtain a Magnetisation Curve which makes the 36Ω voltage-drop line a tangent. The 200 rev/min O.C.C. is a suitable starting point and is plotted. It is seen to be cut at the bottom by the 35Ω line, which therefore is not quite tangential. The required O.C.C. must be at a lower speed and 400 × 0.45 could be tried to give values of 9, 34.2, 54.9, 70.2, etc. If plotted this is seen to be approximately correct and the answer would be 180 rev/min (approx.).

CHAPTER 13

1. 7.5kW = 7500W This is the output.

Input is $\frac{7500 \times 100}{85} = 8824\text{W}$

$$\text{Motor current} = \frac{8824}{110} = 80.2\text{A}$$

$$E_b = V - I_a (R_a + R_{se})$$
$$= 110 - (80.2 \times 0.12)$$
$$= 100.4\text{V} \quad \text{Ans.}$$

2. Power input = 90 000W

$$\text{Current input to motor} = \frac{90\,000}{500} = 180\text{A}$$

$$\text{Shunt-field current } I_f = \frac{500}{100} = 5\text{A}$$

$$\text{Armature current} = 180 - 5 = 175\text{A}$$

$$\text{Back e.m.f.} = E_b = V - I_aR_a$$
$$= 500 - (175 \times 0.1)$$
$$= 482.5\text{V} \quad \text{Ans.}$$

3. No load. Back e.m.f. $E_{b0} = V - I_{a0}R_a$

$$= 460 - (10 \times 0.025)$$
$$= 459.75\text{V} \quad \text{Ans.}$$

Full load. Back e.m.f. $E_{b1} = V - I_{a1} R_a$

$$= 460 - (300 \times 0.025)$$
$$= 452.5\text{V} \quad \text{Ans.}$$

4. (a) Current in 1 parallel path $= \frac{40}{4} = 10\text{A}$

Current in 1 conductor = current per parallel path

So force on 1 conductor is given by BIl

$$= 1.2 \times 10 \times 0.4$$
$$= 4.8\text{N} \quad \text{Ans.}$$

$$\text{Torque of 1 conductor} = \frac{4.8 \times 0.3}{2}$$
$$= 0.72\text{N m} \quad \text{Ans.}$$

(b) Total torque due to all conductors

$$= 240 \times 0.72$$
$$= 172.8\text{N m} \quad \text{Ans.}$$

(c) Power output $= \dfrac{2 \times \pi \times 800 \times 172.8}{60}$ joules/sec

$= \dfrac{6.28 \times 80 \times 172.8}{6}$

$= 14\,469\text{W} = 14.5\text{W}$ Ans.

5. On full load $E_{b1} = 220 - (25 \times 0.2) = 220 - 5 = 215\text{V}$
Also, since torque is unchanged, and $T_1 = T_2$
and since $T \propto \Phi I_a$ or $T = k\Phi I_a$, we can write:

$$k\Phi_1 I_{a1} = k\Phi_2 I_{a2}$$

But $\Phi_2 = 0.9\Phi_1$

so $\Phi_1 I_{a1} = 0.9\Phi_1 I_{a2}$

or $I_{a2} = \dfrac{I_{a1}}{0.9} = \dfrac{25}{0.9} = 27.77\text{A}$

Thus $E_{b2} = 220 - (27.77 \times 0.2)$
$= 214.446\text{V}$

Also since $\dfrac{E_{b2}}{E_{b1}} = \dfrac{k\Phi_2 N_2}{k\Phi_1 N_1}$

$\therefore N_2 = \dfrac{E_{b2} \times \Phi_1 \times N_1}{E_{b1} \times \Phi_2} = \dfrac{214.446 \times \Phi_1 \times 725}{215 \times 0.9\Phi_1}$

$= 804$ rev/min Ans.

6. As a generator. 50kW at 250V gives a load current of

$I_L = \dfrac{50\,000}{250} = 200\text{A}$

Field current $I_f = \dfrac{250}{50} = 5\text{A}$

$\therefore$ Armature current $I_a = I_L + I_f = 200 + 5 = 205\text{A}$

General voltage $E = V + I_a R_a +$ brush voltage drop

or $E = 250 + (205 \times 0.02) + 2$

$= 256.1\text{V}$

As a motor. Input current $I_L = \dfrac{50\,000}{250} = 200\text{A}$

Field current I_f (as before) $= \dfrac{250}{50} = 5\text{A}$

$\therefore$ Armature current $I_a = I_L - I_f = 200 - 5 = 195\text{A}$

Back e.m.f. $E_b = V - I_a R_a -$ brush voltage drop

$= 250 - (195 \times 0.02) - 2$

or $E_b = 244.1$

Again since E and E_b are proportional to flux and speed then,

$E = k\Phi N$ and $E_b = k\Phi N$

Thus $\dfrac{E}{E_b} = \dfrac{5 \times 400}{5 \times N}$ or $\dfrac{256.1}{244.1} = \dfrac{400}{N}$

$$\text{or } N = \frac{400 \times 244.1}{256.1} = 382 \text{ rev/min} \quad \text{Ans.}$$

Note. Since flux can be assumed to be proportional to the field ampere-turns, and hence the exciting current, then the current value has been substituted for the flux values Φ.

7. On no load. $I_f = \dfrac{105}{90} = 1.17\text{A}$ $I_{L0} = 3.5\text{A}$

$$I_{a0} = 3.5 - 1.17 = 2.33\text{A}$$
$$E_{b0} = 105 - (2.33 \times 0.25)$$
$$= 104.42\text{V}$$

On full load. Output $= 3 \times 1000$ watts

$$\text{Input} = \frac{3 \times 1000 \times 100}{82}$$
$$= 3660\text{W}$$

Input line current $I_{L1} = \dfrac{3660}{105} = 34.86\text{A}$

$$I_{a1} = 34.86 - 1.17 = 33.7\text{A}$$
$$E_{b1} = 105 - (33.7 \times 0.25)$$
$$= 96.57\text{V}$$

Now $E \propto \Phi N$ or $E = k\Phi N$

and we can write $\dfrac{E_{b0}}{E_{b1}} = \dfrac{k\Phi_0 N_0}{k\Phi_1 N_1}$

$$\text{or } N_0 = \frac{E_{b0} \times \Phi_1 \times N_1}{E_{b1} \times \Phi_0} = \frac{104.42 \times 1.17 \times 1000}{96.57 \times 1.17}$$
$$= 1080 \text{ rev/min} \quad \text{Ans.}$$

Again, since T is constant and $T \propto \Phi I_a$, we can write

$$T_2 = k\Phi_2 I_{a2} \text{ and } T_1 = k\Phi_1 I_{a1}$$

or $\dfrac{T_2}{T_1} = \dfrac{k\Phi_2 I_{a2}}{k\Phi_1 I_{a1}}$ But $T_2 = T_1$ and $\Phi_2 = \Phi_1$

$\therefore I_{a2} = I_{a1} = 33.7\text{A}$

If R is the added resistance to reduce speed then

$$E_{b2} = 105 - 33.7\,(R + 0.25)$$

Also since $E_b \propto \Phi$ and N, since flux is constant

then $\dfrac{E_{b2}}{E_{b1}} = \dfrac{800}{1000}$ or $E_{b2} = 96.57 \times \dfrac{800}{1000}$

Back e.m.f. (at reduced speed) $= 77.26\text{V}$

$$\text{Thus } 77.26 = 105 - 33.7\,(R + 0.25)$$
$$= 105 - 33.7R - 8.43$$
$$\text{or } 33.7R = 105 - 85.69$$
$$R = \frac{19.31}{33.7} = 0.57\Omega \quad \text{Ans.}$$

Note. As for No 6. Flux being proportional to field current, I_f is substituted for Φ.

8. Cold condition

$$I_{f0} = \frac{230}{200} = 1.15\text{A}$$
$$I_{a0} = 50 - 1.15 = 48.85\text{A}$$
$$E_{b0} = 230 - (48.85 \times 0.2)$$
$$= 220.23\text{V}$$

Hot condition

$$\text{Temp rise} = 60 - 15 = 45°\text{C}$$
$$\therefore R_{a1} = 0.2\left[1 + \left(\frac{0.4}{100} \times 45\right)\right]$$
$$= 0.236\Omega$$
$$\text{Similarly } R_{f1} = 200\left[1 + \left(\frac{0.4}{100} \times 45\right)\right]$$
$$= 200 + (200 \times 45 \times 0.004)$$
$$= 236\Omega$$
$$\text{So } I_{f1} = \frac{230}{230} = 0.975\text{A}$$
$$\text{So } I_{a1} = 50 - 0.975 = 49.025\text{A}$$
$$E_{b1} = 230 - (49.025 \times 0.236)$$
$$= 218.43\text{V}$$

Again since $E_b \propto \Phi$ and N and since $\Phi \propto I_f$ we can write $E_{b0} = kI_{f0}N_0$ and $E_{b1} = kI_{f1}N_1$

$$\text{or } \frac{N_1}{N_0} = \frac{E_{b1}}{E_{b0}} \times \frac{I_{f0}}{I_{f1}} \quad \text{or } N_1 = \frac{E_{b1} \times I_{f0} \times N_0}{E_{b0} \times I_{f1}}$$
$$\text{and } N_1 = \frac{218.43 \times 1.15 \times 1000}{220.23 \times 0.975}$$
$$= 1160 \text{ rev/min} \quad \text{Ans.}$$

9. (a) $V = E_b + I_aR_a$ or $E_b = V - I_aR_a$

$$= 230 - (200 \times 0.35)$$
$$\text{and } E_b = 160\text{V}$$
$$\text{Also since } E_b = \frac{Z\Phi N}{60} \times \frac{P}{A} \text{ then } N = \frac{E_b}{Z\Phi} \times \frac{60A}{P}$$
$$\text{or } N = \frac{160 \times 60 \times 2}{294 \times 0.025 \times 4}$$
$$= 653 \text{ rev/min Ans.}$$

(b) Again Torque is given by:

$$T = 0.159 \times Z\Phi I_a \frac{P}{A} \text{ newton metres}$$
$$= 0.159 \times 294 \times 0.025 \times 200 \times \frac{4}{2}$$
$$= 467.5\text{N m} \quad \text{Ans.}$$

10. $I_{f0} = \dfrac{230}{104.5} = 2.2\text{A} \quad I_{a0} = 5 - 2.2 = 2.8\text{A}$

Also since $I_{f1} = 2.2\text{A}$ then $I_{a1} = 50 - 2.2 = 47.8\text{A}$

Again $E_{b1} = 230 - (47.8 \times 0.4) - 2$

$= 208.88\text{V}$

And $E_{b0} = 230 - (2.8 \times 0.4) - 2$

$= 226.88\text{V}$

(a) Since $\dfrac{E_{b1}}{E_{b0}} = \dfrac{k\Phi_1 N_1}{k\Phi_0 N_0}$

Then $N_0 = \dfrac{E_{b0}N_1}{E_{b1}}$ assuming constant flux.

or $N_0 = \dfrac{226.88 \times 600}{208.88} = 648$ rev/min Ans.

(b) At 600 rev/min $E_{b1} = 208.88\text{V}$

Assuming a constant flux, then for 500 rev/min

$E_{b2} = 208.88 \times \frac{5}{6}$.

$\therefore E_{b2} = 174.07\text{V}$

$\therefore$ The voltage across the armature has to be reduced by

$230 - 174.07 = 55.93\text{V}$

or since $V = E_b + I_a(R_a + R) + 2$

then $V - E_b = I_a(R_a + R) + 2$

or $55.93 = I_a R_a + I_a R + 2$

so $I_a R = 55.93 - 2 - (47.8 \times 0.4)$

$= 34.81$

$R = \dfrac{34.81}{47.8} = 0.73\Omega$ Ans.

(c) Under the new condition, Φ is to be altered,—hence I_f

$\therefore \dfrac{E_{b3}}{E_{b1}} = \dfrac{k\Phi_3 N_3}{k\Phi_1 N_1}$ or $E_{b3} = 230 - (30 \times 0.4) - 2$

$= 216\text{V}$

$\dfrac{216}{208.88} = \dfrac{I_{f3} \times 750}{2.2 \times 600}$

So $I_{f3} = \dfrac{216 \times 2.2 \times 600}{208.88 \times 750}$

$= 1.82\text{A}$

Thus current—and therefore flux, is to be reduced to

$\dfrac{1.82}{2.2} = 0.827 = 82.7$ per cent Ans.

CHAPTER 14

1. Anode current (amperes) $= \dfrac{\text{power dissipated (watts)}}{\text{anode to cathode voltage (volts)}}$

or $I_a = \dfrac{P}{V_a} = \dfrac{336 \times 10^{-3}}{160} = 2.1 \times 10^{-3}$ amperes

Thus anode current = 2.1mA Ans.

2. The d.c. resistance of the valve, for the condition, is given

$= \dfrac{\text{anode to cathode voltage}}{\text{anode current}}$

thus d.c. resistance, $R = \dfrac{58}{6 \times 10^{-3}} = 9.666 \times 10^3$ ohms or 9.67Ω

Power dissipated is given by: $P = V_a I_a$ or $P = I_a^2 R$

Thus $P = 58 \times 6 \times 10^{-3} = 348 \times 10^{-3}$ watts or 348mW Ans.

or $P = (6 \times 10^{-3})^2 \times 9.666 \times 10^3$
$= 347.76 \times 10^{-3}$ watts or 348mW Ans.

3. Resistance, $r_a = \dfrac{1}{\text{conductance}}$

or $r_a = 1 \Big/ \dfrac{1.64 \times 10^{-3}}{8.13} = \dfrac{8.13 \times 10^3}{1.64}$
$= 4.957 \times 10^3$ ohms = 4.96kΩ Ans.

4. A.C. resistance $= \dfrac{\text{change in anode voltage}}{\text{change in anode current}}$

or $r_a = \dfrac{129 - 75}{(22 - 12)10^{-3}}$

$= 5.4 \times 10^3$ ohms or 5.4kΩ Ans.

D.C. resistance, R (for 75V condition)

$= \dfrac{75}{12 \times 10^{-3}} = 6.25 \times 10^3$ ohms

= 6.25kΩ Ans.

D.C. resistance, R (for 129V condition)

$= \dfrac{129}{22 \times 10^{-3}} = 5.864 \times 10^3$ ohms

= 5.86kΩ Ans.

5. The characteristic appears to be the straight line between the 10.5 and 4.2mA values and, over this region, the a.c. resistance would be:

$$r_a = \frac{\delta V_a}{\delta I_a} = \frac{25 - 15}{(10.5 - 4.2)10^{-3}} = 1.59\text{k}\Omega \quad \text{Ans.}$$

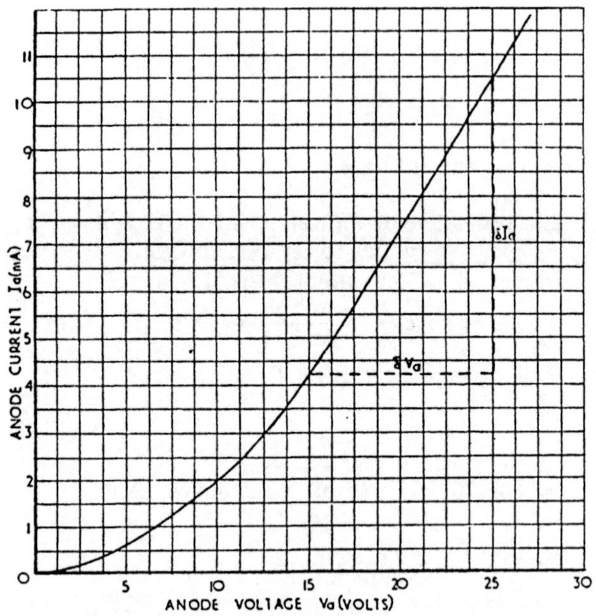

Fig 30

6. The characteristic is plotted as shown and the load line is drawn in thus:
   i. Assume $I_a = 0$. Then voltage on anode would be 60V to give point A.
   ii. Assume the valve resistance to be of minimum value. The anode current would then be $\frac{60}{300} = \frac{1}{5} = 0.2\text{A}$ or 200mA.

Point B is thus obtained and AB gives the load line for a load-resistance value of 300Ω.

The point of intersection with the characteristic is P and the standing current is 100mA. Ans.

The power dissipated in the load resistor

$= (100 \times 10^{-3})^2 \times 300$ watts

$= 3\text{W}$ Ans.

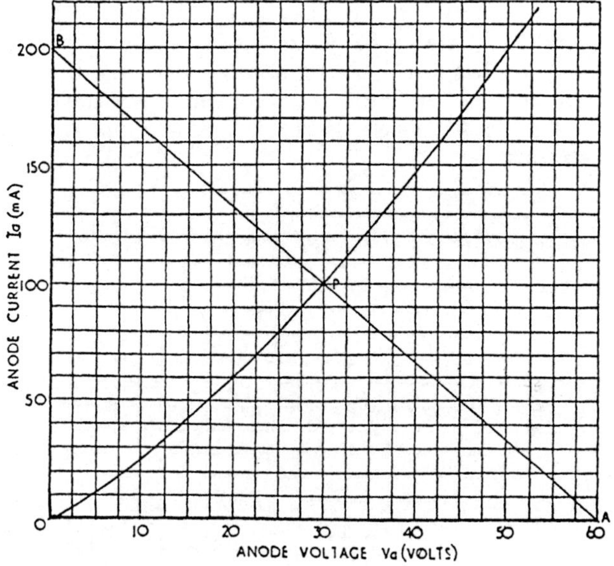

Fig 31

Alternatively (from the load); for point P, $V_a = 30$V
$\therefore$ Voltage dropped across load resistor

$$= V - V_a = 60 - 30 = 30\text{V}$$

Power dissipated = voltage across resistor × anode current

$$= 30 \times 100 \times 10^{-3} = 3\text{W} \quad \text{Ans.}$$

7. Table of results. Since $I_a = V_a^{3/2}$ then:

Anode voltage (V)	0	1	2	3	4	5
Anode current (mA)	0	1	$\sqrt{8}$	$\sqrt{27}$	$\sqrt{64}$	$\sqrt{125}$
,, ,, (mA)	0	1	2.83	5.2	8	11.2
Load voltage (V)	0	1	2.83	5.2	8	11.2
Supply voltage (V)	0	2	4.83	8.2	12	16.2

Specimen calculation, for the 2V anode voltage condition:
$I_a = V_a^{3/2}$. So $I_a = 2^{3/2} = (2^3)^{1/2} = \sqrt{8} = 2.83$mA

The supply voltage for this condition

$$= \text{load voltage drop} + \text{valve voltage drop}$$
$$= I_aR + V_a = (2.83 \times 10^{-3} \times 10^3) + 2$$

or $V = 4.83$ volts.

The deduced valve static and dynamic characteristics are plotted, as shown below, with I_a to a V_a and V base. From the latter, the value of anode current, for an applied voltage of 8V, is seen to be 5.1mA.

Note that this result could have been obtained by drawing a load line on the static characteristics for a resistor of 1kΩ and a supply voltage of 8V. This is shown dotted and the current is seen to be 5.1mA. For this problem, since the dynamic characteristic has been deduced the load-line method is unnecessary.

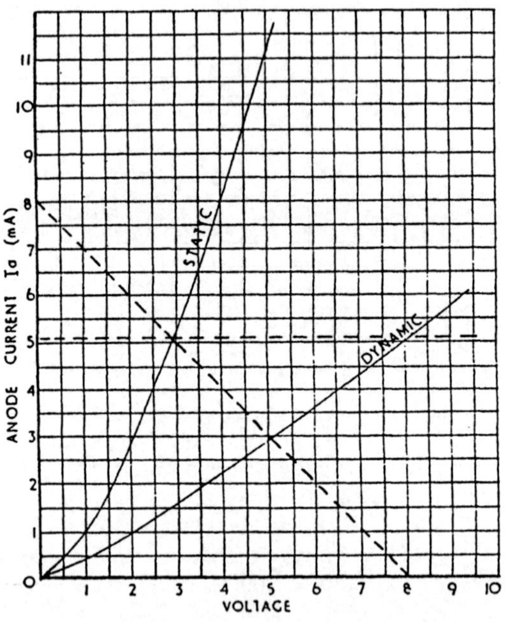

Fig 32

8. The problem is solved by plotting the valve characteristic and drawing the load line for the resistive circuit. Solution will be assisted, if the circuit is drawn out.

Consider the valve on open-circuit *ie* non-conducting. Then the current through the resistive circuit would be

$$\frac{240}{(10 + 50)10^3} = 4\text{mA}$$

and the voltage drop across the anode resistor would be

$$4 \times 10^{-3} \times 10 \times 10^3 = 40\text{V}$$

∴ The p.d. between anode and cathode of the valve, for zero I_a would be 240 − 40 = 200V

This gives a value for point A on the load line.

Next consider the valve 'short-circuiting' *ie* with no p.d. across it. The full 240V would be applied across the anode resistor and the current would be $\frac{240}{10 \times 10^3} = 24\text{mA}$.

This is for a zero anode-voltage condition and gives point B on the load line. Join AB and the intersection point shows a value of 60V applied to the valve and an anode current of some 17mA. Ans.

As a check, we have: For 60V across the valve a current of $\frac{60}{50 \times 10^3} = 1.2\text{mA}$ through the 50kΩ shunt circuit. With 60V across the valve, 240 − 60 = 180V would be dropped across the anode resistor *ie* a current of $\frac{180}{10 \times 10^{-3}}$

= 18mA

∴ Valve current = 18 − 1.2 = 16.8mA, as indicated by the graph. Ans.

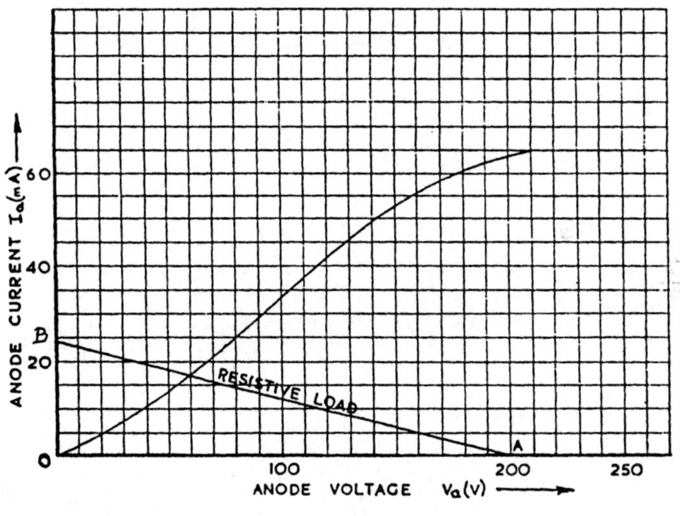

Fig 33

9. In general $g_m = \frac{\delta I_a}{\delta Vg}$ for a constant anode voltage.

$$g_m = \frac{(18 - 10) \times 10^{-3}}{-1.25 - (-3)}$$

$$g_m = 4.58\text{mA/V} \quad \text{Ans.}$$

10. a.c. resistance $r_a = \frac{\delta V_a}{\delta I_a}$ (V_g constant)

$$= \frac{(120 - 80)}{(10 - 6.2) \times 10^{-3}}$$

$$r_a = 10.53\text{k}\Omega$$

$$\text{Mutual conductance } g_m = \frac{\delta I_a}{\delta V_a} \text{ (V}_a \text{ constant)}$$

$$= \frac{(10 - 4) \times 10^{-3}}{-1.3 - (-3.8)}$$

$$g_m = 2.4 \text{ mA/V}$$

$$\text{Amplification Factor } \mu = g_m \times r_a$$

$$= 2.4 \times 10^{-3} \times 10.53 \times 10^3$$

$$\mu = 25.27 \quad \text{Ans.}$$

CHAPTER 15

1. Plot the graph from the given data.

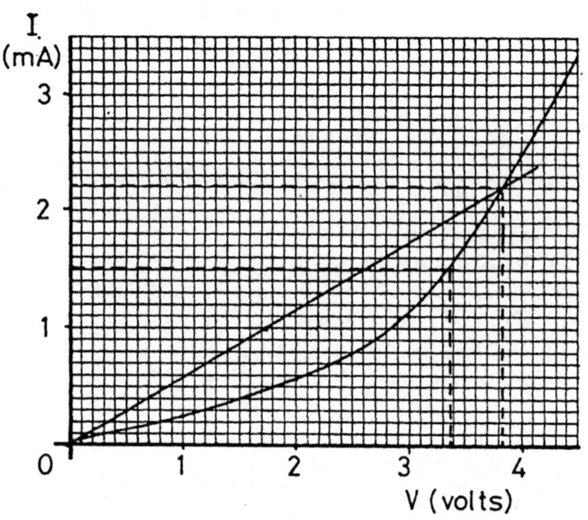

Fig 34

(a) When current is 1.5mA:

$$\text{Forward d.c. resistance} = \frac{3.35}{1.5 \times 10^{-3}}$$
$$= 2.23 \times 10^{3}\Omega$$
$$= 2.23\text{k}\Omega \quad \text{Ans.}$$

(b) Construct d.c. resistance line through origin whose slope is 1.75kΩ.

Thus when d.c. resistance = 1.75kΩ
Anode voltage = 3.83 volts Ans.
Anode current = 2.2mA Ans.

2. $P = VI$, where V is the forward voltage of the diode for a current of 5A.

$\therefore P = 0.7 \times 5 = 3.5\text{W}$ Ans.

The rate of heat dissipation per °C rise is 0.1J/s or 0.1W

Thus the rate of heat dissipation for a 70°C rise is

$0.1 \times 70 = 7\text{W}$

If the voltage drop is constant at 0.7V then the diode current is given by $I_a = \dfrac{7}{0.7} = 10\text{A}$ Ans.

3. Since a number of examples involving the use of a load line have now been illustrated, this problem should present no difficulties. Thus:

(a) For a no anode-current condition, $I_a = 0$ amperes and $V_a = 1$ volt. Thus Point A is obtained.

For a zero diode-resistance condition,

$$I_a = \frac{\text{applied voltage}}{\text{load resistance}}$$

or $I_a = \dfrac{1}{100} = 0.01\text{A}$ or 10mA for $V_a = 0$ volts (across diode)

Point B is then obtained.

The resulting load line intersects with the curves to give points X and Y with the required answers.

At 25°C the load current is 3.4mA

At 60°C the load current is 4.1mA

The current variation would be 0.7mA Ans.

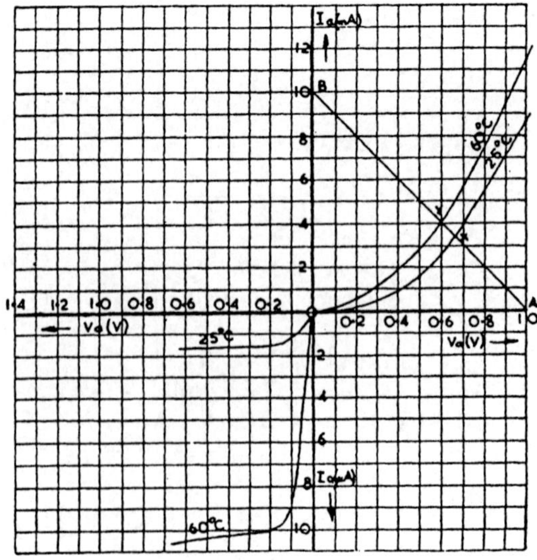

Fig 35

(b) D.C. resistance for point X $= \dfrac{0.66}{3.4 \times 10^{-3}}$

$= 194\Omega$ Ans.

D.C. resistance for point Y $= \dfrac{0.59}{4.1 \times 10^{-3}}$

$= 144\Omega$ Ans.

4. (a)Transformer output voltage Vi $= \dfrac{230}{12} = 19.167V$ (R.M.S. value)

Peak value of Vi $= \sqrt{2} \times 19.167$

$= 27.11$ volts

Peak value of d.c. output Vo $= 27.11 - 0.75$

$= 26.36$ volts Ans.

(b) P.I.V. Rating of diode $= 2 \times 26.36$

$= 52.72$ volts minimum.

Ans.

5. (a) $Vp = \sqrt{2}Vi = \sqrt{2} \times 50 = 70.71$ volts. Ans.
   (b) Since the rectified supply is half wave,

time for discharge $t = \dfrac{1}{f}$

Thus $t = \dfrac{1}{60} = 16.67$ milliseconds

$Q = CVr = It$

$Vr = \dfrac{It}{C} = \dfrac{350 \times 10^{-3} \times 16.67 \times 10^{3}}{1800 \times 10^{-6}}$

Ripple Voltage $Vr = 3.24$ volts Ans.

6. Refer to Figs 232 and 233 for relevant diagram.
   (a) Peak Value $Vm = \sqrt{2}\, V_{R.M.S.}$

$= \sqrt{2} \times 40$

$Vm = 56.57$ volts

Thus the capacitor charges up to 56.67 volts.
Hence the total p.d. across two diodes in series $= 2Vm$

$= 113.14V$

P.I.V. per diode $= 56.57$ volts Ans.

(b) Refer to Fig 235 for relevant diagram.

$Q = CVr = It$

$\therefore Vr = \dfrac{It}{C} = \dfrac{250 \times 10^{-3} \times 0.01}{2000 \times 10^{-6}}$

$= 1.25$ volts

$\therefore$ Ripple Voltage $= 1.25$ volts Ans.

7.

$$Vm = \sqrt{2}\, V_{R.M.S.} = \sqrt{2} \times 50 = 70.71 \text{ volts}$$

Since output voltage is doubled:

Output voltage is 141.42 volts Ans.

8. Refer to Figs 239 and 240

Using the lowest input voltage Vi = 25 − 5V = 20V

When Vi = 20V and I_L = 50mA; I_Z = 1mA

$$\therefore I = I_Z \times I_L$$
$$= 1 + 50\text{mA} = 51\text{mA}$$

(a)
$$Vo = Vi - IR = V_Z + I_Z R_Z$$
$$20 - (51 \times 10^{-3})R = 9.7 + (1 \times 10^{-3}) \times 12$$
$$R = \frac{20 - 9.7 - (12 \times 10^{-3})}{(51 \times 10^{-3})}$$
$$\therefore R = 201.725\Omega \quad \text{Ans.}$$

(b)
$$\text{Stabilised Voltage } Vo = Vi - IR$$
$$= 20 - (51 \times 10^{-3} \times 201.725)$$
$$= 9.712 \text{ volts} \quad \text{Ans.}$$
$$\text{alternatively } Vo = Vz + IzRz$$
$$= 9.7 + (1 \times 10^{-3} \times 12)$$
$$= 9.712 \text{ volts} \quad \text{Ans.}$$

9. (a) When Vi = 30 volts and I_L = 50mA

$$Vo = Vi - IR$$
$$\text{and } I = I_Z + I_L$$
$$\therefore Vi - R(Iz + I_L) = Vz + IzRz$$
$$30 - [201.725(Iz + 50) \times 10^{-3}] = 9.7 + (12Iz \times 10^{-3})$$
$$213.725Iz \times 10^{-3} = 10.214$$
$$\therefore Iz = 47\text{mA} \quad \text{Ans.}$$

(b)
$$Vo = 9.7 + (47 \times 10^{-3} \times 12)$$
$$\text{Stabilised Voltage} = 10.273 \text{ volts} \quad \text{Ans.}$$

(c)
$$Pz = VoIz$$
$$\therefore Pz = 0.48\text{W} \quad \text{Ans.}$$

10. $I = I_L + Iz = 60 + 1 = 61$ mA

$$Vi = 25 \text{ volts (R.M.S.)}$$
$$Vm = \sqrt{2}\, Vi = \sqrt{2} \times 25 = 35.36 \text{ volts}$$

(a)
$$t = \frac{1}{2f} = \frac{1}{2 \times 50} = 0.01 \text{ seconds}$$
$$Q = CVr = It$$
$$Vr = \frac{It}{C} = \frac{61 \times 10^{-3} \times 0.01}{2000 \times 10^{-6}}$$
$$\therefore \text{Ripple Voltage} = 0.305 \text{ volts} \quad \text{Ans.}$$

(b) Total voltage across diode = Vm + Capacitor Voltage
= 35.36 + 35.36V
= 70.72 volts

∴ P.I.V. Rating of each diode = $\frac{70.72}{2}$

= 35.36 volts Ans.

(c) For stable operation the diode must carry 1mA

Thus $I = I_Z + I_L = 1 + 60 = 61\text{mA}$

$Vo = Vz + IzRz = Vi - IRs$

$24 + (1 \times 10^{-3} \times 14) = 35.36 - (61Rs \times 10^{-3})$

$$Rs = \frac{35.36 - (24 + 0.014)}{61 \times 10^{-3}}$$

$Rs = 186\Omega$ minimum Ans.

(d) Power rating is based upon maximum zener current.

$Vi = Vz + Iz\,(Rz + Rs) \quad (I_L = 0)$

$35.36 = 24 + Iz\,(14 + 186)$

$$Iz = \frac{35.36 - 24}{200} = 56.8\text{mA}$$

$Vo = Vz + IzRz$

$= 24 + (56.8 \times 10^{-3} \times 14) = 24.8$ volts

$Pz = VoIz$

$= 24.8 \times 56.8 \times 10^{-3}$

Power Rating of Zener Diode = 1.41 watts Ans.

CHAPTER 16

1. Draw the circuit diagram
   Let I_3 amperes = current in battery towards junction C
   ,, I_1 ,, = ,, ,, CD away from ,, ,,
   ,, I_2 ,, = ,, ,, CA ,, ,, ,, ,,
   Note. There is no current between AD $\therefore$ I_2 flows in AB and I_1 in DB.
   Applying Kirchhoff's laws $I_3 = I_1 + I_2$
   Also point A is at the same potential as point D
   $\therefore$ Voltage drop in AB = voltage drop in DB or $3I_2 = 6I_1$
   giving $I_2 = 2I_1$
   Also voltage drop in CA = voltage drop in CD or $8I_2 = xI_1$.
   Substituting in the above $8 \times 2I_1 = xI_1$ or $x = 16\Omega$ Ans.
   Also, if R = the resistance between points C and B with no current in Y
   then $\frac{1}{R} = \frac{1}{11} + \frac{1}{22} = \frac{2+1}{22}$ or $R = \frac{22}{3} = 7.33\Omega$

   Thus $I_3 = \frac{10}{1 + 7.33} = 1.2\text{A}$ Ans.

2. Let I_1 in current in battery X and I_2 = current in battery Y
   Applying Kirchhoff's laws, we can build up the equations
   $8 = 1.5I_1 + 6(I_1 + I_2) = 7.5I_1 + 6I_2$ . . . (a)
   and $4 = 3I_2 + 6(I_1 + I_2) = 6I_1 + 9I_2$ . . . (b)
   Solving (a) and (b) we have $24 = 22.5I_1 + 18I_2$
   $8 = 12I_2 + 18I_2$

   Subtracting, $16 = 10.5I_1$ or $I_1 = \frac{16}{10.5} = 1.524\text{A}$

   Also from (b) $9I_2 = 4 - 6I_1 = 4 - 6 \times 1.524$
   $= -5.144$
   or $I_2 = \frac{-5.144}{9} = -0.571\text{A}$
   Thus the assumed direction for current I_2 was wrong and battery Y is being charged.
   Current in battery X = 1.524A (discharge) Ans.
   ,, ,, ,, Y = 0.571A (charge) Ans.
   Total current = 1.524 − 0.571 = 0.953A in 6Ω resistor
   So terminal voltage = 6 × 0.953 = 5.72V Ans.

3. (1) By Maxwell's Method. Refer to diagram below (Fig 36).

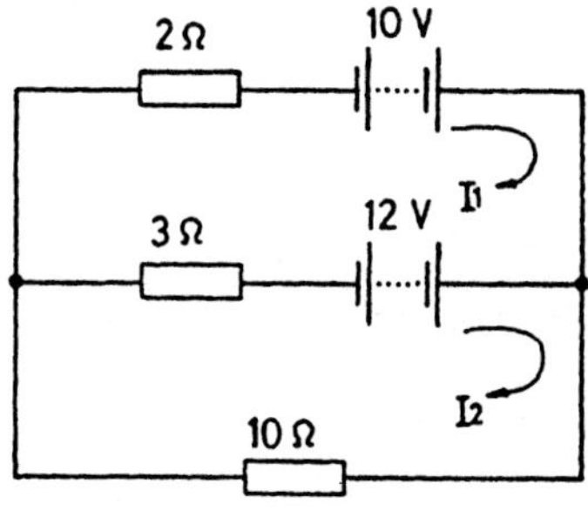

Fig 36

Equating voltage drops to e.m.fs. then:

$2I_1 + 3(I_1 - I_2) = 10 - 12$ or $5I_1 - 3I_2 = -2$

$3(I_2 - I_1) + 10I_2 = 12$ or $-3I_1 + 13I_2 = 12$

giving $15I_1 + 9I_2 = -6$

and $-15I_2 + 65I_2 = 60$

Adding, $56I_2 = 54$ or $I_2 = \frac{54}{56} = 0.964\text{A}$

So $5I_1 = -2 + 3 \times 0.964 = -2 + 2.892 = 0.892$

or $I_1 = \frac{0.892}{5} = 0.1784\text{A}$

Current I_2 in 10Ω resistor = 0.964A Ans.

,, I_1 in 10V battery = 0.1784A = 0.18A Ans.

,, $I_2 - I_1$ in 12V ,, = 0.964 − 0.1784 = 0.7876A
= 0.79A Ans.

(2) By Superposition of Current method. See the attached diagrams (Figs 37a and 37b).

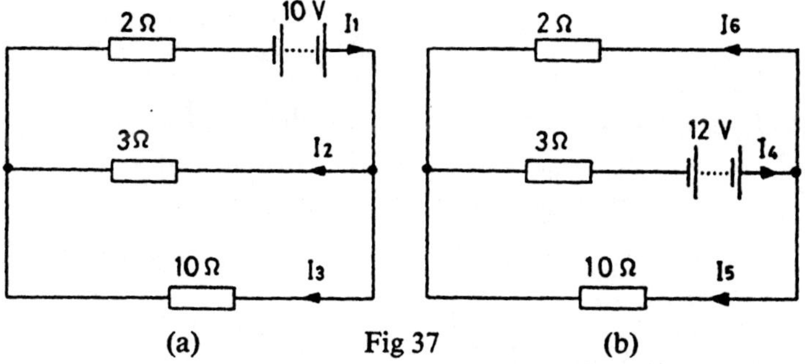

(a) Fig 37 (b)

(a) Only 10V battery is effective. If R = resistance of parallel circuit then $\frac{1}{R} = \frac{1}{3} + \frac{1}{10} = \frac{10 + 3}{30}$ or $R = \frac{30}{13} = 2.315\Omega$

Total resistance $= 2.315 + 2 = 4.315\Omega$

So current $I_1 = \frac{10}{4.315} = 2.325\text{A}$

Voltage drop across $R = 2.315 \times 2.325$

Current $I_2 = \frac{2.315 \times 2.325}{3} = 1.783\text{A}$

,, $I_3 = \frac{5.35}{10} = 0.535\text{A}$

(b) Only 12V battery is effective. If R = resistance of parallel circuit

then $\frac{1}{R} = \frac{1}{2} + \frac{1}{10} = \frac{5 + 1}{10}$ thus $R = 1.666\Omega$

Total resistance $= 3 + 1.666 = 4.666\Omega$

So current $I_4 = \frac{12}{4.66} = 2.575\text{A}$

Voltage drop in parallel section $= 2.575 \times 1.666$
$= 4.29\text{V}$

Current $I_5 = \frac{4.29}{10} = 0.429\text{A}$

Current $I_6 = \frac{4.29}{2} = 2.145\text{A}$

So current in 12V battery $= I_4 - I_2$
$= 2.575 - 1.783$ amperes
$= 0.79\text{A}$ Ans.

,, ,, 10V ,, $= I_1 - I_6$
$= 2.325 - 2.145$ amperes
$= 0.18\text{A}$ Ans.

Current in 10Ω resistor $= I_3 + I_5$
$= 0.535 + 0.429$ amperes
$= 0.964\text{A}$ Ans.

4. Let the currents be as shown in the diagram (Fig 38).

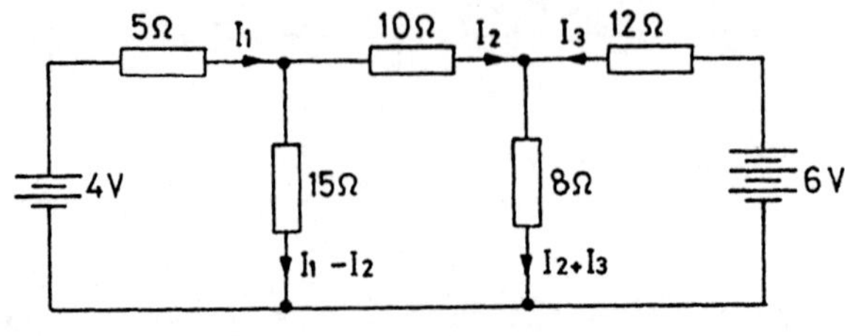

Fig 38

Taking the mesh formed by the 4V battery, the 5Ω, 10Ω and 12Ω resistors and the 6V battery.

$$4 - 6 = 5I_1 + 10I_2 - 12I_3$$

$$\text{or } -2 = 5I_1 + 10I_2 - 12I_3 \qquad \ldots \text{(a)}$$

Taking the mesh formed by the 4V battery, the 5Ω resistor and 15Ω resistor,

$$4 = 5I_1 + 15(I_1 - I_2)$$

$$\text{or } 4 = 20I_1 - 15I_2 \qquad \ldots \text{(b)}$$

Taking the mesh formed by the 6V battery, the 12Ω resistor and the 8Ω resistor.

$$-6 = -12I_3 - 8(I_2 + I_3)$$

$$\text{or } 6 = 20I_3 + 8I_2 \qquad \ldots \text{(c)}$$

Multiply (a) by 4. $\quad -8 = 20I_1 + 40I_2 - 48I_3$

Subtract (b) $\quad 4 = 20I_1 - 15I_2$

giving $\quad -12 = 55I_2 - 48I_3 \qquad \ldots$ (d)

Multiply (c) by 2.4. $\quad 14.4 = 48I_3 + 19.2I_2$

Adding (d) $\quad -12 = -48I_3 + 55I_2$

$$2.4 = 74.2I_2$$

$$\text{Thus } I_2 = \frac{2.4}{74.2} = 0.0324\text{A}$$

Substituting for I_2 in (c)

then $6 = 20I_3 + 8 \times 0.0324$ or $6 = 20I_3 + 0.2592$

$$\text{giving } I_3 = \frac{6 - 0.2592}{20} = 0.287\text{A}$$

Current in 8Ω resistor $= I_2 + I_3 = 0.0324 + 0.287$

$= 0.3194\text{A}$ or 0.32A Ans.

5. With only the 4V battery effective as shown by the diagram (Fig 39).

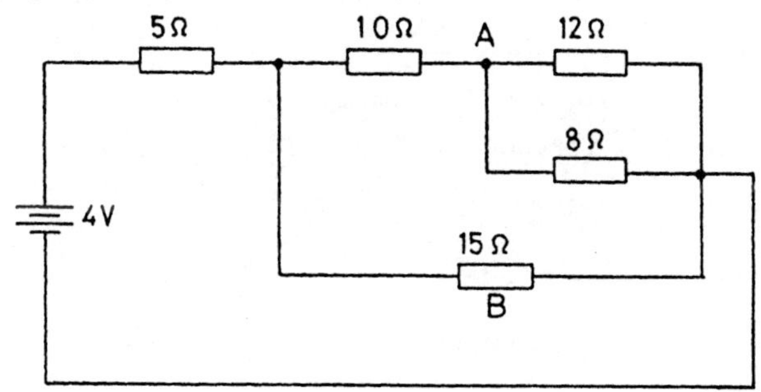

Fig 39

Let R be the resistance of the 12Ω and 8Ω resistors in parallel

$$\text{Then } \frac{1}{R} = \frac{1}{12} + \frac{1}{8} = \frac{2 + 3}{24} \text{ or } R = 4.8\Omega$$

Resistance of branch A = 10 + 4.8 = 14.8Ω
Resistance of branches A and B in parallel = R_{AB}

Then $\frac{1}{R_{AB}} = \frac{1}{14.8} + \frac{1}{15} = \frac{29.8}{222}$ or $R_{AB} = \frac{222}{29.8} = 7.43\Omega$

Total resistance of circuit = 5 + 7.43 = 12.43Ω

$$\text{Circuit current} = \frac{4}{12.43} = 0.322\text{A}$$

Voltage drop across A = 0.322 × 7.43 = 2.59V

$$\text{Current in A} = \frac{2.59}{14.8} = 0.1615\text{A}$$

Voltage drop across 12 and 8Ω resistors = 0.1615 × 4.8
= 0.775V

Current in 8Ω resistor = $\frac{0.775}{8} = 0.097$A

With only the 6V battery effective, as shown by the diagram (Fig 40).

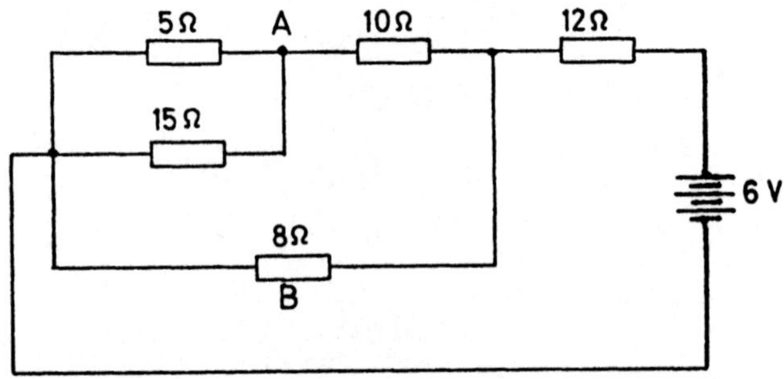

Fig 40

Let R be the resistance of the 5Ω and 15Ω resistors in parallel

Then $\frac{1}{R} = \frac{1}{5} + \frac{1}{15} = \frac{3 + 1}{15}$ or $R = \frac{15}{4} = 3.75\Omega$

Resistance of branch A = 10 + 3.75 = 13.75Ω
Resistance of branches A and B in parallel = R_{AB}

Then $\frac{1}{R_{AB}} = \frac{1}{13.75} + \frac{1}{8} = \frac{8 + 13.75}{110}$

or $R_{AB} = \frac{110}{21.75} = 5.06\Omega$

Total resistance of circuit = 12 + 5.06Ω = 17.06Ω

Circuit current = $\frac{6}{17.06} = 0.352$A

Voltage drop across branch A = 0.352 × 5.06 = 1.78V

Current in 8Ω resistor = $\frac{1.78}{8} = 0.2225$A

Current due to 4V battery
= 0.097A from junction of 10Ω and 12Ω resistors to battery – ve terminal
Current due to 6V battery
= 0.2225A from junction of 10Ω and 12Ω resistors to battery – ve terminal
These currents can be added, thus:
Current in 8Ω resistor = 0.3195A or 0.32A Ans.

6. Consider the top junction or apex of the network (Fig 41).

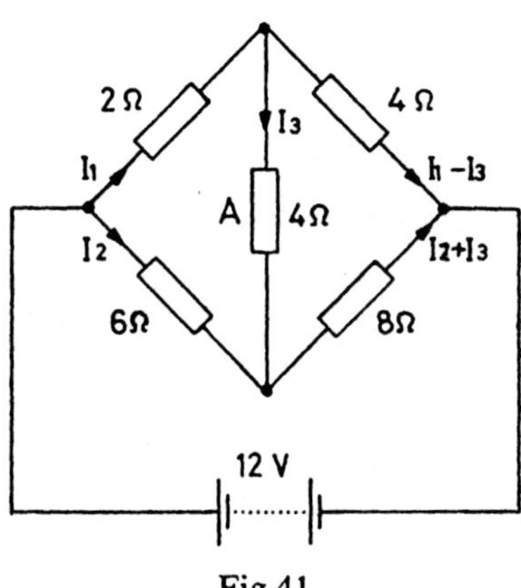

Fig 41

Let I_1 amperes flow in the 2Ω resistor towards this junction
Let I_3 ,, ,, ,, resistor A away from this junction
Let I_2 ,, ,, ,, 6Ω resistor in the direction left to right
Then using Kirchoff's laws, we can build up the following equations.

Left-hand mesh $2I_1 + 4I_3 - 6I_2 = 0$ . . . (a)
Top circuit $2I_1 + 4(I_1 - I_3) = 12$ . . . (b)
Right-hand mesh $4(I_1 - I_3) - 8(I_2 + I_3) - 4I_3 = 0$ . . . (c)
Using (a) and (b)

$$2I_1 + 4I_3 - 6I_2 = 0$$
$$6I_1 - 4I_3 = 12$$

Adding $8I_1 - 6I_2 = 12$ or $I_2 = \dfrac{8I_1 - 12}{6}$

$$= 1.33I_1 - 2$$

Substituting in (a)

$$2I_1 + 4I_3 - 6(1.33I_1 - 2) = 0$$
$$2I_1 + 4I_3 - 7.98I_1 + 12 = 0$$
$$4I_3 - 5.98I_1 = -12 \qquad \ldots (d)$$

Substituting in (c)

$4I_1 - 4I_3 - 8I_2 - 8I_3 - 4I_3 = 0$

$4I_1 - 8I_2 - 16I_3 = 0$

$4I_1 - 8(1.33_1 - 2) - 16I_3 = 0$

$4I_1 - 10.64I_1 + 16 - 16I_3 = 0$

$-6.64I_1 + 16I_3 = -16$

or $6.64I_1 + 16I_3 = 16$ . . . (e)

Multiplying (d) by 4 and solving with (d) and (e)

$16I_3 - 23.921I_1 = -48$

$16I_3 + 6.64I_1 = 16$

Subtracting

$-30.56I_1 = -64$ or $I_1 = 2.09\text{A}$

using (b)

$6I_1 - 12 = 4I_3$ or $I_3 = 1.5I_1 - 3$

thus $I_3 = (1.5 \times 2.09) - 3 = 3.14 - 3 = 0.14\text{A}$

Current is 0.14A (downwards) Ans.

7. $X_A = 2\pi fL = 2 \times \pi \times 50 \times 0.1 = 31.4\Omega$

$Z_A^2 = 30^2 + 31.4^2 = 1886$

Also $X_B = \dfrac{10^6}{2 \times \pi \times 50 \times 30} = \dfrac{10^3}{9.42}$ ohm

or $Z_B^2 = 0^2 + \dfrac{10^6}{9.42^2} = \dfrac{10^6}{9.42^2}$

Then $G_A = \dfrac{30}{1886} = 0.0159\text{S}$ $B_A = \dfrac{-31.4}{1886} = -0.016\ 45\text{S}$

$$G_B = \frac{0}{\frac{10^6}{9.42^2}} = 0\text{S} \quad B_B = \frac{\frac{10^3}{9.42}}{\frac{10^6}{9.42^2}} = \frac{9.42}{10^3} = 0.009\ 42\text{S}$$

Joint $G = 0.0159\text{S}$ Joint $B = -0.007\ 03\text{S}$

$Y = \sqrt{0.0159^2 + 0.007\ 03^2}$

$= 0.0174$ siemens

$= 1.74 \times 10^{-2}$ siemens

$Z = \dfrac{10^2}{1.74} = 57.6\Omega$ Joint Impedance $= 57.6\Omega$ Ans.

The equivalent resistance $= G \times Z^2 = 0.0159 \times 57.6^2$
$= 52.8\Omega$

The equivalent reactance $= B \times Z^2 = 0.007\ 03 \times 57.6^2$
$= 23.32\Omega$

Circuit current $= \dfrac{230}{57.6} = 3.99\text{A}$

Power $= 3.99^2 \times 52.8 = 844\text{W}$ Ans.

8. Branch 1 of section AB

$$X_C = \frac{10^6}{2 \times \pi \times 50 \times 50} = 63.5\Omega$$

and $Z_1^2 = 60^2 + 63.5^2 = 3600 + 4050 = 7650$

Branch 2 of section AB

$X_L = 2 \times \pi \times 50 \times 0.25 = 78.5\Omega$

and $Z^2 = 60^2 + 78.5^2 = 9750$

Thus $G_1 = \frac{60}{7650} = 0.007\ 85S$

$G_2 = \frac{60}{9750} = \underline{0.006\ 15S}$

Total $G = 0.008\ 31S$

Also $B_1 = \frac{63.5}{7650} = 0.008\ 31S$

$B_2 = -\frac{78.5}{9750} = \underline{-\ 0.008\ 06S.}$ (Note the $-$ve sign)

Total $B = -\ 0.000\ 25S$

$Y = 10^{-2}\sqrt{1.4^2 + 0.025^2} = 0.014S$

and $Z = \frac{1}{0.014} = 71.5\Omega$

Equivalent $R = 71.5\Omega$ Equivalent $X = 0\Omega$

Total $R = 100 + 71.5 = 171.5\Omega$

,, $X = 94.2 + 0 = 94.2\Omega$

Note $X_3 = 2 \times \pi \times 50 \times 0.3 = 94.2\Omega$

,, $Z = \sqrt{171.5^2 + 94.2^2}$

$= 197\Omega$

Current $= \frac{500}{197} = 2.54A$

Impedance of section BC $= \sqrt{100^2 + 94.2^2}$

$\therefore Z_{BC} = 137.4\Omega$

and voltage drop $= 2.54 \times 137.4$

$= 348.99V$

Thus 349V will be the voltage across section BC Ans.

9. Branch A $X_L = 2 \times \pi \times 50 \times 0.03$

$= 9.42\Omega$

and $Z_A^2 = 3^2 + 9.42^2 = 97.74$

Branch B $X_C = \frac{10^6}{2 \times \pi \times 50 \times 400} = 7.8\Omega$

and $Z_B^2 = 100^2 + 7.8^2 = 10\ 060$

Branch C $X = X_L - X_C$

$$= (2 \times \pi \times 50 \times 0.02) - \frac{10^6}{2 \times \pi \times 50 \times 300}$$

$$= 6.28 - 10.6 = -4.32\Omega$$

$$Z_C^2 = 7^2 + 4.32^2 = 67.66$$

Then $G_A = \dfrac{3}{97.74} = 0.0307\text{S}$ $\quad B_A = \dfrac{-9.42}{97.74} = -0.0964\text{S}$

$G_B = \dfrac{100}{10\ 060} = 0.009\ 94\text{S}$ $\quad B_B = \dfrac{7.8}{10\ 060} = 0.000\ 776\text{S}$

$G_C = \dfrac{7}{67.66} = 0.103\text{S}$ $\quad B_C = \dfrac{4.32}{67.66} = 0.0637\text{S}$

Total $G = 0.1436\text{S}$ $\quad$ Total $B = -0.0318\text{S}$

$$Y = 10^{-1}\sqrt{1.436^2 + 0.318^2}$$
$$= 0.1471\text{S}$$
$$Z = \frac{1}{0.1471} = 6.798\Omega$$

Thus circuit impedance $Z = 6.798\Omega$ Ans.
Equivalent resistance $R = G \times Z^2 = 0.1436 \times 6.798^2$
$= 6.5\Omega$ Ans.
Equivalent reactance (inductive, since total B is $-$ ve)
$= X_L = B \times Z^2 = 0.0318 \times 6.798^2$
$= 1.48\Omega$ Ans.

10. Circuit A $Z_A = R_A = 20\Omega \qquad X_A = 0\Omega$
Circuit B $X = 2 \times \pi \times 50 \times 0.05$ or $X_B = 15.70\Omega$

$$Z_B^2 = 5^2 + 15.7^2 = 271.49$$

Circuit C $X = \dfrac{10^6}{2 \times \pi \times 50 \times 50}$

or $X_C = 63.65\Omega$

Then $G_A = \dfrac{R_A}{Z_A^2} = \dfrac{20}{20^2} = 0.05\text{S}$

$$G_B = \frac{R_B}{Z_B^2} = \frac{5}{271.5} = 0.0184\text{S}$$

$$G_C = \frac{R_C}{Z_C^2} = \frac{0}{63.65^2} = 0\text{S}$$

Thus total $G = 0.0684\text{S}$

Similarly $B_A = \dfrac{X_A}{Z_A^2} = \dfrac{0}{20^2} = 0\text{S}$

$$B_B = \frac{X_B}{Z_B^2} = -\frac{15.7}{271.5} = -0.0578\text{S}$$

$$B_C = \frac{X_C}{Z_C^2} = \frac{1}{63.65} = 0.0157\text{S}$$

$$\text{Total } B = \overline{-\ 0.0421\text{S}}$$

Hence $Y = 10^{-2}\sqrt{6.84^2 + 4.21^2}$
$= 0.0805\text{S}$

Current, $I = VY = 100 \times 0.0805 = 8.05\text{A}$ Ans.

Circuit power factor is given by $\frac{G}{Y}$ and is lagging, since B is minus or net inductive.

$$\therefore \cos\phi = \frac{0.0684}{0.0805} = 0.85 \text{ (lagging)} \quad \text{Ans.}$$

SELECTION OF TYPICAL EXAMINATION QUESTIONS

SECOND CLASS

1. A thin rectangular plate 350mm by 250mm is totally covered on both sides with nickel 0.12mm thick in 8.25h. The current required would, if supplied to a voltameter, cause 0.0805kg of silver to be deposited in 1h. If the E.C.E. of nickel is 304×10^{-9}kg/C, calculate the density of nickel in kg/m^3, if the E.C.E. of silver is taken as 1118×10^{-9}kg/C.

2. A 110V d.c. lighting system comprises six 150W and 40 60W lamps. Calculate the inductance of a choke-coil of negligible resistance which, when placed in series with this system, would enable it to be operated on 230V, 50Hz mains.

3. The resistance of the armature, field coils and starter of a 220V shunt motor are 0.2, 165 and 9.8Ω respectively, the field being connected across the first stud of the starter and an armature terminal. Calculate (a) the field current at the instant of starting; (b) the field current when running, and (c) the total current taken by the motor at the instant of starting, considering the armature as stationary.

4. If the instantaneous value of a current is represented by $i = 70.7 \sin 520t$, calculate the current's (a) maximum value, (b) r.m.s. value, (c) frequency, (d) instantaneous value 0.0015s after passing through zero.

5. A 220V furnace uses a current of 4.5A to melt 5.5kg of lead in 12 min. If the initial and melting temperatures of the lead are 16°C and 327°C respectively, its specific heat is 0.1278kJ/kg°C, and its latent heat is 22.72kJ/kg, calculate the furnace efficiency.

6. The maximum value of a sinusoidal current wave is 170A. Find graphically the instantaneous current value after 0.001s, 0.003s, 0.006s, and 0.008s after zero and increasing positively. Take the frequency as 50Hz.

7. The open-circuit voltage of a cell, as measured by a voltmeter of 100Ω resistance, was 1.5V, and the p.d. when supplying current to a 10Ω resistance was 1.25V, measured by the same voltmeter. Determine the e.m.f. and internal resistance of the cell.

8. An alternating current series circuit consists of a coil A that has an inductance of 0.3H and negligible resistance and a resistor B of 100Ω. The supply voltage is 200V with a frequency of 50Hz. Determine (a) the impedance of the circuit, (b) the current flowing, (c) the power factor.

9. What is meant by the term 'back e.m.f.' as applied to an electric motor? A 40kW, 220V shunt motor has a full-load efficiency of 90 per cent, an armature resistance of 0.075Ω and a shunt-field resistance of 55Ω. When 'at starting', the starter handle is moved onto the first stud, it is desired to limit the current through the armature to 1.5 times the value which it has when the motor is on full load. What must be the total value of the starting resistance? If, on overload, the speed falls to 90 per cent of its normal full-load value, what would be the armature current? Neglect the effect of armature reaction.

10. Determine the instantaneous value of a sinusoidal e.m.f. of frequency 50Hz and 100V maximum value, 0.001 seconds after it has passed through its zero value. Determine also the time that elapses before the voltage reaches 50 per cent of its maximum value.

11. State Lenz's law. An iron ring is wound with a coil of 84 turns and carries 0.015mWb of residual magnetism. When the coil is excited, the magnetic flux increases to 0.3mWb in 0.12s. Calculate the average value of the e.m.f. which will be self-induced in the exciting coil while the flux is increasing, and state the direction in which it will act relative to the supply voltage, giving reasons.

12. Find the impedance and power factor of an a.c. circuit consisting of two pieces of apparatus in series. Piece A has a resistance of 2Ω and inductive reactance of 14Ω, and piece B has a resistance of 10Ω and a capacitive reactance of 6Ω.

13. Define the temperature coefficient of resistance of a conductor. Name a conductor which has a negative temperature coefficient. When first switched on, the field winding of a 200V shunt motor takes 2A. After running for two hours the field current is observed to have decreased to 1.7A, the supply voltage having remained constant and the setting of the shunt regulator not having been altered. If the ambient air temperature is 15°C, calculate the average temperature rise within the windings. Temperature coefficient of resistance of copper is 0.004 28 at 0°C.

14. A 440V single-phase motor is rated at 7.5kW and operates at a power factor of 0.8 (lagging) with an efficiency of 88 per cent. Find the current taken from the supply.

15. Explain how the value of a current can be measured by the deposition of a metal from a copper sulphate solution. Find the quantity of electricity (in coulombs) which will deposit 0.01kg of copper from a solution of copper sulphate. What current would be required if this process took one hour? What mass of silver would be deposited from a silver nitrate solution if the same current flowed for the same length of time? E.C.E. of copper = 330 × 10^{-9}kg/C. E.C.E of silver = 1118 × 10^{-9}kg/C.

16. A coil consumes 300W when the voltage is 60V d.c. On an a.c. circuit the consumption is 1200W when the voltage is 130V. What is the reactance of the coil?

17. A motor has four poles, its armature is 0.36m in diameter and has 720 conductors whose effective lengths are 0.3m. The flux density of the field under the poles is 0.7T. Each conductor carries 30A. If the armature is turning at 680 rev/min, find the torque in newton metres and the power developed if only two-thirds of the conductors are effective.

18. Define the average value and r.m.s. value of an alternating quantity. Calculate the average r.m.s. value for the stepped half wave given.

Time (ms)	0-10,	10-20,	20-30,	30-40,	40-50,	50-60,	60-70
Steady current (A)	2	4	6	8	6	4	2

19. Two 200V lamps are connected in series across a 400V supply. One lamp is 75W, the other is 40W. What resistance would have to be connected in the circuit so that each lamp gives its correct illumination? What would be the power loss in the resistance?

20. A series circuit consists of a capacitor of 50μF and a coil of inductance 1.5H and resistance 300Ω. Find the total impedance when working on a 50Hz supply. Find whether the current leads or lags the voltage.

21. A 6V battery is connected to points AC of a Wheatstone Bridge ABCD. Sides AB, BC and AD of the bridge have resistance values of 1.5Ω, 2Ω and 1Ω respectively. If the galvanometer is connected across BD and shows no deflection, find the value of resistance of side DC.

22. Explain the term 'power factor'. An alternator supplies 560kW at a power factor of 0.7 (lagging). What extra power would be available if the power factor is increased to 0.8 (lagging) for the same kVA output?

23. The armature winding of a six-pole, lap-wound generator is made up from wire 250m long and 7mm^2 cross-sectional area. If the specific resistance of copper is 1.7×10^{-8}Ωm, find the resistance of the armature.

24. 100V a.c. is applied to a circuit of 3Ω resistance and 4Ω reactance. Find (a) the current in the circuit, (b) the active e.m.f. (resistance voltage drop) of the circuit, (c) the e.m.f. of self-inductance (reactive voltage drop).

25. A generator is over-compounded and designed to supply the circuit load at 220V. If the machine terminal voltage is 220.2V when the load current is 10A, what will be the machine terminal voltage when the load is 600A?

26. A coil of 125Ω impedance has a resistance of 100Ω when connected across a 50Hz supply. Find its inductance. If the impedance falls to 120.6Ω when the frequency is varied, find the new frequency value.

27. 4.5 litres of fresh water at 17°C is heated to boiling point in 15 min. If the heater is 80 per cent efficient and the supply voltage is 220V, find the current taken from the mains and the resistance of the heater. Take the density of water as 1kg/litre and the specific heat capacity as 4.2kJ/kg°C.

28. An alternating voltage of r.m.s. value 100V is applied to a circuit with negligible resistance and an inductive reactance of 25Ω. Determine the r.m.s. variation of current flowing. Show graphically the variation of current and voltage during one cycle of applied voltage. What is the value of the current when the voltage is at its maximum value?

29. A generator has eight brush-arms, each with six brushes, 30mm long and each with a bearing surface of 30mm by 20mm. The current density is 0.054A/mm^2 in the brushes. Find the sectional area of the cables, if the leads to the switchboard are each 9.2m and the current density must not exceed 1.0A/mm^2. Find the power lost in the brushes and cables. The resistivity for carbon and copper is 2550×10^{-8} and 1.7×10^{-8}Ωm respectively.

30. A coil is connected in series with a capacitor of 60μF across a 200V, 50Hz supply. The current is 3A and the power absorbed is 144W. Calculate (a) the p.d. across the capacitor, (b) the resistance and inductance of the coil, (c) the power factor of the coil, and (d) the power factor of the whole circuit.

31. An iron conductor and an aluminium conductor are connected in parallel to a supply. The iron conductor is 10 per cent longer than, and half the diameter of, the aluminium conductor. Given that the ratio of the resistivities of iron to aluminium is 40 to 13, find the ratio of the currents in the two conductors.

32. Two currents I_1 = 14.14A and I_2 = 8.5A with a phase difference of 30°, are fed into a common conductor. Find the resultant current and the heating effect in joules when it passes through a resistor of value 4Ω for a period of 2 minutes.

33. A six-pole d.c. generator has 498 conductors, the e.m.f. per conductor being 1.5V and the current in each 100A.

Find the e.m.f. and current output of the armature, if it is (a) lap wound, (b) wave wound.

34. A circuit takes a current of 10A from 220V, 50Hz mains at a power factor of 0.866 (lagging). Find the value of the current when the voltage is (a) passing through its maximum value and (b) 0.005 seconds later.

35. The filament of an electric lamp is 600mm long and 0.04mm in diameter. The resistivity of the material when cold is $65 \times 10^{-8}\Omega m$. If the resistance of the wire when hot is 5 times the resistance when it is cold, what would be the working current taken by the lamp when placed in a circuit having a supply voltage of 110V?

36. An inductance coil has a resistance of 19.5Ω and when connected to a 220V, 50Hz supply, the current passing is 10A. Find the inductance of the coil.

37. State Faraday's and Lenz's Laws of Electromagnetic Induction. A 4-pole, 250V motor has its armature removed in order to test the continuity of the field windings which are connected in series and consists of 2000 turns each. What is the average e.m.f. induced when the current is switched off, if the flux falls from 0.026Wb to 0.001Wb in 0.2s?

38. A choke when connected across 206V a.c. mains, passed a current of 10A and dissipates 500W. If it is connected in series with apparatus having an impedance of 5Ω and a capacitive reactance of 4Ω, find the impedance and power factor of the complete circuit.

39. A piece of copper wire is bent to form a circle and another piece of the same wire is placed across to form a diameter, all the junctions being electrically connected. If the resistance of the straight wire is 2Ω, find the total current flowing when a p.d. of 220V is applied across the junctions.

40. The average value of a sinusoidal waveform is 125A and the frequency is 60Hz. Calculate the first time from zero when the instantaneous value of the current is 95A. Find also the r.m.s. value of the current.

41. A 15kW motor of efficiency 90 per cent is supplied at 240V by a 2-wire system. The supply cables are 500m long and are of diameter 5mm. Find the current taken by the motor, the voltage at the supply point and the efficiency of the distribution system. Take the specific resistance of copper as $1.7 \times 10^{-8}\Omega$m.

42. If the impedance of a circuit is 20Ω, the resistance is 16Ω and the inductance 0.047 75H, find the frequency of the supply.

43. Differentiate clearly between the kilowatt and the kilowatt-hour. A heater with an efficiency of 85 per cent develops 10MJ in 30 minutes at 200V. Find the energy consumption in kilowatt hours and the current taken. Find also the length of wire in the element if its resistance is 0.26 ohm per metre.

44. An electric heater of resistance 6.5Ω is connected in series with a choke of inductance value 0.1H. If the mains frequency is 50Hz, find the voltage to be applied to the arrangement in order to maintain 110V across the heater. If the frequency was increased by 5 per cent, keeping the applied voltage constant, find the voltage across the heater.

45. When a coil is connected to a 440V, 50Hz supply the current is 15A, and when connected to 440V, 60Hz the current falls to 13A. Calculate the resistance and the inductance of the coil.

46. A resistor of ohmic value 3Ω is connected in series with a coil of inductance 0.1H and resistance 1Ω. If 100V at a frequency of 50Hz is applied to the circuit, find the current flowing.

47. A 500V, d.c. shunt motor has a full-load armature current of 20A. Three per cent of the input power is dissipated as heat in the armature. What would be the current on starting if 500V is applied across the armature. Find also the value of starting resistance required to limit the starting current to twice the full-load current.

48. Find the total effective reactance of a 50Hz circuit made up from a coil of inductance 100mH, in series with a

capacitor of 20μF. If the coil has a resistance of 10Ω, find the impedance of the circuit.

49. The armature resistance of a 200V, shunt motor is 0.4Ω and the no-load armature current is 2A. When fully loaded and taking an armature current of 50A, the speed is 1200 rev/min. Find the no-load speed and state the assumption made in the calculation.

50. A coil takes 5A when connected across a 220V, 50Hz supply. The power taken is 450W. Calculate (a) the power factor, (b) the impedance, resistance, reactance and inductance of the coil.

SOLUTIONS TO TYPICAL SECOND CLASS EXAMINATION QUESTIONS

1. When used in the voltameter: $m = zIt$

or $I = \dfrac{0.0805}{1118 \times 10^{-9} \times 3.6 \times 10^3} = 20\text{A}$

When used for plating:

Area of coating $= 350 \times 250 \times 2 = 175\ 000\text{mm}^2$

Volume ,, $= 175 \times 10^3 \times 12 \times 10^{-2}$

$= 21\ 000\text{mm}^3 = 21 \times 10^{-6}\text{m}^3$

Also mass of nickel deposited

$= 304 \times 10^{-9} \times 20 \times 8.25 \times 3600$

$= 0.181$ kilogrammes

So density $= \dfrac{0.181}{21 \times 10^{-6}}\ \text{kg/m}^3$

$= 8600\text{kg/m}^3$ Ans.

2. Total wattage $= (6 \times 150) + (40 \times 60)$

$= 3300\text{W}$

System current $= \dfrac{3300}{110} = 30\text{A}$

Impedance required on 230V a.c. $= \dfrac{230}{30} = 7.66\Omega$

Resistance of lamps $= \dfrac{110}{30} = 3.66\Omega$

Reactace of choke coil $= \sqrt{7.66^2 - 3.66^2}$

$= 6.73\Omega$

Inductance $= \dfrac{6.73}{2 \times \pi \times 50}$

$= 0.0214\text{H}$ Ans.

3. (a) Field current at instant of starting $= \dfrac{220}{165} = 1.33\text{A}$ Ans.

(b) When running, the starter resistance is inserted into the field circuit by virtue of the position of the contact arm. Field-circuit resistance $= 165 + 9.8 = 174.8\Omega$

Field current when running $= \dfrac{220}{174.8} = 1.26\text{A}$ Ans.

(c) Armature-circuit resistance, at instant of starting
$= 0.2 + 9.8 = 10\Omega$

Armature current $= \frac{220}{10} = 22A$

Total current taken by motor $= 22 + 1.33 = 23.33A$

Ans.

4. (a) Maximum value of current $= 70.7A$ Ans.
(b) The current is sinusoidal
$\therefore$ r.m.s. value $= 0.707 \times$ maximum value
$= 0.707 \times 70.7 = 49.94A$ Ans.
(c) $i = 70.7 \sin (520 \times 0.0015)$ is in the form $i = I_m \sin \omega t$ where $\omega = 520$ radians/second. Also ω equals $2\pi f$ or $2\pi f = 520$
Thus $f = \frac{520}{2 \times \pi} = 82.8Hz$ Ans.
(d) Again $i = 70.7 \sin 2\pi ft$, and if degrees are used for the angle, then
$i = 70.7 \sin (2 \times 180 \times 82.8 \times 0.0015)$
$= 49.65A$ Ans.

5. Energy input to furnace $= \frac{220 \times 4.5}{1000} \times \frac{12}{60}$ kilowatt hour
$= 0.198kW\ h$
$= 0.198 \times 3600$ kilojoules
$= 712.8kJ$
Energy received (by lead)
$= (5.5 \times 311 \times 0.1278) + (5.5 \times 22.72)$ kilojoules
$= 343.7kJ$
Energy put out out by furnace into lead $= 343.7kJ$
Efficiency $= \frac{343.7}{712.8} = 0.482 = 48.2$ per cent Ans.

6. Plot a sine wave on graph paper with a maximum value of 170A and a base of 0.01s. A frequency of 50Hz gives the time of a half wave as 0.01 seconds.

The wave can be plotted from a phasor of length equal to 170A or by use of tables to obtain ordinates. Thus for 30° the instantaneous value or ordinate would be $170 \times \sin 30° = 170 \times 0.5 = 85A$. So for 45° $i = 170 \times 0.707 = 120.2A$. For 60° $i = 147.02A$. For 90° $i = 170A$, etc.

Answers from the deduced waveform:
When time $t = 0.001s$, $i = 57.5A$.
When time $t = 0.003s$, $i = 132A$.

When time $t = 0.006$s, $i = 162$A.
When time $t = 0.008$s, $i = 102$A.

7. Let E = the e.m.f. of the cell and R_i = the internal resistance.
With the voltmeter across the cell terminals only,

Current taken by voltmeter $= \frac{1.5}{100} = 0.015$A

Then $E = 1.5 + (0.015) \times R_i)$ . . . (a)

With the voltmeter and resistor across the cell terminals,

Current taken by resistor $= \frac{1.25}{10} = 0.125$A

Current taken by voltmeter $= \frac{1.25}{100} = 0.0125$A

Current supplied by cell $= 0.125 + 0.0125 = 0.1375$A

Thus $E = 1.25 + 0.1375R_i$ . . . (b)

Solving (a) and (b) then $E = 1.25 + 0.1375R_i$
and $E = 1.5 + 0.015R_i$

Subtracting $0 = -0.25 + 0.1225R_i$

or $R_i = \frac{0.25}{0.1225} = 2.04\Omega$ Ans.

and $E = 1.5 + (0.015 \times 2.04) = 1.5 + 0.031$
Thus cell e.m.f. $= 1.531$V Ans.

8. $R_A = 0\Omega$ $X_A = 2\pi fL = 2 \times \pi \times 50 \times 0.3$
$= 94.2\Omega$
$R_B = 100\Omega$ $Z = \sqrt{100^2 + 94.2^2}$
$= 137.4\Omega$ Ans.

Current $I = \frac{200}{137.4} = 1.45$A Ans.

Power factor, $\cos\phi = \frac{R}{Z} = \frac{100}{137.4} = 0.73$ (lagging) Ans.

9. Output $= 40 \times 1000$ watts

Input $= 40 \times 10^3 \times \frac{100}{90}$ watts

$= 44\,444$W

Input current $= \frac{44\,444}{220} = 202.02$A

Shunt-field current $= \frac{220}{55} = 4$A

Armature current $= 202.02 - 4 = 198.2$A
Armature starting current $= 198.2 \times 1.5 = 297.3$A

Resistance of armature circuit $= \frac{220}{297.3} = 0.74\Omega$

Resistance to be added $= 0.74 - 0.075 = 0.665\Omega$ Ans.

On normal load $E_b = 220 - (198.2 \times 0.075)$ volts $= 205.13\text{V}$

On 90 per cent speed $E_{b1} = 0.9 \times 205.13 = 184.62\text{V}$

$\therefore$ Armature voltage drop $= 220 - 184.62 = 35.38\text{V}$

Armature current $= \frac{35.38}{0.075} = 471.7\text{A}$ Ans.

10. Substituting in $v = V_m \sin 2\pi ft$

$v = 100 \times \sin(2 \times 180 \times 50 \times 0.001)$

$= 30.9\text{V}$ Ans.

If $v = 50\text{V}$, then $50 = 100 \sin(2 \times 180 \times 50 \times t)$

and $\frac{50}{100} = \sin(2 \times 180 \times 50 \times t)$

or $\frac{1}{2} = \sin\theta$, whence $\theta = 30°$

Thus $30 = 18 \times 10^3 t$ or $t = \frac{3}{18 \times 10^2}$

$= 0.166 \times 10^{-2}$ seconds

The required time $= 1.66\text{ms}$ Ans.

11. $E_{av} = \frac{N(\Phi_2 - \Phi_1)}{t}$ volts

$= \frac{84(3 \times 10^{-4} - 1.5 \times 10^{-5})}{0.12}$

$= 0.1995\text{V}$ or 199.5mV Ans.

The induced voltage will oppose the applied supply voltage, thus reducing the rate of current growth.

12. Total resistance of circuit $= 2 + 10 = 12\Omega$

Total reactance ,, ,, $= 14 - 6 = 8\Omega$ (inductive)

From impedance relationship:

$$Z = \sqrt{12^2 + 8^2} = 14.4\Omega$$

The circuit impedance is 14.4Ω Ans.

$\cos\phi = \frac{R}{Z} = \frac{12}{14.4} = 0.832$

Thus power factor $= 0.832$ (lagging) Ans.

13. Carbon has a negative temperature coefficient.

Resistance of field (cold) $= R_1 = \frac{200}{2} = 100\Omega$

Resistance of field (hot) $= R_2 = \frac{200}{1.7} = 117.64\Omega$

Then $100 = R_0 (1 + 0.004\,28 \times 15)$
and $117.64 = R_0 (1 + 0.004\,28 \times T)$

Dividing $\frac{117.64}{100} = \frac{1 + 0.0024\,28T}{1 + 0.0642}$

or $1.1764 \times 1.0642 = 1 + 0.004\,28T$

and $1.2519 - 1 = 0.004\,28T$

$\therefore T = \frac{0.2519}{0.004\,28} = 58.85°C$

Temperature rise of winding $= 58.85 - 15°C$
$= 43.85°C$ Ans.

14. Motor output $= 7.5 \times 1000 = 7500W$

Motor input $= 75000 \times \frac{100}{88} = 8523W$

But $VI \cos \phi = P \therefore I = \frac{P}{V \cos \phi} = \frac{8523}{440 \times 0.8}$ amperes

or $I = \frac{8523}{352} = 24.21A$

Thus motor current = 24.2A Ans.

15. 330×10^{-9} kilogrammes of copper are deposited by 1 coulomb

$\therefore$ 0.01kg of copper will require $\frac{0.01}{330 \times 10^{-9}}$

$= 30\,303C$ Ans.

If time was 1 hour, since $Q = It$

$\therefore I = \frac{Q}{t} = \frac{30\,303}{3600}$

or current required = 8.418A Ans.

Mass of silver would be in proportion to the E.C.E.

So mass $= \frac{0.01 \times 1118 \times 10^{-9}}{330 \times 10^{-9}}$

= 0.033 87kg or 33.87g Ans.

16. When on d.c. $P = \frac{V^2}{R} \therefore R = \frac{V^2}{P} = \frac{60 \times 60}{300} = 12\Omega$

When on a.c. $P = 1200W$ also $P = I^2R$

So $I^2 = \frac{1200}{12} = 100$ or $I = 10A$

The impedance Z of the circuit $= \frac{V}{I} = \frac{130}{10} = 13\Omega$

So reactance $X = \sqrt{13^2 - 12^2}$
$= 5\Omega$

Thus reactance of coil $= 5\Omega$ Ans.

17. Force on 1 conductor is given by $F = BIl$ newtons
or $F = 0.7 \times 30 \times 0.3 = 6.3$N

No of conductors in the field at any instant $= \frac{2}{3} \times 720$
$= 480$

Total force $= 480 \times 6.3 = 3024$N
Torque = force × radius $= 3024 \times 0.18$ newton metres
$= 544.32$
So torque exerted = 544Nm Ans.

Power developed $= \frac{2\pi NT}{60}$

$= \frac{2 \times \pi \times 680 \times 544}{60}$

$= 38.7$kW Ans.

18. If plotted, this waveform will be found to be made up of seven rectangular blocks, the mid-ordinates of which are 2, 4, 6, etc, as given.

$\therefore$ Average value $= \frac{2 + 4 + 6 + 8 + 6 + 4 + 2}{7} = \frac{32}{7}$

$= 4.57$A Ans.

Also r.m.s. value $= \sqrt{\frac{2^2 + 4^2 + 6^2 + 8^2 + 6^2 + 4^2 + 2^2}{7}}$

$= \sqrt{\frac{4 + 16 + 36 + 64 + 36 + 16 + 4}{7}}$

$= \sqrt{\frac{176}{7}} = \sqrt{25.14} = 5$A Ans.

19. 75W lamp. $I = \frac{75}{200} = 0.375$A

40W lamp. $I = \frac{40}{200} = 0.2$A

With lamps in series 40W lamp will only pass 0.2A
$\therefore$ (0.375 − 0.2) amperes must be passed through a shunt resistor connected across the 40W lamp. This resistor is also to be suitable for 200V, and its resistance value must

$= \frac{200}{0.175} = 1143\Omega$ Ans.

Power loss in this resistor $= 200 \times 0.175 = 35$W Ans.

20. Here $L = 1.5$H so $X_L = 2\pi fL = 2 \times 3.14 \times 50 \times 1.5$
$= 471\Omega$

Also $C = 50\mu F$ so $X_C = \frac{1}{2\pi fC} = \frac{10^6}{2 \times \pi \times 50 \times 50}$

$= 63.7\Omega$

Total reactance $X = X_L - X_C \therefore 471 - 63.7 = 407.3\Omega$ (inductive)

Impedance of circuit, $Z = \sqrt{R^2 + X^2} = \sqrt{300^2 + 407.3^2}$
$= 506\Omega$

The circuit impedance is 506 ohms Ans.
Current will lag the voltage because the circuit is inductive. Ans.

21. With no current in BD, points B and D are at the same potential, and since A is common, then:
P.D. across AB = p.d. across AD.
Similarly since C is common, then:
P.D. across BC = p.d. across DC.
Let I_1 = current in top branch and I_2 = current in bottom branch.
Then $I_1 \times 1.5 = I_2 \times 1$ and $I_1 \times 2 = I_2 \times R_X$
Written as a ratio $\frac{R_X \times I_2}{1 \times I_2} = \frac{2 \times I_1}{1.5 \times I_2}$

Thus $R_X = \frac{2}{1.5} = 1.33\Omega$ Ans.

22. Since $P = VI \cos \phi$, then $VI = \frac{P}{\cos \phi}$

or kilovolt amperes (S) = $\frac{\text{kilowatts } (P)}{\text{power factor } (\cos \phi)}$

Thus $S = \frac{560}{0.7} = 800\text{kVA}$

If this kVA or S value is to be maintained
New $P = 800 \times 0.7 = 640\text{kW}$
Thus extra power available = $640 - 560 = 80\text{kW}$ Ans.

23. Here $\varrho = 1.7 \times 10^{-8}\Omega\text{m}$. $l = 250\text{m}$. A = $7 \times 10^{-6}\text{m}^2$.
Then $R = \frac{\varrho l}{A} = \frac{1.7 \times 10^{-8} \times 250}{7 \times 10^{-6}} = 0.607\Omega$

This would be the resistance of the length of wire. but this is a lap-wound generator, with six parallel paths in the armature. Thus resistance of 1 parallel path $= \frac{0.607}{6}$
$= 0.101\ 16\Omega$

But there are 6 paths in parallel so the equivalent resistance is one-sixth of the above $= \dfrac{0{:}101\ 16}{6}$

$= 0.016\ 86\Omega$ Ans.

24. Impedance of circuit $= \sqrt{3^2 + 4^2}$

$= 5\Omega$

Thus $Z = 5\Omega$

(a) Current $I = \dfrac{V}{Z} = \dfrac{100}{5} = 20\text{A}$ Ans.

(b) Resistive voltage drop $V_R = IR = 20 \times 3 = 60\text{V}$ Ans.

(c) Reactive voltage drop $V_X = IX = 20 \times 4 = 80\text{V}$ Ans.

25. When the load current is 10A, the voltage drop in the cables, between machine terminals and load = 220.2 − 220 = 0.2V

By proportion the voltage drop for 600 amperes

$$= 0.2 \times \frac{600}{10} = 12\text{V}$$

If load voltage is still to be 220V, the terminal voltage would need to be 220 + 12 = 232V.

Thus machine terminal voltage = 232V Ans.

26. Reactance X_L at 50Hz $= \sqrt{Z^2 - R^2} = 100\sqrt{1.25^2 - 1^2}$

$= 75\Omega$

Also $X_L = 2\pi fL \therefore L = \dfrac{75}{2 \times \pi \times 50}$

$= 0.24\text{H}$ Ans.

At new frequency $X_L = \sqrt{120.6^2 - 100^2}$

$= 67.4\Omega$

So $\dfrac{\text{new frequency}}{50} = \dfrac{67.4}{75}$

or new frequency $= 67.4 \times \dfrac{50}{75}$

$= 44.93$ hertz

The new frequency value would be 45Hz Ans.

27. Mass of water = volume × density = 4.5 × 1 = 4.5kg

Heat received by water = 4.5 × 4.2 (100 − 17)

= 18.9 × 83 kilojoules = 1569kJ

Electrical energy supplied to heater $= 1569 \times \dfrac{100}{80}$

$= 1961\text{kJ}$

$$\text{Power rating of heater} = \frac{\text{energy}}{\text{time}} = \frac{1961}{15 \times 60}$$
$$= 2.18\text{kW}$$

$$\text{Current taken from mains} = \frac{2180}{220} = 9.9\text{A}$$

Mains current = 10A (approx) Ans.

$$\text{Resistance of heater} = \frac{220}{9.9} = 22.2\Omega \quad \text{Ans.}$$

28. Here $R = 0\Omega$ and $Z = X = 25\Omega$

$$\therefore \text{ r.m.s. value of current } I = \frac{V}{Z} = \frac{100}{25} = 4\text{A}$$

The graphical solution consists of a sinusoidal voltage wave with a sinusoidal current wave lagging it by 90°, since the circuit is wholly inductive. Thus when voltage is maximum current is zero. When voltage has fallen to zero, the current has risen to its maximum value and as voltage rises to its negative maximum the current falls to zero.

When V is a maximum, current value is zero Ans.

29. Bearing surface of 1 brush = 30 × 20 square millimetres
= 600mm²

With a current density of 0.054A/mm², the current carried by one brush = 600 × 0.054 = 32.4A

With 8 brush arms, there are 4 positive and 4 negative brush arms. Also since there are 6 brushes per arm, the number of brushes in parallel carrying current into or out of the armature = 4 × 6 = 24 brushes, or total current carried by brushes = 32.4 × 24 = 777.6A.

Current display density in the cable is limited to 1.6A/mm².

$$\therefore \text{ Cable is to carry 777.6A} = \frac{777.6}{1.6} = 486\text{mm}^2 \quad \text{Ans.}$$

$$\text{Resistance of cable} = \frac{\varrho l}{A} = \frac{1.7 \times 10^{-8} \times 9.2 \times 2}{486 \times 10^{-6}}$$
$$= 0.643 \times 10^{-3} \text{ ohms}$$

$$\text{Voltage drop in cable} = 777.6 \times 0.643 \times 10^{-3}$$
$$= 0.5\text{V}$$

Power loss in cable = 777.6 × 0.5 = 388.8W Ans.

Resistance of the brushes is given by $R = \frac{\varrho l}{A}$ where l is the length of a +ve plus a −ve brush and A is the area of half the total number of brushes.

Thus $R = \frac{2550}{10^8} \times \frac{2 \times 30 \times 10^{-3}}{600 \times 10^{-6} \times 24}$

$= 0.106 \times 10^{-3}$ ohms

Power loss in the brushes is given by I^2R

$= 777.6^2 \times 0.106 \times 10^{-3}$

$= 64.1$ watts

Thus power loss in brushes = 64.1W Ans.

30. $X_C = \frac{10^6}{2\pi fC} = \frac{10^6}{2 \times \pi \times 50 \times 60} = 53\Omega$

If Z is the circuit impedance then $Z = \frac{V}{I} = \frac{200}{3} = 66.6\Omega$

Now only resistance is responsible for power dissipation, then $P = I^2R$ or $144 = 3^2R$. Thus $R = \frac{144}{9} = 16\Omega$

For this circuit X is made up of inductive reactance X_L and capacitive reactance X_C which nullify each other.

The resultant reactance $X = \sqrt{Z^2 - R^2}$

$= \sqrt{66.6^2 - 16^2}$

$= 64.7\Omega$

But X_C alone is 53Ω $\therefore X_L = 64.7 + 53 = 117.7\Omega$

Thus (a) p.d. across capacitor = $53 \times 3 = 159$V Ans.

(b) Resistance of coil = $R = 16\Omega$ Ans.

Inductance of coil $= \frac{X_L}{2\pi f} = \frac{117.7}{100 \times \pi}$

$= 0.38$H Ans.

(c) Impedance of coil $Z_L = \sqrt{16^2 + 117.7^2}$

$= 119\Omega$ Ans.

Power factor of coil $= \cos \phi_L = \frac{R}{Z_L} = \frac{16}{119}$

$= 0.1346$ (lagging) Ans.

(d) Power factor of circuit $= \cos \phi = \frac{R}{Z}$

$= \frac{16}{66.6} = 0.245$ (lagging) Ans.

Note. Both power factors are lagging, the circuit being net inductive.

31. The equations for the iron and aluminium conductors can be written as $R_i = \frac{\varrho_i l_i}{A_i}$ and $R_a = \frac{\varrho_a l_a}{A_a}$

$\therefore \quad \frac{R_i}{R_a} = \frac{\varrho_i L_i}{A_i} \Big/ \frac{\varrho_a l_a}{A_a} = \frac{\varrho_i l_i A_a}{\varrho_a l_a A_i}$

But $l_i = 1.1\ l_a$

and $d_i = \frac{1}{2}d_a \therefore A_i = \frac{A_a}{4}$ since area $\propto$ diameter^2.

So $\frac{R_i}{R_a} = \frac{40 \times 1.1 \times l_a \times A_a \times 4}{13 \times l_a \times A_a}$

$= \frac{13.54}{1}$

Thus, since the resistance ratio of the iron to the aluminium wires are 13.54 to 1, and as the wires are in parallel, the currents in the wires are in the ratio Iron : Aluminium = 1 : 13.54. Ans.

32. Resultant current $I = \sqrt{I_1^2 + I_2^2 + (2 \times I_1 \times I_2 \times \cos 30)}$
$= \sqrt{14.14^2 + 8.5^2 + (2 \times 14.14 \times 8.5 \times 0.866)}$
$= 21.92$A

Resultant current = 21.92A Ans.
Power dissipated = $I^2R = 21.92^2 \times 4$ watts
Energy at heat = I^2Rt joules
$= 21.92^2 \times 4 \times 2 \times 60$
= 230 602J Ans.

33. (a) Lap-wound. A = P = 6

Conductors in series = $\frac{\text{total conductors}}{\text{parallel paths}} = \frac{498}{6} = 83$

E.m.f. of 1 parallel path = e.m.f. of machine
$= 1.5 \times 83 = 124.5$V Ans.

Current per parallel path = current in 1 conductor
= 100A

Current of 6 paths in parallel = 6×100 = 600A Ans.

(b) Wave-wound. A = 2

Conductors in series = $\frac{\text{total conductors}}{\text{parallel paths}} = \frac{498}{2} = 249$

E.m.f. of 1 parallel path = e.m.f. of machine
$= 1.5 \times 249 = 373.5$V Ans.

Current per parallel path = current in 1 conductor
= 100A

Current of two paths in parallel = 2×100 = 200A Ans.

34. The power factor of this circuit is 0.866 (lagging) or $\cos \phi = 0.866$ (lagging) and $\phi = 30°$, where ϕ is the angle of lag between the voltage and the current—the latter lagging the former.

The values of 10A and 220V as given, can be assumed to be r.m.s. values. So the maximum value of current is given by $I = 0.707\, I_m$ (here I_m is the maximum value). Sine-wave working is assumed.

or $I_m = \dfrac{10}{0.707} = 14.14\text{A}$

Also $V = 0.707\, V_m$ (maximum value). Sine-wave working is assumed.

or $V_m = \dfrac{220}{0.707} = 311.08\text{V}$

The voltage and current can be written as

$v = V_m \sin \omega t$

and $i = I_m \sin (\omega t - \phi)$ ϕ is in radians

At a frequency of 50Hz, time for 1 cycle $= \dfrac{1}{50}$ seconds

(a) When the voltage is at a maximum, the time is for $\frac{1}{4}$ cycle

or $t = \dfrac{1}{4 \times 50} = 0.005$ seconds

Current at this instant is given by substituting in

$i = 14.14 \sin (2\pi 50 \times 0.005 - \phi)$

or $i = 14.14 \sin (2 \times 180 \times 50 \times 0.005 - 30)$. π and ϕ in degrees

$= 14.14 \sin (90 - 30)$

$= 14.14 \sin 60° = 12.25\text{A}$ Ans.

(b) At an instant 0.005s later t would be 0.01s

$\therefore i = 14.14 \sin (2 \times 180 \times 50 \times 0.01 - 30)$

or $i = 14.14 \sin (180 - 30) = 14.14 \sin 150°$

$i = 14.14 \sin 30° = 7.07\text{A}$ Ans.

35. Resistance of filament (cold)

$$R = \frac{\varrho l}{A} = \frac{65 \times 10^{-8} \times 600 \times 10^{-3} \times 4}{\pi \times (0.04 \times 10^{-3})^2}$$

or $R = 310.35\Omega$

$\therefore$ Resistance when hot $= 5 \times 310.35$ ohms

Working current $= \dfrac{110}{5 \times 310.35}$ amperes

$= 0.071\text{A}$ Ans.

36. From the information given, the impedance Z of the circuit is $= \dfrac{V}{I}$ or $Z = \dfrac{220}{10} = 22\Omega$

The resistance R is 19.5Ω. Therefore the reactance X is obtained from $X = \sqrt{Z^2 - R^2} = \sqrt{22^2 - 19.5^2}$

$= 10.15\Omega$

Also $X = 2\pi fL \therefore L = \dfrac{10.15}{2 \times \pi \times 50}$

$= 0.032\text{H}$ Ans.

37. From Faraday's law.

$$E_{av} = \frac{N(\Phi_1 - \Phi_2)}{t} \text{ volts}$$

$$= \frac{2000 \times 4(2.6 \times 10^{-2} - 0.1 \times 10^{-2})}{0.2}$$

Induced e.m.f. = 1000V or 1kV Ans.

38. Z of coil $= \dfrac{206}{10} = 20.6\Omega$

Also since $P = I^2R$, then R of coil $= \dfrac{500}{10^2} = 5\Omega$

Thus reactance X of coil $= \sqrt{Z^2 - R^2}$

or $X = \sqrt{20.6^2 - 5^2}$

$= 19.98\Omega$

Z of additional apparatus $= 5\Omega$

X ,, ,, $= 4\Omega$ (capacitive)

$\therefore R$,, ,, $= \sqrt{5^2 - 4^2} = 3\Omega$

Total resistance of circuit $= 5 + 3 = 8\Omega$

Total reactance ,, $= 19.98 - 4 = 15.98\Omega$

Note. The inductive and capacitive reactances have been subtracted.

Total impedance of circuit $= \sqrt{8^2 + 15.98^2}$

$= 17.88\Omega$ Ans.

Since $\cos\phi = \dfrac{8}{17.88} = 0.44$ (lagging)

Thus power factor = 0.44 (lagging), since circuit is net inductive Ans.

39. Since the circuit is built up from wire of the same material and sectional area, then the resistance of various parts of the circuit are proportional to length.

The resistance of the diameter = 2 ohms

$\therefore$ The resistance of the circumference $= \pi d = 2\pi$ ohms.

The resistance of $\frac{1}{2}$ circumference $= \pi$ ohms

The circuit is made up of a diameter and two $\frac{1}{2}$ circumferences in parallel. $\therefore$ if R is the circuit resistance

$$\frac{1}{R} = \frac{1}{2} + \frac{1}{\pi} + \frac{1}{\pi} = \frac{\pi + 2 + 2}{2\pi} = \frac{\frac{22}{7} + 4}{2 \times \frac{22}{7}}$$

$$\text{or } \frac{1}{R} = \frac{\frac{22 + 28}{7}}{\frac{44}{7}} = \frac{50}{44} = \frac{25}{22}$$

$$\text{or} \quad R = \frac{22}{25} = 0.88\Omega$$

With 220V applied across R, the current would be

$$= \frac{220}{0.88} = 250\text{A} \quad \text{Ans.}$$

40. For a sine wave;

Average value = 0.6365 × maximum value

$$\therefore \quad \text{Maximum value of current } I_m = \frac{\text{av value}}{0.6365} = \frac{125}{0.6365}$$

$$\text{or } I_m = 196.4\text{A}$$

Also as current is sinusoidal, then $i = I_m \sin 2\pi ft$

$$\text{or } 95 = 196.4 \sin (2 \times 180 \times 60 \times t)$$

$$\text{and } \sin (18 \times 12 \times 10^2 \times t) = \frac{95}{196.4} = 0.483$$

Now from sine tables, the sine of 28.9° is 0.483

$$\therefore 28.9 = 216 \times 10^2 t \text{ or } t = \frac{28.9 \times 10^{-2}}{216}$$

$$\text{Thus } t = 0.001\ 34\text{s.}$$

First time from zero = 0.001 34s or 1.34ms Ans.

R.M.S. value of current = 0.707 × 196.4

= 138.86A Ans.

41. Output of motor = 15kW

$$\text{Input to motor} = \frac{15}{0.9} \text{ kilowatts} = 16\ 666\text{W}$$

$$\text{Input current or current in cables} = \frac{16\ 666}{240}$$

$$= 69.44\text{A} \quad \text{Ans.}$$

Resistance of cable is given by $\frac{\varrho l}{A}$

$$\text{or } R = \frac{1.7 \times 10^{-8} \times 500 \times 2}{\frac{\pi}{4} \times (5 \times 10^{-3})^2}$$

$$= 0.866\Omega$$

Voltage drop in cable = 69.44 × 0.866 = 60.14V

Input voltage at supply cables = 240 + 60.14 = 300.14V

Ans.

$$\text{Efficiency of distribution} = \frac{\text{Power output from cables}}{\text{Power input to cables}}$$
$$= \frac{240 \times 69.44}{300.14 \times 69.44}$$
$$= 0.799 = 79.9 \text{ per cent} \quad \text{Ans.}$$

42. Here $Z = 20\Omega$ and $R = 16\Omega \therefore X = \sqrt{20^2 - 16^2}$
$= 12\Omega$

Also $X = 2\pi fL = 2 \times \pi \times f \times 0.047\ 75 = 12$

$$\text{Thus } f = \frac{12}{2 \times \pi \times 0.047\ 75}$$
$$= 40\text{Hz}$$

The frequency of the supply is 40Hz Ans.

43. Output of heater = 10MJ
also 1kW h = $3600 \times 1000 = 36 \times 10^5$ joules

$$\text{Now energy output of heater} = \frac{10 \times 10^6}{36 \times 10^5}$$
$$= 2.78\text{kW h}$$

$$\text{Energy input} = \frac{\text{output}}{\text{efficiency}} = \frac{2.78}{0.85} = 3.27\text{kW h} \quad \text{Ans.}$$

$$\text{Power input} = \frac{\text{energy}}{\text{time}} = \frac{3.27}{0.5} = 6.54\text{kW}$$

$$\text{Also, since } P = I^2R = I \times R \times I = V \times \frac{V}{R} = \frac{V^2}{R}$$

$$\therefore R = \frac{V^2}{P}$$

$$\text{Thus } R = \frac{200^2}{6540} = 6.12\Omega$$

$$\text{Length of element at } 0.26\Omega/\text{m} = \frac{\text{resistance}}{\text{ohms per metre}} = \frac{6.12}{0.26}$$
$$= 23.54 \quad \text{Ans.}$$

$$\text{Current taken} = \frac{P}{V} = \frac{6540}{200} = 32.7 \quad \text{Ans.}$$

44. If 110V is maintained across the heater, the current would be $\frac{110}{6.5} = 16.92\text{A}$

At 50Hz reactance of choke $= 2\pi fL$
$= 2 \times \pi \times 50 \times 0.1$
$= 31.4\Omega$

Impedance Z of complete circuit $= \sqrt{6.5^2 + 31.4^2}$
$= 32.07\Omega$

Applied voltage for 16.92A = 16.92 × 32.07 = 542.6V
Ans. If frequency rises 5 per cent, reactance rises 5 per cent.
New reactance = 3.14 × 1.05 = 32.97Ω
New impedance = $\sqrt{6.5^2 + 32.97^2}$
= 33.6Ω

Circuit current = $\frac{542.6}{33.6}$ = 16.15A

Voltage across heater = 16.15 × 6.5 = 104.97
= 105V (approx) Ans.

45. At 50Hz $Z = \frac{440}{15} = 29.33\Omega$

At 60Hz $Z = \frac{440}{13} = 33.85\Omega$

In general $Z^2 = R^2 + X_L^2$ or $R^2 = Z^2 - X_L^2$
and $X_L = 2\pi fL$
Hence $R^2 = Z^2 - (2\pi fL)^2$
$\therefore \quad R^2 = 29.33^2 - (2\pi \times 50 \times L)^2$
$R^2 = 860.24 - (100\pi L)^2$
also $R^2 = 33.85^2 - (2\pi \times 60 \times L)^2$
$R^2 = 1145.56 - (120\pi L)^2$
$860.44 - (100\pi L)^2 = 1145.56 - (120\pi L)^2$
$(120\pi L)^2 - (100\pi L)^2 = 1145.56 - 860.24$
$L = 0.08\text{H}$ Ans.

Thus $R^2 = Z^2 - X_L^2$
$= 29.33^2 - (2\pi 50 \times 0.08)^2$
$R = 15.12\Omega$ Ans.

46. Total resistance of circuit = 3 + 1 = 4Ω
Reactance of circuit is given by $X = 2\pi fL$
or $X = 2 \times \pi \times 50 \times 0.1$
$= 31.4\Omega$

Circuit impedance $Z = \sqrt{4^2 + 31.4^2}$
= 31.65Ω

Current flowing = $\frac{100}{31.65}$ = 3.16A Ans.

47. Neglecting the field current, since it will be small
Input power = VI
= 500 × 20 = 10 000W

3 per cent of the input power = $\frac{3}{100} \times 10\,000$ = 300W

This is dissipated as heat in the armature, *ie* it is a copper or I^2R loss.

$$\therefore I^2R_a = 300 \text{ or } 20^2R_a = 300$$

$$\text{and } R_a = \frac{300}{400} = 0.75\Omega$$

The starting current with only armature resistance to limit the armature current

$$= I_{as} = \frac{500}{0.75} = 666.66\text{A} \quad \text{Ans.}$$

Twice full load current = 20 × 2 = 40A

∴ Total resistance required in the armature circuit to limit starting current to 40A $= \frac{500}{40} = 12.5\Omega$

Since $R_a = 0.75\Omega$

Series resistance would be 12.5 − 0.75 = 11.75Ω Ans.

48. Inductive reactance $X_L = 2\pi fL$

$$= 2 \times \pi \times 50 \times 100 \times 10^{-3}$$

$$\text{or } X_L = 31.4\Omega$$

$$\text{Capacitive reactance } X_C = \frac{10^6}{2\pi fC} = \frac{10^6}{2 \times \pi \times 50 \times 20}$$

$$= 159.23\Omega$$

$$\text{Total effective reactance } X = 159.23 + 31.4$$

$$\text{or } X = 127.83\Omega \text{ (capacitive)} \quad \text{Ans.}$$

$$\text{Impedance of circuit } Z = \sqrt{R^2 + X^2}$$

$$= \sqrt{10^2 + 127.83^2}$$

$$= 128.3\Omega \quad \text{Ans.}$$

49. On no-load: $E_{b0} = V - I_aR_a = 200 - (2 \times 0.4)$

$$= 200 - 0.8 = 199.2\text{V}$$

On full-load: $E_{b1} = 200 - (50 \times 0.4) = 200 - 20$

$$= 180\text{V}$$

As this is a shunt motor, constant field current and therefore the same constant flux can be assumed for the no-load and full-load conditions.

Since $E_{b0} = k\Phi_0N_0$ and $E_{b1} = k\Phi_1N_1$ and $\Phi_0 = \Phi_1$

then $\frac{E_{b0}}{E_{b1}} = \frac{k\Phi_0N_0}{k\Phi_1N_1}$ or $N_0 = \frac{N_1E_{b0}}{E_{b1}}$

This gives $N_0 = 1200 \times \frac{199.2}{180} = 1328$ rev/min Ans.

50. Impedance Z of coil $= \frac{220}{5} = 44\Omega$

Also Power $P = I^2R$ or $R = \frac{P}{I^2} = \frac{450}{5^2}$

Hence $R = 18\Omega$

Also since $Z^2 = \sqrt{R^2 + X^2}$ then $X = \sqrt{Z^2 - R^2}$

or $X = \sqrt{44^2 - 18^2}$

Hence $X = \sqrt{1612}$

$= 40.14\Omega$

(a) Power factor $\cos\phi = \frac{R}{Z} = \frac{18}{44} = 0.41$ (lagging) Ans

(b) Impedance $Z = 44\Omega$ Resistance $R = 18\Omega$

Reactance $X = 40.14\Omega$

Inductance L is given by: $X = 2\pi fL$

So $L = \frac{X}{2 \times \pi \times f} = \frac{40.14}{2 \times \pi \times 50}$

Inductance, $L = 0.127\text{H}$ Ans.

SELECTION OF TYPICAL EXAMINATION QUESTIONS

FIRST CLASS

1. The rudder motor of a Ward-Leonard type steering gear is a compound-wound machine, details of which are given below. From this information find: (a) armature current, (b) the torque developed by the motor, (c) output power of motor. Armature — volts 90, resistance 0.0288Ω, rev/min 370. *IR* drop over armature may be taken as 8.5V. Shunt field — separately excited from 110V supply, resistance 65Ω, turns per pole 1000. Series field — separately excited by prime mover line current of 325A, turns per pole 11. Torque Nm = (armature current (A) × flux (kWb) per pole × 10^{-4} × 6.8) − 15.

M.M.F. per pole (At)	3000	3500	4000	4500	5000	5500	6000
Flux per pole (kWb)	2500	2700	2860	2990	3100	3190	3275

2. A 500V installation consists of a synchronous motor taking 50kW working in parallel with a load of 90kW having a power factor of 0.6 (lagging). If the power factor of the combined load is 0.8 (lagging), find the power factor and reactive kVA of the motor.

3.

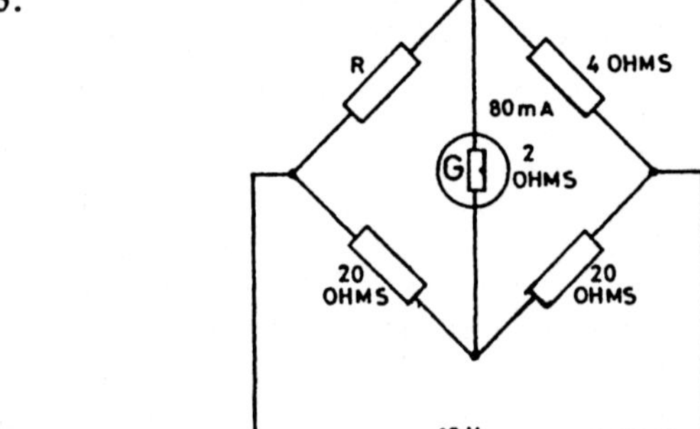

3. The electrical circuit of an explosive gas sampling device is shown above. The resistance R is heated by the gas being ignited and causes a current of 80mA to flow through the galvanometer. If the element is of platinum, to what temperature has it been raised if the instrument was originally balanced at 15.55°C? (Increase in resistance per ohm per °C for platinum = 0.002 43Ω).

4. A 175kVA, 6600/440V, single-phase transformer has an iron loss of 2.75kW. The primary and secondary windings have resistances of 0.4Ω and 0.0015Ω respectively. Calculate the efficiency on full load when the power factor is 0.9.

5. A shunt-wound generator has the following open-circuit characteristic. If the actual field-resistance value is half that of the critical field resistance, above which the machine will fail to excite, find the open-circuit voltage. The e.m.f. when the generator is operating at a load of 200A, falls to 135V. Find the terminal voltage and the armature resistance.

Field current (A)	0.5	1.0	2.0	3.0	4.0	5.0
Open-circuit voltage (V)	55	90	133	160	179	193

6. What are the units of length, mass and force in the SI system of units?

A current of 12A produces a magnetic flux of 0.4mWb in a coil of 60 turns. What e.m.f. would be induced in the coil if the current of 12A was reversed in 25 milliseconds? What is the inductance of the coil in henrys?

7. A two-wire ring main, 2km long, is supplied at a point 'X' with 220V. At a point 'Y' situated 400m from 'X' there is a load of 110A. The resistance of 1km of single, main conductor is 0.032Ω. Calculate the current in each section of the main, and the voltage at the load.

8. In a supply, the voltage and current vary sinusoidally at a frequency of 50Hz, the r.m.s. values being 311.2V and 70.7A respectively, at a power factor of 0.866 (lagging). Plot on one chart the graphs, for a single cycle, of the voltage and current in their correct relative positions. From them, derive the instantaneous values of voltage and current at 3, 6, 11 and 18 milliseconds from the time that the voltage passed through zero, in the course of it increasing positively.

9. The moving coil of a permanent-magnet voltmeter is made of copper and has a resistance of 5Ω when at 20°C. The instrument is connected in series with a resistance, the value of which is 995Ω at 20°C. Calculate the percentage error, high or low, of the reading at 50°C, if the series resistance is made of (a) copper and (b) manganin. Take the temperature coefficient of copper as 0.004 28/°C at 0°C and that of manganin as zero.

10. A parallel circuit consists of a branch A of resistance 10Ω, inductance 38mH and capacitance 312.8μF, and a branch B of negligible resistance and inductive reactance of 12Ω. Derive graphically, the total current and its phase angle, if the system is connected to a 440V, 50Hz supply.

11. The field windings of a motor comprise 8 coils connected in series, each coil having 1200 turns. The flux linked with each coil is 0.05Wb when the current is 5A. Calculate the inductance of the field circuit and the value of the average e.m.f. induced, if the current was cut off in 50ms.

12. A 550kVA, 50Hz, single-phase transformer has 1875 and 75 turns in the primary and secondary windings respectively. If the secondary voltage is 220V, calculate (a) primary voltage, (b) primary and secondary currents, (c) maximum value of flux.

13. The armature and field resistances of a 220V shunt motor are 0.25 and 110 ohms respectively and, when running on no load, the motor takes 6A. Calculate the losses attributable to iron, friction and windage and, assuming this value to remain constant on all loads, determine the efficiency when the current supplied is 62A.

14. Explain the principles underlying the necessity for the introduction of the term 'power factor' when considering a.c. machinery.

An alternator is supplying a load of 560kW at a power factor of 0.7 (lagging). If apparatus is installed that raises the power factor to 0.8 (lagging), calculate the increase in power available for the same kVA loading.

15. A battery consisting of 48 cells, having an internal resistance of 0.02 ohm per cell, is to be charged so that the

e.m.f. per cell is raised from 1.8 to 2.2V. Determine the maximum number of carbon lamps of a parallel bank, which may be switched on in series with the circuit so that the current from the 200V mains does not exceed 10A at the commencement of charging.

If the circuit remains unaltered, calculate the current flowing through the battery when the charge is approaching completion. Each lamp is marked 110V, 100W.

16. A heater unit of negligible inductance has a resistance of 6.5Ω and is intended for use with 100V mains. For what 50Hz voltage would it be suitable when placed in series with an external apparatus, of negligible resistance, having an inductance of 0.01H? If the frequency rises by 5 per cent and this voltage remains constant, what would be the resulting change of voltage at the heater terminals?

17. The magnetic field in the air gap of a two-pole motor has a flux density of 0.8T. The armature is wound with 246 conductors, each of 400mm effective length, mounted at 150mm effective radius, and at full load each conductor carries a current of 20A. Assuming that the actual torque produced is equivalent to that due to two-thirds of the number of conductors cutting the lines of force at right-angles, find (a) the torque in newton metres, and (b) the shaft power developed at 500 rev/min.

18. State briefly, the meaning of the expressions 'star-connected' and 'delta-connected' as applied to three-phase a.c. practice. What is the ratio of the maximum line voltage to the maximum phase voltage in each case.

Determine the line current taken by a 440V, three-phase, star-connected motor having an output of 45kW at 0.88 (lagging) power factor and an efficiency of 93 per cent.

19. A battery is to consist of a number of cells connected in series. Each cell has an e.m.f. of 1.5V and an internal resistance of 0.5Ω. The external load has a resistance of 100Ω and requires approximately 2W for satisfactory operation. Determine how many cells will be required.

20. A battery comprises six cells each of e.m.f. 2.2 volts and internal resistance 0.1Ω. Determine the value of the load resistance connected when the battery delivers maximum power and evaluate this maximum power.

21. The earth-lamps on a main switchboard comprise two 240V 60W lamps connected in the usual manner. The potential difference at the busbars is 220V. Damage by sea water occurs to a distribution cable so that the insulation resistance to earth is reduced to 16Ω and 6Ω for +ve and −ve cables respectively. Find by calculation (a) which of the two lamps burns the brighter; and (b) the additional load on the generators occasioned by the fault. The resistance of the cables and the ship's structure may be neglected, and that of the lamps taken as constant at the value corresponding to the 60W rating.

22. If an alternator supplies the following loads: (a) 200kW lighting load at unity power factor, (b) 400kW induction-motor load at 0.8 (lagging), power factor, (c) 200kW synchronous-motor load, find the power factor of the synchronous-motor load, to give an overall power factor of 0.97 (lagging).

23. A d.c. shunt-wound machine is run as a motor, being supplied with 55kW at 220V when its speed is 500 rev/min. Find the speed at which this machine should be driven to generate an output of 55kW with a terminal potential difference of 220V. The resistance of the armature is 0.02Ω and that of the field, which is the same for each case, is 110Ω.

24. A single-phase, 50Hz transformer, has a core with a square cross-section, each side being 270mm. The transformation ratio is 3500/440V, and the maximum flux density in the core is not to exceed 1.4T. Find the number of turns of the windings required if the frequency is 50Hz.

25. Explain the purpose of interpoles and state their magnetic polarity relative to the main poles of both generators and motors.

A 200V, long-shunt compound-wound generator has a full-load output of 20kW. The various resistances are as follows; armature (including brush contact) 0.15Ω, series field 0.025Ω, interpole field 0.028Ω, shunt field (including the field-regulator resistance) 115Ω. The iron losses at full load are 780W, and the friction and windage losses 590W. Calculate the efficiency at full load.

26. A 500V, three-phase alternator supplies a balanced delta-connected load in parallel with a balanced star-connected load. The delta load is 30kW at a power factor of 0.92 (leading) and the star load is 40kW at a power factor of 0.85 (lagging). Calculate the line current and the power factor of the supply.

27. A 440V shunt motor takes an armature current of 30A at 700 rev/min. The armature resistance is 0.7Ω. If the flux is suddenly reduced 20 per cent, to what value will the armature current rise momentarily? Assuming unchanged resisting torque to motion, what will be the new steady values of speed and armature current? Sketch graphs showing armature current and speed as functions of time during the transition from initial to final, steady-state conditions.

28. A load takes 250A at 240V and a power factor of 0.8 (lagging) from a 50Hz supply. The supply cable is then operating at its full rating. Find graphically, the additional power which can be supplied, without the cable exceeding its rating, when a 800μF capacitor is connected across the load.

29. Two shunt generators work in parallel. Each has an armature resistance of 0.015Ω and a shunt-field resistance of 85Ω. Machine A is excited so that its e.m.f. is 600V, while the other machine B is excited so that its e.m.f. is 620V. What is the output of each machine when they jointly supply a load of 2500A and what is the busbar voltage?

30. Illustrate what is meant by phase difference in an a.c. circuit, by both wave and phasor diagrams. Three currents which can be represented by: $i_A = 19 \sin \theta$ amperes

$$i_B = 35 \sin\left(\theta - \frac{\pi}{2}\right) \text{ amperes}$$

$$\text{and } i_C = 42 \sin\left(\theta - \frac{5\pi}{6}\right) \text{ amperes}$$

flow in the same conductor. Find graphically the maximum value of the resultant current, its r.m.s. value and its phase position relative to the 42A (maximum value) current?

31. A 230V motor, which normally develops 10kW at 1000 rev/min with an efficiency of 85 per cent, is to be used as a

generator. The armature resistance is 0.15Ω and the shunt-field resistance is 220Ω. If it is driven at 1080 rev/min and the field current is adjusted to 1.1A, by means of the shunt regulator, what output in kW could be expected as a generator, if the armature copper loss was kept down to that when running as a motor?

32. Two coils are connected in parallel across a 220V, 60Hz supply. At the supply frequency, their impedances are 16Ω and 25Ω respectively, and their resistances are 3Ω and 7Ω respectively. Find the current in each coil, the total current and the total power. Draw a complete phasor diagram for the system.

33. A shunt-wound generator has a magnetisation-curve given by the figures below. The total resistance in the field circuit is 20Ω and the armature resistance is 0.02Ω. With the machine on load, estimate the e.m.f. generated and the armature current when the terminal voltage of the machine is 140V.

Field current (I_f)—amperes	1.2	2.8	5.0	7.0	7.7	9.0	11.0
Generated e.m.f. (E)—volts	46	88	126	149	154	162	168

34. A 12-pole, three-phase, delta-connected alternator runs at 600 rev/min and supplies a balanced star-connected load. Each phase of the load is a coil of resistance 35Ω and inductive reactance 25Ω. The line terminal voltage of the alternator is 440V. Determine (a) frequency of supply, (b) current in each coil, (c) current in each phase of the alternator, (d) total power supplied to the load.

35. A 400V shunt motor has an armature resistance of 0.45Ω and a shunt-field resistance of 450Ω. When run on no load, the current taken from the supply is 2.8A. Calculate the output power and efficiency of the machine when the current it takes from the mains is 35A.

36. A coil of 100Ω resistance and 0.1H inductance is connected in series with a 0.1μF capacitor to a 230V variable frequency a.c. supply. Calculate the resonant frequency and the p.d. across the capacitor at resonance.

37. Three d.c. generators connected in parallel, each supply a load of 640A to a set of 220V busbars. The e.m.f. of one

generator is raised from 230V to 235V. If the load and the resistances are constant, determine the current supplied from each generator and the voltage at the busbars.

38. A balanced delta-connected load and a balanced star-connected load are connected in parallel to a 220V, three-phase supply. The delta-connected load takes a total power of 50kW at a power factor of 0.75 (lagging), and the star-connected load, 40kW at a power factor of 0.62 (leading). Calculate the power, volt amperes and power factor of the supply.

39. A section of supply cable AB 1km long has a fault to earth such that, when end B is disconnected, the resistance measurement from end A to earth is 5Ω. When end A is disconnected, the resistance reading from end B to earth is 3Ω. The length of the cable AB has a resistance of 4Ω when intact. Find the distance of the fault from end A.

40. A coil of 0.84H inductance and 50Ω resistance is connected in series with a capacitor of 14μF capacitance. (a) find the frequency for resonance and the potential differences across the capacitor, across the coil and across the two, when a current of 5A at this frequency, is flowing, (b) find the three potential differences when the same current flows at 60Hz.

41. A series-connected d.c. motor has a field and armature resistance of 0.1Ω and runs at 600 rev/min when taking a full-load current of 100A from a 210V supply. Calculate the speed of the motor when the torque is reduced 75 per cent.

42. Each phase of a star-connected load consists of a resistor of 14Ω in parallel with a 400μF capacitor. Calculate the line current, power and power factor when the above load is connected to a 440V, 60Hz, three-phase supply. What power would be dissipated in the load, if it is reconnected in delta?

43. A ring-main, 900m long, is supplied at a point A at a p.d. of 220V. At a point B, 240m from A, a load of 45A is drawn from the main, and at a point C, 580m from A, measured in the same direction, a load of 78A is taken from the main. If the resistance of the main (lead and return) is 0.25Ω per

kilometre, calculate the current which will flow in each direction round the main from the supply point A and the potential difference across the main, at the load where it is lowest.

44. A non-inductive coil of 6Ω resistance is connected in parallel with an inductive coil of 3Ω resistance and 9Ω impedance at 50Hz. If a potential difference of 110V is applied to the terminals, find the current in each coil and in the mains. If a capacitor of 600μF is connected in parallel with the coils of the above example, calculate the total current.

45. A shunt 400V motor has an armature resistance of 0.3Ω and a shunt-field circuit resistance of 300Ω. Calculate the speed attained by the motor when the current taken from the supply is (a) 30A and (b) 50A. The open-circuit characteristic of the machine, when driven as a generator at 1500 rev/min, is given by the following figures:

Field current (A)	0.25	0.5	0.75	1.0	1.25	1.5
Open-circuit voltage (E)	220	398	515	598	638	662

Neglect the effects of armature reaction.

46. A three-phase transformer has 560 turns on the primary and 42 turns on the secondary. The primary windings are connected to a line voltage of 6.6kV. Calculate the secondary line voltage when the transformer is connected (a) Star-Delta, (b) Delta-Star.

47. Two batteries, of e.m.f.s 220V and 225V and internal resistances of 0.2Ω and 0.3Ω respectively, are connected in parallel to supply a load resistor of 10Ω. Find the current supplied by each battery and the terminal voltage.

48. A kettle, when connected to a 220V d.c. supply, boils 1 litre of water from 11°C in 3.5 min. Calculate the percentage time difference when the water is boiled, by connecting the kettle to a 220V, 50Hz a.c. supply. The inductance of the element of 0.05H. Take 1 litre of water to have a mass of 1kg and its specific heat capacity as 4.2kJ/kg°C.

49. In a shunt motor the four field coils are connected in series. Each coil is wound to give 750 ampere-turns, the

length of each turn being 450mm. At the safe working temperature, there are 45 watts dissipated at each coil. If the supply voltage is 220V, find (a) the field current, (b) the diameter of the wire, and (c) the length of wire in each coil. Take the resistivity of copper as $2.0 \times 19^{-8}\Omega m$.

50. An eight-pole alternator running at a speed of 720 rev/min supplies current to synchronous and induction motors with 48 poles. Calculate the frequency and speed of rotation of the motors if the induction motor runs with 2 per cent slip.

SOLUTIONS TO TYPICAL FIRST-CLASS EXAMINATION QUESTIONS

1. This problem is not as involved as it would appear, but requires an understanding of the electrical connections of a Ward-Leonard system.

Armature-resistance voltage drop 8.5V $R_a = 0.0288\Omega$

$$\therefore I_a = \frac{8.5}{0.0288} = 295.15\text{A} \quad \text{Ans.}$$

$$\text{Shunt-field current } I_{sh} = \frac{110}{65} = 1.692\text{A}$$

Shunt ampere-turns/pole = 1.692 × 1000 = 1692
Series ampere-turns/pole = 325 × 11 = 3575
Total ampere-turns/pole = 1692 + 3575 = 5267

Plot the graph. From this Φ = 3150 kilowebers per pole.

From the given expression:

$$T = 295.15 \times 3150 \times 10^{-4} \times 6.8 - 15$$
$$= 617.2\text{Nm} \quad \text{Ans.}$$

$$\text{Output power} = \frac{2\pi NT}{60} = \frac{2 \times \pi \times 370 \times 617.2}{60}$$
$$= 24\text{kW} \quad \text{Ans.}$$

2. Total active power of combined load, $P = 50 + 90$
$= 140$kW

Total apparent power of combined load, $S = \dfrac{140}{\cos\phi} = \dfrac{140}{0.8}$
$= 175$kVA

Total reactive power of combined load, $Q = 175 \times 0.6$
$= 105$kVAr

Active power of original load, $P_1 = 90$kW

Apparent power of original load, $S_1 = \dfrac{90}{\cos\Phi} = \dfrac{90}{0.6}$
$= 150$kVA

Reactive power of original load, $Q_1 = 150 \times 0.8$
$= 120$kVAr

Thus the reactive power is reduced by 120 − 105 = 15 kVAr and this kVAr figure is that of the motor, Q_2 operating at a leading power factor. Thus reactive power Q_2 of motor = 15kVAr Ans.

Also apparent power of motor, $S_2 = \sqrt{50^2 + 15^2}$
$= 52.2\text{kVA}$

Power factor of motor, $= \dfrac{P_2}{S_2} = \dfrac{50}{52.2} = 0.96$ (leading) Ans.

3. When cold and the 'bridge is balanced' B and D are at the same potential. If R increases in resistance value, current in branch ABC falls and the potential of point B falls below D until, for the 'hot' condition being considered B is $2 \times 80 \times 10^{-3} = 160 \times 10^{-3} = 0.16\text{V}$ below D.

Currents will flow as shown in the diagram and, as a solution by Kirchoff's laws will be made for the hot condition, the values I_1 and I_2 are allocated.

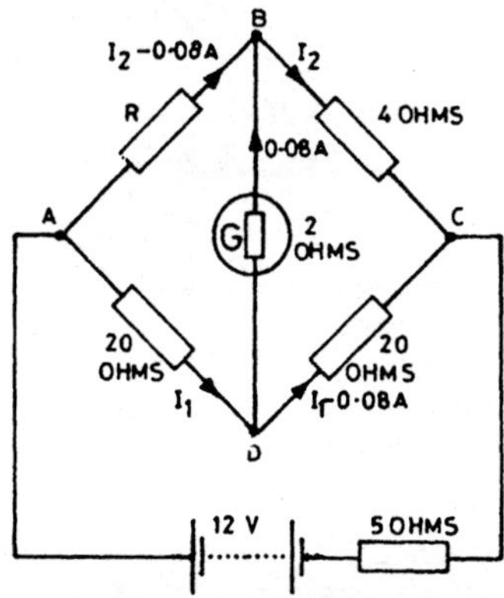

When 'cold' and at balance; value of $R = 4\Omega$
When 'hot': consider mesh BCD

Then $4I_2 - 20(I_1 - 0.08) + (2 \times 0.08) = 0$
or $4I_2 - 20I_1 + 1.6 + 0.16 = 0$
and $4I_2 - 20I_1 + 1.76 = 0$
whence $I_2 - 5I_1 + 0.44 = 0$ ... (a)

Consider circuit ADC and battery:

Then $20I_1 + 20(I_1 - 0.08) + 5(I_2 + I_1 - 0.08) = 12$
or $20I_1 + 20I_1 - 1.6 + 5I_2 + 5I_1 - 0.4 = 12$
and $45I_1 + 5I_2 - 2 = 12$
whence $45I_1 + 5I_2 = 14$ ... (b)

Solving (a) and (b) then: $45I_1 + 5I_2 = 14$
and $-45I_1 + 9I_2 = -3.96$
Adding $14I_2 = 10.04$
or $I_2 = \dfrac{5.02}{7} = 0.717\text{A}$

Again $5I_1 = I_2 + 0.44 = 0.717 + 0.44 = 1.157\text{A}$

and $I_1 = \dfrac{1.157}{5} = 0.2314\text{A}$

Applying Kirchhoff's law to circuit ABC and battery. Then:

$R(0.717 - 0.08) + 4(0.717) + 5(0.717 + 0.2314 - 0.08) = 12$

or $R(0.637) + 2.868 + 5(0.8684) = 12$

and $0.637R + 2.868 + 4.342 = 12$

whence $0.637R = 12 - 7.21$

Thus $0.637R = 4.79$ or $R = \dfrac{47.9}{6.37} = 7.52\Omega$

Let T_2 = the hot temperature and T_1 = the cold temperature = 15.55°C. The temperature coefficient, as given, is assumed to be for a base of 0°C. Thus $\alpha = 0.002\,43/°\text{C}$.

Thus $\dfrac{7.52}{4} = \dfrac{R_0(1 + 0.002\,43 \times T_2)}{R_0(1 + 0.002\,43 \times 15.55)}$

or $7.52(1 + 0.037\,79) = 4(1 + 0.002\,43T_2$

whence $7.52 + 0.2842 = 4 + 0.009\,72T_2$

$7.8042 - 4 = 0.009\,72T_2$

or $T_2 = \dfrac{3.8042}{0.009\,72}$

$= 391.37°\text{C}$ Ans.

4. Although the transformer has not been dealt with in any detail in this book, this problem can be worked from first principles.

Thus efficiency $(\eta) = \dfrac{\text{output}}{\text{input}} = \dfrac{\text{output}}{\text{output + losses}}$

In a transformer there are no rotational losses and the expression can be written as:

$$\eta = \frac{\text{output (kW)}}{\text{output (kW)} + \text{copper loss (kW)} + \text{iron loss (kW)}}$$

Thus $\eta = \dfrac{\text{kVA} \cos\phi}{\text{kVA} \cos\phi + P_C + P_{Fe}}$

Here $P_{Fe} = 2.75\text{kW}$

Now primary current $= \dfrac{175\,000}{6600} = 26.51\text{A}$

Primary copper loss $= \dfrac{26.51^2 \times 0.4}{1000}$

$= 0.2812\text{kW}$

Similarly, secondary current $= \dfrac{175\,000}{440} = 397.7\text{A}$

Secondary copper loss $= \dfrac{397.7^2 \times 0.0015}{1000}$

$= 0.237\text{kW}$

$$\text{Thus } \eta = \frac{175 \times 0.9}{175 \times 0.9 + (0.2812 + 0.237) + 2.75}$$
$$= 0.98$$

Efficiency on full load = 98 per cent. Ans.

5. The O.C. characteristic is plotted as shown and a tangent drawn. The resistance value obtained from this tangent is the critical resistance. Thus consider a field current of 1A. The voltage value for this current, reading off the critical-resistance line is 120V. The critical resistance is therefore $\frac{120}{1} = 120\Omega$.

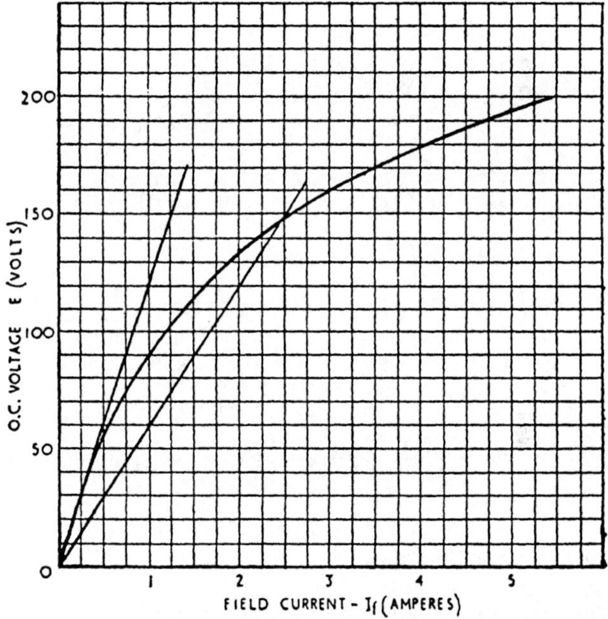

According to the given data, the field resistance is $\frac{120}{2} = 60\Omega$ and the required O.C. voltage is obtained, by taking any field-current value, say 2A, and finding the voltage value for this current. Thus $2 \times 60 = 120$V. Plot this value and join the point to the origin to obtain the field voltage-drop line. The intersection with the O.C.C. gives the required voltage = 146V Ans.

Again, from the O.C.C., an e.m.f. of 135V requires a field current of 2.1A. It must be remembered that the field resistance has not been altered and is 60Ω. The voltage necessary to maintain 2.1A through this field resistance is the terminal voltage.

$\therefore$ Terminal voltage = 2.1 × 60 = 126V Ans.

Armature voltage drop = 135 − 126 = 9V

Also $9 = I_a R_a$

$$\therefore \quad R_a = \frac{9}{200 + 2.1} = 0.045\Omega \quad \text{Ans.}$$

Note. I_a = line current + shunt-field current.

6. Using the expression $E_{av} = \dfrac{N(\Phi_2 - \Phi_1)}{t}$

$$\text{We have } E_{av} = \frac{60\,[(0.4 \times 10^{-3}) - (-0.4 \times 10^{-3})]}{25 \times 10^{-3}} \text{ volts}$$

or E_{av} = 1.92V. Ans.

Also $E_{av} = L \times$ average rate of change of current

or $1.92 = L \times \dfrac{I}{t}$ Thus $1.92 = L \times \dfrac{12}{12.5 \times 10^{-3}}$

Note. The time taken for the current to fall to zero has been taken.

$$\text{so, } L = \frac{1.92 \times 12.5 \times 10^{-3}}{12} = 0.002\text{H}$$

The inductance of the coil is 2mH Ans.

L can also be found from first principles, since the inductance value can be determined from the flux linkages per ampere

$$\text{Thus } L = \frac{N\Phi}{I} = \frac{60 \times 0.4 \times 10^{-3}}{12}$$

$= 0.002$H Ans.

7. Let I be the current in the short section of the ring, *ie* in the 400m length. Therefore $(110 - I)$ is the current in the 2000 − 400 = 1600m length.

Resistance of 400m of cable (double conductor)

$$= \frac{0.032 \times 800}{1000} = 0.0256\Omega$$

Resistance of 1600m of cable

$= 0.0256 \times 4 = 0.1024\Omega$

Since points X and Y are connected by both sections of the ring, it follows that the voltage drop in the short-section = the voltage drop in the longer section

or $I \times 0.0256 = (110 - I) \times 0.1024$.

and $0.0256I = 0.1024 \times 110 - 0.1024I$

or $0.128I = 0.1024 \times 110$

$\therefore I = 88$A

Current in shorter section = 88A Ans.
Current in longer section = 110 − 88 = 22A Ans.
Voltage at load = 220 − (88 × 0.0256)
= 217.75V Ans.

8. Cos ϕ = 0.866. $\therefore$ ϕ = 30°C. Thus current lags voltage by 30°.

$$\text{Maximum value of voltage} = \frac{311.2}{0.707} = 440\text{V}$$

$$\text{Maximum value of current} = \frac{70.7}{0.77} = 100\text{A}$$

$$\text{Time for 1 cycle} = \frac{1}{50}\text{ second} = 0.02\text{s}$$

Time for $\frac{1}{2}$ cycle = 0.01s

Since voltage is sinusoidal, then $v = 440 \sin(2 \times 180 \times 50 \times t)$

When t = 0.001s, v = 440 sin (2 × 180 × 50 × 0.001)
= 440 sin 18° = 136V
When t = 0.002s, v = 440 sin 36° = 259V
,, = 0.003s, v = 440 sin 54° = 356V
,, = 0.004s, v = 440 sin 72° = 418V
,, = 0.005s, v = 440 sin 90° = 440V
,, = 0.006s, v = 418V, *etc, etc*

The voltage wave is plotted to a time base t, as shown. Similarly for the current wave.

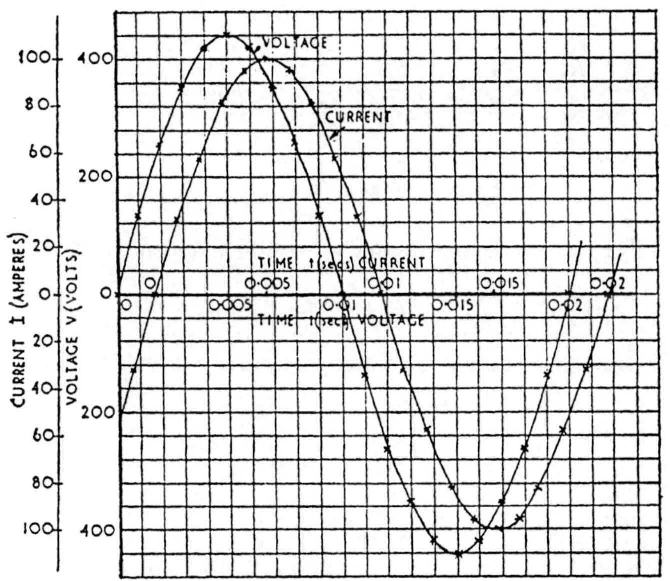

Since $i = 100 \sin (2 \times 180 \times 50 \times t)$
When $t = 0.001\text{s}, i = 100 \sin 18° = 30.9\text{A}$
,, $= 0.002\text{s}, i = 100 \sin 36° = 58.8\text{A}$
,, $= 0.003\text{s}, i = 100 \sin 54° = 80.9\text{A}$
,, $= 0.004\text{s}, i = 100 \sin 72° = 95.1\text{A}$
,, $= 0.005\text{s}, i = 100 \sin 90° = 100\text{A}$
,, $= 0.006\text{s}, i = 95.1\text{A}$, *etc, etc.*

Draw the current wave to a new scale with its zero displaced from that of the voltage time scale by 0.001 66s. Use the ordinates derived above for this new time scale.

Note. If a time of 90° = 0.005s, 30° = 0.001 66s.

Read off the required answers from the original voltage-time scale on both the voltage and current waveforms.

Thus when
$t = 3 \times 10^{-3} = 0.003\text{s}$, $v = 355\text{V}$ and $i = 38\text{A}$ Ans.
$t = 0.006\text{s}$, $v = 415\text{V}$ and $i = 96\text{A}$ Ans.
$t = 0.011\text{s}$, $v = -140\text{V}$ and $i = 25\text{A}$ Ans.
$t = 0.018\text{s}$, $v = -260\text{V}$ and $i = 93\text{A}$ Ans.

9. (a) Assume the meter indicates correctly at 20°C when the resistance is 1000Ω. Let I_1 be the meter current at this temperature. Also let R_2 and T_2 be the resistance and temperature at 50°C and let R_1 and T_1 be the resistance and temperature at 20°C.

Then from the resistance to temperature relation,

$$\frac{R_2}{R_1} = \frac{R_0(1 + \alpha T_2)}{R_0(1 + \alpha T_1)} \quad \therefore R_2 = \frac{1000(1 + \alpha 50)}{(1 + \alpha 20)} \text{ ohms}$$

$$\text{or } R_2 = \frac{1000(1 + 0.004\ 28 \times 50)}{1 + 0.004\ 28 \times 20}$$

$$\text{and } R_2 = 1118.2\Omega$$

$$\text{Meter current } I_2 \text{ at } 50°\text{C} = \frac{V}{118.2} \text{ where } I_1 = \frac{V}{1000}$$

The voltage across the meter can be assumed constant.

$$\therefore \quad \frac{I_2}{I_1} = \frac{1000}{1118.2} \text{ or } I_2 = 0.894I_1 \text{ amperes}$$

Now meter deflection is proportional to current

$$\therefore \text{Percentage error} = \frac{\text{difference in readings}}{\text{true reading}} \times 100$$

$$\text{or } \frac{I_1 - I_2}{I_1} \times 100 = \frac{I_1 - 0.894I_1}{I_1} \times 100$$

$$\text{Thus percentage error} = \frac{0.106I_1}{I_1} \times 100$$

$$= 10.6 \text{ per cent (low)} \quad \text{Ans.}$$

(b) If the series resistor is made from manganin, its value at 50°C does not vary and the resistance of the 5Ω copper coil is given by R_2

$$\therefore \quad \frac{R_2}{R_1} = \frac{R_0(1 + 0.004\,28 \times 50)}{R_0(1 + 0.004\,28 \times 20)}$$

or $R_2 = 5.58\Omega$

The resistance of the meter at t_2 is now 995 + 5.58 ohms = 1000.58Ω

New current $I_2 = \dfrac{V}{1000.58}$ or as before $I_1 = \dfrac{V}{1000}$ amps.

$\therefore \dfrac{I_2}{I_1} = \dfrac{1000}{1000.58}$ So new current $I_2 = 0.9994 I_1$

∴ Percentage error, as deduced before,

$$= \frac{I_1 - 0.9994 I_1}{I_1} \times 100 = \frac{1 - 0.9994}{1} \times 100$$

or percentage error = 0.0006 × 100 = 0.06 per cent (low) Ans.

10. Branch A. Inductive reactance, $X_L = 2\pi f L$

or $X_L = 2 \times \pi \times 50 \times 38 \times 10^{-3}$
$= 11.932\Omega$

Capacitive reactance, $X_C = \dfrac{1}{2\pi f C}$

or $X_C = \dfrac{10^6}{2 \times \pi \times 50 \times 312.8}$
$= 10.16\Omega$

Resultant reactance, $X = X_L - X_C$
$= 11.932 - 10.16$
or $X = 1.772\Omega$ (inductive)

Impedance of Branch A, $Z_A = \sqrt{10^2 + 1.772^2}$
$= 10.17\Omega$

Current $I_A = \dfrac{V}{Z_A} = \dfrac{440}{10.17} = 43.2\text{A}$

$\cos \phi_A = \dfrac{10}{10.17} = 0.98$ (lagging)

$\sin \phi_A = \dfrac{1.772}{10.17} = 0.174$

Branch B. $R = 0 \quad X_L = 12\Omega$

$\therefore Z_B = 12\Omega \quad I_B = \dfrac{440}{12} = 36.66\text{A}$

$\cos \phi_B = \dfrac{0}{Z_B} = 0 \quad \sin \phi_B = 1$

Solving mathematically: $I_a = (43.2 \times 0.98) + (36.66 \times 0)$
$= 42.3 + 0 = 42.3\text{A}$

$$I_r = (43.2 \times 0.174) + (36.66 \times 1)$$
$$= 7.52 + 36.66 = 44.18\text{A}$$

Resultant current

$$I = \sqrt{42.3^2 + 44.18^2}$$
$$= 61\text{A} \quad \text{Ans.}$$

$$\text{Power factor, } \cos\phi = \frac{42.3}{61} = 0.694 \text{ (lagging)} \quad \text{Ans.}$$

Note. The problem has been solved mathematically, but the graphical solution is made by choosing a suitable current scale and, with voltage as reference, *ie* along the horizontal, draw I_A lagging the voltage by an angle whose cosine is 0.98. Next draw I_B lagging the voltage by 90°. Complete the parallelogram. Draw the longer diagonal, measure this to scale and also the angle of lag. The cosine of this angle will give the power factor of the circuit. The answers should check with those obtained above.

11. Using the expression:
Inductance (in henries) = flux-linkages per ampere

$$\text{Then } L = \frac{N\Phi}{I} = \frac{8 \times 1200 \times 50 \times 10^{-3}}{5}$$
$$= 96\text{H} \quad \text{Ans.}$$

$$\text{Also } E_{av} = L \times \text{rate of change of current}$$
$$= 96 \times \frac{5}{50 \times 10^{-3}}$$
$$= 9600\text{V}$$

Thus the value of induced e.m.f. is 9.6kV Ans.

12. (a) Although no work has been done in this book on the transformer, it can be stated that the voltages induced in the primary and secondary windings are in direct proportion to the turns.

$$\text{Thus } \frac{V_1}{V_2} = \frac{N_1}{N_2} \text{ or } V_1 = V_2 \times \frac{N_1}{N_2}$$

$$\text{Hence } V_1 = 220 \times \frac{1875}{75} = 220 \times 25$$

and primary voltage = 5500V or 5.5kV Ans.

(b) The kVA rating applies equally to the primary and secondary sides.

$$\text{Thus primary current} = \frac{\text{kVA rating} \times 1000}{V_1}$$
$$= \frac{550 \times 1000}{5500} = 100\text{A} \quad \text{Ans.}$$

$$\text{Secondary current} = \frac{\text{kVA rating} \times 1000}{V_2}$$
$$= \frac{550 \times 1000}{220}$$
$$= 2500\text{A} \quad \text{Ans.}$$

(c) This part of the question is answered by a knowledge of the e.m.f. formula for the transformer. This is developed in Volume 7 but can be memorised.

Thus $V_1 = 4.44\Phi_m f N_1$ volts.
or $5500 = 4.44 \times \Phi_m \times 50 \times 1875$

$$\text{Thus } \Phi_m = \frac{5500}{2.22 \times 100 \times 1875}$$
$$= 13.2 \times 10^{-3} \text{ webers}$$

Maximum value of flux = 13.2mWb Ans.

13. Since $I_f = \frac{220}{110} = 2\text{A}$, then $I_a = I_L - I_f = 6 - 2 = 4\text{A}$

Input power = $220 \times 6 = 1320\text{W}$

Copper loss (armature), $I_a^2 R_a = 4^2 \times 0.25 = 16 \times \frac{1}{4} = 4\text{W}$

Copper loss (field), $I_f^2 R_f = 2^2 \times 110 = 440\text{W}$

Total copper loss = 444W

Rotational loss = 1320 − 444 = 876W. These are the losses attributable to iron, friction and windage. Ans.

When the current is 62A, the input is $220 \times 62 = 13\,640\text{W}$

The output = input − losses (all values in watts)
= 13 640 − copper losses − rotational losses
= $13\,640 - (60^2 \times 0.25 + 2^2 \times 110) - 876$
= 13 640 − (900 + 440) − 876
= 13 640 − 2216 = 11 424W

So efficiency $= \frac{11\,424}{13\,640} = 0.837$ or 83.7 per cent Ans.

14. A basic definition of power factor is that: it is the ratio of the 'active power' to the 'apparent power' being expended in a circuit.

$$\text{Thus power factor, } \cos\phi = \frac{\text{active power}}{\text{apparent power}} = \frac{P}{S}$$

or active power = apparent power × power factor

For the problem:

$P = S\cos\phi$ or kW = kVA × $\cos\phi$

Hence 560 = kVA × 0.7

Thus apparent power supplied $= \frac{560}{0.7} = 800\text{kVA}$

With the power factor increased to 0.8 and the apparent power kept constant, the new 'active power' = 800 × 0.8 = 640kW.

The increase in active power would be:

$$640 - 560 = 80\text{kW} \quad \text{Ans.}$$

15. The equation for the charging circuit is $V = E_b + IR_i + IR$ where E_b is the battery e.m.f., I the current, R_i the battery internal resistance, R the resistance of the lamp bank and V the supply voltage.

At commencement of charge;

$200 = 10R + 10R_i + E_b$ here $R_i = 48 \times 0.02\text{m} = 0.96\Omega$

$E_b = 48 \times 1.8 = 86.4\text{V}$

$\therefore 200 = 10R + (10 \times 0.96) + 86.4$

or $10R = 200 - 9.6 - 86.4 = 200 - 96 = 104$

$\therefore R = 10.4\Omega.$

Now a lamp is rated at 110V,100W

Resistance of a lamp $= \frac{V^2}{P} = \frac{110^2}{110} = 121\Omega$

If 'n' lamps are used in parallel

then $\frac{1}{R} = \frac{1}{121} + \frac{1}{121} + \frac{1}{121} \ldots n$ times

$\therefore \frac{1}{R} = \frac{n}{121}$ or $n = \frac{121}{R} = \frac{121}{10.4}$

$= 11.63$, say 12

Thus number of lamps = 12 lamps in parallel at start. Ans.

At start of charge:

If $R = \frac{121}{12} = 10.08$ ohms, the current is obtained

from $200 = (I_1 \times 10.08) + (I_1 \times 0.96) + 86.4$

or $I_1 = \frac{200 - 86.4}{11.04} = 10.29\text{A}$

At end of charge:

If 12 lamps are in parallel, then $R = 10.08\Omega$.

So $200 = (I_2 \times 10.08) + (I_2 \times 0.96) + E_b$. Here I_2 is the final charging current value and E_b is the final back e.m.f.

Then $E_b = 48 \times 2.2 = 105.6\text{V}$

and $200 - 105.6\text{V} = 11.04I_2$

Thus $I_2 = \frac{94.4}{11.04} = 8.55\text{A}$

Current at end of charge = 8.55A Ans.

16. Resistance of heater $= 6.5\Omega$ Working voltage 100V

$\therefore$ Rated current $= \dfrac{100}{6.5} = 15.38\text{A}$

Reactance of choke, $X_L = 2\pi fL$

or $X_L = 2 \times \pi \times 50 \times 0.01 = 3.14\Omega$

Impedance of circuit, $Z = \sqrt{6.5^2 + 3.14^2}$

or $Z = 7.22\Omega$

Circuit voltage would be $= 15.38 \times 7.22 = 111.04\text{V}$

Applied voltage should be 111.04V to give 100V on heater. Ans.

If frequency rose to $50 + \left(\dfrac{5}{100} \times 50\right) = 52.5\text{Hz}$, X_L would rise in proportion.

$\therefore$ New reactance $X_{L1} = 3.14 \times \dfrac{52.5}{50}$

$= 3.297\Omega$

New impedance $Z_1 = \sqrt{6.5^2 + 3.297^2}$

or $Z_1 = 7.288\Omega$

New heater current $= \dfrac{V}{Z_1} = \dfrac{111.04}{7.288} = 15.23\text{A}$

New voltage across heater terminals $= 15.23 \times 6.5$

$= 98.995\text{V}$

Change of voltage $= 100 - 98.995 = 1.005\text{V}$ Ans.

17. Force on one conductor:

$F = BIl$ newtons

$= 0.8 \times 20 \times 400 \times 10^{-3}$ newtons

$= 6.4\text{N}$

Force due to all active conductors

$= \dfrac{2}{3} \times 246 \times 6.4 = 164 \times 6.4$

$= 1050\text{N}$

(a) Torque produced $= 1050 \times 150 \times 10^{-3}$

$= 157.5\text{Nm}$ Ans.

(b) Shaft power developed $= \dfrac{2\pi NT}{60}$ watts

$\dfrac{2 \times \pi \times 500 \times 157.5}{60}$

$= 8.24\text{kW}$ Ans.

18. Output from motor $= 45\text{kW}$

Input to motor $= \dfrac{45 \times 1000 \times 100}{93}$ watts

$= 48.4\text{kW}$

But 3-phase power (watts) $= \sqrt{3}VI \cos\phi$

$$\therefore\ 48.4 \times 10^3 = \sqrt{3} \times 440 \times I \times 0.88$$

$$I = \frac{48.4 \times 10^3}{1.732 \times 440 \times 0.88} = 72\text{A} \quad \text{Ans.}$$

19. Let n = the no of cells in series.

Battery e.m.f. = $n \times 1.5$ volts

Battery internal resistance = $n \times 0.5$ ohms

Also since $P = I^2R$, then for the load: $2 = I^2 \times 100$

$$\text{or } I = \frac{\sqrt{2}}{10} = 0.1414\text{A}$$

Also for the circuit, current I is given by:

$$I = \frac{1.5n}{100 + 0.5n} \text{ or } \frac{1.5n}{100 + 0.5n} = 0.1414$$

$$\therefore\ 1.5n = 14.14 + 0.0707n$$

$$\text{or } 1.4293n = 14.14$$

$$\text{giving } n = 10 \text{ (approx)} \quad \text{Ans.}$$

$$\text{As a check: } I = \frac{1.5 \times 10}{100 + (0.5 \times 10)} = 0.143\text{A}$$

$$\text{So power dissipated} = 0.143^2 \times 100 = 2.045\text{W}$$

or the given rating, 2W (approx)

20. Total internal resistance = $6 \times 0.1\Omega = 0.6\Omega$. For maximum power; external load resistance and internal battery resistance are equal.

Thus Load Resistance = 0.6Ω Ans.

$$\text{Load current } I = \frac{E}{R + R_i} = \frac{6 \times 2.2}{0.6 + 0.6} = 11\text{A}$$

$$\therefore \text{ Maximum power } P = I^2R = 11^2 \times 0.6 = 72.6\text{W} \quad \text{Ans.}$$

An alternative solution is possible using Calculus.

$$I = \frac{E}{R + R_i} \text{ and } P = I^2R$$

$$\therefore\ P = \left(\frac{E}{R + R_i}\right)^2 R = \frac{E^2R}{R^2 + 2RR_i + R_i^2}$$

$$= \frac{E^2}{R + 2R_i + R_i^2R^{-1}}$$

$$= \frac{13.2^2}{R + 1.2 + 0.6^2R^{-1}}$$

$$\therefore\ P = \frac{174.24}{R + 1.2 + 0.36R^{-1}}$$

P is maximum when the denominator is minimum. Differentiate denominator with respect to R.

$$\text{Thus } \frac{d(R + 1.2 + 0.36R^{-1})}{dR} = 1 + 0 - 0.36R^{-2}$$
$$= 0 \text{ for min. value}$$
$$1 + 0 - 0.36R^{-2} = 0$$
$$\therefore R = 0.6\Omega$$

The 2nd differential determines whether R is maximum or minimum value.

$$\frac{d^2(R + 1.2 + 0.36R^{-1})}{dR^2} = \frac{d(1 + 0 - 0.36R^{-2})}{dR}$$
$$= 0.72R^{-3}$$

Since the second differential is positive, R is a minimum value.

Thus when $R = 0.6\Omega$ P is a maximum value. Ans.

Calculate power as shown previously.

21. Current taken by a lamp at full voltage

$$= \frac{60}{240} = \frac{1}{4} = 0.25\text{A}$$

Resistance of 1 earth lamp $= \dfrac{240}{0.25} = 960\Omega$

Let R_A be the resistance between the +ve line and earth.

$$\text{Then } \frac{1}{R_A} = \frac{1}{960} + \frac{1}{16} = \frac{1 + 60}{960} = \frac{61}{960}$$
$$\text{So } R_A = \frac{960}{61} = 15.74\Omega$$

Let R_B be the resistance between the −ve line and earth.

$$\text{Then } \frac{1}{R_B} = \frac{1}{960} + \frac{1}{6} = \frac{1 + 160}{960} = \frac{161}{960}$$
$$\text{So } R_B = \frac{960}{161} = 5.96\Omega$$

∴ Total resistance between +ve and −ve line

$$= 15.74 + 5.96 = 21.7\Omega$$

Current flowing between the +ve and −ve line

$$= \frac{220}{21.7} = 10.14\text{A}$$

Voltage across +ve line and earth $= 10.14 \times 15.74$ volts $= 159.6$V

,, ,, −ve ,, ,, ,, $= 10.14 \times 5.96$ volts $= 60.43$V

(a) The +ve lamp, *ie* lamp connected between the +ve line and earth will burn the brighter. Ans.

(b) Current passed by the fault $= \dfrac{220}{15.74 + 5.96}$
$= 10A$ (approx)

Additional load on generator $= 220 \times 10 = 2200W$
$= 2.2kW$ Ans.

22. Load (a) $P_a = 200kW \quad \cos\phi_a = 1 \quad \sin\phi_a = 0$
$S_a = 200kVA \quad Q_a = S_a \sin\phi_a = 0kVAr$
,, (b) $P_b = 400kW \quad \cos\phi_b = 0.8 \quad \sin\phi_b = 0.6$
$S_b = P_b/\cos\phi_b = \dfrac{400}{0.8} = 500kVA$
$Q_b = S_b \sin\phi_b = 500 \times 0.6$
$= 300kVAr$
,, (c) $P_c = 200kW$

Total active power of loads, $P = 200 + 400 + 200 = 800kW$

Total apparent power of all loads, $S = \dfrac{P}{\cos\phi}$
$= \dfrac{800}{0.97} = 824.74kVA$

Also $\cos\phi = 0.97$. $\therefore \phi = 14°4'$ and $\sin\phi = 0.234$
So reactive power of all loads, $Q = S \sin\phi = 824.74 \times 0.243$
$= 200.4kVAr$

Thus lagging kVAr value is reduced by $300 - 200.4 = 99.6$
This must therefore, be the leading reactive power Q_c of the synchronous motor. Apparent power rating of motor
$S_c = \sqrt{200^2 + 99.6^2}$
$= 223.3kVA$

Power factor of motor, $\cos\phi_c = \dfrac{200}{223.3} = 0.89$ (leading)
Ans.

23. Motor condition.

Input power = 55kW Line current $= \dfrac{55\,000}{220} = 250A$

Field current $= \dfrac{220}{110} = 2A$

Armature current $= 250 - 2 = 248A$
Back e.m.f. $= 220 - (248 \times 0.02) = 220 - 4.96$
$= 215.04V$

Generator condition.

Output power = 55kW Line current $= \dfrac{55\,000}{220} = 250A$

Field current $= \frac{220}{110} = 2A$ Armature current $= 250 + 2$
$= 252A$

Generated e.m.f. $= 220 + (252 \times 0.02)$
$= 225.04V$

As e.m.f. is proportional to speed, then $\frac{225.04}{215.04} = \frac{N_2}{500}$

$$\text{or } N_2 = \frac{225.04}{215.04} \times 500$$

Thus generator speed $= 1.0464 \times 500$
$= 523.3$ rev/min   Ans.

24. If $B_m = 1.4T$ and area of core $= 0.27 \times 0.27 = 0.0729m^2$
Then $\Phi_m = 1.4 \times 0.0729 = 0.102Wb$
Substituting in the formula:

$$V_1 = 4.44\Phi_m f N_1 \text{ or } N_1 = \frac{3500}{4.44 \times 0.102 \times 50} \text{ turns}$$

$$\therefore N_1 = \frac{3500}{2.22 \times 10.2} = 154.7 \text{ turns}$$

or primary turns $= 155$ (approx)

Secondary turns $= \frac{440}{3500} = \frac{N_2}{155}$

$$\text{or } N_2 = 155 \times \frac{44}{350} = 19.5 \text{ turns}$$

Thus secondary turns $= 20$ (approx)   Ans.

25. Load current of generator $= \frac{20\,000}{200} = 100A$

Shunt-field current $= \frac{200}{115} = 1.74A$

Armature current $= 101.7A$

Copper loss of armature circuit $= 101.7^2(0.15 + 0.025 + 0.028) = 2.1kW$

Copper loss in shunt field $= 1.74 \times 200 = 348W = 0.348kW$

Total copper loss $= 2.1 + 0.348 = 2.448kW$

$$\text{Efficiency} = \frac{\text{output}}{\text{input}} = \frac{\text{output (kW)}}{\text{output (kW)} + \text{losses (kW)}}$$

$$= \frac{20}{20 + \text{copper loss} + \text{iron loss} + \text{rotational loss}}$$

$$= \frac{20}{20 + 2.448 + 0.78 + 0.59}$$

$$= 0.839$$

Thus full-load efficiency $= 83.9$ per cent   Ans.

26. Delta-connected load.

Active power, P_1 = 30kW at a power factor or 0.92 (leading)

$\therefore$ Apparent power, $S_1 = \frac{30}{0.92} = 32.61\text{kVA}$

$\cos\phi = 0.92 \quad \therefore \phi = 22°56' \quad \sin\phi = 0.3896$

So the reactive power, $Q_1 = 32.61 \times 0.3896 = 12.7\text{kVAr}$

Star-connected load.

Active power, P_2 = 40kW at a power factor of 0.85 (lagging)

Apparent power, $S_2 = \frac{40}{0.85} = 47.1\text{kVA}$

$\cos\phi = 0.85 \quad \therefore \phi = 31°37' \quad \sin\phi = 0.5242$

Thus reactive power, $Q_2 = 47.1 \times 0.5242 = -24.7\text{kVAr}$

Note. – ve sign given to the lagging reactive power value to distinguish it from the leading reactive power value of the other load.

Total active power on alternator, $P = P_1 + P_2$
$= 30 + 40 = 70\text{kW}$

Total reactive power loading, $Q = Q_1 + Q_2$
$= 12.7 - 24.7$
$= -12\text{kVAr}$

So apparent power, loading on alternator is:

$S = \sqrt{70^2 + 12^2}$
$= 71.06\text{kVA}$

Again three-phase kilovolt amperes is given by $\frac{\sqrt{3}VI}{1000}$

Hence $\frac{1.732 \times 500 \times I}{1000} = 71.06$

or $I = \frac{71.06 \times 2}{1.732} = 82.1\text{A}$

Line current = 82.1A Ans.

Supply power factor $= \frac{P}{S} = \frac{70}{71.06} = 0.98$ (lagging) Ans.

The lagging condition is determined from the resultant – ve sign of the total reactive power value.

27. Original conditions:

$E_{b1} = 440 - (30 \times 0.7) = 419\text{V}$

Original flux condition Φ_1

Final flux condition:

$\Phi_2 = 0.8\Phi_1$. Assuming no speed change then since generated e.m.f. is proportional to flux new E_b value = $0.8 \times 419 = 335.2$V

and momentary current is given by $\frac{V - E_b}{R_a}$ amperes

$$\text{or } I_a = \frac{440 - 335.2}{0.7} = 149.7\text{A} \quad \text{Ans.}$$

If final torque condition T_2 = original torque T_1 and since torque is proportional to flux and armature current, then

$$T = k\Phi I_a$$

Thus we can write $T_1 = T_2$ or $k\Phi_1 I_{a1} = k\Phi_2 I_{a2}$

$$\text{and } I_a = \frac{\Phi_1 I_{a1}}{\Phi_2} = \frac{\Phi_1 I_{a1}}{0.8\Phi_1} = \frac{3.0}{0.8} = 37.5\text{A}$$

New armature current will be 37.5A Ans.

New back e.m.f. $= 440 - 37.5 \times 0.7$
$= 413.75$V

Also since $E_b \propto \Phi N$

$$\text{we can write } \frac{E_{b2}}{E_{b1}} = \frac{k\Phi_2 N_2}{k\Phi_1 N_1} \text{ or } \frac{413.75}{419} = \frac{0.8\Phi_1 N_2}{\Phi_1 700}$$

$$\text{whence } N_2 = \frac{413.75 \times 700}{419 \times 0.8} = 864.4 \text{ rev/min} \quad \text{Ans.}$$

28. Original power transmitted by supply cable is given by:

$$P_1 = VI\cos\phi = \frac{240 \times 250 \times 0.8}{1000} = 48\text{kW}$$

Current taken by a 800μF capacitor is given by:

$$I_C = \frac{V}{X_C} \text{ where } X_C = \frac{1}{2\pi fC} = \frac{10^6}{2 \times \pi \times 50 \times 800}$$

$$\text{Thus } X_C = \frac{10^2}{25.12} = 3.98 = 4\Omega \text{ (approx)}$$

$$\text{and capacitor current} = \frac{240}{4} = 60\text{A}$$

There are now two currents, I_L = 250A, lagging the voltage by 36°44′ (Note cos 36°44′ = 0.8) and I_C = 60A, leading the voltage by 90°. The problem calls for a graphical solution. Draw a voltage ordinate horizontally. Choose a suitable current scale and, from the origin, *ie* left-hand point of the voltage ordinate, draw I_C, to scale, vertically upwards. Next from the origin draw I_L to scale below the voltage ordinate by 36°44′. Complete the parallelogram for the current vectors, draw and measure the resultant current for magnitude and phase. This part of the problem is worked out here mathematically.

$$\text{Thus } I = \sqrt{I_L^2 - I_C^2 - 2I_L I_C \cos\theta}$$
$$= \sqrt{250^2 + 60^2 - (2 \times 250 \times 60 \times \cos 53°16')}$$
$$= 219.4\text{A} \quad \text{Ans.}$$

Assuming the same power of 48kW was being supplied, the new power factor of the supply would be obtained from:

$$P_2 = VI \cos \phi_2 \text{ or } \cos \phi_2 = \frac{48\,000}{240 \times 219.4}$$
$$= 0.91 \text{ (lagging)}$$

The question is not clear as to the final loading conditions but it is assumed that this power factor condition of 0.91 (lagging) is maintained. The maximum current would be limited to 250A and the power transmitted would be, P_2.

$$\text{Thus } P_2 = \frac{240 \times 250 \times 0.91}{1000} \text{ kilowatts}$$
$$= 54.6\text{kW}$$

Additional power, $P_2 - P_1 = 54.6 - 48 = 6.6\text{kW}$ Ans.

29. Although the parallel working of generators has not been considered in this book, this problem can be solved by a direct application of Kirchoff's laws. Thus, let I_A be the current output from machine A and I_B the current output from machine B. Let V = the common busbar or terminal voltage.

Then $I_A + I_B = 2500\text{A}$ and two voltage equations can be built up as follows, since $I_{aA} = I_A + \frac{V}{85}$ and $I_{aB} = I_B + \frac{V}{85}$

$$\text{Thus } V = 600 - 0.015 I_{aA}$$
$$\text{and } V = 620 - 0.015 I_{aB}$$
$$\text{or } V = 600 - 0.015\left(I_A + \frac{V}{85}\right)$$
$$\text{or } V = 600 - 0.015\left(I_B + \frac{V}{85}\right)$$
$$\text{giving } V = 600 - 0.015 I_A - \frac{0.015V}{85} \quad \ldots \text{(a)}$$
$$V = 620 - 0.015 I_B - \frac{0.015V}{85} \quad \ldots \text{(b)}$$

Subtracting (a) from (b),

$$0 = 20 - 0.015 I_B + 0.015 I_A \text{ or } 0 = 20 - 0.015(I_B - I_A)$$
$$\text{and } 0 = 20 - 0.015[I_B - (2500 - I_B)]$$
$$= 20 - 0.015[I_B - 2500 + I_B]$$
$$= 20 - 0.03 I_B + 37.5$$
$$\text{Thus } I_B = \frac{57.5}{0.03} = 1916.66\text{A} \quad \text{Ans.}$$
$$I_A = 2500 - 1916.66 = 583.34\text{A} \quad \text{Ans.}$$
$$\text{From (a) } V = 600 - (583.34 \times 0.015) - \frac{0.015V}{85}$$

$$= 600 - (5.8334 \times 1.5) - \frac{0.015V}{85}$$

$$\text{or } V = 600 - 8.75 - \frac{0.015V}{85}$$

$$\text{Whence } V + \frac{0.015V}{85} = 591.25$$

$$\text{giving } \frac{85V + 0.015V}{85} = 591.25$$

$$\text{or } 85.015V = 591.25 \times 85$$

$$\text{and } V = \frac{591.25 \times 85}{85.015}$$

Thus $V =$ 590.75 Volts Ans.

$$\text{Output of machine A} = \frac{583.34 \times 590.75}{1000} = 344.6\text{kW} \quad \text{Ans.}$$

$$\text{,, \quad ,, \quad B} = \frac{1916.66 \times 590.75}{1000} = 11.32\text{kW} \quad \text{Ans.}$$

Note. Although the kW rating of B is correct for the values given in the question, it is high for the accepted sizes of d.c. machine.

30. The current can be written as:

$$i_A = 19 \sin \theta$$

$$i_B = 35 \sin \left(\theta - \frac{180}{2}\right) \text{ or } i_B = 35 \sin (\theta - 90)$$

$$\text{and } i_C = 42 \sin \left(\theta - \frac{900}{6}\right) \text{ or } i_C = 42 \sin (\theta - 150)$$

A graphical solution is required. Use i_A as reference. Choose a suitable current scale and draw OA horizontally from left to right to equal 19A. Draw OB vertically downwards, *ie* 90° behind OA and make a length to represent 35A. Finally, draw OC, 42A long 150° behind A, *ie* it will be 30° below the horizontal and should be drawn from right to left. Complete the parallelogram for OA and OB, draw the resultant and call this OX. Now use OX with OC to complete a new parallelogram and draw the resultant OY. Measure OY to scale and this gives the required answer.

Multiply the answer by 0.707 and a further answer is obtained. Finally measure the angle between OC and OY. The required phase position is thus obtained and OY will lead OC. A mathematical solution for the above problem is set out below.

Resolving into horizontal and vertical components we have:

$$I_H = 19 \cos 0 + 35 \cos 90 - 42 \cos 30$$

$= (19 \times 1) + (35 \times 0) - (42 \times 0.866)$
$= 19 + 0 - 36.372 = -17.372\text{A}$
$I_V = 19 \sin 0 - 35 \sin 90 - 42 \sin 30$
$= (19 \times 0) - (35 \times 1) - (42 \times 0.5)$
$= 0 - 35 - 21 = -56\text{A}$
The resultant $I = \sqrt{17.372^2 + 56^2}$
$= 58.64\text{A}$ Ans.

The phase angle can be obtained from

$$\cos \phi = -\frac{17.372}{58.64} = -0.29 \quad \phi = 73°8'$$

The resultant is thus $(73°8' - 30°) = 43°8'$ ahead of the third current of maximum value = 42A Ans.
The required r.m.s. value of the resultant = 41.46A Ans.

31. Motor output = 10kW

,, input $= 10 \times \frac{100}{85} = 11.76\text{kW} = 11\,760\text{W}$

Current taken from supply $= \frac{11\,760}{230} = 51.31\text{A}$

Shunt-field current, $I_f = \frac{230}{220} = 1.045\text{A}$

Armature current, $I_a = 51.13 - 1.045 = 50.085\text{A}$
Back e.m.f., $E_b = 230 - (50.1 \times 0.15)$
$= 222.5$ volts

As a generator, speed is increased and flux is increased in proportion to the shunt-field current.

$$\therefore E = 222.5 \times \frac{1080}{1000} \times \frac{1.1}{1.045}$$
$$= 252.98\text{V}$$

Also since armature copper loss is the same as for the motor, armature current must be the same = 50.1A.

$\therefore$ Terminal voltage, $V = 252.98 - (50.1 \times 0.15)$
$= 245.47\text{V}$

Output current $= 50.1 - 1.1 = 49\text{A}$

So output $= \frac{245.47 \times 49}{1000} = 12.03\text{kW}$ Ans.

32. Let I_A = the current in the first coil, then $I_A = \frac{220}{16}$
$= 13.75\text{A}$ Ans.

$$\cos \phi_A = \frac{3}{16} = 0.187 \quad \phi_A = 79°13' \quad \sin \phi_A = 0.9824$$

I_B = the current in the second coil, then $I_B = \frac{220}{25}$
$= 8.8\text{A}$ Ans.

$\cos \phi_A = \frac{7}{25} = 0.28 \quad \phi_B = 73°44' \quad \sin \phi_B = 0.96$

Total active components, $I_a = (13.75 \times 0.187)$
$+ (8.8 \times 0.28)$ amperes
$= 2.57 + 2.464 = 5.034\text{A}$

Total reactive components. $I_r = (-13.75 \times 0.9824)$
$- (8.8 \times 0.96)$ amperes
$= -13.51 - 8.448$
$= -21.96\text{A}$

resultant current. $I = \sqrt{5.034^2 + 21.96^2}$ amperes
$= 22.53\text{A}$ Ans.

$$\cos \phi = \frac{5.034}{22.53} = 0.223 \text{ (lagging)}$$

$$\text{Power, } P = \frac{220 \times 22.53 \times 0.233}{1000} \text{ kilowatts} = 1.105\text{kW}$$

Ans.

The phasor diagram for this problem, can be considered as one of the basics for a parallel circuit, and has already been illustrated several times.

33. The magnitude of the e.m.f. generated on open circuit is determined from the point where the field voltage-drop line intersects the O.C.C.

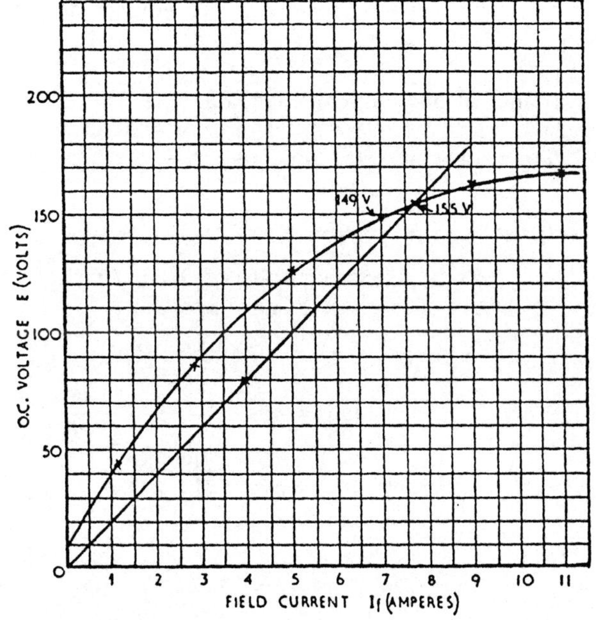

Plot the O.C.C. and the field voltage-drop line by obtaining a typical value. Thus, for a field current of 4A, the voltage drop would be 4 × 20 = 80V. Plot this point. Draw a line from the origin through this point and produce the line to cut the O.C.C. The induced e.m.f. is then 155V.

The voltage drop across the field, is also the terminal voltage of the generator and from a graph a terminal voltage of 140V is obtained when the field current is 7A, *ie* read horizontally for 140V on the voltage scale to determine the corresponding field current value on the field voltage line. This value is 7A. For this value of field current however, the e.m.f. generated is 149V. Ans.

Also $V = E - I_a R_a$ $\quad \therefore 140 = 149 - (I_a \times 0.02)$ or

$0.02 I_a = 149 - 140$

Giving an armature current, $I_a = \dfrac{9}{0.02} = 450\text{A}$ Ans.

34. A fundamental formula for the alternator and an a.c. supply is:

$$f = \frac{PN}{120} \text{ where } P = \text{the number of poles.}$$

(a) Therefore $f = \dfrac{12 \times 600}{120} = 60\text{Hz}$ Ans.

(b) Since the load is balanced, the voltage across each phase, $V_{ph} = \dfrac{440}{\sqrt{3}}$ volts

Impedance of one phase of load, $Z_{ph} = \sqrt{35^2 + 25^2}$ ohms $= 43.01\Omega$

For a star-connected load, current in one phase of load,

I_{ph} = Line current, I

or $I_{ph} = \dfrac{440}{\sqrt{3} \times 43.01} = 5.91\text{A}$

Thus current in a coil = 5.91A Ans.

(c) Current in each phase of alternator,

or $I_{ph} = \dfrac{I}{\sqrt{3}} = \dfrac{5.91}{\sqrt{3}} = 3.41\text{A}$ Ans.

(d) Power factor of load $= \cos\phi = \dfrac{R}{Z} = \dfrac{35}{43.01}$

$= 0.81$ (lagging).

and total power of load $= \sqrt{3}VI\cos\phi$

$= \dfrac{\sqrt{3} \times 440 \times 5.91 \times 0.81}{1000}$ kilowatts

$= 3.65\text{kW}$ Ans.

35. No load

Input to motor = 400 × 2.8 = 1120W

Shunt-field current, $I_f = \frac{400}{450} = 0.89\text{A}$

Armature current, $I_a = 2.8 - 0.89 = 1.91\text{A}$

Copper losses = field copper loss + armature copper loss

$= (400 \times 0.89) + (1.91^2 \times 0.45)$

$= 356 + 1.64 = 357.64\text{W}$

Rotational loss of machine = input − copper losses

or $P_R = 1120 - 357.64 = 762.36\text{W}$

On load

Input to motor = 400 × 35 = 14 000W

Losses on load = copper loss + rotational loss

Copper losses, P_{Cu} = field copper loss + armature copper loss

$= (400 \times 0.89) + [(35 - 0.89)^2 \times 0.45]$ watts

Note. Load armature current = line current − field current

Thus, $P_{Cu} = 356 + (34.11^2 \times 0.45)$

$= 356 + 523 = 879\text{W}$

Total losses = 879 + 762.36 = 1641.36W = 1.6414kW

Output of motor = 14 − 1.6414 = 12.358kW

$$\text{Efficiency} = \frac{\text{output}}{\text{input}} = \frac{12.358}{14} = 0.882$$

or η = 88.2 per cent Ans.

36. At resonance $2\pi fL = \frac{1}{2\pi fC}$

$$\therefore f = \frac{1}{2\pi\sqrt{LC}}$$

$$= \frac{1}{2\pi\sqrt{0.1 \times 0.1 \times 10^{-6}}}$$

Resonant Frequency = 1592 Hz. Ans.

At resonance there is no resultant reactance

ie R = Z

$$\therefore I = \frac{V}{R} = \frac{230}{100}$$

$$= 2.3\text{A}$$

$$X_c = \frac{1}{2\pi fC} = \frac{1}{2\pi \times 1592 \times 0.1 \times 10^{-6}}$$

$$= 1000\Omega$$

p.d. across capacitor = $IX_c = 2.3 \times 1000$

= 2300 volts. Ans.

37. Before adjusting the e.m.f. of one of the machines, the e.m.f. of each machine is considered to be 230V. The busbar voltage is 220V. Voltage drop due to resistance R of one machine and cables up to busbars = 230 − 220 = 10V. Current = 640A.

$\therefore$ Resistance R of machine and cables $= \frac{10}{640} = \frac{1}{64}$ ohms

Let V = the busbar voltage under the new condition.

Then current supplied by generator No 1 $= \dfrac{235 - V}{\frac{1}{64}}$ amps

$= 64(235 - V)$

Current supplied by generator No 2 = $(230 - V)64$ amps

" " " " " 3 = $(230 - V)64$ amps

Power supplied by the three machines

$= V(235 - V)64 + V(230 - V)64 + V(230 - V)64$ watts

$= 64(235V - V^2) + 2 \times 64(230V - V^2)$

Now the original power supplied by three machines

$= 3 \times 640 \times 220$ watts = 422 400W or 422.4kW

$\therefore 3 \times 640 \times 220 = 64(235V - V^2) + 2 \times 64(230V - V^2)$

and $6600 = 235V - V^2 + 460V - 2V^2$

or $-3V^2 + 695V = 6600$

Thus $V^2 - 231.66V + 2200 = 0$

Solving for V using the quadratic formula:

$$V = \frac{231.66 \pm \sqrt{231.66^2 - 4 \times 2200}}{2}$$

$$= \frac{231.66 \pm 211.9}{2} = \frac{443.56}{2} = 221.78\text{V}$$

Thus busbar voltage would be 221.8V Ans.

Current of Machine No 1 $= \dfrac{235 - 221.78}{\frac{1}{64}}$

$= 13.22 \times 64 = 846$A Ans.

Current of Machine No 2 $= \dfrac{230 - 221.78}{\frac{1}{64}}$

$= 8.22 \times 64 = 526$A Ans.

Current of Machine No 3 $= \dfrac{230 - 221.78}{\frac{1}{64}}$

$= 8.22 \times 64 = 526$A Ans.

Check. Power supplied = $(846 + 526 + 526)221.78$

$= 1898 \times 221.78$ watts

= 421kW (approx).

Using the other root of the quadratic equation we have:

$$V = \frac{231.66 - 211.9}{2} = \frac{19.76}{2} = 9.88\text{V}$$

Current of Machine No 1 = $(235 - 9.88) \times 64$
= $225.12 \times 64 = 14\,408$A

Current of Machine No 2 = $(230 - 9.88) \times 64$
= $220.12 \times 64 = 14\,088$A

Current of Machine No 3 as for No 2.

The above conditions though theoretical, would relate to a busbar voltage of 9.88V and could be imagined as the result of a 'short-circuit' at the busbars, where the power of 422kW could be assumed to be dissipated.

38. Delta load

Active power, P_1 = 50kW $\cos\phi = 0.75$
Hence $\phi = 41°25'$ and $\sin\phi = 0.6615$

Apparent power, $S_1 = \frac{50}{0.75}$
= 66.66kVA

Reactive power, $Q_1 = 66.66 \times 0.6615$
= -44.1kVAr (lagging)

Star load

Active power, P_2 = 40kW
$\cos\phi = 0.62$ $\phi = 51°41'$ and $\sin\phi = 0.7846$

Apparent power, $S_2 = \frac{40}{0.62} = 64.5$kVA

Reactive power, $Q_2 = 64.5 \times 0.7846 = 50.6$kVAr (leading)

Total power, $P = P_1 + P_2 = 50 + 40$
= 90kW Ans.

Total reactive power, $Q = Q_1 + Q_2 = -44.1 + 50.6$
= 6.5kVAr (leading)

Apparent power, $S = \sqrt{90^2 + 6.5^2}$
= 90.23kVA Ans.

Power factor = $\frac{P}{S} = \frac{90}{90.23} = 0.988$ (leading) Ans.

39. Let the resistance of the fault be R ohms
Since the cable resistance itself is 4 ohms,
then (Resistance of end A to earth $- R$) + (Resistance of end B to earth $- R$) = 4

or $5 - R + 3 - R = 4$ or $8 - 2R = 4$
giving $2R = 4$ or $R = 2\Omega$

It can be assumed that for a cable, the resistance is proportional to length.
Then from end A to fault = 5Ω and the fault resistance is 2Ω
∴ cable length is such as to have a resistance of 3Ω.

Thus fault must be $\frac{3}{4} \times 1000 = 750$ metres from end A Ans.

40. (a) At resonance $X_L = X_C$ or $2\pi fL = \frac{1}{2\pi fC}$

Thus resonant frequency $f^2 = \frac{1}{(2\pi)^2 CL}$

or $f^2 = \frac{10^6}{(2 \times \pi)^2 \times 14 \times 0.84}$

and $f^2 = 2150.5$

Thus $f = 46.38$ Hz. Ans.

Also $X_L = 2\pi fL = 2 \times \pi \times 46.38 \times 0.84$
$= 244.89\Omega$

For a current flow of 5A
Voltage drop across the inductive reactance of the coil
$= 5 \times 244.89 = 1224.45$V

Voltage drop across resistance of coil $= 5 \times 50$
$= 250$V

Voltage drop across capacitance = 1224.45V Ans.
Voltage drop across coil $= \sqrt{250^2 + 1224.5^2}$
$= 1250$V Ans.

Voltage drop across coil and capacitor = voltage across resistance only $= 5 \times 50 = 250$V Ans.

(b) At 60Hz, since X_L is proportional to frequency

Then $X_L = 244.89 \times \frac{60}{46.38}$
$= 316.8\Omega$

Voltage drop across inductive reactance $= 5 \times 316.8$
$= 1584$V

Voltage drop across resistance = 250V
Again X_C is inversely proportional to frequency

$\therefore X_C = \frac{10^6}{2\pi \times 60 \times 14}$
$= 189.3\Omega$

Voltage across capacitor $= 5 \times 189.3 = 946.5$V Ans.
Voltage across resultant reactance $= 1584 - 946.5$
$= 637.5$V

Voltage across resistance = 250V

$\therefore$ Supply voltage $= \sqrt{637.5^2 + 250^2}$
$= 684.5\text{V}$ Ans.
Voltage across coil $= \sqrt{1584^2 + 250^2}$
$= 1604\text{V}$ Ans.

41. The torque of a motor $T \propto \Phi I_a$ and since, for a series motor, $\Phi \propto I_f$ and $I_f = I_a = I_L$ then $T \propto I_L^2$ or $T = kI_L^2$

On full load:

$$E_{b1} = V - I_a R$$
$$= 210 - (100 \times 0.1) = 200\text{V}$$

When load torque is reduced to 25 per cent of full-load value, then $T_2 = T_1 \times \frac{1}{4}$ where, T_2 is the new torque and T_1 the original torque.

$$\text{Then } \frac{T_2}{T_1} = \frac{kI^2_{L2}}{kI^2_{L1}} \text{ or } \frac{I^2_{L2}}{I^2_{L1}} = \frac{1}{4}$$

$$\therefore I_L{}^2{}_2 = \frac{I_L{}^2{}_1}{4} \text{ or } I_{L2} = \frac{I_{L1}}{2} = \frac{100}{2} = 50\text{A}$$

On new load condition

$$E_{b2} = 210 - (50 \times 0.1) = 205\text{V}$$

As $E_b \propto \Phi N$ then $\frac{E_{b2}}{E_{b1}} = \frac{k\Phi_2 N_2}{k\Phi_1 N_1}$ Thus $\frac{205}{200} = \frac{50 \times N_2}{100 \times 600}$

Note that field current $I_f = I_a = I_L$ is now substituted for flux, since $\Phi \propto I_f$.

$$\text{Thus } N_2 = \frac{205}{200} \times 600 \times 2 = 1230 \text{ rev/min.}$$

Speed of motor on 75 per cent load torque = 1230 rev/min Ans.

42. Reactance X_C of capacitor $= \frac{1}{2\pi fC} = \frac{10^6}{2 \times \pi \times 60 \times 400}$
$= 6.63\Omega$

Since the load is balanced:

$$\text{The voltage across a phase, } V_{ph} = \frac{440}{\sqrt{3}} = 254.5\text{V}$$

Current in the resistor $I_R = \frac{254.5}{14}$
$= 18.18\text{A}$, in phase with V_{ph}

Current in the capacitor $I_C = \frac{254.5}{6.63}$
$= 38.38\text{A}$, leading V_{ph} by 90°

Let I_{ph} = the resultant of 18.18A and 38.38A which are in quadrature.

$\therefore I_{ph} = \sqrt{18.18^2 + 38.38^2}$

$= 42.45$A. This is also the line current since the load is star connected $\therefore I = 42.45$A Ans.

Power factor of load $= \cos\phi = \frac{I_R}{I_{ph}} = \frac{18.18}{42.45}$

$= 0.43$ (leading) Ans.

Power of load, $P = \sqrt{3}\ VI \cos\phi$

$= \sqrt{3} \times 440 \times 42.45 \times 0.43$ watts

$= 13.9$kW Ans.

If the load is in delta, the current per phase would rise in the proportion of $\frac{440}{254.5}$ or $= \sqrt{3} \times$ original I_{ph}

$= \sqrt{3} \times 42.45$ amperes

The line current would be $\sqrt{3}$ times this new phase current.

$\therefore$ New $I = \sqrt{3} \times \sqrt{3} \times 42.45 = 3 \times 42.45 = 127.35$A

The power factor of the load would remain the same, so new power, $P = \sqrt{3}\ VI \cos\phi = \sqrt{3} \times 440 \times 127.35 \times 0.43$ watts

$= 41.7$kW Ans.

43. Total load $= 45 + 78 = 123$A. Length BC is $(580 - 240) =$ 340m. Let the current in the remaining 320m section AC be I amperes;

Then current in section AC $= I$ amperes

,, ,, ,, ,, AB $= (123 - I)$ amperes

,, ,, ,, ,, BC $= (123 - I - 45)$ amperes

$= (78 - I)$ amperes

Resistance of section AC $= \frac{0.25}{1000} \times 320 = 0.08\Omega$

,, ,, ,, AB $= \frac{0.25}{1000} \times 240 = 0.06\Omega$

,, ,, ,, BC $= \frac{0.25 \times 340}{1000} = 0.085\Omega$

By Kirchhoff's laws, the voltage drops in either section of the main feeding the load at C, are equal.

$\therefore I \times 0.08 = (123 - I)\ 0.06 + (78 - I)\ 0.085$

or $8I = 6\ (123 - I) + 8.5\ (78 - I)$

giving $8I = 738 - 6I + 663 - 8.5I$

or $22.5I = 738 + 663 = 1401$

Thus current I in section AC $= \frac{1401}{22.5} = 62.27$A Ans.

Current in section AB $= 123 - 62.27 = 60.73$A Ans.

Current in section BC $= 78 - 62.27 = 15.73$A Ans.

P.D. at point C $= 220 - (62.27 \times 0.08)$

$= 215$V

,, ,, ,, B $= 220 - (60.73 \times 0.06)$

$= 216.36\text{V}$

P.D. at load C is lowest $= 215\text{V}$ Ans.

44. Let non-inductive coil of 6Ω be designated branch A.

Then $I_A = \frac{110}{6} = 18.33\text{A}$ Ans. $\cos\phi_A = 1$ $\sin\phi_A = 0$

Let inductive coil of impedance 9Ω be designated branch B

Then $I_B = \frac{110}{9} = 12.22\text{A}$ Ans.

$\cos\phi_B = \frac{3}{9} = 0.33$ (lagging) $\sin\phi_B = 0.943$

Resolving in active and reactive components

$$\begin{aligned} I_a &= I_A\cos\phi_A + I_B\cos\phi_B \\ &= 18.33 \times 1 + 12.22 \times 0.33 = 18.33 + 4.033 \\ &= 22.363\text{A} \end{aligned}$$

$$\begin{aligned} \text{and } I_r &= -I_A\sin\phi_A - I_B\sin\phi_B \\ &= -18.33 \times 0 - 12.22 \times 0.943 = -0 - 11.52 \\ &= -11.52\text{A} \\ \therefore I &= \sqrt{22.36^2 + 11.52^2} \\ &= 25.2\text{A} \end{aligned}$$

Thus current taken from mains is 25.2A Ans.

With capacitor of 600μF connected in parallel:

$$X_C = \frac{1}{2\pi fC} = \frac{10^6}{2 \times \pi \times 50 \times 600} = 5.3\Omega$$

$$\text{Current } I_C = \frac{110}{5.3} = 20.75\text{A}$$

Resolving as before, active component I_a remains the same but reactive component $I_r = 20.75 - 11.52 = 9.23\text{A}$ and is now vertically upwards, *ie* leading V by 90°.

Resultant current $I = \sqrt{22.36^2 + 9.23^2}$
$= 24.2\text{A}$ Ans.

45. Here $R_f = 300\Omega$. $\therefore I_f = \frac{400}{300} = 1.33\text{A}$

$I_L = 30\text{A}$. $\therefore I_a = 30 - 1.33 = 28.67\text{A}$

And for the first load condition:

$$\begin{aligned} E_b &= 400 - (28.67 \times 0.3) \\ &= 391.4\text{V} \end{aligned}$$

With the data given and the O.C.C. plotted, it will be seen that with 1.33A field current, the e.m.f. generated at 1500

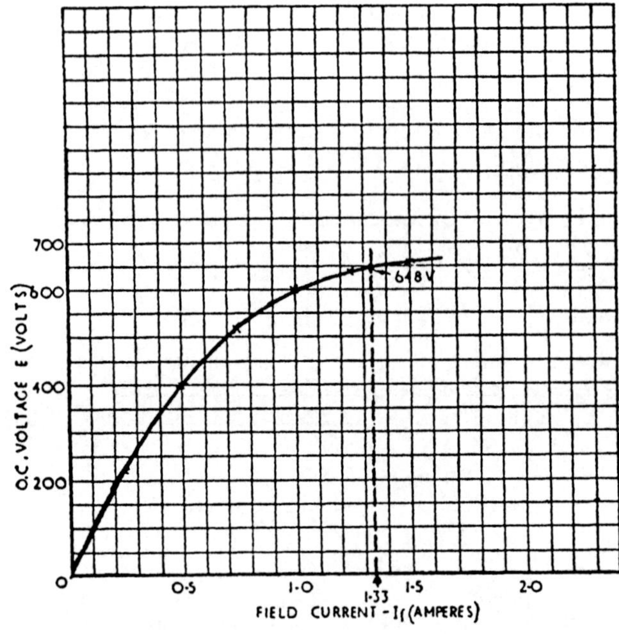

rev/min is 648V. Therefore for the same field current and for 391.4V to be generated, since $E \propto N$ the speed will be:

$$\frac{1500 \times 391.4}{648} = 906 \text{ rev/min} \quad \text{Ans.}$$

For the second load condition $\quad I_a = 50 - 1.33 = 48.67\text{A}$

The E_b is now $= 400 - (48.67 \times 0.3)$
$= 385.4\text{V}$

As before, the new speed condition is given by $\dfrac{1500 \times 385.4}{648}$

$= 892$ rev/min Ans.

46. The turns ratio per phase of the transformer are 560 to 42 or $\frac{560}{42} = \frac{13.33}{1}$. The voltages per phase will be in the same proportion.

(a) With the transformer connected; Primary in star and Secondary in delta; then:

Primary voltage per phase $= \dfrac{6600}{\sqrt{3}}$ volts

Secondary ,, ,, ,, $= \dfrac{6600}{\sqrt{3} \times 13.33}$

$= 286\text{V}$

But for a delta connection, line voltage = phase voltage
or $V = V_{ph} = 286\text{V}$ Ans.

(b) With the transformer connected; Primary in delta and Secondary in star, then:
Primary voltage per phase = 6600V

Secondary ,, ,, ,, $= \dfrac{6600}{13.33}$

$= 495.4\text{V}$

But for a star connection, line voltage $= \sqrt{3}$ phase voltage or $V = \sqrt{3}V_{ph} = \sqrt{3} \times 495.4$
$= 858\text{V}$ Ans.

47. Let V = the terminal voltage, I_1 the current given by the 220V battery and I_2 the current given by the 225V battery. The following equations can then be set down.

$220 - 0.2I_1 = V$ . . . (a)

$225 - 0.3I_2 = V$ . . . (b)

and $10(I_1 + I_2) = V$ . . . (c)

Subtracting (a) from (b) we have:

$225 - 220 - 0.3I_2 + 0.2I_1 = 0$

Thus $5 = 0.3I_2 - 0.2I_1$ or $I_2 = \dfrac{5 + 0.2I_1}{0.3}$

Using (b) and (c) we can write:

$225 - 0.3I_2 = 10I_1 + 10I_2$ and substituting for I_2

$$225 - 0.3\left(\frac{5 + 0.2I_1}{0.3}\right) = 10I_1 + 10\left(\frac{5 + 0.2I_1}{0.3}\right)$$

Thus $225 - 5 - 0.2I_1 = 10I_1 + \dfrac{50 + 2I_1}{0.3}$

and $(220 \times 0.3) - 0.06I_1 = 3I_1 + 50 + 2I_1$

or $66 - 50 = 3I_1 + 2I_1 + 0.06I_1$

Hence $16 = 5.06I_1$ $I_1 = \dfrac{16}{5.06} = 3.16\text{A}$ Ans.

Also $I_2 = \dfrac{5 + 0.2I_1}{0.3} = \dfrac{5 + 0.632}{0.3} = 18.77\text{A}$ Ans.

and $V = 10(I_1 + I_2) = 10(3.16 + 18.77)$
$= 219.3\text{V}$ Ans.

48. Heat received by the water = mass × temperature rise × specific heat capacity

$= 1 \times (100 - 11) \times 4.2$ kilojoules

$= 373.8\text{kJ}$

∴ Energy put into water = 373.8kJ and assuming an efficiency of 100 per cent for the kettle, then;

Energy taken from mains = 373 800J

With a d.c. supply, the current taken would be $\frac{373\ 800}{220 \times 3.5 \times 60}$

or $I = 8.09\text{A}$

and resistance of kettle element $= \frac{220}{8.09} = 27.19\Omega$

On a.c., the reactance X of the kettle becomes effective.

$\therefore X = 2\pi fL = 2 \times \pi \times 50 \times 0.05 = 15.7\Omega$

Hence $Z = \sqrt{27.19^2 + 15.7^2}$

$= 31.4\Omega$

Thus on an a.c. supply, current taken would be $\frac{220}{31.4}$

or $I = \frac{22}{3.14} = 7.06\text{A}$

The input power would be given by $P = I^2R$

or $P = 7.06^2 \times 27.19 = 1355\text{W}$

The time taken to produce 373 800J is given by:

$$t = \frac{373\ 800}{1355} = 275.7\text{s}$$

The percentage time difference between heating with a.c. as against d.c. would be $\frac{276 - 210}{210} = 0.314$

$= 31.4$ per cent Ans.

49. (a) Total power dissipation of coils $= 4 \times 45 = 180\text{W}$

If applied voltage = 220V, then field current $= \frac{180}{220}$

or $I_f = 0.818\text{A}$ Ans.

(c) No. of turns per field coil $= \frac{750}{0.818} = 916.86$

Length of turn $= 450\text{mm}$

$\therefore$ Length of wire $= 450 \times 916.86 = 412\ 587\text{mm}$

$= 412.6\text{m}$ Ans.

(b) Since $R = \frac{\varrho l}{A}$ and $R = \frac{220}{4} \times \frac{11}{9}$

$= 67.22\Omega$

Then $A = \frac{\varrho l}{R} = \frac{2 \times 10^{-8} \times 412.6}{67.22} = 12.3 \times 10^{-8}\ \text{m}^2$

or $12.3 \times 10^{-8} = \frac{\pi d^2}{4}$ where d = diameter of wire

Then $d^2 = \frac{12.3 \times 10^{-8} \times 4}{\pi}$

giving $d^2 = \frac{15.62}{10^8}$ or $d = \frac{\sqrt{15.62}}{10^4} = \frac{3.95}{10^4}$ metres

or diameter of wire $= 3.95 \times 10^{-4} \times 10^3\text{mm}$

$= 0.395\text{mm}$ Ans.

50. The frequency of the e.m.f. produced by an alternator is proportional to the number of poles and the speed at which they are rotated. The relationship can be expressed by:

$f = \frac{pN}{60}$ where f = the frequency in hertz = the number of pairs of poles and N = speed in rev/min.

$$\therefore \text{ frequency of the supply } = f = \frac{4 \times 720}{60} = 48\text{Hz}$$ Ans.

The machine speed is also referred to as the synchronous speed, and if a supply of 48Hz was fed back into the alternator and it was allowed to 'motor', it would rotate at a synchronous speed of 720 rev/min.

Synchronous motors are constructed like alternators and the relation between speed and frequency is directly applicable. Thus for the motor in question

$$N = \frac{60 \times f}{p} = \frac{60 \times 48}{24}$$

or speed of synchronous motor = 120 rev/min Ans.

An induction motor because of its construction and principle of operation, does not run at synchronous speed but 'slips' this speed by an amount determined by the design and loading of the machine. The amount of 'slip' is usually expressed as a fraction or percentage.

$$\text{Thus percentage slip} = \frac{\text{syn. speed} - \text{rotor speed}}{\text{syn. speed}} \times 100\%$$

or $s = \frac{N_s - N}{N_s} \times 100$, where s is the % slip.

For the problem, N_s = synchronous speed = 120 rev/min
N = the speed of the motor
and s = 2 per cent

$$\text{Thus } 2 = \left(\frac{120 - N}{120}\right) \times 100$$

$$\text{or } 0.02 = \frac{120 - N}{120}$$

giving $2.4 = 120 - N$ or $N = 120 - 2.4 = 117.6$ rev/min Ans.

The above could be deduced more easily thus:

With 2 per cent slip, if synchronous speed was 100 rev/min, then rotor speed would be 98 rev/min. Accordingly, if synchronous speed was 120 rev/min, then rotor speed would be

$$= \frac{98 \times 120}{100} = 117.6 \text{ rev/min}$$ Ans.

INDEX

N

O

P

Q

R

S

REED'S MARINE ENGINEERING SERIES

Vol. 1 MATHEMATICS
Vol. 2 APPLIED MECHANICS
Vol. 3 APPLIED HEAT
Vol. 4 NAVAL ARCHITECTURE
Vol. 5 SHIP CONSTRUCTION
Vol. 6 BASIC ELECTROTECHNOLOGY
Vol. 7 ADVANCED ELECTROTECHNOLOGY
Vol. 8 GENERAL ENGINEERING KNOWLEDGE
Vol. 9 STEAM ENGINEERING KNOWLEDGE
Vol. 10 INSTRUMENTATION AND CONTROL SYSTEMS
Vol. 11 ENGINEERING DRAWING
Vol. 12 MOTOR ENGINEERING KNOWLEDGE

REED'S ENGINEERING KNOWLEDGE FOR DECK OFFICERS
REED'S MATHS TABLES AND ENGINEERING FORMULAE
REED'S MARINE DISTANCE TABLES
REED'S OCEAN NAVIGATOR
REED'S SEXTANT SIMPLIFIED
REED'S SKIPPERS HANDBOOK
REED'S COMMERCIAL SALVAGE PRACTICE
REED'S MARITIME METEOROLOGY
SEA TRANSPORT – OPERATION AND ECONOMICS

These books are obtainable from all good Nautical Booksellers
or direct from:

THOMAS REED PUBLICATIONS
The Barn, Ford Farm
Bradford Leigh
Bradford-on-Avon
Wiltshire BA15 2RP
United Kingdom

Tel: 01225 868821
Fax: 01225 868831

Email: sales@abreed.demon.co.uk